T0091496

Springer Proceedings in Physics 283

Indexed by Scopus

The series Springer Proceedings in Physics, founded in 1984, is devoted to timely reports of state-of-the-art developments in physics and related sciences. Typically based on material presented at conferences, workshops and similar scientific meetings, volumes published in this series will constitute a comprehensive up-to-date source of reference on a field or subfield of relevance in contemporary physics. Proposals must include the following:

- Name, place and date of the scientific meeting
- A link to the committees (local organization, international advisors etc.)
- Scientific description of the meeting
- List of invited/plenary speakers
- An estimate of the planned proceedings book parameters (number of pages/articles, requested number of bulk copies, submission deadline).

Please contact:

For Americas and Europe: Dr. Zachary Evenson; zachary.evenson@springer.com
For Asia, Australia and New Zealand: Dr. Loyola DSilva; loyola.dsilva@springer.com

Chengmin Liu
Editor

Proceedings of the 23rd Pacific Basin Nuclear Conference, Volume 1

PBNC 2022, 1–4 November, Beijing & Chengdu, China

Set 2

Springer

Editor
Chengmin Liu
Nuclear Power Institute of China
Chengdu, China

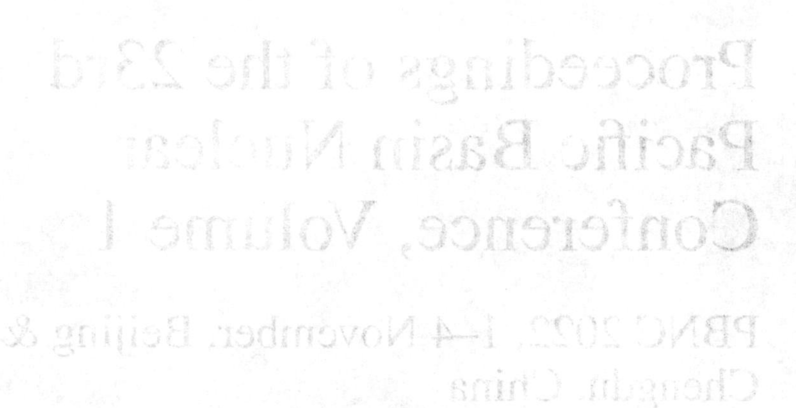

ISSN 0930-8989 ISSN 1867-4941 (electronic)
Springer Proceedings in Physics
ISBN 978-981-99-1022-9 ISBN 978-981-99-1023-6 (eBook)
https://doi.org/10.1007/978-981-99-1023-6

This Springer imprint is published by the registered company Springer Nature Singapore Pte Ltd.
The registered company address is: 152 Beach Road, #21-01/04 Gateway East, Singapore 189721, Singapore

Acknowledgment

Thanks to all the members of the PBNC 2022 Conference Committees

COMMITTEES

Title	Name
Conference Chairman	Jianfeng YU Shoujun WANG
Steering Committee	
Chairs	Qi LUO
Co-Chairs	Yanfeng SHEN Tao ZHANG Hongzao LU Gang WAN Zhi WANG Corey McDaniel
Members	Fengxue WANG、Kamal Verma、Mimi Holland Limbach、Jorge Spitalnik、Vladimir Kuchinov、Benhamin Rouben、George Christidis、Kazuaki Matsui、Masahito Kinoshita、Armando Gomez、Ozemoyah Peter、Javier Palacios Javier、John Harries、Tomofumi Yamamoto、Faridah Idris
Organizing Committee	
Chairs	Jianqiao LIU Conglin WANG
Co-Chairs	Dazhu YANG Jiashu TIAN Keli GAO Yongjin FENG
Members	Fengxue WANG、Lixin SHEN、Xiaoguang LIU 、Ning SHEN、Laisheng YANG、Hong TANG、Aubrey Whittington
Chair Assistant	Ruoshan XU、Yanyan ZHU、Siyan LIU、Dan DENG、Li YANG、Yan QUAN
Technical Program Committee	
Chairs	Chengmin LIU
Executive Chair	Hongxing YU
Members	Danrong SONG、Zhou ZHOU、Wei SHI、Hangzhou ZHANG、Tao ZHANG、Wei YI、Guanghui SU、Yizhe LIU、Xiaoyong WU、Yue ZUO、Jiang HU、Yaming LI、Suyuan YU
Technical Secretariat	Wenjie LI、Bowen QIU

Contents

Research on the Radioactive Waste Discharge Permit Regulation

Li Huang, Ting Liu(✉), Yun Fan, Meng Chang, and Yiman Dong

Nuclear and Radiation Safety Center, MEE, Beijing, China
liuting@chinansc.cn

Abstract. For the release of radioactive waste, the regulatory body must establish or approve operational limits and conditions relating to public exposure, including approved discharge limits. The parties concerned must ensure that the release of radioactive waste and radioactive substances into the environment is managed in accordance with the instrument of ratification. Discharge authorization procedures, including the establishment of emission authorizations and emission allowances, the establishment and use of dose constraints, emission characteristics, and the circumstances of exposure used in the establishment of emission allowances, etc., According to the legal requirements of our country, at present, all units in the nuclear industry that produce and discharge radioactive waste gases and liquids into the Environment shall rely on the examination and approval of prior environmental impact reports or environmental impact assessment documents, the permit mode of total discharge control for the examination and approval of the designed nuclide discharge of radioactive waste gas and liquid. At present, China's radioactive contamination emissions are not included in the management of emission permits, nuclear and radiation-related key industries and units are not included in the "Fixed Source Emission Permits Classified Management Directory" management. Whether it is feasible to carry out the pollution discharge permit system in the nuclear and radiation industry of our country, in the theoretical system, there are already relevant laws and regulations of our country to guarantee or can guarantee, however, there are still some problems in the actual operation, such as insufficient research on radiation environmental quality objectives,further work is needed to promote the full implementation of the discharge permit system in the nuclear industry.

Keywords: Radioactive Waste · Discharge · Limit Value · Control Value · Permit · License

1 Introduction

The discharge of radioactive waste must be regulated and restricted by the state. Article 41 of my country's "Radioactive Pollution Prevention and Control Law" stipulates that units that generate radioactive waste gas and waste liquid discharge into the environment radioactive waste gas and waste liquid that meet the national standards for the

C. Liu (Ed.): PBNC 2022, SPPHY 283, pp. 589–595, 2023.
https://doi.org/10.1007/978-981-99-1023-6_51

prevention and control of radioactive pollution, shall report to the environmental protection administrative director who has examined and approved the environmental impact assessment documents. The department applies for the discharge of radionuclides and regularly reports the discharge measurement results.

In January, China issued regulations on the management of pollution permits, 2021 that enterprises, institutions and other producers and operators under the management of pollution permits should apply for pollution permits in accordance with the regulations, no pollutants shall be discharged. According to the factors such as the amount of pollutants produced, the amount of pollutants discharged and the degree of impact on the environment, etc., the discharge units shall be subject to the classified management of discharge permits. However, the nuclear and radiation industries are not explicitly included in the current "Classified management list of emission permits for fixed sources of pollution", and there are no specific requirements for emission permits in China's current Local ordinance regulations, therefore, the industry has not implemented a licensing system.

2 Relevant Regulations of Chinese Standards

2.1 GB18871 "Basic Standard for Protection Against Ionizing Radiation and for the Safety of Radiation Source"

Article 8.6.2 stipulates that radioactive waste liquid shall not be discharged into ordinary sewers, unless it is confirmed by the review and management department that it is low-level waste liquid that meets the following conditions, it can be directly discharged into ordinary sewers with a flow rate greater than 10 times the discharge amount, and Each discharge should be recorded:

a) The total activity emitted per month does not exceed 10 ALI_{min} (ALI_{min} is the smaller of the ingestion and inhalation ALI values corresponding to occupational exposure, and the specific value can be obtained according to the provisions of B1.3.4 and B1.3.5);

b) The activity of each discharge shall not exceed 1 ALI_{min} and shall be flushed with not less than 3 times the amount of water discharged after each discharge.

2.2 GB6249 "Environmental Radiation Protection Regulations for Nuclear Power Plants"

Article 6 "Dose constraint value and emission control value in operating state" stipulates that the effective dose to any individual in the public caused by the radioactive substances released to the environment from all nuclear power reactors at any site must be less than the dose constraint value of 0.25 mSv per year.

The operating organization of the nuclear power plant shall formulate the dose management target values for airborne radioactive effluent and liquid radioactive effluent respectively according to the dose constraint value approved by the audit and management department. The nuclear power plant must implement the control of the total annual emission of radioactive effluents for each reactor. For a reactor with a thermal power of 3000 MW, the control value is as follows (Tables 1 and 2).

Table 1. Control values of airborne radioactive effluents

	Light water reactors	Heavy water reactors
Inert gas	$6 * 10^{14}$ Bq/a	
Iodine	$2 * 10^{10}$ Bq/a	
Particles (half-life > 8 days)	$5 * 10^{10}$ Bq/a	
14C	$7 * 10^{11}$ Bq/a	$1.6 * 10^{12}$ Bq/a
3H	$1.5 * 10^{13}$ Bq/a	$4.5 * 10^{14}$ Bq/a

Table 2. Control values for liquid radioactive effluents

	Light water reactors	Heavy water reactors
3H	$7.5 * 10^{13}$ Bq/a	$3.5 * 10^{14}$ Bq/a
14C	$1.5 * 10^{11}$ Bq/a	$2.0 * 10^{11}$ Bq/a (Except 3H)
Other nuclides	$5.0 * 10^{10}$ Bq/a	

3 Radioactive Waste Discharge Permits in China

At present, the discharge of radioactive pollutants in my country has not been included in the management of pollutant discharge permits for the time being, and relevant key industries and units have not been included in the "List of Fixed Pollution Source Pollutant Discharge Permit Classification Management" management.

According to the requirements of the Pollution Discharge Law, at present, all units in my country that generate and discharge radioactive waste gas and waste liquid into the environment implement a permit system based on construction permits, safety permits, mining permits, etc., with pre-environmental impact reports or environmental impact assessment documents. Based on the approval of radioactive waste gas and waste liquid, the total emission control permit mode is approved for the design of radioactive waste gas and waste liquid emission.

When submitting an environmental impact report to the ecological environment department, the unit producing radioactive waste gas and liquid should also submit an application for the estimated discharge of radioactive waste gas and waste liquid, including determining the characteristics and activities of the radionuclides to be discharged and possible discharges where and how, and the exposure doses to key populations of the public that may be caused by the planned discharge. According to the discharge situation of other facilities in the area where the emission unit is located, the ecological environment department will allocate a certain emission share to the unit according to the public dose limit standard. The approved emission limits for nuclear power plants and research reactors must be included in the operating limits and conditions. The nuclear technology utilization unit generally adopt emission and clean decontamination models that conform to national standards, while uranium mines adopt emission limits for corresponding nuclides based on environmental standards, and associated mines currently do not have relevant standards, the corresponding limits of national sewage discharge standards can be used.

A nuclear power plant unit is a million kilowatt-class pressurized water reactor nuclear power unit. According to the approval of the original Ministry of Environmental Protection "Approval of the Environmental Impact Report of a Nuclear Power Plant (Operation Stage)" and other documents, as well as the monitoring of the effluent of the nuclear power plant, the annual emission control value and actual value of the radioactive effluent of the unit in a certain year as follows (Table 3):

Table 3. Annual emission control value and actual value of radioactive effluents from 1/2 unit of a nuclear power plant

		6249 control value (single unit)	A nuclear power plant (1/2 units control value)	A nuclear power plant (1/2 units actual value)	Proportion
Airborne effluent	Inert gas	6.00E+14	1.37E+14	4.54E+11	0.33%
	Iodine	2.00E+10	1.18E+09	1.15E+07	0.97%
	Particles	5.00E+10	1.31E+08	1.56E+06	1.19%
	14C	7.00E+11	7.81E+11	3.00E+11	38.38%
	3H	1.50E+13	9.90E+12	7.38E+11	7.46%
Liquid effluent	3H	7.50E+13	9.90E+13	4.58E+13	46.27%
	14C	1.50E+11	5.87E+10	5.87E+09	9.99%
	Other nuclides	5.00E+10	5.56E+10	6.16E+08	1.11%
	Total concentration	1000Bq/L	900Bq/L (all)	180Bq/L	–
Dose		0.25 mSv (all units)		0.000452 mSv (all units)	0.18%

The annual emissions of Airborne Carbon 14 and liquid tritium were higher than the annual control value, and the emissions of other airborne and liquid radioactive effluents were far lower than the annual control value. The annual maximum individual effective dose of effluents released to the public was $4.52 * 10^{-7}$ Sv, and the residents receiving the maximum individual dose were adults within 1–2 km north-northeast (NNE) of the plant site, about 0.18% of the dose-limiting value of 0.25 mSv/a.

For radioactive contamination, the monthly and quarterly monitoring results of the water samples near the outfall comply with the first-class Standard in Table 4 of the comprehensive wastewater discharge standard (GB8978-1996), and there is no obvious change compared with the outfall. Monthly monitoring and analysis items include dissolved oxygen, residual chlorine, conductivity, Ph value, hexavalent chromium, total chromium, orthophosphate, hydrazine. Quarterly monitoring items include Boron, iron, anions, sulphates, lithium, sodium, nickel and oils.

Table 4. Annual emission control value and actual value of radioactive effluents from 3/4 unit of a nuclear power plant

		6249 control value (single unit)	A nuclear power plant (3/4 units control value)	A nuclear power plant (3/4 units actual value)	Proportion
Airborne effluent	Inert gas	6.00E+14	9.11E+13	7.04E+11	0.77%
	Iodine	2.00E+10	6.45E+08	4.95E+06	0.77%
	Particles	5.00E+10	7.14E+07	1.00E+06	1.40%
	14C	7.00E+11	7.65E+11	2.31E+11	30.18%
	3H	1.50E+13	5.63E+12	4.98E+11	8.85%
Liquid effluent	3H	7.50E+13	6.30E+13	3.28E+13	52.11%
	14C	1.50E+11	5.62E+10	5.46E+09	9.72%
	Other nuclides	5.00E+10	4.00E+10	7.24E+08	1.81%

3.1 Nuclear Technology Utilization Units

My country's nuclear technology utilization units basically have very few waste gas emissions, and the amount is also very small. The main radioactive pollutants in the management of pollutant discharge licenses are mainly concentrated in waste liquids, and in various industries, waste liquid discharges are mainly medical-related units. Although there are many types of radionuclides used in medical institutions, most of them are of low toxicity and short half-life. However, in the investigation and research on the application status of nuclear medicine projects in Guangdong Province, it was found that the total β of the wastewater at the outlet of the decay pool of the hospital with a large amount of iodine-131 was greater than the "Water Pollutant Discharge Standard for Medical Institutions" (GB18466-2005) the specified emission limits.

3.2 Associated Minerals

At present, my country has no associated radioactive mine radiation environmental safety standard guidelines and associated mine development and utilization project effluent discharge limit. From the actual discharge situation, the discharge concentration of wastewater may be higher than the relevant limit in the "Integrated Wastewater Discharge Standard" (GB8978-1996), and the radon concentration in some places will also be higher than the limit, but the resulting dose is basically low The public individual effective dose limit for the development and utilization of mineral resources is 1 mSv/a.

3.3 Situation Analysis

After investigation, various local provinces in my country have carried out relevant explorations in the implementation of the radioactive pollutant discharge permit system.

Article 18 of the "Administrative Measures for the Prevention and Control of Radioactive Pollution in Sichuan Province" promulgated by Sichuan Province in 1999 stipulates: "Any unit that discharges radioactive waste water and waste gas into the environment must be monitored by a statutory monitoring agency; It is prohibited to discharge radioactive waste water and waste gas into the environment without a pollution discharge permit or beyond the regulations of the pollution discharge permit." But this method has been abolished at present, and there is no such relatedness in the newly issued Regulations on the Prevention and Control of Radiation Pollution in Sichuan Province in 2016. Regulation. At present, radioactive pollutant discharge permits have also been set up in the ecological environment examination and approval of Shandong and other provinces, but there has been zero application status.

Whether it is feasible to implement a pollutant discharge permit system in my country's nuclear and radiation industry, in terms of theoretical system, there are relevant laws and regulations in my country to guarantee or can be guaranteed, but in terms of practical operation, there are problems such as insufficient research on radiation environmental quality objectives.

The ultimate purpose of the implementation of the pollutant discharge permit system is to reduce the discharge of pollutants, thereby improving the quality of the environment. However, in the nuclear and radiation industry, there is a lack of relevant standards and research on how to link the allowable amount of radioactive pollutant discharge with environmental quality goals.

According to the pollution discharge permit management regulations, three modes of management are stipulated according to factors such as pollutant generation, discharge, and degree of impact on the environment, which are divided into key management, simplified management and registration management. The corresponding degree of environmental impact is divided into three categories. In the nuclear and radiation industry, it is generally based on the natural background, and the 1mSv equivalent dose limit for humans is used to control the effluent discharge of relevant units, but when it comes to the degree of environmental impact It is difficult to define the size of the industry and the impact on the ecological environment, which is also a problem that affects the determination of which mode of supervision the relevant industry should be.

4 Conclusions

In general, the conditions for implementing a pollutant discharge permit system in accordance with the requirements of the pollutant discharge permit regulations in my country's nuclear and radiation industries are not yet mature. However, in the medical and associated mining industries where management is relatively weak at present, a pilot registration management model can be considered, and It is possible to explore more scientific and reasonable emission standards and management models based on practice, laying the foundation for the comprehensive implementation of the emission permit system in the nuclear and radiation industry. At the same time, it is recommended to establish a radiation environment quality index system, improve the standard system, and strengthen the building of monitoring capabilities and professional supervision teams. Carry out research on the relationship between the discharge of radioactive pollutants and environmental quality objectives, to provide sufficient theoretical basis for the implementation of key management, simplified management, and registration management of pollutant discharge permits, and to achieve the goal of improving environmental quality and ensuring the health of personnel with pollutant discharge permits.

References

1. Li, D., et al.: Talking about the difficulties and countermeasures of implementing the pollutant discharge permit system in my country. Environ. Manage. China (5) (2016)
2. Liang, Z., et al.: Legislative positioning and legislative demands of my country's pollutant discharge permit system—legal reflections on the establishment of pollutant discharge permit management regulations (Continued). Environ. Impact Assess. (2) (2018)
3. Li, Y., et al.: Study on the radiation environmental impact of the development and utilization of the Bayan Obo Rare Earth Mine. Rare Earths (4) (2020)
4. Jia, P., et al.: Research on the setting system of pollution discharge permit under the background of ecological civilization system reform . J. Harbin Inst. Technol. Sci. (Soc. Sci.) (1) (2020)
5. Wang, S., et al.: The experience of the management of pollutant discharge permits in the United States. Oil Gas Field Environ. Protect. (1) (2017)

Preparation and Shielding Performance of Gamma Ray Shielding Composite Materials Based on 3D Printing Technology

Yulong Li, Chengxin Li(✉), Danfeng Jiang, Feng Liu, Xiajie Liu, and Li Li

China Nuclear Power Technology Research Institute, Shenzhen, Guangdong, China
cikerlee416@163.com

Abstract. Excessive gamma-rays will be emitted when a nuclear power plant is under the refueling overhaul, leading to a certain number of hotspots. To meet the shielding requirements of these hotspots of complex components, a nylon-tungsten shielding composite material was developed by laser selective sintering 3D printing technology. The effects to shielding performance of 3D printing shielding materials were emphatically studied for two preparation processes (including mechanical mixing method and coating method) of 3D printing composite powders. Experimental results show that the nylon-tungsten shielding composite material with tungsten content of 70–85% was obtained by 3D printing technology, which realizes the manufacture of mold-free customized and bonded shielding materials. The shielding material prepared by 3D printing technology by coating method is better than that by mechanical mixing method in shielding performance. When the mass ratio of tungsten powders is 80%, the linear attenuation coefficient can reach 0.32, which is the best formula of the composite material and can be used for shielding of complex components. All these results lay a theoretical foundation for the engineering application of 3D printing shielding materials.

Keywords: 3D printing · Shielding materials · Performance study

1 Introduction

During the operation of a PWR nuclear power plant, many radiation hotspots may appear on different operating pipelines and equipments in the control area. Therefore, it is necessary to adopt certain radiation shielding measures to reduce the external dose for working staff near the radiation hotspots.

At present, it is common to use sheet lead or lead apron for radiation shielding in the domestic nuclear power plants [1, 2], but there are many problems for this shielding way. Sheet lead is messy and difficult to clean. Sheet lead and lead apron are soft, which is easy to cause deformation and affect the stability of radiation protection. This shielding method does not meet the seismic performance requirements of nuclear power plants, and hinders the operation and maintenance of instruments and valves. In addition, most of the wrapped shielding sheet lead and lead apron need to be removed when the refueling overhaul is finished for the nuclear power plant. Repeating installations and

C. Liu (Ed.): PBNC 2022, SPPHY 283, pp. 596–608, 2023.
https://doi.org/10.1007/978-981-99-1023-6_52

removals not only consumes a lot of sheet lead and lead apron, but also causes high dose on the working staff. According to the management requirements of the control area of nuclear power plants, it is necessary to develop a new shielding composite material which is customized and completely fitting with the component, and innovate the traditional shielding mode of nuclear power plants.

At present, research on the new type of shielding materials has aroused more and more attention. Shin et al. [3] developed a BN/high-density polyethylene composite material with good neutron shielding performance; A kind of rubber-based flexible shielding materials has been developed by British ITW Company for radiation shielding in nuclear industry and the nondestructive testing field, which has been applied in the Size-well B nuclear power plant in Britain; Fu Ming et al. [4] developed a flexible shielding material with styrene butadiene rubber (SBR) for matrix; Zhuo Mingchuan et al. prepared PA6-tungsten shielding composites with different densities by melt extrusion method [5]; Ahmed et al. Studied the gamma-ray shielding characteristics of tungsten-silicone rubber composites, which can be used as manufacturing materials for protective clothing such as gloves and jackets [6]; Samantha et al. used 3D printing technology to prepare resin-based neutron and gamma-ray shielding materials, of which the content of functional filler Bi can reach 40% or the content of ^{10}B can reach 60% [7].

It is complex and diverse for equipment structure at radiation hotspots in nuclear power plants. Radiation protection composite materials based on polymer materials are generally produced by injection molding or compression molding. There are many disadvantages for preparing shielding materials with complex structures by traditional manufacturing method, such as a long manufacturing cycle, high mold cost, low mold utilization rate and easy to cause lead pollution. Therefore, it is necessary to develop a new gamma-ray shielding composite material based on advanced manufacturing technology, in order to meet the requirements of customized coated shielding materials for complex equipments in nuclear power plants.

3D printing technology is an additive manufacturing technology, which was born in 1980s. With this technology, rapid and free molding can be realized, and polymer materials with complex structure can be manufactured conveniently, quickly and at low cost. In this paper, a new type of nylon-tungsten shielding composite materials is prepared by 3D printing technology, in which thermoplastic nylon is used as matrix material, and tungsten is used as functional filler. Through the research of powder mixing process and shielding performance, the best material formula is obtained, and the shielding protection problem of complex components in nuclear power site is solved effectively.

2 3D Printing Radiation Protection Material Design

3D printing is a new type of additive manufacturing technologies compared with the current traditional manufacturing technologies (such as turning, milling, planning, grinding, etc.). This technology is a rapid manufacturing technology by which materials can be fusioned in one-time. Based on digital model files, it uses powdery metal or plastic and other bindable materials to construct three-dimensional objects by printing layer by layer and superimposing different shapes [8]. Compared with traditional manufacturing process, 3D printing has some advantages as followed: (1) Rapid free forming without mold; (2) Short manufacturing cycle and low cost; (3) near net shape and even net shape of complex structure; (4) Full digitalization; (5) multi-materials arbitrary composite manufacturing.

At present, the most widely used radiation protection materials are those containing the elements with high atomic numbers such as lead and tungsten. Lead is toxic and has a "Pb weak absorption zone" for rays with energy between 40–88 keV, which is easy to produce secondary bremsstrahlung radiation. In addition, lead has poor structural strength and is not resistant to high temperature. It is commonly used as lead containers, movable screens, lead bricks, etc. Tungsten has high density and atomic number. As an ideal shielding material, it has many advantages, such as good ray shielding effect and no secondary electron radiation produced [9]. It is an environmentally friendly radiation protection material with the strong ability of ray protection. However, tungsten has the characteristics of high melting temperature, high strength, poor plasticity and toughness, so it is difficult to process it into parts with complex shapes. Nylon is one of the engineering materials with the characteristics of excellent mechanical properties, low cost, easy processing and widely used, especially commonly used as one kind of 3D printing materials. It is a practical and effective way to solve the above problems by using nylon powders as substrate and the filler of tungsten powders to prepare nylon-tungsten composite products with complex shapes, which is based on 3D printing technology. Nylon-tungsten composites belong to the environmentally friendly materials. It is particularly important that the parts with complex shapes, which are difficult to be formed with pure tungsten, can be easily produced by using 3D printing technology with nylon-tungsten composites. Based on the above analysis, a new nylon-tungsten gamma-ray shielding composite was prepared by using 3D printing technology with thermoplastic nylon as the matrix and tungsten powders as the functional filler, which can meet the customized shielding requirements of the complex components in the control area of nuclear power plant.

3 Materials and Method

3.1 Raw Materials and Forming Process

The median particle size of tungsten particles with 99.99% pure is 15–20 μm, while the median particle size of nylon powders is 50–55 μm.

Composite materials are composed of polymer materials and metal materials. 3D printing technology that can both print nonmetallic materials and metal materials mainly includes fused deposition modeling (FDM), stereo lithography apparatus (SLA) and selective laser sintering (SLS). Table 1 shows the comparison of three process types of 3D printing technology. Compared with SLA and FDM, SLS has the following advantages: (1) SLS process is applicable to a wide range of materials, and the powder laying process adopted by SLS process is more suitable for mixed materials printing with great different density such as nylon powder and tungsten powder, which can ensure the dispersion and distribution of tungsten powder particles in nylon matrix; (2) SLS process is better than SLA and FDM process in forming precision and forming speed; (3) The SLS printing environment is generally a sealed space with inert gases such as argon as the protective gas. There is no risk of toxic and harmful substances being released during the printing process; (4) SLS does not need supporting materials, and is more suitable for the preparation of complex structural parts. Comprehensively considering material applicability, precision, forming efficiency and other factors, SLS printing process is chosen as the best one of the 3D printing process of shielding composite materials.

Table 1. Comparison of Process Types

Process	Precision/mm	Volume molding rate/$cm^3 \cdot h^{-1}$	Support	Applicable materials	Printing environment
SLA	±0.1	1000–2000	need	Photohardening resin	seal
FDM	±0.3	500–1000	need	Thermoplastics	open
SLS	±0.1	2000–3000	no need	Thermoplastics, Low melting point metal powder, ceramic powder	Sealing, protection gas

Figure 1 shows the molding principle of the SLS process of shielding composite materials. When the laser acts on the mixed powder of nylon and tungsten, the mixed powder absorbs heat, and its temperature gradually increases. When the temperature reaches the melting temperature (T_R) of nylon powder, the nylon powder changes from a solid state at room temperature to a liquid viscous flow state. As the temperature increases, the viscosity of the melt decreases, but its fluidity increases. It is easy to contact with the surrounding particles of tungsten powders. After cooling, the melt nylon powder solidifies and binds together. The particles of tungsten powders are uniformly dispersed in the nylon matrix without any change. The parts of the shielding material which are molded by laser selective sintering, mainly bond tungsten powder particles together through the adhesion of nylon. The adhesion value is determined by the cohesion and adhesion force of liquid nylon. Cohesion force refers to the force between the molecules of nylon binder itself, that is, the strength of nylon. Adhesion force is the force between the particles of nylon and tungsten powders, that is, the force of nylon adhering to the surface of tungsten powders.

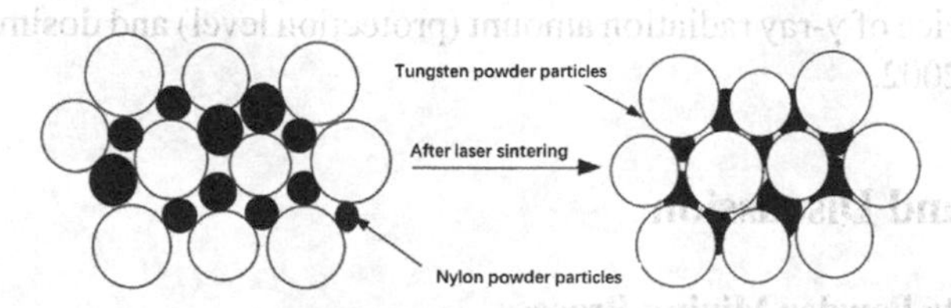

Fig. 1. SLS Forming Principle of Shielding Composites Material

Figure 2 shows the process flow of the shielding material prepared by 3D printing. The specific preparation process is as follows: Firstly, a certain quality of nylon powder and tungsten powder are mixed according to the design ratio by the mixing process; Secondly, the fully mixed nylon and tungsten powders were put into the laser selective sintering 3D printing equipment. After preheating the mixture and parameter settings, including laser power, scanning speed, scanning spacing and scanning mode, researchers input the model size of the printing part, and start printing; Finally, the sample was taken out for post-treatment. The powder adhered to the sample surface was removed, and then the sample was polished to make its surface bright, as shown in Fig. 3. During the preparation of 3D printing composite shielding materials, the batching design, mixing process and SLS sintering process are the most important factors, which affect the shielding performance, mechanical properties and forming accuracy of the shielding material.

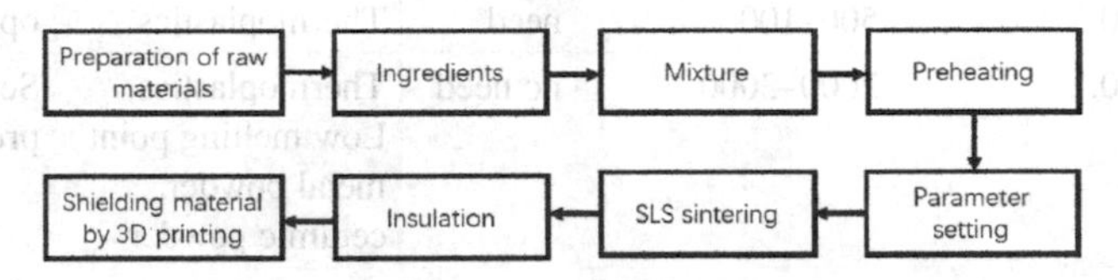

Fig. 2. 3D Printing Shielding Material Preparation Process

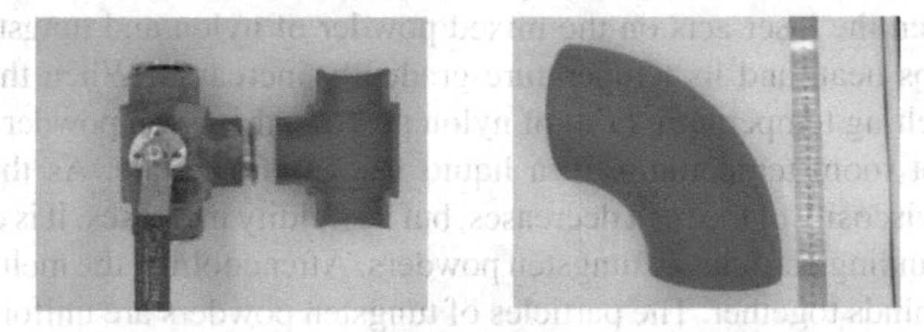

Fig. 3. 3D Printing Nylon-Tungsten Composite Shielding Material

3.2 Test and Analysis

The gamma shielding performance test of radiation protection composite materials was conducted with 137Cs (0.662 meV). The radiation decay rate of the sample was tested by standard device of γ-ray radiation amount (protection level) and dosimeter, according to GBZT 147-2002.

4 Results and Discussion

4.1 Composite Powder Mixing Process

At present, there are two main methods for preparing composite powders suitable for SLS printing, including the mechanical mixing method and the film coating method.

4.1.1 Mechanical Mixing Method

Mechanical mixing method is to mix polymer powders and filler powders mechanically in the mixer for three-dimensional movement, high speed kneading machine and other mixing equipments. The tungsten powder and nylon powder with a certain mass ratio were put into the double-motion mixer and mixed for 30 min. The surface appearance of the composite powder is shown in Fig. 4 (a). The mechanical mixing method is simple in process, low in equipment requirements, and economical. However, when the particle size of filler powders is very small (for example, less than 10 microns), or when the proportion of the filler powders (for example, metal powders) is much larger than that of polymer, it is difficult to disperse inorganic filler particles evenly in the polymer matrix by mechanical mixing method. The powder particles are easy to segregate during transportation and SLS powder laying, which leads to the existence of non-uniform distribution of filler particle aggregates in SLS formed parts, resulting in the decrease of capability of the products.

4.1.2 Film Covering Method

The coating method is to coat the polymer material on the outer surface of the filler powders to form a kind of composite powders with a polymer coating. The steps of preparing the shielding composite material powder by the film coating method are as follows: putting nylon powders, tungsten powders, coupling agent, leveling assistant into a stainless steel reaction kettle in a certain proportion, sealing and injecting nitrogen for protection, slowly heating up to about 150 °C, so that nylon powders are completely dissolved in solvent, then cooling the kettle to room temperature at a certain rate under vigorous stirring, and solid-liquid separation is carried out to obtain precipitated coated composite powders. After vacuum drying, crushing and sieving the obtained aggregates, the composite material containing nylon coated tungsten powders with suitable particle size distribution can be obtained. The surface appearance of the powder is shown in Fig. 4(b). In the coated powder, the filler and polymer matrix are mixed evenly without the segregation phenomenon in the process of transportation and powder spreading.

a) Preparation of powder by mechanical mixing

b) Preparation of powder by film coating

Fig. 4. Nylon-tungsten composite powder

According to the characteristics of the two composite powder mixing processes, nylon can be coated on the surface of tungsten powders uniformly by coating method.

The particle size of the nylon is significantly reduced during melting, coating on the surface of tungsten powder particles to achieve adhesion effect, so that the dispersion of tungsten powders in nylon matrix is more uniform.

4.2 Shielding Performance Research

4.2.1 Theoretical Shielding Efficiency of Composites

In the study of γ-ray shielding materials, the shielding performance of the materials was simulated in advance, which facilitated formulation design and adjustment. WPA1, WPA2, WPA3, WPA4, WPA5 and WPA6 were defined as different proportions of tungsten and nylon. The material proportions are listed in Table 2. The simulation of 90% (mass ratio) of tungsten powder ratio was not carried out because the sample could not be formed in the machine.

Table 2. Proportions of shielding composite materials

Material type	Raw material ratio
WPA1	60%W + 40%PA
WPA2	65%W + 35%PA
WPA3	70%W + 30%PA
WPA4	75%W + 25%PA
WPA5	80%W + 20%PA
WPA6	85%W + 15%PA

The linear attenuation coefficient model of the shielding composites was established. The theoretical shielding coefficients of the composites were calculated by MCNP (Monte Carlo N Particle Transport Code) program. MCNP is a general software package developed by Los Alamos National Laboratory (LANL) based on Monte Carlo method for computing neutrons, photons, electrons or coupled neutrons, photons, electrons transport problems in complex three-dimensional geometry. It can greatly save research funds and time by simulating shielding performance of materials with MCNP. The calculation model is shown in Fig. 5. The source is a unidirectional point source, which is 1m away from the detection point and 0.46 m away from the material surface. The material size is 20 cm (length) × 20 cm (width) × 1 cm (thickness). The attenuation of the material to 137Cs source was simulated, and the source energy was 0.662 meV. The radiation intensity before and after the material was recorded by F1 card. The linear attenuation coefficient μ of the composite material was obtained according to the formula (2).

All the test results were converted into linear attenuation coefficients in order to compare the test results of different formulations. The attenuation rate and linear attenuation coefficient were calculated as follows:

When gamma-rays pass through the shielding material, they will not be completely absorbed, but attenuated (intensity decreases). The attenuation relationship is shown in

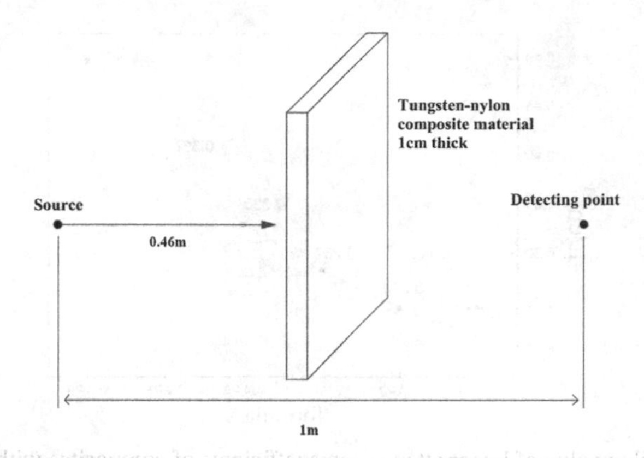

Fig. 5. Calculation model of linear attenuation coefficients of composite materials

Eq. (1):

$$R_x = R_0 e^{-\mu x} \tag{1}$$

where R_x is the dose rate after passing through a certain thickness x of shielding material;
R_0 is the dose rate before shielding;
X is the thickness of the shielding material;
and μ is the linear attenuation coefficient of the shielding material.
Formula (1) is converted to the following Eq. (2):

$$\mu = -\frac{-Ln\left(\frac{R_x}{R_0}\right)}{x} \tag{2}$$

The simulation values of linear attenuation coefficients of the shielding composites with different formulas were calculated by formula (2), which are shown in Fig. 6. The γ-ray shielding performance of composites increases exponentially with the increase of tungsten powder content. When the proportion of tungsten powders is low (<70%), the theoretical linear attenuation coefficient of the composite is low. Thus the ingredients with tungsten powder ratio between 70 and 85% were selected for 3D printing experiment.

4.2.2 Effect of Powder Ratio on Shielding Performance of Composites

According to the simulation results of linear attenuation coefficients of composite materials in Fig. 6 and combined with the powder mixing process of composite materials, the sample preparation and testing schemes of different formulas are set in Table 3. The linear attenuation coefficients of the samples with different formulas are obtained through shielding performance tests and conversion results by formula (2).

The shielding effect of the samples with different contents of tungsten powders was tested. The results are shown in Fig. 7: (1) The shielding performance of the composite is obviously improved when the mass fraction of tungsten powders increases from 70% to 80%. When the mass fraction of tungsten powders increases from 80% to 85%, the

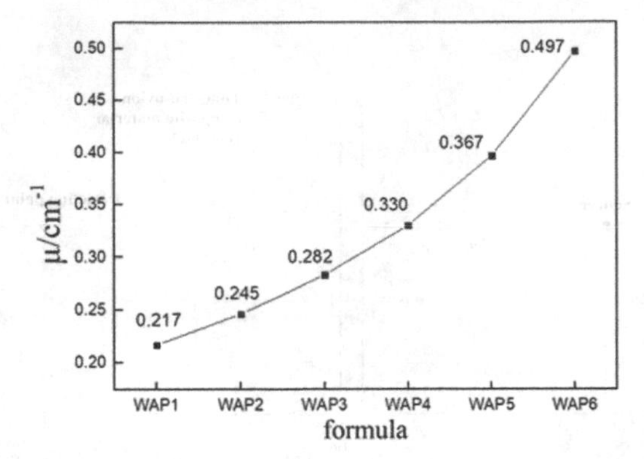

Fig. 6. Analog value of linear attenuation coefficients of composites without formula

Table 3. Test values of attenuation coefficients of samples with different formulations

Serial number	Tungsten mass fraction/%	Mixing method	γ-ray shielding rate/%	Linear attenuation coefficient/μ/cm^{-1}
1	70	Coated mixture	21.7	0.24
2	80	Mechanical mixing	21.5	0.24
3	80	Coated mixture	27.6	0.32
4	85	Coated mixture	25	0.29

shielding performance of the composite decreases. When the mass fraction of tungsten powders increases from 70% to 80%, the laser selective sintering of composites is in good condition, and there is no obvious defect in the samples. When the mass fraction of tungsten powders increases from 80% to 85%, due to the nylon content of the composite material is only 15%, cracks and warping occur in some areas of the sample during laser selective sintering, resulting in the final measured shielding performance of the composite material being lower than that of the composite material with the mass fraction of tungsten powders of 80%. (2) For mechanical mixing process, the linear attenuation coefficient of tungsten powders with 80% mass fraction is obviously lower than that of coated tungsten powders with 80% mass fraction. (3) The linear attenuation coefficient of the sample coated with film is lower than the simulation value of Monte Carlo under any tungsten powder mass ratio, which shows that there is still much room for improvement in the 3D printing process, and the uniform dispersion of tungsten powders in the matrix needs further improvement. When using the film covered mixing process, the best formula is that the mass fraction of tungsten powders and nylon matrix are 80% and 20% respectively.

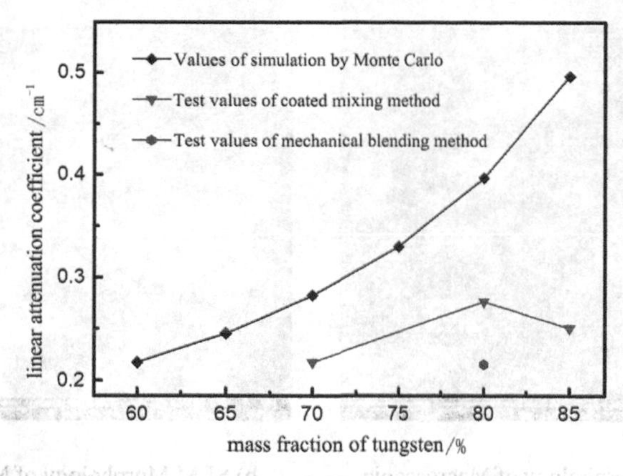

Fig. 7. Comparison of shielding capability of 3D printing shielding materials with different formulas

The main factor affecting the shielding performance of the shielding composite is the content of tungsten powders in the composite, followed by the dispersion of tungsten powders in the matrix. According to the above results, the higher the content of tungsten powders, the better the shielding performance of qualified samples. Under the condition of same mass fraction, the more uniform the dispersion of tungsten powders in matrix, the better the shielding performance of the material. The mixing process of composite powders has a great influence on the dispersion of tungsten powders in matrix.

4.2.3 Effect of Powder Mixing Process on Shielding Performance of Composite Materials

In order to further explore the reasons for the difference of shielding performance of samples under the two mixing process conditions, SEM-scan of 3D printing products of mechanical mixing and coated mixing was carried out, as shown in Fig. 8. It can be found that the macroscopic fracture surface of mechanical mixture presents obvious lines, showing typical characteristics of brittle fracture, while the microscopic fracture presents agglomeration of tungsten powders and only nylon powders in some areas (as shown in the box area in Fig. 5 (e)); In contrast, the surface of macroscopic fracture is smooth and the distribution of tungsten powder in the matrix is uniform in microstructure This also verified the results of the shielding performance experiment, and the 3D printing shielding composite products under the condition of mulching and mixing technology showed better shielding performance. The composite powder prepared by coating method has good sintering capability, high bonding strength and high forming precision, and the shielding composite powder is well coated, and almost no exposed tungsten powder is found.

The essential attribute of 3D printing radiation protection composite materials lies in its shielding performance, which is mainly reflected in the mass fraction and dispersion state of tungsten particles in the composite material. Excellent powder mixing and manufacturing process can ensure the uniform dispersion of tungsten powders in the matrix under high mass fraction and ensure the shielding performance.

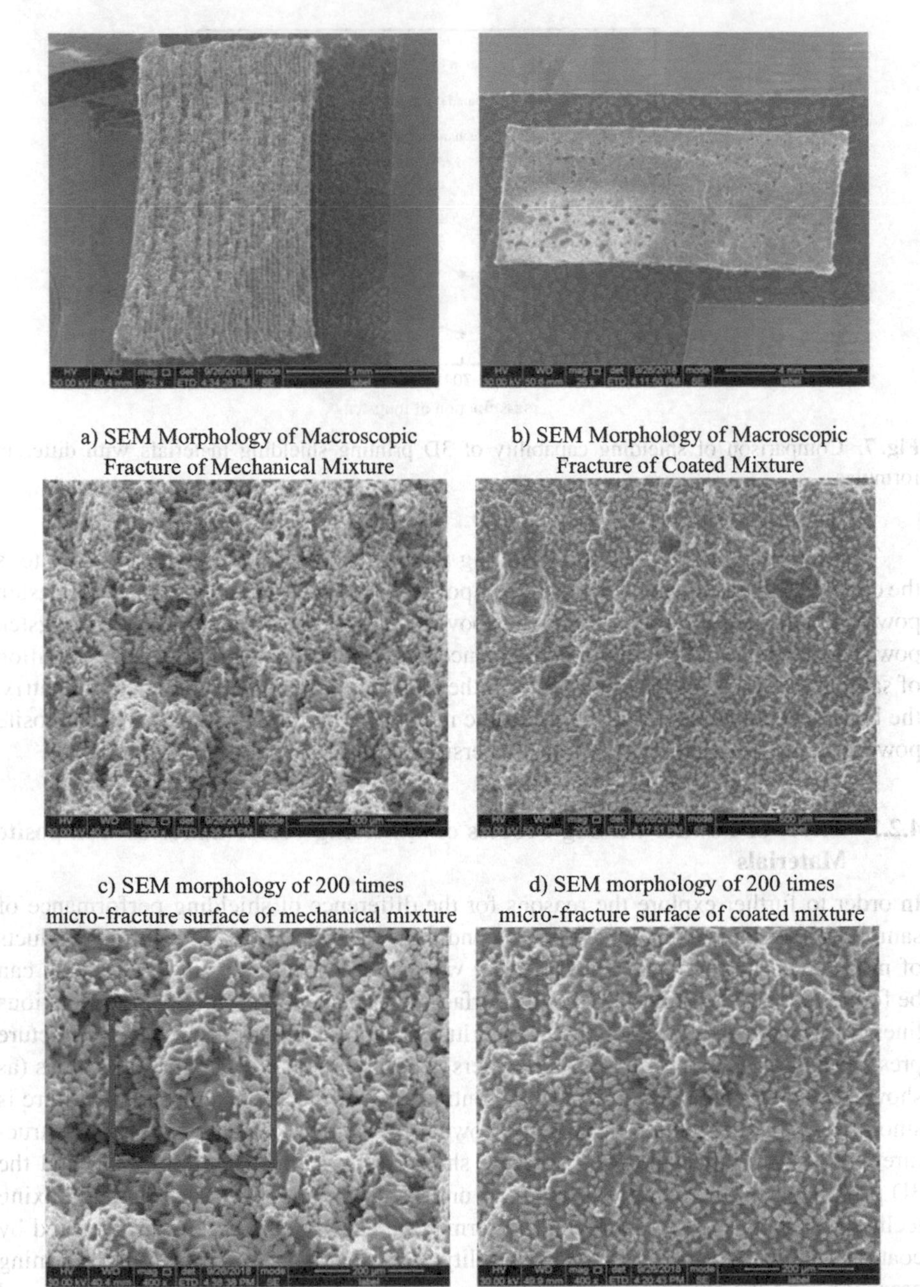

a) SEM Morphology of Macroscopic Fracture of Mechanical Mixture

b) SEM Morphology of Macroscopic Fracture of Coated Mixture

c) SEM morphology of 200 times micro-fracture surface of mechanical mixture

d) SEM morphology of 200 times micro-fracture surface of coated mixture

e) SEM morphology of 400 times micro-fracture surface of mechanical mixture

f) SEM morphology of 400 times micro-fracture surface of coated mixture

Fig. 8. Fracture scan photo

5 Conclusions

In this paper, a nylon-tungsten gamma-ray shielding composite material was developed by using laser selective sintering 3D printing technology, and two kinds of 3D printing composite powders were prepared by mechanical mixing method and coating method, respectively. The performance of the shielding materials was studied, and the conclusions are as follows:

(1) The laser selective sintering 3D printing technology adopting the technology route of powder spreading is more suitable for the manufacture of customized and laminated shielding materials for complex components in nuclear power plants;
(2) The coating method is preferred in the mixing process of the powders of the 3D printing shielding material composite. Compared with the mechanical mixing method, the product by the former method has better shielding performance. Through 400 times SEM scanning micro-fracture morphology, tungsten powders are still uniformly dispersed in nylon matrix;
(3) By carrying out performance research on different ratios of shielding materials, the best formula of nylon-tungsten 3D printing shielding materials was obtained. By shielding performance test, the linear attenuation coefficient of the shielding material reached 0.32 at 80% of tungsten powder mass ratio, which is the best formula and can be used for shielding of complex components;
(4) 3D printing technology can conveniently, quickly and low-costly prepare customized shielding composite materials to meet the need of special-shaped parts such as valves, elbows, large and small heads in the control area of nuclear power plants. This method has practical application value, and is a powerful technical supplement to the mold manufacturing process.

    3D printing shielding composite materials can effectively solve the radiation protection problem of complex components in the control area of nuclear power plants. This study lays a theoretical foundation for the engineering application of 3D printing shielding composite materials.

References

1. Gu, H.: The effect of reasonable shielding on reducing the site radiation level. Daya Bay Nuclear Power **4**, 36–38 (2003). (in Chinese)
2. Han, Y., Yu, W., Chen, F., et al.: Investigation of temporary radiation shielding facilities in Nuclear Power Plants. Nucl. Electron. Detect. Technol. **36**(12), 1263–1267 (2016). (in Chinese)
3. Singh, V.P., Badiger, N.M., Chanthima, N., et al.: Evaluation of gamma-ray exposure buildup factors and neutron shielding for bismuth borosilicate glasses. Radiat. Phys. Chem. **98**(1), 14–21 (2014)
4. Fu, M., Wang, Y., Li, F.: Preparation of SBR rubber based flexible shield material. Nucl. Power Eng. **34**(6), 165–168 (2012). (in Chinese)
5. Zhuo, M., Li, Y., Xiao, Y., et al.: Investigation on properties of PA6/Tungsten shielding composite. Eng. Plast. Appl. **39**(3), 23 (2011). (in Chinese)
6. Ahmed, B., Shah, G.B., Malik, A.H., et al.: Gamma-ray shielding characteristics of flexible silicone tungsten composites. Appl. Radiat. Isotopes **155**, 1–7 (2019)

7. Talley, S.J., Robison, T., Long, A.M., et al.: Flexible 3D printed silicones for gamma and neutron radiation shielding. Radiat. Phys. Chem. **188**, 1–12 (2021)
8. Yang, Y., Song, C.: Technology Roadmap of Guangdong Additive Manufacturing (3D Printing) Industry. South China University of Technology Press, pp. 1–2 (2017)
9. Lin, Q., Yang, Y., He, Y.: Simulation and verification of γ-multi-layer shielding with Monte Carlo method. Nucl. Phys. Rev. **27**(2), 182–186 (2010)

Dynamic Response Analysis of Floating Nuclear Power Plant Containment Under Marine Environment

Jialin Cui[1,2], Lijuan Li[3], Meng Zhang[1(✉)], Hongbing Liu[1,2], and Xianqiang Qu[1,2]

[1] Yantai Research Institute and Graduate School, Harbin Engineering University, Yantai, China
{cuijialin001,zhangmeng}@hrbeu.edu.cn

[2] College of Shipbuilding Engineering, Harbin Engineering University, Harbin, China

[3] Nuclear Power Institute of China, Chengdu, China

Abstract. Floating nuclear containment is in a harsher environment than conventional onshore nuclear containment. In view of the Marine environment under the condition of floating nuclear power plant containment structure safety, combined water dynamics and structural mechanics, considering the containment response under random movement of hull in the Marine environment, the influence of the containment structure load calculation, thus checking containment when working in pile structure safety, provide theoretical basis for the safe operation of floating nuclear power plants. In this paper, taking a floating nuclear power plant as an example, ANSYS 2021R1, Workbench, Fluent and other software of finite element analysis are used to conduct fatigue simulation of floating nuclear power plant. The time course curve of the 6-dof motion of the ship's center of gravity is obtained, then, a remote displacement method is adopted to transfer the hull motion to the containment vessel to realize the numerical simulation of the containment vessel movement with the hull, thus to solve maximum normal stress and strain, the maximum load component of containment bearing under the action of Marine environmental load is obtained. The results show that the maximum stress and strain of the vessel increase obviously in the moving state compared with the static state of the vessel, which indicates that the random motion response of the vessel must be considered in the structural safety analysis of the floating nuclear power plant containment.

Keywords: floating nuclear power plant · Ocean circulation · containment · structural loads · Random motion response

1 Introduction

With the continuous adjustment and optimization of China's energy structure and the continuous promotion of the strategy of becoming a maritime power, traditional fossil energy and emerging energy such as wind, wave and solar energy are increasingly difficult to meet the energy needs brought about by China's coastal oil and gas resources and island development. Therefore, as a clean, efficient and flexible location of offshore

C. Liu (Ed.): PBNC 2022, SPPHY 283, pp. 609–623, 2023.
https://doi.org/10.1007/978-981-99-1023-6_53

nuclear power generation technology, the national government pays more and more attention to it. Marine floating nuclear power station is a mobile floating offshore platform equipped with nuclear reactor and power generation system. It is the organic combination of mobile small nuclear power station technology with ship and ocean engineering technology. In floating nuclear power plants, a sealed steel containment structure is usually installed around the reactor and other auxiliary power generation structures to protect the normal operation of the reactor and the external environment. Compared with the traditional onshore nuclear power plant containment, the environment and load borne by the small steel containment (including support) of floating nuclear power plant at sea are very different, especially the complexity of the Marine environment leads to more complex load borne by the containment (including support). Therefore, in order to ensure the safe operation of floating nuclear power plants and protect the surrounding personnel and the external environment from nuclear radiation damage, it is urgent to carry out researches on mechanical analysis and safety evaluation technology of steel containment of floating nuclear power plants in marine environment.

The floating nuclear power plant containment vessel is not only subjected to huge vertical and horizontal loads, but also inevitably subjected to wind load, wave load and current load. Therefore, it is particularly important to analyze the dynamic response of the floating nuclear power plant containment vessel in the Marine environment. Reissner [1] studied the vibration characteristics of rigid circular foundation plate under vertical load, and proposed and verified the feasibility of elastic half-space theory in vibration research of foundation and foundation. Choprah [2] proposed the dynamic substructure method, which made numerical calculation effectively applied in this field. Lysmer et al. [3] proposed lumped parameter method, which laid the foundation for structural dynamic response analysis. Gazetas [4] and Mrakis et al. [5] proposed the calculation and analysis method of pile-soil-structure dynamic interaction, and provided empirical expressions of stiffness coefficients and damping coefficients. Fan Min et al. [6] conducted nonlinear seismic response analysis and research, and the results showed that the soil-pile-structure interaction system would affect the dynamic characteristics of the structure, resulting in the extension of the natural vibration period and the increase of damping of the system. Wang et al. [7] used hydrodynamic model to simulate the evolution of wind, wave and tide under 32 typhoon events in Bohai Bay from 1985 to 2014, and used two-dimensional Gumbel Logistic model to establish the joint distribution of wave and storm surge in Bohai Sea. De Waal and Van Gelder [8] established the joint distribution of extreme wave height and period through Copula function. Michele et al. [9] used two-dimensional Copula function to analyze the frequency of effective wave height, storm duration, storm direction and storm interval of ocean storms, and established the joint probability distribution between pairs. Xu et al. [10] established the two-dimensional joint distribution of storm surge height and effective wave height. Dong Sheng et al. [11] established the joint distribution of the annual maximum wave height and the corresponding wind speed of a jacket platform for 24 years based on Archimedean Copula function, combined with the response of the offshore platform, and found that considering the joint effect, the response of the offshore platform could be reduced in the same return period. Chen Minglu et al. [12] conducted hydrodynamic analysis and wave load prediction for semi-submersible offshore platforms. Zhou Sulian et al. [13] studied the mooring system design of deep-water semi-submersible platform.

Due to the complex wind, wave and current environment under the action of Marine environment of floating nuclear power plant containment, the research on dynamic response characteristics of floating nuclear power plant containment is insufficient and almost no reports have been reported. Therefore, it is of great significance to carry out dynamic response analysis of floating nuclear power plant containment under Marine environment conditions, and to explore the variability and ultimate load effect of short-term time history analysis, so as to understand dynamic response characteristics of floating nuclear power plant containment under Marine environment.

The main work of this paper is to establish a hydrodynamic analysis model, using Ansys Workbench HD software module for frequency domain analysis of the platform and obtain the hydrodynamic parameters of the platform. The anchor chain model was added to the platform model to control the six degrees of freedom movement of the platform under the action of wave and flow. The HR module of Ansys Workbench software was used to conduct time-domain analysis of the platform to obtain the time-history curve of the platform motion response. The structural dynamic response analysis of the containment vessel of floating nuclear power plant under the action of wind load, wave load and current load is carried out, which provides important reference for the safe operation of floating nuclear power plant.

2 Theoretical Basis of Potential Flow

2.1 Small Scale Member

The stress of offshore floating structures in waves is studied. The stress of offshore structures is the most important topic in the field of offshore engineering, in which the wave force of piles is the basis of the stress of offshore structures. The method proposed by Morison et al. in 1950 is used to calculate wave force for small components, that is, structures whose diameter is smaller than the wavelength of the incident wave. Morison equation is basically an empirical formula, which takes wave particle velocity, acceleration and cylinder diameter as parameters to calculate the wave force in each depth of water, and then obtains the wave force along the length of the column.

Morrison et al. believed that the horizontal wave force acting on any height of the cylinder included two components:

$$f_H = f_D + f_I \tag{1}$$

Its magnitude is in the same mode as the drag force exerted on the column by unidirectional steady water flow, that is, it is proportional to the square of the horizontal velocity of the wave water point and the projection area of the unit column height perpendicular to the wave direction. The difference is that the wave water points oscillate periodically, and the horizontal velocity is positive and negative, so the drag force on the cylinder is also positive and negative:

$$f_D = \frac{1}{2} C_D \rho A u_x |u_x| \tag{2}$$

$$f_I = \rho V_0 \frac{du_x}{dt} + C_m \rho V_0 \frac{du_x}{dt} = C_M \rho V_0 \frac{du_x}{dt} \tag{3}$$

In the engineering design of floating buildings and piled offshore platforms, one of the main problems to be solved is to determine the movement, stress and deformation of these structures under the action of external forces such as wave and wind. We can regard these structures as a dynamic system, the wave action is called the input of the system, and the movement, stress and deformation of the structure are called the output of the system.

Remember this transformation as:

$$y(t) = K[x(t)] \tag{4}$$

For different systems and different inputs, the operator K may have different forms. According to the different operators, dynamic systems can be divided into linear systems and nonlinear systems. An operator is a linear system if it has the following properties, and its operator is denoted by L.

a. Superposition property

$$L[x_1(t) + x_2(t)] = L[x_1(t)] + L[x_2(t)] = y_1(t) + y_2(t) \tag{5}$$

b. The constant α can be removed from the operator

$$L[\alpha x(t)] = \alpha L[x(t)] = \alpha y(t) \tag{6}$$

therefore:

$$L\left[\sum_i \alpha_i x_i(t)\right] = \sum_i \alpha_i y_i(t) \tag{7}$$

It's called the principle of linear superposition, which means that the response of a linear system to inputs is equal to the sum of the responses of the inputs acting independently.

The following operations are linear transformations:

$$y(t) = \frac{\mathrm{d}}{\mathrm{d}t}x(t),\ y(t) = \int_0^T x(t)\mathrm{d}t,\ y(t) = \varphi(t)x(t) \tag{8}$$

In Formula (1-179), $\varphi(t)$ is A non-random function. The above types are linear secondary operations. If a certain function is added, they are called linear non-secondary operations, such as:

$$y(t) = \frac{\mathrm{d}}{\mathrm{d}t}x(t) + \varphi(t) \tag{9}$$

Systems whose operators do not conform to the above conditions are called nonlinear systems. Linear systems are often encountered in practical work, and some nonlinear systems can be linearized within a certain range.

2.2 Large Scale Member

Ship hydrodynamic problems can be solved by frequency domain method. The frequency-domain method is based on the assumption that the wave-ship interaction has lasted for quite a long time, the initial disturbance of the incident wave and the transient influence of the initial rocking of the ship have disappeared, and the fluid motion in the field has reached a steady state. In this case, if the incident wave is harmonic, then the ship's motion is also harmonic (the encounter frequency must be the changing frequency), and the steady-state solution can be obtained in the frequency domain.

Due to the action of waves, the ship has six degrees of freedom besides constant speed forward motion. Assuming that the motion of the six degrees of freedom is small, the ship's center of gravity G point can be at. The three linear displacements (swing, roll, and heave) and the three angular displacements (roll, pitch, and yaw) around the G point in the O-XYZ coordinate system are represented. In the stable state, its displacement vector will be regarded as the harmonic quantity with the encounter frequency field as the changing frequency:

$$\{\eta(t)\} = \{\eta\}e^{i\omega t} = (\eta_1\ \eta_2\ \eta_3\ \eta_4\ \eta_5\ \eta_6)^T e^{i\omega t} \tag{10}$$

According to rigid body dynamics, the ship motion equation with the center of gravity G as the center of moment can be expressed as:

$$[M]\left\{\ddot{\eta}(t)\right\} = \{F(t)\} = \{F\}e^{i\omega t} \tag{11}$$

For the convenience of calculation, the fluid loads acting on the hull are divided into two parts: hydrostatic loads due to changes in the position of the ship's relative hydrostatic equilibrium and hydrodynamic loads dependent on wave and ship motion. Hydrostatic load comes from the contribution of hydrostatic pressure change caused by ship movement, which can be directly given by ship statics as follows:

$$\left\{F^S(t)\right\} = -[C]\{\eta(t)\} \tag{12}$$

Among them, only 5 items of hydrostatic coefficient $C_{ij},\ i,j = 1,2,\cdots,6$ are not zero, they are:

$$\begin{cases} C_{33} = \rho g A \\ C_{35} = C_{53} = -\rho g S_y \\ C_{44} = \rho g \forall h_x \\ C_{55} = \rho g \forall h_y \end{cases} \tag{13}$$

In Formula (13), A and Sy are the waterplane area of the ship and the static moment to the Y-axis, $\forall$ is the drainage volume of the ship, h_x and h_y are the transverse and longitudinal metacentric heights of the ship respectively. According to the different dimensions of the mathematical model used to solve the flow field, it can be divided into three dimensional method and two dimensional method (slice method).

The above equation shows that the free surface condition under low speed has the same form as that under zero speed.

If zero velocity radiation potential ϕ_j^0 and ϕ_j^U additional velocity potential ϕ_j^U are defined, let them satisfy continuity equation [*L*] of the fixed solution, bottom condition [*D*], distant radiation condition [*R*], free surface condition and object surface condition [*S*] defined by the following formula:

$$\begin{cases} \frac{\partial}{\partial n}\phi_j^0 = n_j, (j = 1, 2, ...6)\ \textit{It's on plane S} \\ \frac{\partial}{\partial n}\phi_j^U = m_j, (j = 1, 2, ...6)\ \textit{It's on plane S} \end{cases} \tag{14}$$

For the above fixed solution problem, the disturbance potential ϕ_j and its gradient $\nabla\phi_j(j = 1 \sim 7)$ can be determined by using appropriate numerical solution method. Introducing differential operator:

$$\frac{d}{dt} \equiv \frac{\partial}{\partial t} - U\frac{\partial}{\partial x} = i\omega - U\frac{\partial}{\partial x} \tag{15}$$

According to the linearized Bernoulli equation, the hydrodynamic pressure after deducting the change of hydrostatic pressure is as follows:

$$p(x, y, z, t) = -\rho\frac{d}{dt}\left[\phi_T(x, y, z)e^{i\omega t}\right] \tag{16}$$

3 Wave Load and Structural Response in Frequency Domain

3.1 Wave Load in Frequency Domain

A floating nuclear power plant with a total length of 229.8 m and a total weight of 83,200 tons is taken as the numerical simulation object, and the model is modeled by the APDL module in Ansys 2021R1. The finite element model of hull structure is constructed by Shell181 and BEAM188 elements. The mesh size of the bottom and supporting part of the containment is 0.1 m, the mesh size of the upper part of the containment is 0.2 m, and the mesh size of the rest of the containment is 0.8 m. The total number of the whole ship elements is about 1.56 million. The finite element model of the built floating nuclear power plant and containment vessel is shown in the figure below (Figs. 1 and 2).

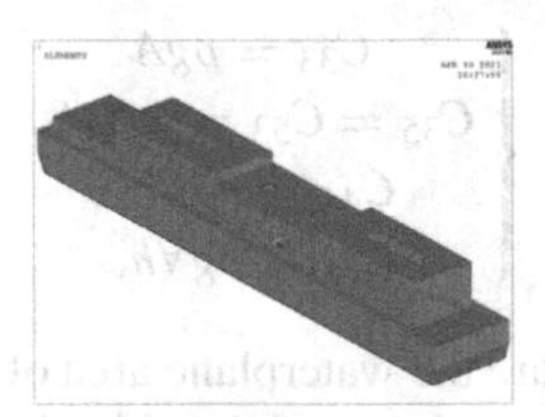

Fig. 1. Overall finite element model of floating nuclear power plant structure

In general, the hydrodynamic response of large floating body structure is a linear system. Therefore, when calculating the hydrodynamic response of the containment vessel of floating nuclear power plant under random waves, regular waves can be used

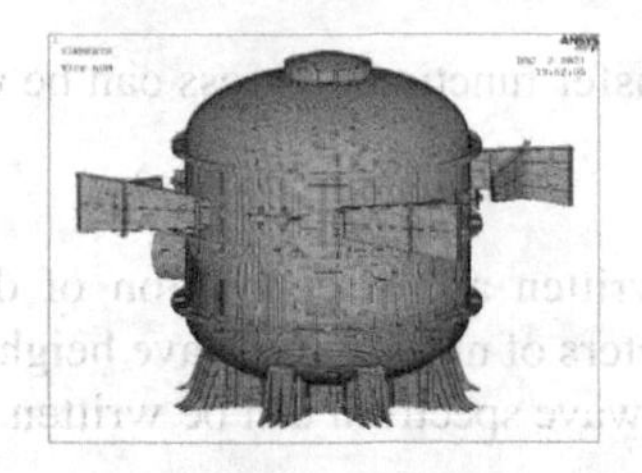

Fig. 2. Finite element model of floating nuclear power plant containment

to calculate first, and the calculated response can be divided by wave amplitude. In this way, the RAO of the floating nuclear power plant containment vessel can be obtained, which preliminarily reflects the hydrodynamic performance of the containment vessel, and then the HD module in Ansys Workbench is used for frequency domain analysis. As the floating nuclear power plant in this paper is symmetric about X-axis and Y-axis, the dynamic response caused by waves within the range of 0°–180° to the containment vessel of floating nuclear power plant is mainly analyzed. A wave direction is set every 15°, which is divided into 13 wave directions in total. The schematic diagram of wave incidence Angle is shown in the figure below (Fig. 3).

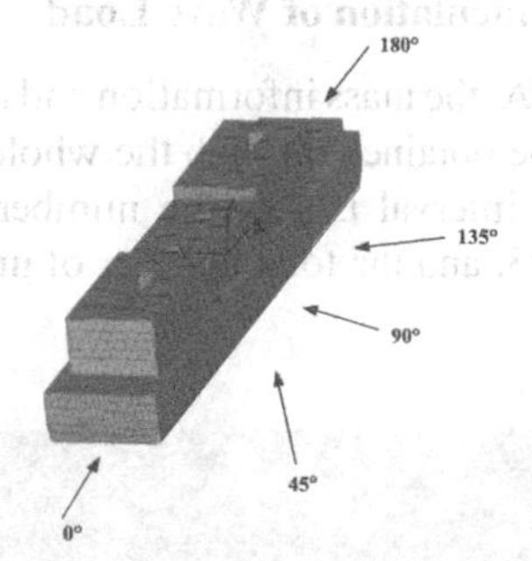

Fig. 3. Diagram of wave incidence Angle

The wave load calculation and structure analysis in this paper are based on linear theory. Under this condition, if the wave is a stationary random process, so is the alternating stress obtained by transformation. According to random process theory, the power spectral density of the above two stationary random processes has the following relationship:

$$G_{XX}(\omega) = |T(\omega)|^2 G_{NN}(\omega) \tag{17}$$

For A linear system composed of ships and waves, the stress response follows the characteristics of the linear system, and the synthetic stress can be written as $\sigma = \sigma_C + i\sigma_S$. In actual calculation, it is necessary to process the loads generated by regular waves of unit amplitude with different frequencies according to the real and imaginary parts respectively to obtain the corresponding response σ_c and σ_s, and then synthesize it into $\sigma_A(\omega_e)$. Thus, the transfer function of stress can be written as:

$$H_\sigma(\omega_e) = \sigma_A(\omega_e) \tag{18}$$

P-m spectrum can be written as the expression of different parameters. If it is expressed by the two parameters of meaningful wave height Hs and mean zero-crossing period *Tz*, the expression of wave spectrum can be written as follows:

$$G_{\eta\eta}(\omega) = \frac{H_S^2}{4\pi}(\frac{2\pi}{T_Z})^2\omega^{-5}\exp(-\frac{1}{\pi}(\frac{2\pi}{T_Z})^4\omega^{-4}) \tag{19}$$

In the analysis, the actual response frequency should be the encounter frequency A, and its relationship with wave frequency A is as follows:

$$\omega e = \omega(1 + \frac{2\omega U}{g}\cos\theta) \tag{20}$$

Therefore, the response spectrum of stress can be expressed as:

$$G_{XX}(\omega_e) = |H_\sigma(\omega_e)|^2 G_{\eta\eta}(\omega_e) \tag{21}$$

3.2 Structural Response Calculation of Wave Load

In the pre-processing of AQWA, the mass information and moment of inertia information of the hull structure need to be obtained through the whole ship finite element analysis. The incident wave direction interval is 15°, the number of wet surface units of the hydrodynamic model is 20403, and the total number of units is 38559, as shown in the figure below (Fig. 4).

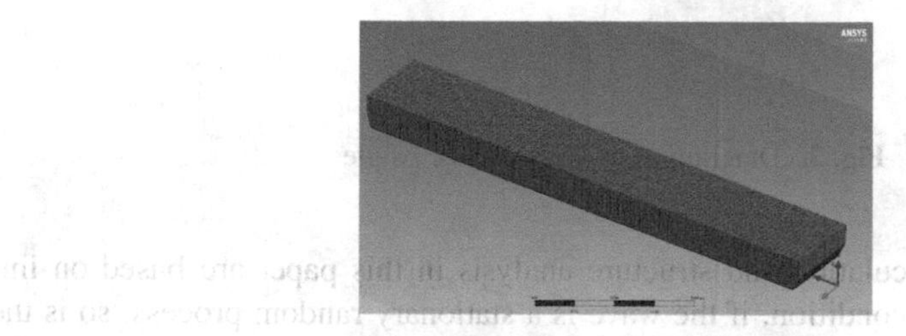

Fig. 4. Hydrodynamic model

The calculated frequencies of waves in the frequency domain were 0.01592 Hz–0.27 Hz, and 48 calculated frequency points were interpolated at equal intervals, totaling 50 calculated frequency points.

In AQWA calculating unit amplitude structural response under the action of amplitude, according to the main control parameters (RY), namely the total longitudinal bending moment, derived the relationship between the frequency and phase, and then according to the wave height (2 m), wave Angle of incidence, frequency and phase extraction wet surface wave pressure and the ship's hull acceleration, and loads it into structural response of the hull computation, as shown in the figure below (Figs. 5 and 6):

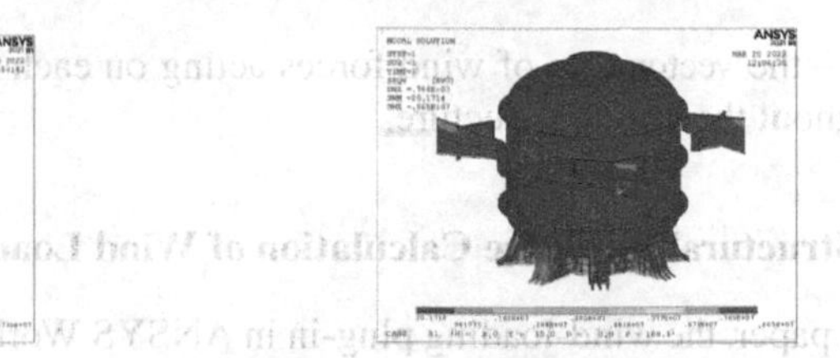

Fig. 5. Stress cloud of containment at wave frequency 0.06139 Hz

Fig. 6. Stress cloud of containment at wave frequency 0.06688 Hz

4 Wind Load and Structural Response

4.1 Wind Load

The wind load rapid loading plug-in currently in use is based on API 4F specification: "Drilling and Workover Structure Specification", 2008 edition. Typical structural analysis is derrick, fan pile leg and jacket platform. The API Wind load Quick loading plugin is limited to API 4F specification profiles and methods and is not intended for general use. Other wind codes (like ASCE 7-05/7-10) are not included in this plugin.

This plug-in has the following advantages:

(1) Suitable for different geometric types. No matter solid, shell or beam structure can take advantage of this plug-in.
(2) Enable load step selection, allowing multiple wind load conditions.
(3) Directly implement API 4F specification, allowing factor coverage.
(4) Wind load can be applied to the leeward side.
(5) The actual windward surface can be detected.

The total wind force on the structure is estimated by the vector sum of the wind force acting on individual components and accessories, as shown in the following formula:

$$F_{\mathrm{m}} = 0.00338 \times K_i \times V_z^2 \times C_s \times A \tag{22}$$

$$F_t = G_f \times K_{sh} \times \Sigma F_m \tag{23}$$

In Eqs. (3-1) and (3-2):

Fm – The force of the wind perpendicular to the vertical axis of a single member, or to the surface of the wind wall, or to the projected area of the appendage.

Ki – a factor of the inclination Angle φ between the longitudinal axis of a single member and the wind.

Vz – Local wind speed at altitude Z.

Cs – shape coefficient.

A – The projected area of A single member is equal to the length of the member multiplied by its projected width with respect to the normal wind component.

Gf – Gust effect factor, used to explain spatial coherence.

Ksh – the conversion factor for the total shielding of a member or accessory and the variation of airflow around the end of the member or accessory.

Ft – the vector sum of wind forces acting on each individual member or accessory throughout the drilling structure.

4.2 Structural Response Calculation of Wind Load

In this paper, the wind-loading plug-in in ANSYS Workbench is used. Firstly, the structural model of floating nuclear power plant is fixed rigidly and released inertia, and then constant Wind load is carried out on the Wind receiving surface of floating nuclear power plant (Wind speed 50.7 = 98.56 knot, Wind direction: 13, 0° to 180° with wind direction set at 15° intervals), at the same time, then export the result file, and perform post-processing in classic ANSYS.

The wind speed is 50.7 m/s, and the maximum equivalent stress of the containment is 8.15 MPa. The maximum stress occurs at the junction between the upper support and the bulkhead, as shown in the figure below (Figs. 7, 8 and 9).

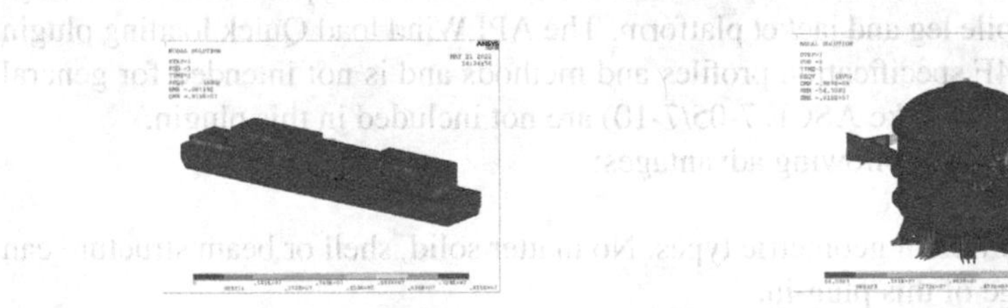

Fig. 7. Ship - wide stress cloud

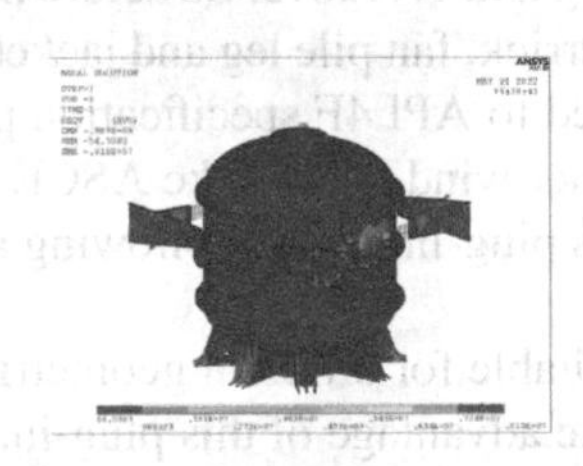

Fig. 8. Containment stress cloud map

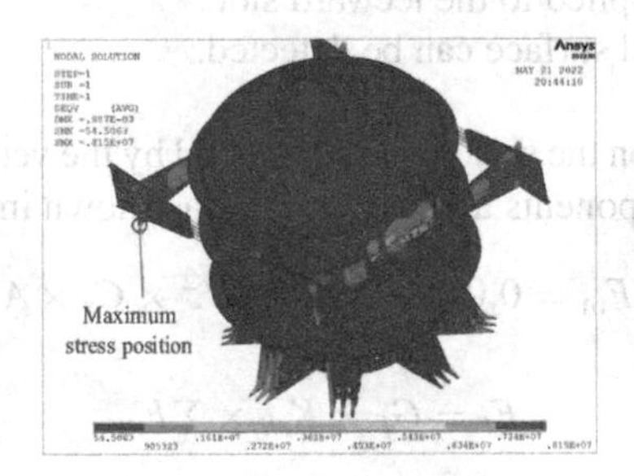

Fig. 9. Maximum stress and position of containment under wind load

5 Flow Load and Structural Response

5.1 Current Loading

Ocean currents can be caused by many factors, such as local stationary currents caused by ocean circulation, tidal currents caused by periodic changes in the gravitational pull of the sun and moon on the Earth, differences in the density of ocean water, and the action of wind. It should be pointed out that the speed of wind on the sea surface is about 3% of the speed at 10 m above the sea surface. Tidal currents have an important influence on the flow field in some restricted waters. Tidal currents in restricted waters generally have a speed of 2–3 m/s and a maximum of 10 m/s.

For floating bodies, the flow of the ocean surface is of greatest concern to us. However, for the mooring system at sea, the distribution of water flow along the water depth is also our concern. For the designers, the maximum limit flow encountered during the operation of floating body is the most important factor affecting the design, so the actual measurement and monitoring of water flow velocity is essential. Since the velocity and direction of the water flow change slowly, we can approximately consider the water flow to be steady.

The action of water flow on floating body can be divided into the following two parts:

(1) Viscosity effect. Viscosity resistance due to frictional effects, and differential pressure resistance. For blunt body, the friction resistance can be ignored, and the pressure difference resistance is mainly.

(2) Influence of potential flow. The lift effect caused by the ring volume and the drag effect caused by the free surface effect are small in comparison.

Using flow force coefficient to estimate the flow load on the surface ship floating body:

The flow force/moment can be calculated by the following formula, in which the flow force coefficients are usually determined by model test methods.

$$\left.\begin{aligned} X_c &= \frac{1}{2}\rho v_c^2 C_{X_c}(\alpha_c) A_{TS} \\ Y_c &= \frac{1}{2}\rho v_c^2 C_{Y_c}(\alpha_c) A_{LS} \\ N_c &= \frac{1}{2}\rho v_c^2 C_{N_c}(\alpha_c) A_{LS} L \end{aligned}\right\} \tag{24}$$

Similar to the wind load calculation of ship type floating body, the key to the flow load of ship type floating body is how to obtain the flow load coefficient, which is mainly obtained by model test.

For the calculation of flow load of large oil tankers, OCIMF based on model test data gives flow force coefficient curves of two different bow forms, full load and ballast, which have good reference value and are widely used in engineering design and analysis of mooring ships.

Remery and Van Oortmerssen carried out an experimental study in the MARIN Tank to test the flow loads on tanker models of different profiles and sizes. Since ships are mostly slender bodies, the axial flow load is mainly caused by frictional resistance. If the axial flow velocity is small, this resistance is difficult to measure, and it is not accurate to predict the axial resistance of real ships from model tests, because the scale effect is very obvious.

The axial flow load is important for anchored vessels. The force can be estimated based on the frictional resistance of the plate. The following formula is recommended by ITTC:

$$X_c = \frac{0.075}{(\log_{10}(R_N - 2))} \cdot \frac{1}{2}\rho V_c^2 \cos\alpha_c \cdot |\cos\alpha_c| \cdot S \tag{25}$$

$$\mathrm{Re} = \frac{|\cos\alpha_c|V_c \cdot L}{\upsilon} \tag{26}$$

For tanker, it is generally not a problem to estimate the transverse force and yawing moment of the real ship. For the transverse flow of ocean current in an oil tanker, the ship can be regarded as a blunt body. Because the bilge radius of the hull is relatively small, it can be considered that the flow separation situation in the model and the real ship is consistent, and the transverse force and the bow rolling moment can be considered independent of the Reynolds number.

In MARIN pool, the lateral force and bow rolling moment are expanded into Fourier series through experiments.

$$C_{Y_c}(\alpha_c) = \sum_{1}^{n} b_n \cdot \sin(n \cdot \alpha_c) \tag{27}$$

$$C_{N_c}(\alpha_c) = \sum_{1}^{n} c_n \cdot \sin(n \cdot \alpha_c) \tag{28}$$

The above formula can be applied in deep water, but for shallow water, the lateral force and bow torque coefficients need to be multiplied by a correction factor. In practice, the free surface effect is small for deep water with flow rates of 3kn.

5.2 Structural Response Calculation of Flow Load

In this paper, Workbench Fluent and classical ANSYS 2021R1 are used to realize flow field analysis and structural response calculation. The viscous flow model, namely K-Omega (2EQN) SST model, is selected for flow field analysis. A THREE-DIMENSIONAL flow field model is established in Workbench DM module, and the flow field size is 500 × 600 × 27.6 (m), as shown in the figure below (Fig. 10):

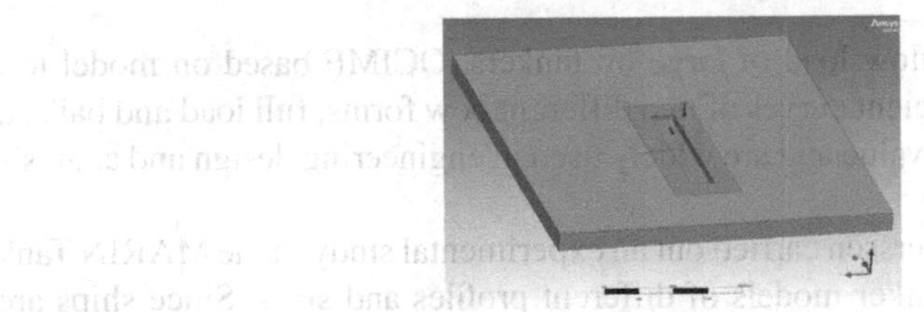

Fig. 10. 3d model of flow field

The mesh of fluid domain is divided in Workbench mesh module. The mesh size of near-field fluid domain is 1.5 m, and that of far-field fluid domain is 3.5 m. The flow field and structural stress (75° flow direction as an example) corresponding to flow velocity of 0.83 m/s are shown as follows (Fig. 11):

At a flow rate of 0.83 m/s and a flow direction of 75°, the maximum equivalent stress of the containment fourth strength is 1.15 MPa. The maximum stress occurs at the arc transition of the equipment base inside the containment vessel, as shown below (Fig. 12).

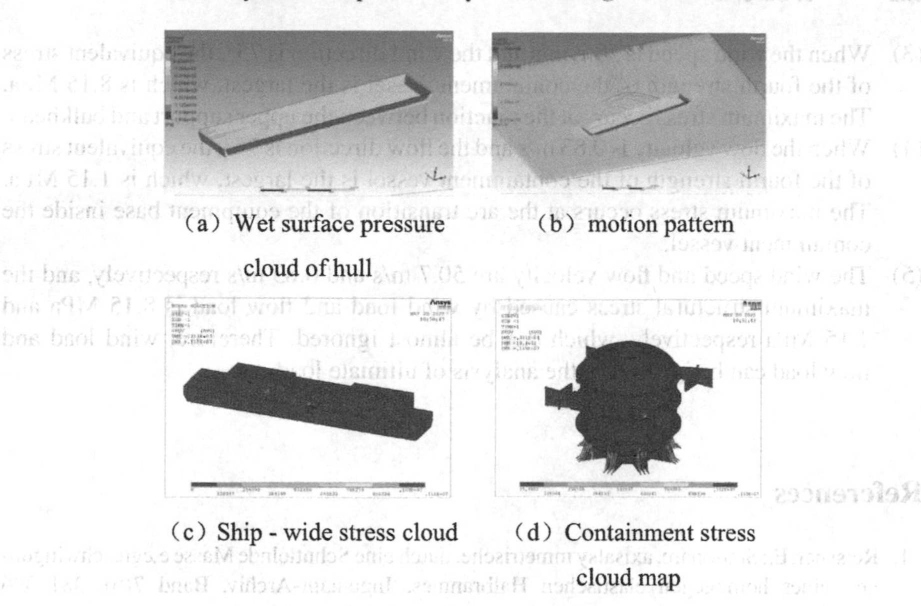

(a) Wet surface pressure cloud of hull (b) motion pattern

(c) Ship - wide stress cloud (d) Containment stress cloud map

Fig. 11. Calculation results of 75° flow direction

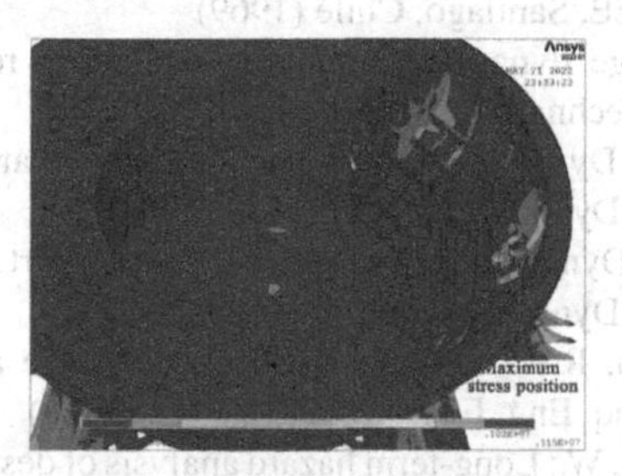

Fig. 12. Maximum stress position

6 Conclusions

(1) Under the action of wave load in the frequency domain, the maximum load component (absolute value) of the containment support is as follows: $F_X = 2.68 \times 10^5 N$ (wave direction 60°), $F_Y = 5.89 \times 10^6 N$ (wave direction 90°), $F_Z = 1.16 \times 10^6 N$ (wave direction 90°), $M_X = 3.93 \times 10^7 N \cdot m$ (wave direction 90°), $M_Y = 1.93 \times 10^6 N \cdot m$ (wave direction 75°), $M_Z = 1.13 \times 10^5 N \cdot m$ (wave direction 105°). This part of the resultant force only includes the structural response caused by waves, not the structural response caused by static equilibrium.

(2) Under the action of wave load in the frequency domain, the maximum equivalent stress of the containment vessel appears when the wave direction is 90° and the frequency is 0.09433 Hz, and the size is 98.3 MPa. The maximum stress occurs at the junction between the upper support and bulkhead.

(3) When the wind speed is 50.7 m/s and the wind direction is 75°, the equivalent stress of the fourth strength of the containment vessel is the largest, which is 8.15 MPa. The maximum stress occurs at the junction between the upper support and bulkhead.
(4) When the flow velocity is 0.83 m/s and the flow direction is 75°, the equivalent stress of the fourth strength of the containment vessel is the largest, which is 1.15 MPa. The maximum stress occurs at the arc transition of the equipment base inside the containment vessel.
(5) The wind speed and flow velocity are 50.7 m/s and 0.83 m/s respectively, and the maximum structural stress caused by wind load and flow load is 8.15 MPa and 1.15 MPa respectively, which can be almost ignored. Therefore, wind load and flow load can be ignored in the analysis of ultimate load.

References

1. Reissner, E.: Stationare, axisalsy mmetrische, dutch eine Schuttelnde Masse e,egte. chwingungen eines homogegen elastischen Halbranmes. Ingenianr-Archiv, Band **7**(6), 381–396 (1936)
2. Chopra, A.K., Perumalswami, P.R.: Dam foundation interaction during earthquakes. In: Proceedings of 4th WCEE, Santiago, Chile (1969)
3. Gou, M.: Pile-soil-bridge dynamic interaction in seismic response analysis of Bridges. Qingdao University of Technology, Shandong (2011)
4. Gazetas, G., Mrakis, N.: Dynamic pile-soil-pile interaction part I: analysis of axial vibration. Earthquake Eng. Struct. Dynam. **20**(2), 115–132 (1991)
5. Mrakis, N., Gazetas, G.: Dynamic pile-soil-pile interaction part II: analysis of lateral vibration. Earthquake Eng. Struct. Dynam. **21**(2), 145–162 (1992)
6. Fan, M., Xie, M.Y., Bu, R.F.: Nonlinear seismic response analysis of soil-pile-structure interaction system. Earthq. Eng. Eng. Vib. **5**(3), 6–12 (1985)
7. Wang, Y., Mao, X., Jiang, W.: Long-term hazard analysis of destructive storm surges using the ADCIRC-SWAN model: a case study of Bohai Sea, China. Int. J. Appl. Earth Obs. Geoinf. **73**, 52–62 (2018)
8. de Waal, D.J., van Gelder, P.H.A.J.M.: Modelling of extreme wave heights and periods through copulas. Extremes **8**(4), 345–356 (2005)
9. De Michele, C., Salvadori, G., Passoni, G., et al.: A multivariate model of sea storms using copulas. Coast. Eng. **54**(10), 734–751 (2007)
10. Xu, H., Xu, K., Lian, J., et al.: Compound effects of rainfall and storm tides on coastal flooding risk. Stoch. Env. Res. Risk Assess. **33**(7), 1249–1261 (2019)
11. Dong, S., Zhai, J.J., Tao, S.S.: Combined statistical analysis of wind and wave based on Archimedean Copula function. J. Ocean Univ. China (Nat. Sci. Edn.) **44**(10), 134–141 (2014)
12. Chen, M.L., Ji, C.Y., Liu, Z.: Hydrodynamic analysis and wave load prediction for semi-submersible offshore platforms. Mar. Technol. **31**(4), 68–70 (2012)
13. Zhou, S.L., Nie, W., Bai, Y.: Design and research on mooring system of deep water semi-submersible platform. Ship Mech. **14**(5), 495–502 (2010)

Research on Process Diagnosis of Severe Accidents Based on Deep Learning and Probabilistic Safety Analysis

Zheng Liu[1(✉)] and Hao Wang[2]

[1] CNNC Key Laboratory on Severe Accident in Nuclear Power Safety, CNPE, Beijing, China
liuzheng.2009@tsinghua.org.cn

[2] Department of Engineering Physics, Tsinghua University, Beijing, China

Abstract. Severe accident process diagnosis provides data basis for severe accident prognosis, positive and negative effect evaluation of Severe Accident Management Guidelines (SAMGs), especially to quickly diagnose Plant Damage State (PDS) for operators in the main control room or personnel in the Technical Support Center (TSC) based on historic data of the limited number of instruments during the operation transition from Emergency Operation Procedures (EOPs) to SAMGs. This diagnosis methodology is based on tens of thousands of simulations of severe accidents using the integrated analysis program MAAP. The simulation process is organized in reference to Level 1 Probabilistic Safety Analysis (L1 PSA) and EOPs. According to L1 PSA, the initial event of accidents and scenarios from the initial event to core damage are presented in Event Trees (ET), which include operator actions following up EOPs. During simulation, the time uncertainty of operations in scenarios is considered. Besides the big data collection of simulations, a deep learning algorithm, Convolutional Neural Network (CNN), has been used in this severe accident diagnosis methodology, to diagnose the type of severe accident initiation event, the breach size, breach location, and occurrence time of the initial event of LOCA, and action time by operators following up EOPs intending to take Nuclear Power Plant (NPP) back to safety state. These algorithms train classification and regression models with ET-based numerical simulations, such as the classification model of sequence number, break location, and regression model of the break size and occurrence time of initial event MBLOCA. Then these trained models take advantage of historic data from instruments in NPP to generate a diagnosis conclusion, which is automatically written into an input deck file of MAAP. This input deck originated from previous traceback efforts and provides a numerical analysis basis for predicting the follow-up process of a severe accident, which is conducive to severe accident management. Results of this paper show a theoretical possibility that under limited available instruments, this traceback and diagnosis method can automatically and quickly diagnose PDS when operation transit from EOPs to SAMGs and provide numerical analysis basis for severe accident process prognosis.

Keywords: Process Diagnose · Process Traceback · EOP · SAMG · Deep Learning · PSA · ET · MAAP

C. Liu (Ed.): PBNC 2022, SPPHY 283, pp. 624–634, 2023.
https://doi.org/10.1007/978-981-99-1023-6_54

1 Introduction

The application of the concept of defense in depth is the primary means of preventing accidents in a nuclear power plant and mitigating the consequences of accidents if they do occur [1]. If an accident occurs at a nuclear power plant, to restore safety, two types of accident management guidance documents are typically used: EOP for preventing fuel rod degradation, and SAMG for mitigating significant fuel rod degradation when a severe accident is imminent [2]. And the relationship between different components of an Accident Management Programme (AMP) is illustrated in Fig. 1. Preventive accident management EOPs integrate actions and measures needed to prevent or delay severe damage to the reactor core. Mitigatory accident management SAMGs refers to those actions or measures which become necessary if the preventive measures fail and severe core damage occurs or is likely to occur [3].

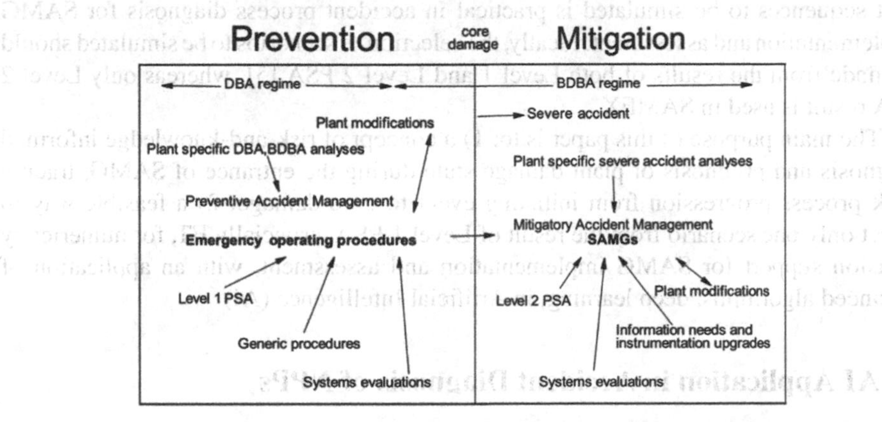

Fig. 1. Relationship between EOPs and SAMGs

Severe accident management guidelines should be comprehensively Verified and Validated [4] (V&V). Implementing and enhancing the existing SAMG program in NPPs is an important post-Fukushima activity [5]. Expert judgment, simulators, field training, tabletop exercises, emergency drill, and exercise and analysis are current practices related to SAMG V&V. Moreover, informing SAMG and actions through analytical simulation is a practical and commendable practice [5].

While numerical simulation is scenario/event based from initiating event to core damage, and even to fission product release to the off-site environment. There is a great challenge that SAMGs take an integrated, symptom-and-knowledge-based approach [5], and that means, SAMGs contain actions to be taken that are based on the values of directly measurable plant parameters [6].

SAMEX, a decision support system, is developed for use in severe accident management following an incident at a nuclear power plant [6]. Risk-informed severe accident risk database management module (RI-SARD), which is a risk-informed accident diagnosis and prognosis module of SAMEX [7], examines (a) a symptom-based diagnosis of a plant damage state (PDS) sequence in a risk-informing way and (b) a PDS sequence-based prognosis of key plant parameter behavior, through a prepared database, SARDB, which stores the data of integrated severe accident analysis code results like MAAP and MELCOR for hundreds of high-frequency scenarios from the plant damage state event tree of a Level 2 PSA. RI-SARD predicts a series of potential severe accident sequences that match the user-specified symptom criteria, and then prioritizes and adds more symptoms to screen down scenarios for prognosis analysis.

The accident process diagnosis and prognosis method, with fast-running severe accident codes and scenario selection, used in SAMEX, fills the gap between symptom-based analysis and event-based analysis. But with the given symptom of key parameters of NPPs, a small set of initiating events and sequences may lead to a similar plant damage state, and this method is not precisely focused on one scenario, which means multiple parallel sub-section simulations should be conducted to provide information for users to select.

Numerically reproducing symptom-based scenarios and selecting a small set of accident sequences to be simulated is practical in accident process diagnosis for SAMG implementation and assessment. Ideally, the selection of scenarios to be simulated should be made from the results of both Level 1 and Level 2 PSA [5], whereas only Level 2 PSA result is used in SAMEX.

The main purpose of this paper is to: 1) a concept of risk-and-knowledge informed diagnosis and prognosis of plant damage state during the entrance of SAMG, tracing back process progression from initiating event to core damage; 2) a feasible way to select only one scenario from the result of Level 1 PSA, especially ET, for numerically decision support for SAMG implementation and assessment, with an application of advanced algorithms, deep learning, in Artificial Intelligence (AI).

2 AI Application in Accident Diagnosis of NPPs

Since the 1980s, with the field of artificial intelligence evolved, kinds of artificial intelligence methodologies are implemented for accident diagnosis in NPPs.

Jaques Reifman [8] provides a comprehensive survey of computer-based diagnostic systems using artificial intelligence techniques that have been proposed for the nuclear industry up to 1997. Two computing tools: expert systems or Artificial Neural Networks (ANNs), are used for artificial intelligence-based systems for process diagnostics. Hybrids of them and a combination of these technologies with numerical quantitative simulation programs have also been proposed.

The early diagnostic approaches were based on expert systems, and then ANNs before the 21st century, which follows chronologically the popularity of these artificial intelligence technologies.

Kinds of diagnosis systems were developed based on an expert system. REACTOR, an expert system for diagnosis and treatment of nuclear reactor accidents, was developed at the beginning of the 1980s in the U.S.A [9]. DISKET, which is based on knowledge engineering in the field of expert systems, has been developed to identify the cause and the type of abnormal transient of a nuclear power plant in the middle of the 1980s in Japan [10]. ADAM, an accident diagnostic, analysis, and management system application for severe accident simulation and management in the 2000s in the U.S.A [11]. Even in the past few years, a decision support system called SEVERA was developed in Europe [12], aimed at supporting the decision-making team during an accident or a training exercise, using decision modeling software DEXi, which is based on if-then rules.

The ANNs method application for diagnosis in NPPs is data-driven with model training and is mainly trained with simulation data of NPPs. Diagnosis result is highly dependent on the quality of simulation results. At the early stage, numerical simulation is not quite competent for accident progression. No systemic application was built based on ANNs, only in some specific domains, ANNs were applied as a methodology basis.

But nowadays, integrated simulation code, MAAP or MELCOR, is advanced involved in the past few decades. And several diagnosis systems were developed. ADAS, using neural networks, was developed in the 2000s in Korea for accident diagnosis and support operator decision-making [13]. A severe accident diagnosis and response support system was developed in China, which uses three diagnostic methods including BP neural network method, SDG expert diagnosis, and artificial diagnosis [14].

Moreover, some other machine learning methods emerged and are applied for accident diagnosis in NPPS. A cascaded support vector regression (CSVR) model is used to predict accident scenarios, accident locations, and accident information [15]. An approach for diagnosis of multiple failures based on dynamic Bayesian networks (DBNs) is proposed to support emergency response in case of an incident [16].

Nowadays, the most popular AI technology is Deep Learning, which is a subset of machine learning and essentially a neural network with three or more layers. Related algorithms in this domain include Convolutional Neural Networks (CNN), Recurrent Neural Networks (RNN), Auto Encoder (AE), Generative Adversarial Network (GAN), etc. These algorisms greatly improve the ability to solve practical issues, such as image recognition, speech recognition, auto driving, language translation, etc. And inspires new prosperity of AI applications.

Although numerical simulation is applied to generate accident data, those AI methods mentioned above do require not a very large dataset to train the model, hundreds or even thousands of simulations may be enough. As a drawback, the diagnosis capability may be limited.

3 Traceback Accident Progression Methodology During Transition from EOP to SAMG

Following the occurrence of initiating event, EOPs are step-by-step procedures for operators in the main control room to put the NPPs into a safe state. Prevention efforts, including safety system function and human performance, with success or failure execution, direct to different branches of the progression path, which are static sets of ETs of L1 PSA. Key procedures in EOPs are simplified as header events of ETs, for example, depressurizing the primary system. This path with the uncertainty of time may produce tens of thousands of scenarios.

According to different SAMG entrance conditions, for example, core exit temperature exceeds 650 °C, TSC is formed to mitigate accident progression from large nuclide release. Plant damage state assessments are required when operation transitions from EOPs to SAMG.

With fast-running simulation code, qualitatively PDS and SAMG action assessment is conducive for accident management. But the obstacle gap between symptom-based SAMG and event-based numerical analysis needs to be filled first.

This paper promotes a traceback method for TSC, which can automatically generate an input deck of simulation code, based on chronological parameter data from initiating event occurrence time to time when SAMG entrance condition is realized. And this method is data-driven, using the good performance of deep learning algorithms to diagnose header events and their time to define a relatively reasonable scenario that confines symptoms on assessment time. It's their advantage to diagnose from big datasets for DL algorithms, while with a little number of datasets, they may diagnose with poor capability.

3.1 Traceback Methodology

The main steps are described as follows:

1) Analysis of event trees and header events selection

First, a review of event trees from publicly available Level 1 PSA is performed to select the main header events for every sequence path which causes core damage.

2) Sequences simulation and database generation

With the previously identified header events delineated for each sequence, using a simulation tool, and with the occurring time uncertainty branch, a database is produced with a large amount of CPU calculation work. The branching method will be explained in the following section. Selection of key parameters as the figure of merits is performed for simulation output.

3) Traceback model training for scenario confinement

Using DL algorithms, classification and regression models are trained with formerly generated scenarios database. This may last for weeks. Different kinds of DL algorithms are tried to be used, and hyperparameters are tuned with effort.

4) PDS diagnose with simulation and real-time instruments indicator

Put chronological real-time instrument data into DL-trained models to diagnose confined scenarios, and then generate an input deck of the simulation tool, ranging from initiating event occurrence time to the time when core exit temperature exceeds 650 °C. This defined scenario is a base scenario for the following accident simulation.

5) Prognosis and SAMG action assessment

Put SAMG action into input deck of base scenario, and following simulation code execution, which runs fast ahead of real-time, will prognosis the following accident progression with/without SAMG action.

3.2 Header Event Branches

According to step 2, header event branching is conducted in dealing with scenario uncertainties.

The third issue described in the report of OECD/NEA [5] is the treatment of simulation uncertainty for assessing SAM actions using an analytical tool. The best-estimate approach is recommended for analytical simulation because a conservative approach may not be of much help and sometimes could even lead to a wrong decision.

Despite code uncertainties, representation uncertainties, numerical inadequacies, user effects, computer/compiler effects, and plant data uncertainties for the analysis of an individual event, branching time is one important uncertainty for sequence definition. Lots of evaluation effort has been done for computer code uncertainty.

While some safety systems function automatically, there is still uncertainty between simulation and real-time accident progression. In addition, automatic safety systems may fail and manual work with delayed time may still be needed to define accident scenarios.

Besides, manual actions following up EOPs may delay branching time for diagnosis time, preparation time, and execution time.

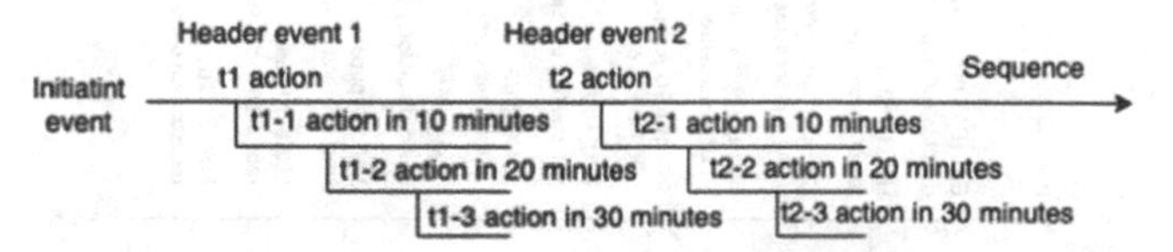

Fig. 2. Event tree branches

Kinds of breaching time for both safety systems and manual actions are simulated for scenarios, as shown in Fig. 2. For one sequence of ETs, header event 1 action should be done at time t1 as action condition is fulfilled in simulation, three branches deviated t1 time, which are 10 min after t1, 20 min after t1 and 30 min after t1. The safety system or manual actions at header event 2 is similar to header event 1, and three branches are formed with different delay times. And another three branches for t2 action are formed for a scenario with t1–2 action in 20 min. So, the permutation of delayed action time for header events makes a very large number of scenarios for one sequence of ETs.

4 Case Study – Medium Break LOCA (MBLOCA) in a Pressurized Water Reactor (PWR)

This section presents an example application of the above-referred traceback methodology to define a suitable input deck of accident progression scenario from initiating event to core exit temperature exceeding 650 °C in no more than one day, which is the basic scenario for SAMG actions implementation assessment.

The initiating event of this case study is the middle break loss of coolant accident in a pressurized water reactor. According to L1 PSA, two ETs are analyzed which differ in the position of break, one break is on one hot leg, and the other one is on one cold leg. Only hog leg ETs is analyzed in this case study.

4.1 Event Tree Analysis and Header Events Selection

The event tree of MBLOCA with a break on the hot leg is depicted in Fig. 3. There are a total of 21 sequences in this ET, 4 sequences result in bringing back to safety status (OK), and the remaining 17 sequences lead to core damage (CD), which cause different plant damage state.

Among these 17 sequences, some of them may not need to be distinguished due to their header events. For example, sequences 7, 8, and 9 are like sequences 11, 12, and 13 respectively. The difference between them is human error in safety actions or the mechanical inherent failure of these safety systems. So, 13 sequences causing CD are screened for the next steps.

Not all of the header events in Fig. 3 are used for each sequence. For example, only E01, H01, and E02 are used for scenario 2. And head events needed for each sequence are analyzed for each sequence.

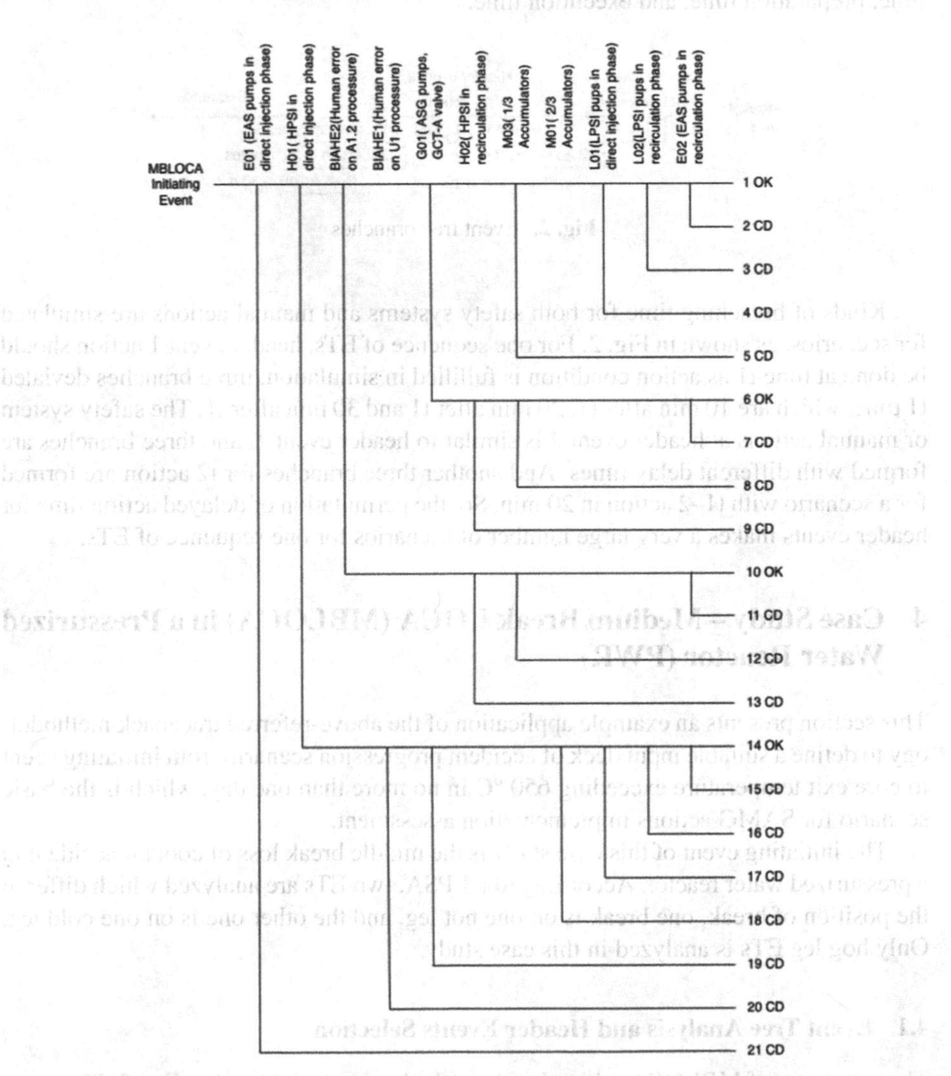

Fig. 3. MBLOCA Event Trees

4.2 Event Tree Scenarios Simulation with Branches

Modular Accident Analysis Program (MAAP) is used as a numerical analysis tool for scenario simulation.

For every one of the screened 13 sequences, 10 different initiating event occurrence times, 10 break sizes, 2 break locations (near/not pressurizer), and 10 branching times for each header event are analyzed and simulated. The branching point is managed in the above-referred method. All these simulation results generate a database of knowledge on accident progression. Some parameters chosen as figure-of-merits (FOMs) are listed in Table 1.

Table 1. FOMs of simulation

No.	Parameter	Unit
1	The collapsed water level in the broken steam generator downcomer	m
2	The collapsed water level in the unbroken steam generator downcomer	m
3	Pressure in broken steam generator	Pa
4	Pressure in unbroken steam generator	Pa
5	Pressure in the primary system	Pa
6	Pressure in pressurizer	Pa
7	Pressure in accumulator	Pa
8	Core exit temperature	K
9	Boiled-up water level measured from the bottom of RPV	m
10	The collapsed water level in the cavity room	m
11	Pressure in cavity room	Pa
12	The temperature of the gas in the cavity room	K
13	The temperature of water in the cavity room	m
…	…	…

4.3 DL Models Training

1D-CNN (one-dimension convolutional neural network) algorithm is used for this modeling task. The knowledge database formally generated is used as training and testing data for classification and regression models.

The classification models include the sequence number of ETs, and break location near or not near the pressurizer.

And regression models include initiating event occurrence time, break size, and branching time for every header event in every sequence.

The result of the training models is listed in Table 2.

Due to the large scale of scenarios, weeks of training time may be needed for this task.

Table 2. Result for sequence 2 of ETs in MBLOCA

No.	Type	Description	Accuracy/Error
1	Classification	Sequence No 2	85.94%
2	Classification	If break on pressurizer side	86.09%
3	Regression	Break size	0.0262 (0.8 quantile)
4	Regression	E01 time	0.2231 (0.8 quantile)
5	Regression	H01 time	0.0944 (0.8 quantile)
6	Regression	E02 time	0.0124 (0.8 quantiles)
7	Regression	Initiating event occurrence time	0.2258 (0.8 Quantile)

4.4 Trace Back Accident Progression

Accident data from the simulator or MAAP is generated for testing 1D-CNN trained models. In the case of real implementation, parameters of instrument indicators are used for gathering chronological data of accident progression.

The task in this section is to automatically generate a defined input deck of MAAP, which relatively accurately depicts the chronological progression from initiating event to core exit temperature exceeding 650 °C. And the methodology is to input chronological data to 1D-CNN trained models and get classification and regression data for each model. These tracked back model data are written to an input deck of MAAP.

Based on this section, a basic scenario for testing accident data is confined to an input deck of MAAP.

4.5 Application of Basic Scenario for SAMG Assessment

With the above-referred tasks, TSC can diagnose plant damage state using an input deck of basic scenarios. For the prognosis of an accident, TSC personnel put related action commands for SAMG actions into this basic input deck, with the calculation result of MAAP, the negative or positive effect of these actions can be foreseeable with fast running code. And the simulation result is used for decision support of TSC.

5 Conclusions

Numerical analysis is a practical tool for SAMG assessment. But the gap between symposium-based SAMG actions and event/scenario-based numerical simulation is the first obstacle to on-time decision support. Although methods of using L2 PSA are provided, the drawback of several screened scenarios are not focused and may generate controversial calculation results.

Event trees of L1 PSA are used for scenarios database generation, and header events selection is performed to generate the skeleton of scenarios.

Uncertainty of scenarios is considered in this traceback method, with different time branching points for each header event of scenarios, and confidence is gained for the following decision-making support.

The method described in this paper is good at automatically generating an input deck for a basic scenario ranging from initiating event to core exit temperature exceeding 650 °C. And the track back result fills the gap between symposium-based SAMG actions and event-based simulations. The prognosis for SAMG assessment can be realized based on this methodology.

References

1. Safety of nuclear power plants: Design. International Atomic Energy Agency, Vienna (2016)
2. Accident management programmes in nuclear power plants. International Atomic Energy Agency, Vienna n.d.
3. Implementation of accident management programmes in nuclear power plants. International Atomic Energy Agency, Vienna (2004)
4. Accident management insights after the Fukushima Daiichi NPP accident report of the CNRA task group on accident management, p. 72 (2014)
5. Th, S.: Informing severe accident management guidance and actions for nuclear power plants through analytical simulation n.d.
6. Park, S.-Y., Ahn, K.-I.: SAMEX: a severe accident management support expert. Ann. Nucl. Energy **37**, 1067–1075 (2010). https://doi.org/10.1016/j.anucene.2010.04.014
7. Ahn, K.-I., Park, S.-Y.: Development of a risk-informed accident diagnosis and prognosis system to support severe accident management. Nucl. Eng. Des. **239**, 2119–2133 (2009). https://doi.org/10.1016/j.nucengdes.2009.06.001
8. Reifman, J.: Survey of artificial intelligence methods for detection and identification of component faults in nuclear power plants. Nucl. Technol. **119**, 76–97 (1997). https://doi.org/10.13182/NT77-A35396
9. Nelson, W.R.: REACTOR: an expert system for diagnosis and treatment of nuclear reactor accidents. In: AAAI (1982)
10. Yokobayashi, M., Yoshida, K., Kohsaka, A., Yamamoto, M.: Development of reactor accident diagnostic system DISKET using knowledge engineering technique. J. Nucl. Sci. Technol. **23**, 300–314 (1986). https://doi.org/10.1080/18811248.1986.9734987
11. Zavisca, M.J., Khatib-Rahbar, M., Esmaili, H., Adam, S.R.: An accident diagnostic, analysis and management system—applications to severe accident simulation and management. In: 10th International Conference on Nuclear Engineering, vol. 2, pp. pp. 131–136. ASMEDC, Arlington, Virginia, USA (2002). https://doi.org/10.1115/ICONE10-22195
12. Bohanec, M., Vrbanić, I., Bašić, I., Debelak, K., Štrubelj, L.: A decision-support approach to severe accident management in nuclear power plants. J Decis Syst **29**, 438–449 (2020). https://doi.org/10.1080/12460125.2020.1854426
13. Lee, S.J., Seong, P.H.: A dynamic neural network based accident diagnosis advisory system for nuclear power plants. Prog. Nucl. Energy **46**, 268–281 (2005). https://doi.org/10.1016/j.pnucene.2005.03.009
14. Chen, P., Xu, W., Yang, F., Liao, Y.: Introduction of three methods used for the nuclear accident diagnosis. In: Decontamination and Decommissioning, Radiation Protection, Shielding, and Waste Management; Mitigation Strategies for Beyond Design Basis Events, vol. 7, p. V007T11A007. American Society of Mechanical Engineers, Shanghai, China (2017). https://doi.org/10.1115/ICONE25-66433

15. Yoo, K.H., Back, J.H., Na, M.G., Hur, S., Kim, H.: Smart support system for diagnosing severe accidents in nuclear power plants. Nucl. Eng. Technol. **50**, 562–569 (2018). https://doi.org/10.1016/j.net.2018.03.007
16. Zhao, Y., Tong, J., Zhang, L., Wu, G.: Diagnosis of operational failures and on-demand failures in nuclear power plants: an approach based on dynamic Bayesian networks. Ann. Nucl. Energy **138**, 107181 (2020). https://doi.org/10.1016/j.anucene.2019.107181

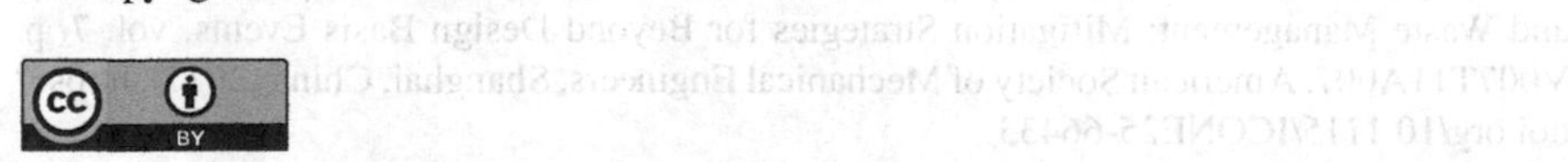

Study on Calculation Method of Corrosion Product Source Term in Lead-Bismuth Fast Reactor Coolant System

Haixia Wan(✉)

China Institute of Atomic Energy, Beijing, China

why_1022@163.com

Abstract. In the coolant system of lead-bismuth fast reactor, the corrosion products produce great occupational radiation dose to the workers, especially in the process of maintenance and repair of nuclear facilities. Therefore, it is very important to accurately calculate the corrosion product source term caused by the reaction between coolant and structural materials. The generation, migration, decay and deposition of corrosion products are described by the corrosion characteristics of lead-bismuth alloy on stainless steel in coolant loop, and the mathematical equilibrium equation is established. The equation is used to calculate the corrosion product source term of the coolant loop in the $20MW_{th}$ lead-bismuth fast reactor and the variation laws of the corrosion product with time are obtained. The results show that the radioactivity of the corrosion products mainly comes from nuclides such as ^{51}Cr, ^{54}Mn, ^{58}Co and ^{60}Co, and the short-lived nuclides such as ^{51}Cr and ^{58}Co decay gradually after shutdown, long-lived nuclides such as 54Mn and ^{60}Co are the main sources of radioactivity.

Keywords: Corrosion Product · Coolant · Lead-Bismuth Alloy · Stainless Steel · Activity

1 Introduction

Lead-bismuth fast reactor is one of the six main types of the fourth generation reactor, which has the characteristics of safety, economy, continuity and nuclear non-proliferation. However, the corrosion ability of lead-bismuth alloy cannot be neglected, and the corrosion products caused by it should not be underestimated. Studies have shown that more than 90% of occupational exposure is caused by corrosion products deposited on the pipe wall from the coolant, which continue to decay and emit gamma rays in the pipe wall or coolant, in particular, some long-life nuclides in the shutdown for a period of time will still cause radiation damage to equipment maintenance workers. Therefore, it is important to study the generation and migration of corrosion products in the alkali metal coolant loop, predict the change and distribution of corrosion products for the radiation shielding design of lead-bismuth fast reactor, the inspection and maintenance of the reactor, and the accident analysis.

C. Liu (Ed.): PBNC 2022, SPPHY 283, pp. 635–642, 2023.
https://doi.org/10.1007/978-981-99-1023-6_55

In view of the important influence of corrosion products on radiation protection, a great deal of research has been carried out at home and abroad, Such as the PWR-GALE program developed in the United States, the PACTOLE program in France, the Nuclear Power Institute, the Suzhou Thermal Engineering Institute, Tsinghua University, the North China Electric Power University Shanghai Jiao Tong University, and Harbin Engineering University. The above calculation program or method, the principle used is basically the same, but the simplification and assumptions used, as well as the specific treatment methods are not the same. The point reactor model is often used in the calculation, which does not take into account the influence of uneven distribution of neutron flux rate and coolant flow time, so its calculation precision is not high. In addition, the existing source item calculation program has some limitations; the calculation system and equipment are also specific, not universal. At present, the study on source term of corrosion products in water-cooled reactor is mainly focused in China, but there is no systematic study on source term of corrosion products for lead-bismuth coolant.

Based on the characteristics of lead-bismuth reactor coolant loop, the corrosion mechanism of lead-bismuth alloy and the corrosion rate of lead-bismuth alloy, the process of corrosion products generation, decay, migration and deposition in the loop are simulated in this paper; the mathematical equilibrium equation is established. The source term of corrosion products of lead-bismuth coolant was calculated by calculating the fast neutron reaction cross section of the reactor. This study provides reference data for related research in China.

2 Corrosion Source

2.1 Impurities in the Coolant

The impurities in the coolant mainly come from the raw materials and the impurities introduced in the operation. The lead-bismuth alloy was synthesized from lead and bismuth in the ratio of 44.5% and 55.5%, the impurities are mainly non-metallic impurities such as oxygen, carbon, and metallic impurities such as calcium. The pipelines and equipment of the reactor loop are processed, welded and cleaned in the process of manufacture, installation and maintenance, this process inevitably leaves behind some dirt, grease, gasoline, metal chips, welding slag, surface oxides and moisture, which is another major cause of contamination and impurities in the coolant system.

2.2 The Structural Material of the Coolant Channel

Lead and bismuth coolants have the characteristics of high melting point, high boiling point, chemical property inactivity and "Negative" cavitation reactivity. Lead and bismuth are chemically inert with fuels, low alloy steel, water and air. The structural material commonly used in contact with lead-bismuth alloys is stainless steel.

By analyzing the main components and impurities in stainless steel, the possible activation reaction types were determined. Finally, the radionuclide types of various corrosion products, the corresponding reaction types and the main sources of structural materials were counted; the results are shown in Table 1. The common active corrosion products in the main circuit are radionuclide as ^{24}Na, ^{51}Cr, ^{56}Mn, ^{59}Fe, ^{58}Co and ^{60}Co, and their initial nuclides are mainly Fe, Cr, Ni, Mn and Co in the structural materials.

Table 1. Source of Corrosion Products

Target Nucleus	Reaction Type	Radiation sources	Half-life	Decay constant, 1/s
^{50}Cr	^{50}Cr (n, γ) ^{51}Cr	^{51}Cr	27.72d	2.89413E−07
^{54}Fe	^{54}Fe (n, p) ^{54}Fe	^{54}Fe	312.5d	2.56721E−08
^{55}Mn	^{55}Mn (n, γ) ^{56}Mn	^{56}Mn	2.587h	7.44263E−05
^{58}Ni	^{58}Ni (n, p) ^{58}Co	^{58}Co	71.3d	1.12518E−07
^{59}Co	^{59}Co (n, γ) ^{60}Co	^{60}Co	5.26a	4.17862E−09
^{60}Ni	^{60}Ni (n, p) ^{60}Co			
^{58}Fe	^{58}Fe (n, γ) ^{59}Fe	^{59}Fe	45.1d	1.77883E−07
^{60}Ni	^{60}Ni (n, p) ^{60}Co			

3 Method of Establishment

3.1 Coolant Migration Process

When calculating the source term of corrosion products in the main loop, there are two general conditions. One is that the materials in contact with the coolant in the main circuit are first corroded and dissolved in the coolant; the other is that the material in the core is first activated by radiation, and then corroded down to dissolve in the coolant. As the coolant flows, some of the activated corrosion products in the main circuit will be deposited on the equipment or pipelines in the main circuit, most of which will be cleaned by the purification system, and some will remain in the main circuit coolant. The generation and migration of corrosion product source term in the main loop can be divided into six processes as follows:

(1) The structural material in the core is activated by irradiation;
(2) The material in contact with the coolant in the main circuit is corroded down and dissolved in the coolant to become the corrosion product;
(3) Non-radioactive corrosion products in the coolant flow with the coolant, in the main circuit migration and balance;
(4) The corrosion products in the coolant are irradiated and activated as they flow through the core;
(5) Transfer and equilibrium of radionuclide in the coolant with the coolant flow;
(6) The radionuclide in the coolant precipitates in the main circuit equipment.

3.2 Computational Model

In the coolant loop, the amount of the target radionuclide of the structural material decreases gradually with neutron irradiation.

$$\frac{dN_1}{dt} = -\overline{\sigma}_1\overline{\phi}_1 N_1$$

$$\frac{dN_2}{dt} = \overline{\sigma}_1\overline{\phi}_1 N_1 - \lambda_2 N_2$$

In the formula, N_1 is the nuclear density of the target nucleus in the structural material; t is the time;N_2 is the nuclear density of the irradiated radionuclide; σ_1 is the average neutron activation cross section of the target nucleus; ϕ_1 is the average neutron flux rate of irradiation; λ_2 is the decay constant of the irradiated radionuclide.

The increase in radionuclide in the primary circuit due to corrosion of the reactor activated materials at constant reactor power can be calculated as follows:

$$R_{ci} = \sum_l \left(\sum_j \frac{f_{ai} \cdot C_{0j} \cdot S_j \cdot N_A}{A_i} \right)$$

$$= \sum_l \left(\sum_j \frac{f_{nll} \cdot f_{sllj} \cdot C_{0j} \cdot S_j \cdot N_A}{A_i} \cdot x_i \right)$$

$$= \sum_l \left[\sum_j \frac{f_{nll} \cdot f_{slj} \cdot C_{0j} \cdot S_j \cdot N_A}{A_i} \cdot \frac{-\overline{\phi} \cdot \overline{\sigma}_l}{\lambda_i} (1 - e^{-\lambda_i tt}) \right]$$

$$= \frac{N_A \cdot \overline{\phi}}{A_i \cdot \lambda_i} \cdot (1 - e^{-\lambda_j \cdot t}) \sum_l \left[\overline{\sigma}_{ll} \cdot \left(\sum_j f_{nnl} \cdot f_{slj} \cdot C_{0j} \cdot S_j \right) \right]$$

In the formula, f_{ai} is the quality share of the radionuclide in the material; C_{0j} is the material corrosion rate of module j; S_j is the corrosion area of module j; N_A is the Amado Avogadro constant; A_i is the atomic weight of radionuclide I; f_{sILj} is the mass share of the nuclide chemical element in the material composition of module j; f_{nII} is the natural abundance of the target nucleus; λ_i is the decay constant of radionuclide i.

When the reactor power is constant, the total nucleon number of the activated corrosion products in the coolant changes with time according to the following equation:

$$\frac{dn_{vit}}{dt} = R_{ci} - \lambda_i \cdot n_{vi}$$

$$= \frac{N_A \cdot \overline{\phi}}{A_i \cdot \lambda_l} (1 - e^{-\lambda_1 \cdot t}) \sum_l \left[\overline{\sigma_{ll}} \left(\sum_j f_{vil} f_{sllj} C_0 S_j \right) \right] - \lambda_i \cdot n_{wt}$$

$$R_{ci} = \frac{N_A \cdot \overline{\phi}}{A_1 \cdot \lambda_l} \sum_l \left[\overline{\sigma_{ll}} \left(\sum_j f_{nll} f_{sllj} C_0 S_j \right) \right]$$

$$\frac{dn_{wl}}{dt} = R_{cl} (1 - e^{-\lambda_i t}) - \lambda_l \cdot n_{wl}$$

If $t \leq t_1$, the analytical solution of the equation is:

$$n_{wi}(t) = \frac{R_{ct}}{\lambda_t} - \frac{R_{ct}\lambda_t t + R_{ct}}{\lambda_l} e^{-\lambda_1 t}$$

if $t > t_1$, The power of the reactor has changed, the average neutron flux rate of the reactor has changed, The solution of the equation is:

$$\mathrm{n_{wt}(t)} = \frac{R_{ci2}}{\lambda_l}\left[1 - \mathrm{e}^{\lambda_i(t_1-t)}\right] - R_{cc2}(t - t_1)\mathrm{e}^{-\lambda_t t} + \frac{R_{ci1}}{\lambda_l}\left[\mathrm{e}^{\lambda_l(t_1-t)} - \mathrm{e}^{-\lambda_l t}\right] - R_{ct1}t_1\mathrm{e}^{-\lambda_l t}$$

3.3 Example Description

The type of this example is a compact pool structure, as shown in Fig. 1; the main parameters are shown in Table 2. Under normal operating conditions, the circulation flow of lead-bismuth medium in the first circuit is as follows: Lead and Bismuth are heated from bottom to top by the core in the internal components of the reactor, and then enter the upper steam generator of the collector chamber through openings in the upper part of the containment vessel, After the heat transfer is completed from the top to the bottom of the primary side of the steam generator and the secondary side, the secondary side outlet is reintegrated into the lower collecting cavity of the side shield, and the inner opening of the side shield returns to the upper collecting cavity from the bottom to the top, Then flow down the annular passage between the side shield and the main vessel and enter the main pump inlet, under the pumping of the main pump, the annular passage between the lower shield and the lower head of the main vessel enters the lower collecting cavity of the core along the flow distribution mechanism, and finally returns to the core.

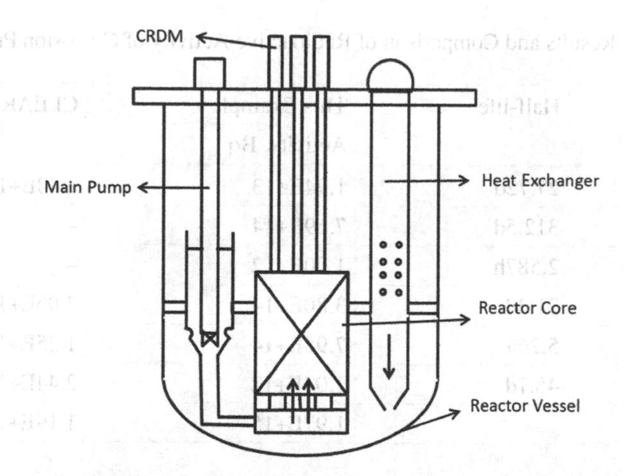

Fig. 1. Schematic Diagram of Pool-type Structure of Lead-bismuth Fast Reactor

Table 2. Main Parameters

Parameters	This Example	CLEAR-1
Thermal Power	20 MW	10 MW
Design Life	30 years	30 years
Loop Number	2	4
Coolant Temperatures	280–485 °C	260–450 °C
Coolant Flow	660 kg/s	529.5 kg/s
Structural Materials	SS316H	SS316L

3.4 Calculation Results and Comparative Analysis

The fast neutron average reaction cross sections for various reactions were calculated using the MCNP code, as shown in Table 3. The radioactivity of the corrosion products in the model coolant is calculated using the formula in Sect. 3.2, and a comparison with the results of the same type of reactor is shown in Table 4.

Table 3. Fast Neutron Average Reaction Cross Section

	^{50}Cr-^{51}Cr	^{54}Fe-^{54}Mn	^{55}Mn-^{56}Mn	^{58}Ni-^{58}Co	^{59}Co-^{60}Co	^{60}Ni-^{60}Co	^{58}Fe-^{59}Fe	^{59}Co-^{59}Fe
barn	0.0123	0.0095	0.0753	0.0098	0.0642	0.0003	0.0019	0.0002

Table 4. Results and Comparison of Radioactive Activity of Corrosion Products

Nuclides	Half-life	This Example	CLEAR-1
		Activity, Bq	
^{51}Cr	27.72d	1.94E+13	7.40E+14
^{54}Mn	312.5d	7.69E+14	–
^{56}Mn	2.587h	1.00E+12	–
^{58}Co	71.3d	3.80E+14	2.06E+14
^{60}Co	5.26a	7.95E+14	1.25E+12
^{59}Fe	45.1d	1.04E+12	2.44E+14
Total		1.97E+15	1.19E+15

As can be seen from Table 4, the total activity of corrosion products in the lead-bismuth coolant circuit is 1.97E+15Bq. Among them, ^{51}Cr, ^{54}Mn, ^{58}Co and ^{60}Co have the largest proportion, but ^{51}Cr and ^{58}Co are short-lived nuclides, which can decay rapidly after shut down for a period of time, and ^{54}Mn and ^{60}Co have longer lifetime, which are the main contributors to the total activity.

Compared with ClEAR-1, the model has same life, but different power, temperature, flow, and structural materials. These factors cause the total amount of corrosion products to be more than ClEAR-1.

4 Conclusions

The sediment source term of corrosion products is the main source of occupational irradiation, and it is also the key and difficult point in the source term analysis of nuclear facilities. Based on the characteristics of lead-bismuth coolant system, the release, migration, decay and deposition of corrosion products in the coolant loop are fully considered, a method for calculating the source term of corrosion products in lead-bismuth coolant loop is developed, The method is used to simulate the coolant loop of lead-bismuth fast reactor, and the source term of corrosion products is calculated. The results show that the source terms of the corrosion products are mainly composed of ^{51}Cr, ^{54}Mn, ^{56}Mn, ^{58}Co, ^{60}Co and ^{59}Fe, among which the long-lived nuclides ^{60}Co and ^{54}Mn are the main contributors to the radioactive activity. The analytical methods and conclusions of this paper can provide theoretical support for relevant domestic research.

References

1. Wu, Y., et al.: Conceptual design of China lead-based research reactor CLEAR-1. Nucl. Sci. Eng. **6** (2014)
2. Chandrasekaran, T., Lee, J.Y., Willis, C.A.: Calculation of Releases of Radioactive Materials in Gaseous and Liquid Effluents from Pressurized Water Reactors. Division of Systems Integration Office of Nuclear Reactor Regulation U.S. Nuclear Regulatory Commission, Washington, D.C (1985)
3. Dacquait, F., Nguyen, F., Martean, H.: Simulations of Corrosion Product Transfer with the PACTOLE V3.2 Code
4. Zhang, C.: Source terms calculation analysis for the reactor and primary coolant system of Qinshan Phase II NPP project. Nucl. Power Eng. **24**, 73–77 (2003)
5. Ding, S., Shangguan, Z., Tao, Y.: Study on computational models of secondary source terms for normal operation of CPR1000 Nuclear Power Plants. Radiat. Protect. **29** (2009)
6. Liu, Y.: Calculation method and computer program for radionuclide concentration in primary circuit of light water reactor. Radiat. Protect. (1986)
7. Liu, Z., et al.: Source team analysis of DORAST code used in AP1000 primary and secondary coolant system. Atomic Energy Sci. Technol. **47**, 625–629 (2013)
8. Wu, M.: Study on Radiation Source Term of Main Coolant in Nuclear Power Plant. Shanghai Jiao Tong University (2005)
9. Yang, Y., et al.: Analysis of dose field in marine reactor cabin. Nucl. Sci. Eng. (2014)
10. Goorley, T., Bull, J., Brown, F., et al.: Release of MCNP 5_RSICC_1.30, Los Alamos National Laboratory (2004)
11. Gang, T.: Study on Radioactive Source Term and Dose Assessment for Lead-Bismuth Cooled Reactor (2013)

BY

Research on ΔI Control Strategy During Rapid Power Reduction

Desheng Meng(✉), Rong Duan, and Zhijun Li

China Nuclear Power Technology Research Institute, Shenzhen, Guangdong, China
rodolphe.mengdesheng@foxmail.com, {duanrong, lizhijun}@cgnpc.com.cn

Abstract. Cold source for the cooling system is provided by CRF pump in nuclear power plant. During failure of CRF pump, a rapid core power reduction to a lower power level is needed, which poses a challenge to ΔI control of the core. Based on the requirements of operating technical specifications, the influence of various factors concerned the ΔI control strategy such as cycle burn-up, low power level and power reduction methods is researched for the rapid power reduction process in a certain balance cycle. Meanwhile, sensitivity analysis is carried out on the action of the control banks for different phases during the rapid power reduction process. Based on the analysis of the main influencing factors, proposal on the rapid power reduction strategy related to the reduced power level, power reduction method and different characteristic burn-up is put forward. The match of the cooling capacity of the CRF pump and the reduced core power is realized, thus ensure the safety and economy performance of reactor core.

Keywords: ΔI · Power Reduction · Control Strategy · Operation Diagram · CRF Pump

1 Introduction

Two circulating water pumps (CRF pumps) are equipped in the three loops of the nuclear power plant in order to provide cooling water and thus to realize the heat removal [1]. During a cold source failure or CRF pump failure, a rapid core power reduction to a lower power level is demanded based on the available CRF pump cooling capacity. During the power reduction process and the stay of the low-power platform, the core axial power distribution is disturbed, and the axial power deviation of the core (ΔI) is relatively hard to control, which is likely to cause a reactor trip. Consequently, it is vital to study the ΔI control strategy in order to achieve a compromise between the CRF pump cooling capacity and ΔI control. It is also of importance for core safety and plant economics.

In the process of rapid power reduction, many problems appear in ΔI control. The power reduction methods in nuclear power plants can be divided into two types: boronization and rod insertion. As for the method of boronization, the power distribution is less perturbed. However, due to the limitation of the boronization speed, the power change rated is relatively low, which is difficult to satisfy the rapid reduce power requirements

C. Liu (Ed.): PBNC 2022, SPPHY 283, pp. 643–655, 2023.
https://doi.org/10.1007/978-981-99-1023-6_56

at some time. Meanwhile, when using the boronization method to reduce power, ΔI is likely to continue increasing. As the power difference is so large until exceeding the right boundary the operation diagram. It will cause a reactor trip. The safety and economy performance of the unit are infected. As far as the way of rod insertion is concerned, the power reduction process can be realized within a short period of time by a rapid introduction of negative reactivity. Nevertheless, the rod insertion method disturbs largely the core power distribution. Meanwhile, The GN rods must be fully extracted out of the core within a certain period of time after the operating technical specifications which strictly restricts on rod insertion time. During the GN rod withdrawal, it is likely for ΔI to exceed the right boundary. In addition, during the rapid power reduction, the core ΔI is also significantly related to the core burnup and the magnitude of the power reduction, that's why the core control strategy research is challenging.

In this paper, the SCIENCE program is utilized to simulate the variation of ΔI in the process of rapid power reduction. The control strategy of ΔI, which involves fuel burnup, low-power platform, and power-reduction methods, is optimized with the help of SOPHORA program. At the same time, the control rod action sensitivity analysis is also completed in this paper. Suggestions for ΔI control strategy in the process of rapid power reduction are given.

2 Research Background of Rapid Power Reduction

Researchers have carried out a lot of explorations and researches on the problem of ΔI control strategy of the rapid power reduction process. Among them, qualitative analysis has been given from the perspective of core characteristic including burn up, burnable absorber, control rod position, fuel type and moderator temperature in a large number of literatures [2–10]. At the same time, power reduction types, such as power reduction to hot shutdown, daily power reduction or augmentation, extended low power operation (namely ELPO), stretch out operation, are also described qualitatively [5]. Nevertheless, literature related to rapid power reduction is relatively scarce, especially the quantitative analysis. The difficulty of ΔI control is mentioned [9] in a case that the reactor realize a reduction of power rapidly by a speed of 50 MW/min due to cooling source failure. Therefore, the power is forced to reduce to below 30% FP in order to prevents load rejection action due to the uncontrollable ΔI. On the basis of previous research, quantitative and in-depth research on the impact of different factors in the process of rapid power reduction is meaningful to the ΔI control and to the matching between the low-power platform and the cooling capacity of the cold source pump, thus ensuring the safety and economy performance of the reactor operation.

3 Research Methods

In order to ensure the safe operation of reactor, the operating diagram limitations on ΔI must be met. In addition, operation technical specifications are also mandatory considering the operation factors. Under the condition of meeting these limitations, SCIENCE and SOPHORA programs are utilized to study the influence of difference factors, such as burnup, low-power platform, power reduction method and control rod action on ΔI control in the process of rapid power reduction.

3.1 Operation Technical Specifications

As one of the main bases for the operator to control the reactor, the operation diagram shows the relationship between the allowable value of ΔI and the relative reactor power. During the rapid power reduction process, the operator must ensure that ΔI satisfies the operation diagram. In addition, the ΔI performance also needs to meet the operation technical specification in which the operation diagram is defined. Related operating technical specifications are listed: when $P_{\Delta Iref}$ is determined and the power of the core is reduced from high power to 50% or less, control rod insertion time is limited to 12 h in any 24 h period in region I and ΔI is limited at region I. In the case of a capacity of a quick return to high power levels, the control rod insertion limitation time can be extended to 24 h. When all the power compensation rods(GN rod) are completely drawn out, ΔI is able to leave the area I after 6 h of stable operation.

3.2 SCIENCE and SOPHORA Software

SCIENCE graphical software package, which invoked by the Copilote graphical user interface is used in this paper for calculations. SMART programs is used for 3D homogenized cores calculation. SOPHORA program is utilized for ΔI control strategy optimization.

3.3 Overall Control Strategy

Under the condition of meeting the demands of operation technical specifications, two periods related to rapid power reduction, which is the period of rapid power reduction and the low power platform is respectively considered in order to evaluate various influencing factors in this article.

Concerned the strategy of rapid power reduction period, the control of ΔI is achieved in two ways. One is to insert the GN rod according to the G9 curve, while the power decrease at a speed of 10MW/min until reaching the low power platform. The other is to increase the boron concentration of the core firstly, at the same time, the power decrease at a rate of 3 MW/min until 80%FP. Then, GN rods is inserted in order to reduce the power at a relatively high speed of 10 MW/min. As for the strategy of low power platform, the core is boronized in order to withdraw all GN rods while R rod is used to control ΔI in order to meet the requirements of operating technical specifications in Sect. 3.1 (the time limitation for rod insertion).

In order to study the ΔI control strategy of reducing power to different low-power platforms at different fuel burnup points, a certain 18-month refueling cycle was selected for research at 150 (BLX), 6000, 11000 (ΔI most negative burnup point), 14000 (MOL), 16000 (80%EOL), 18000MWd/tU and EOL. Low power platform of 50%, 60%, 65%, 70%, 80% FP are studied and the ΔI control strategy is carried out in case of reactor cold source failure.

4 Optimization Results

The ΔI control strategy optimization results of rapid power reduction is researched by mainly focusing on three factors: fuel burnup, reached low power level platform, and the power reduction method. The comprehensive impact evaluation of the three factors is also studied. Then, this paper introduces the sensitivity analysis of the control rod action in the process of rapid power reduction, and gives the ΔI control strategy conclusion of the rapid power reduction process.

4.1 Influence of Fuel Burnup

During the reactor normal operation, the temperature of the lower part of the core is lower than that of the upper part. Due to the negative feedback of temperature, the core ΔI is generally negative. Along with the deepening of burnup, the burnup in the lower part of the core is gradually larger than that of the upper part, therefore, ΔI tends to increase due to the influence of the burnup effect. In this research, after the GN rods are extracted, a larger ΔI is unfavorable for core control. Worse, the xenon oscillation effect manifest more negative effects at EOL than BOL.

In order to study the influence of the fuel burnup effect, the GN rod is inserted to reduce the power to 50% PN during the rapid power reduction period and the ΔI variation is compared at BLX, MOL and EOL. R rod is used to control ΔI only at the low power platform. The variations of ΔI are shown in Fig. 1:

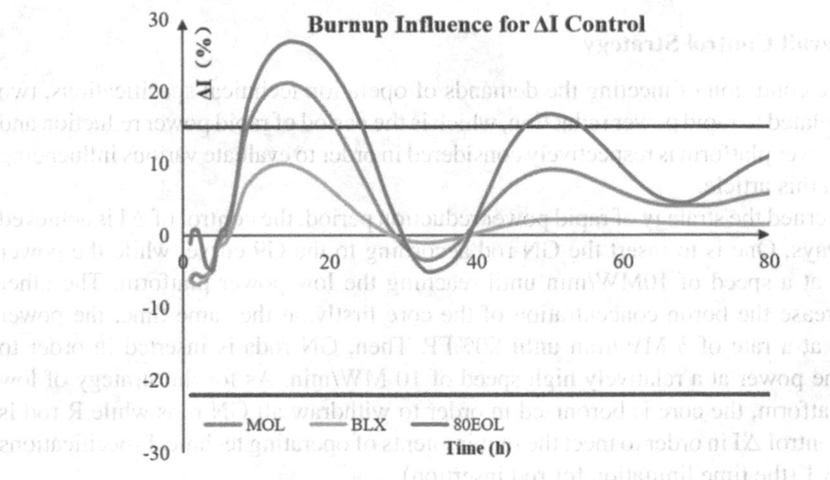

Fig. 1. Burnup influence for ΔI control

It can be seen from Fig. 1 that ΔI exceeds zone I when the GN insertion time is limited to 12h for all three burnup points. Therefore, we can draw that for the whole cycle, it's likely to exceed zone I when the power is reduced to 50%FP rapidly. We also note that ΔI will not exceed the right boundary at operation diagram at BLX; Nevertheless, at MOL and 80% EOL burnup, the xenon oscillates more violently and ΔI tends to be bigger, thus it exceeds the right boundary. ΔI control is influenced significantly by fuel burnup.

4.2 Impact of Low-Power Platforms

In addition to fuel burnup, the low power level reached by the core after a rapid power reduction also has a considerable impact on ΔI control. When the core power is reduced from 100% to 80% FP or more, ΔI is easy to control due to the small variations in power. Additionally, if the core power is reduced to 30% FP or even less, although power distribution go worse, ΔI may still be controllable thanks to the lower power levels. However, as the power is reduced to 30%–80% FP, ΔI is the most difficult to control.

ΔI variations for three low-power platforms (80%, 60% and 50%FP) is compared at burnup of 11000 MWd/tU in this section. R rod is not inserted at the power reduction period. At the low power platform, the core is boronized in order to extract GN rods. Meanwhile, R rod is inserted to control ΔI. The variation of ΔI for the three low-power platforms is shown in the Fig. 2:

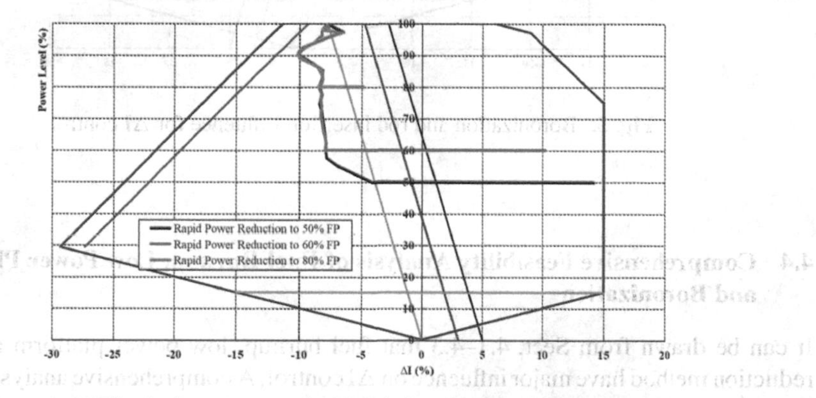

Fig. 2. Low-power platform influence for ΔI control

As shown in Fig. 2, when the power is reduced to a low-power platform which is greater than 50% PN, the greater is the power reduction, the greater is the ΔI oscillation caused by the xenon oscillation. When the power is decreased to the 80% FP, ΔI varies slightly within zone I. While the power is reduced to 60% or 80% FP, ΔI exceeds zone I. Therefore the low-power platform also affects largely the ΔI control.

4.3 Influence of Boronization and Rod Insertion

It can be obtained from Sect. 3.3. That two methods can be used for rapid power reduction: one is direct rod insertion and the other is boronization to 80%FP and then rod insertion. The power drops rapidly in the upper part of the core because of the rod insertion, and thus ΔI becomes more negative. The boronization power reduction method disturbs less to the core ΔI than rod insertion method, and it leads to a more positive ΔI. In order to compare the effects of two power reduction methods on ΔI control, the core ΔI elevation of the two cases is compared.

It can be seen from Fig. 3 that the ΔI performance is slightly improved by adopting the method of boronizing and then inserting the rod. At the same time, studies have shown that if boronization itself is used to reduce power to 50% PN, not only the power change rate is limited, but ΔI is easy to exceeds the right boundary. The ΔI oscillation can be slightly reduced by using the rapid power reduction strategy of boronization and then insertion the control rod.

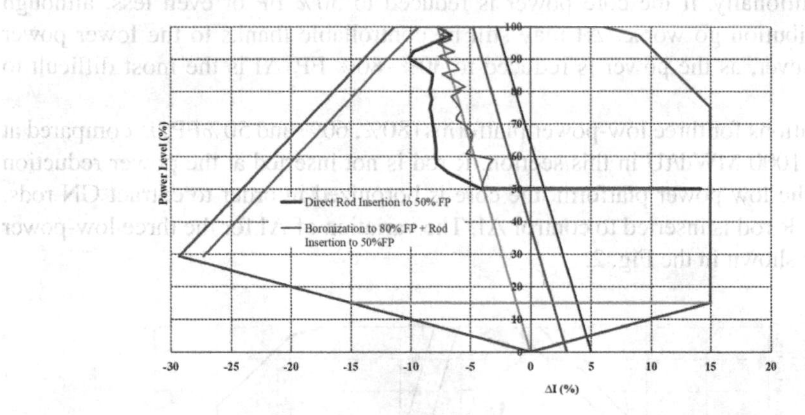

Fig. 3. Boronization and rod insertion influence for ΔI control

4.4 Comprehensive Feasibility Analysis of Fuel Burnup, Low Power Platform and Boronization

It can be drawn from Sect. 4.1–4.3 that fuel burnup, low power platform and power reduction method have major influence on ΔI control. A comprehensive analysis contains the ΔI control strategy optimization at different fuel burnups to different low-power platforms and by using different methods. We also note that the criteria of feasibility for rapid power reduction strategy is that ΔI does not exceed the operation diagram.

The power reduction analysis is based on two assumptions to facilitate the calculation. The first is that, at a higher fuel burnup, if the core power can be reduced to a certain power level without exceeding the right boundary, the core power can also be reduced to that level without exceeding the right boundary at a relatively lower burnup. The other is that, if the direct insertion of control rod method is feasible to reduce the power without exceeding the right boundary, the power reduction method of boronization and then rod insertion will also be feasible. These two assumptions correspond to the conclusions in Sects. 4.1 and 4.3, respectively. On the basis of these two assumptions, the conclusion of whether ΔI is controllable can be obtained for different fuel burnups (the unit of fuel burnup is %EOL), with different power reduction method (use direct rod insertion, boronization without stay at 80%FP and rod insertion, and finally boronization with stay at 80%FP and rod insertion) to different power platform including 80%, 70%, 65%, 60% and 50% FP. The result is shown in Fig. 4:

From Fig. 4, the boundary between controllable and uncontrollable power is obtained. When the fuel burn-up is less than about 50% EOL (about 11GWd/tU), ΔI does not exceed the operation diagram if the power is reduced rapidly to 50%. However, as the

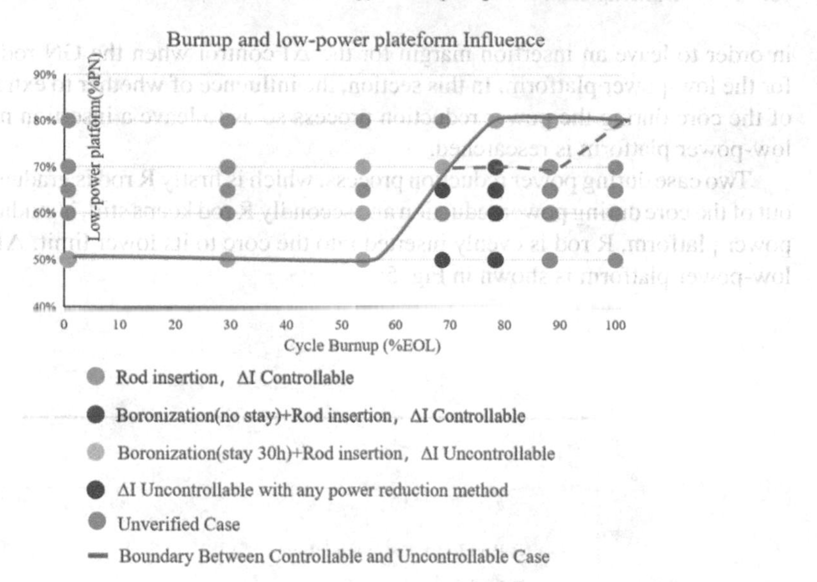

Fig. 4. Combined influence of burn-up, low power platform and power reduction method for the feasibility of rapid power reduction

fuel burn-up increases, in order to prevent ΔI from exceeding the operation diagram, the power reduction magnitude should be gradually reduced. When the fuel burn-up is high enough, a direct rod insertion to reduce the power may cause the ΔI exceeding the right boundary. In this case, the power can be reduced by boronization followed by rod insertion, and a stay of 80%FP may be necessary to avoid ΔI oscillation. The matching of cooling capacity and core power can be achieved with help of the strategy.

5 Sensitivity Analysis of Control Rod Motion

To limit ΔI oscillations during rapid power reduction, the R-rod action can be manually optimized. In addition, after reaching the low power platform, the GN rod is supposed to be raised in 12h, the GN rod action is flexible in time and in speed. Therefore, it is also of importance to carry out the analysis of the influence of the control rod action on ΔI during the rapid power reduction and the low-power platform.

Considering the power reduction method of direct insertion of control rod, according to the time and logical sequence, the sensitivity analysis of control rod action can be divided into: R rod action in power reduction process, GN rod action at low-power platform and R rod action low-power platform. The sensitivity analysis is simulated at a burnup of 80%EOL.

5.1 Sensitivity Analysis of R Rod Action During Power Reduction Process

During the power reduction process, GN rod needs to be inserted according to the known G9 curve, ΔI is generally more negative. In this process, the R rod can be properly raised in order to leave an insertion margin for the ΔI control when the GN rod is extracted for the low-power platform. In this section, the influence of whether to extract R rod out of the core during the power reduction process so as to leave a insertion margin in the low-power platform is researched.

Two case during power reduction process, which is firstly R rod is gradually extracted out of the core during power reduction and secondly R rod keeps still, is studied. At lower-power platform, R rod is evenly inserted into the core to its lower limit. ΔI variation at low-power platform is shown in Fig. 5:

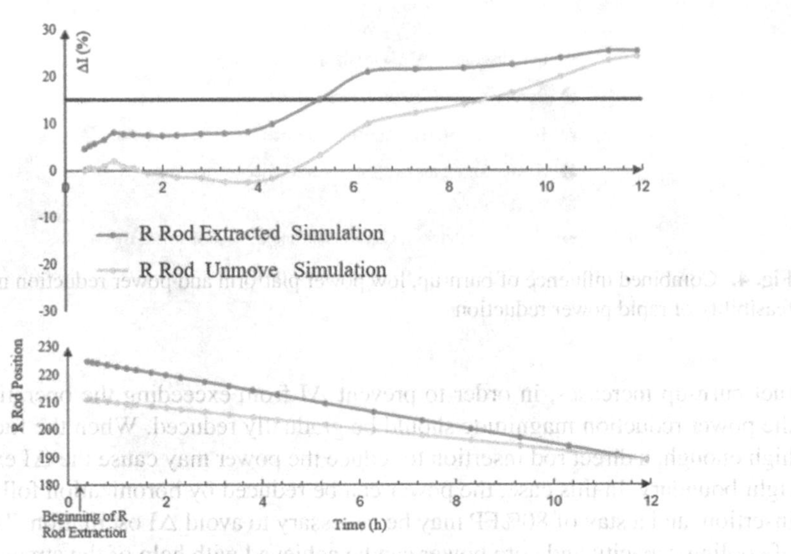

Fig. 5. Sensibility study of for the withdrawal of R bank

It can be drawn from Fig. 5 that in the process of power reduction, an extraction of R rod for more insertion margin has no significant optimization effect on the ΔI control strategy. ΔI raises as the GN rod is extracted, and finally exceeds the right boundary. The value of final ΔI is basically the same in both cases.

5.2 Sensitivity Analysis of GN Rod Motion of Low Power Platform

In order to meet the requirements of the operation technical specifications, the GN rods must be extracted of the core within 12 h. During the low-power platform, when GN rods are extracted, studies of its action include: extraction speed, beginning time and time interval of GN rod withdrawal.

For the analysis of extraction speed of GN rods, three cases of GN rod extraction at low-power platform are simulated. The result of ΔI is shown in Fig. 6:

It can be seen from the above figure that the maximum ΔI is positively correlated with the GN rod extraction rate, but both ΔI are far beyond the right boundary.

For the beginning time of the GN rod extraction, at low-power platform, due to the Xe effect, ΔI tends to decrease in the low-power platform from 0 to 10 h in which is

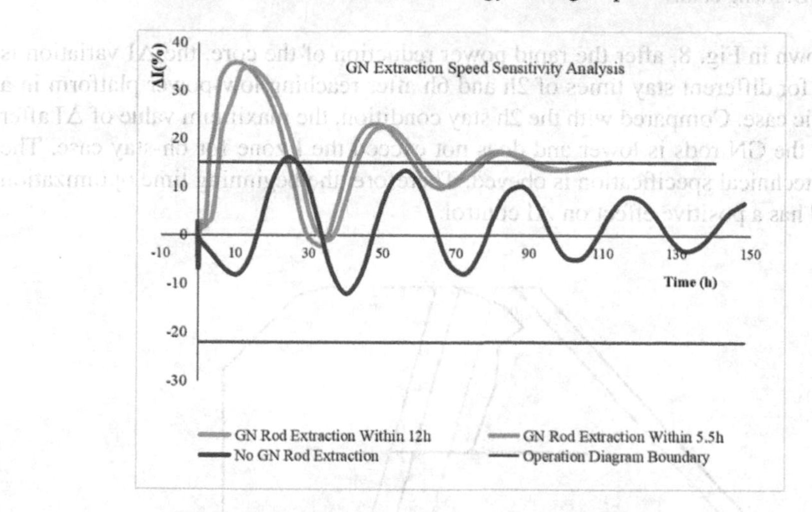

Fig. 6. Influence of the withdrawal speed of GN bank on ΔI

favorable for GN rods extraction. The GN rods are selected to be extracted at 2, 4, 6, and 8 h, and all of the cases are fully extracted at the 12th hour. The R rod keeps still, and the comparison of ΔI changes is shown in Fig. 7:

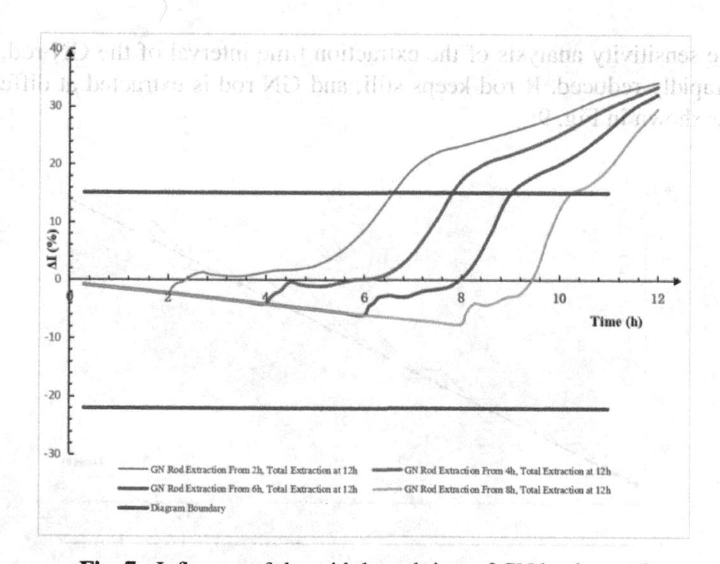

Fig. 7. Influence of the withdrawal time of GN bank on ΔI

From Fig. 7, we observe the changes of ΔI In all four cases, the xenon oscillation causes ΔI to exceed the right boundary. However, by adjusting the time of GN rod withdrawal, the peak value of ΔI can be slightly reduced and the time to exceed right boundary can be delayed.

As shown in Fig. 8, after the rapid power reduction of the core, the ΔI variation is simulated for different stay times of 2h and 6h after reaching low-power platform in a real specific case. Compared with the 2h stay condition, the maximum value of ΔI after extraction the GN rods is lower and does not exceed the I zone for 6h-stay case. The operation technical specification is obeyed. Therefore, the beginning time optimization of GN rod has a positive effect on ΔI control.

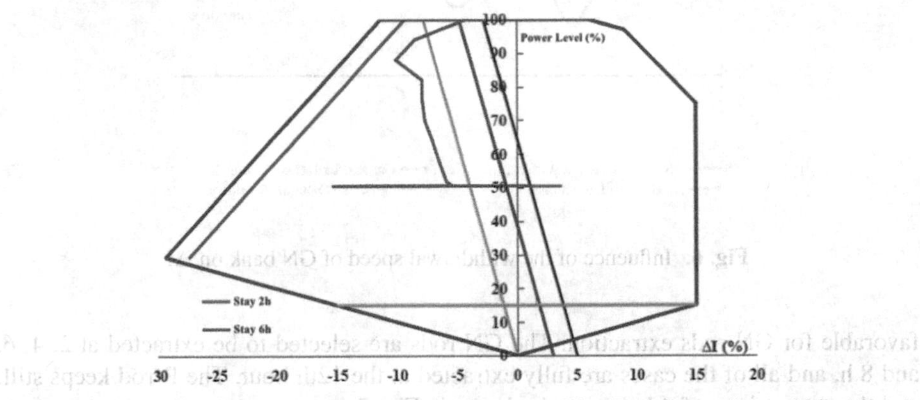

Fig. 8. Influence of the withdrawal time of GN bank on ΔI for Daya Bay unit at middle of cycle

For the sensitivity analysis of the extraction time interval of the GN rod, after the power is rapidly reduced, R rod keeps still, and GN rod is extracted at different time interval, as shown in Fig. 9:

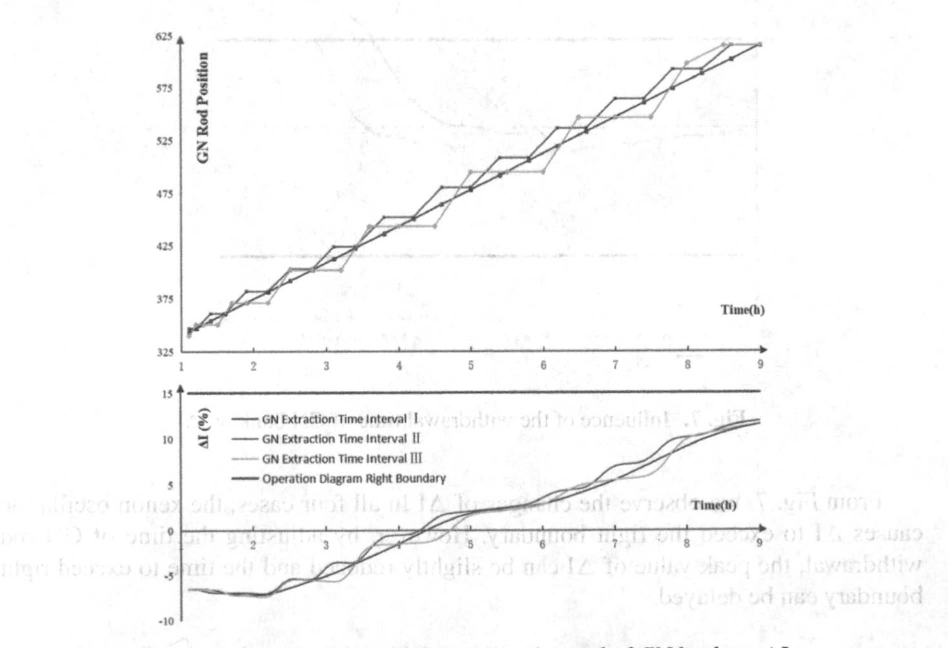

Fig. 9. Influence of the withdrawal time interval of GN bank on ΔI

From the above figure, ΔI attains the same value in the end. The extraction time interval of the GN rod does not affect greatly the variation of ΔI. Consequently, the GN rod can be evenly raised on the power platform.

5.3 Sensitivity Analysis of R-rod Motion of Low-Power Platform

In the low-power platform, during the process of the extraction of GN rod, the R rod needs to be inserted to control ΔI. The R rod can be either evenly inserted to its lower limit, or inserted once at a certain moment or inserted at different time. This section is to study the effect of the R rod action.

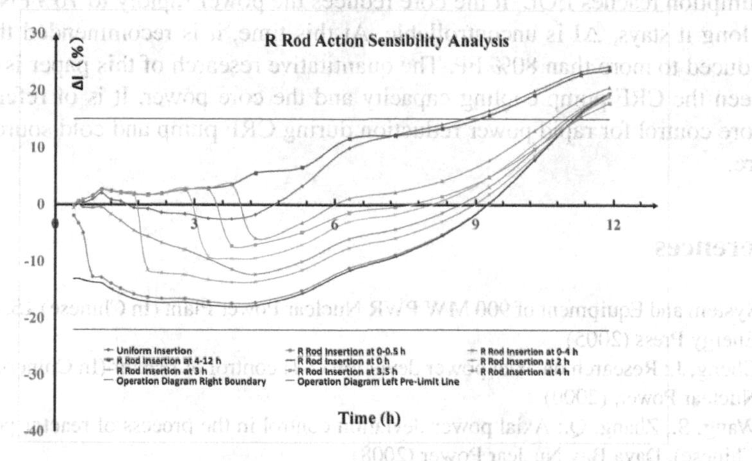

Fig. 10. Influence of the R bank action on ΔI at reduced power level

As can be seen from Fig. 10, the R rod adopts different insertion strategies, none of them can control the ΔI within the right boundary. There is no significant correlation between R rod insertion method and final ΔI value. The later is mainly affected by the final position of R rod (namely the low-low limit).

5.4 Summary of Sensitivity Analysis

In this section, at the 80% EOL burn-up point, the sensitivity analysis of the control rod action to ΔI in the process of rapid power reduction and low-power platform is carried out, including the analysis of the R rod action in the process of the GN rod insertion, the action of the GN rod and R-rod at the low-power platform. We can basically conclude that the control rod motion improves slightly the ΔI performance, but it has no decisive effect on the core control. The fuel burnup, low power platform and power reduction method keeps the main factor affecting ΔI during the rapid power reduction.

6 Conclusion

Based on SCIENCE software, this article describes the effects of different factors such as fuel burnup, low power platform, power reduction method and control rod action on rapid power reduction. Aimed at the three main factors, the comprehensive feasibility analysis is carried out and based on a specific cycle, a rapid power reduction ΔI control strategy is given: when the fuel consumption is less than approximately 50% EOL, ΔI is controllable when the power is quickly reduced to 50% FP. As the fuel burnup deepens, in order to keep ΔI within the operation diagram, the power reduction should be gradually reduced to 70% FP. When the fuel burn-up is high enough (80% EOL), using the direct rod insertion method to reduce the power will cause the ΔI to exceed the right boundary. The power can be reduced by boronization and then rod insertion. When the fuel burnup deepens to 90% EOL, in order to quickly reduce the power to 50% FP, it is recommended to stay for a certain time after boronization. When the fuel consumption reaches EOL, if the core reduces the power rapidly to 70%PN, no matter how long it stays, ΔI is uncontrollable. At this time, it is recommended that the unit be reduced to more than 80% FP. The quantitative research of this paper is carried out between the CRF pump cooling capacity and the core power. It is of reference value for core control for rapid power reduction during CRF pump and cold source response failure.

References

1. System and Equipment of 900 MW PWR Nuclear Power Plant (In Chinese) **15**, 488. Atomic Energy Press (2005)
2. Cheng, J.: Research on axial power deviation (ΔI) control of reactor (In Chinese). Daya Bay Nuclear Power, (2000)
3. Wang, S., Zhang, Q.: Axial power deviation control in the process of reactor power rise (In Chinese). Daya Bay Nuclear Power (2008)
4. Long, S., Gao, Y.: Control of axial power deviation in daily operation of M310 nuclear power unit during load variation (In Chinese). Technol. Innov. Appl. **15**, 12–13 (2019)
5. Liao, Y., Xiao, M., Li, X., Zhu, M.: Axial power difference control strategy and computer simulation for GNPS during stretch-out and power decrease (In Chinese). Nuclear Power Engineering, pp. 297–299 (2004)
6. Duan, C., Wu, X.: Analysis of ΔI control of power reduction at the end of cycle (In Chinese). Shandong Industrial Technology, p. 276 (2016)
7. Wang, B.: Operation control strategy of nuclear power units during ELPO (In Chinese). J. Manag. Sci., 45–46 (2015)
8. Liu, Z., Pan, Y.: Control of axial power deviation during power variation (In Chinese). In: CORPHY 2006, pp. 533–542 (2006)
9. Chen, Z., Chen, C.: Core axial power deviation control at the end of life of PWR nuclear power (In Chinese). Daya Bay Nucl. Power **4**, 7–12 (2011)
10. Zhang, H.: Analysis on axial power distribution and ΔI control in pressurized water reactor

Minor Actinides Transmutation in Thermal, Epithermal and Fast Molten Salt Reactors with Very Deep Burnup

Chunyan Zou, Chenggang Yu, Jun Zhou, Shuning Chen, Jianhui Wu(✉), Yang Zou, Xiangzhou Cai, and Jingen Chen(✉)

Shanghai Institute of Applied Physics, Chinese Academy of Sciences, Shanghai, China
{zouchunyan,wujianhui,chenjg}@sinap.ac.cn

Abstract. Deep burnup and high minor actinides (MA) loading are two alluring features for molten salt reactors (MSR) to incinerate the nuclear wastes. The transmutation capability of minor actinides in MSR is tightly related with the neutron spectrum, the loading of MA and the carrier salt compositions. In this work, three MSR core designs (thermal, epithermal and fast) and two types of salt compositions (Flibe and Flinak) with different solubility limits of transuranic elements are chosen for analyzing the transmutation capability of MA. With a significant mole fraction of MA loading (4% in the Flibe salt and 10% in the Flinak salt) and the continuous MA refueling, MSR acquires an excellent transmutation rate. The specific incineration rates of MA in the thermal, epithermal and fast MSR cores with the Flibe salt are about 167, 185 and 206 kg/GWth/y, respectively. With a larger loading of MA in the Flinak salt, a higher annual incineration rate of MA can be obtained, which are about 170, 206 and 247 kg/GWth/y in thermal, epithermal and fast MSRs, respectively. On the other hand, since there is a preferred neutron economy for the Flibe salt, a higher MA incineration ratio is achieved than that for the Flinak salt. When the neutron spectrum varies from the thermal to fast region, the MA incineration ratio ranges from 0.79 to 0.82 for the Flibe salt and it ranges from 0.75 to 0.81 for the Flinak salt. The transmutation capability of MA in MSR is much higher than that in solid-fueled reactors (~20 kg/GW.y in a PWR), which can provide a feasible way for reducing the current nuclear wastes.

Keywords: Molten Salt Reactor · Minor Actinides · Transmutation Capability · Incineration Rate · Incineration Ratio

1 Introduction

Minor actinides (MA) in the spent fuel of the current pressurized water reactors (PWRs), namely neptunium (Np), americium (Am) and curium (Cm), are the main long-lived radioactive waste in the long term, which is one of the most severe issues associated with sustainable nuclear energy development [1–4]. Until now, transmutation is considered to be an effective way for nuclear waste management and tremendous works have been devoted to achieving a high transmutation capability of MA in different types

C. Liu (Ed.): PBNC 2022, SPPHY 283, pp. 656–672, 2023.
https://doi.org/10.1007/978-981-99-1023-6_57

of reactors, such as PWR, Gas-cooled Fast Reactor (GFR), Lead-cooled Fast Reactor (LFR), Sodium-cooled Fast reactor (SFR) and Accelerator Driven Sub-critical System (ADS) [5–8].

Molten salt reactor (MSR) is an old concept but has been gathered many attentions in recent years due to its special features [9, 10]. One of the most alluring advantages in MSR is that it operates with a liquid fuel, which permits an arbitrary core design and a flexible reprocessing system. Furthermore, the molten salt is capable to dissolve various fissile fuels (eg. enriched uranrium, ^{233}U and transuranium elements (TRUs)), which is convenient to implement different types of fuel cycles in an MSR [11, 12]. A closed nuclear fuel cycle is expected to be realized and the utilization of nuclear fuel can be significantly improved due to the effective burning of minor actinides in an MSR. The MOlten Salt Actinide Recycler & Transmuter (MOSART) was first proposed by the Kurchatov Institute of Russia within the International Science and Technology Center project 1606 (ISTC#1606) with aims to effectively transmute TRUs [13]. A series of studies have been conducted to demonstrate the feasibility of MOSART for reducing TRU radiotoxicity. Afterwards, another fast spectrum MSR concept of the Molten Salt Fast Reactor (MSFR), which was proposed by the Centre National de la Recherche Scientifique (CNRS) to achieve a high thorium breeding ratio, was also applied to investigate the feasibility of TRU transmutation [14]. Recently, some thermal spectrum MSR concepts are also put forward to evaluate the possibility of TRU transmutation [15, 16].

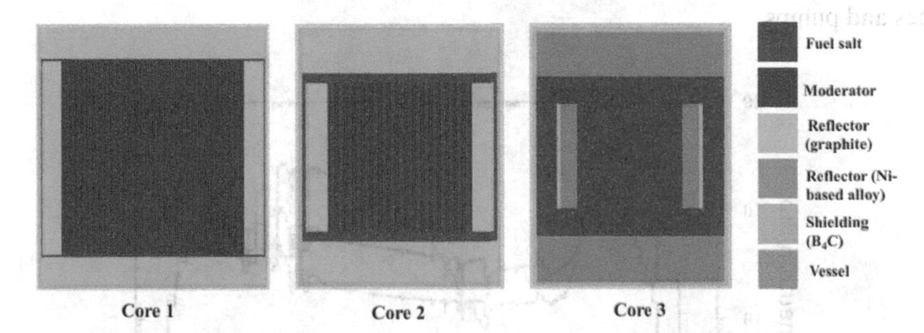

Fig. 1. Geometrical description of the three MSR cores

In an MSR, the MA transmutation capability is tightly related with the neutron spectrum and the MA loading in the core. The dominant transmutation way for MA is varied with the neutron spectrum since the fission and capture cross sections of MA have significant discrepancies with the neutron energy. In addition, different types of carrier salt compositions have various TRU solubility limits, which have a direct influence on the MA loading in MSR. In this paper, it is aimed to evaluate the transmutation behaviors of MA with different neutron spectra and various MA loadings in MSR, which can provide a reference for realizing various transmutation objects for different MA elements. Three typical MSR cores (thermal, epithermal and fast) are proposed to compare the MA transmutation capability. Meanwhile, two typical molten salt compositions (Flibe and FlinaK) which have different TRU solubility limits are selected to analyze the influence of MA loading on the transmutation capability. Furthermore, the radiotoxicity and safety parameters are also analyzed in detail.

A general description of the thermal, epithermal and fast MSRs and the calculation tools is presented in Sect. 2. In Sect. 3, the MA transmutation capabilities in the three MSR cores with two types of molten salts are first analyzed. And then the neutronic performances at different burnups in the thermal, epithermal and fast MSR cores are presented and discussed. The conclusions are given in Sect. 4.

2 General Description of the Geometry Models and Calculation Tools

2.1 Description of the Thermal, Epithermal and Fast MSRs

In the past, a series of MSR core designs ranging from the thermal to fast neutron spectrum have been conducted to achieve a high thorium breeding ratio in our research group [11, 12]. This work extends three typical reactor core configurations, corresponding to thermal, epithermal and fast spectrum cores, respectively. The geometrical descriptions for the three cores are shown in Fig. 1. In an MSR, the power density is a vital parameter to determine the main neutronic behaviors. Therefore, a constant thermal power of 2500 MWth and a constant fuel volume of about 46.2 m^3 are designed for the three cores, respectively. Therein, two thirds of the total fuel volume is located in the core salt channels, the upper and lower plena, and the other one third is located in the heat exchangers, pipes and pumps.

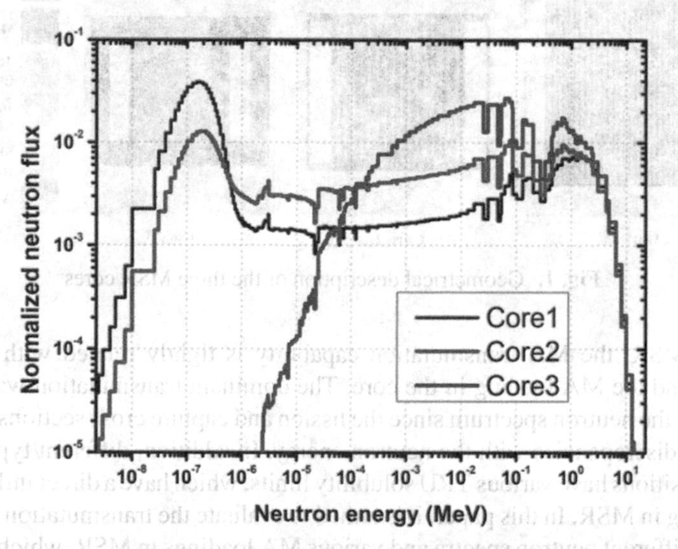

Fig. 2. Neutron spectra of the three cores

The core parameters are detailed in Table 1. For the thermal and epithermal MSR cores, the reactor core is a cylindrical geometry assembled with graphite hexagons and surround by the graphite reflectors. The radii of the fuel salt channel in each graphite hexagon are designed as 3 cm and 7.5 cm to obtain thermal and epithermal spectra, respectively. In the fast MSR, the graphite reflector is replaced by nickel-based alloy to prevent nuclear fissions from being decentralized to reactor's borders. To maintain a constant fuel volume in the core, the dimension for the three MSRs varies from each other, which is designed by keeping the height and the diameter equivalent. The radial reflector with 0.5 m thickness and the axial reflectors with 1.3 m thickness are designed around the core for the three cores to improve the neutron economy. The neutron spectra of the three core configurations fueled with Th-232 and U-233 are shown in Fig. 2, which are shown as typical thermal, epithermal and fast neutron spectra.

Table 1. Main parameters of the three MSR cores

Parameters	Core 1	Core 2	Core 3
Thermal power (MWth)	2500	2500	2500
Fuel volume (m^3)	46.2	46.2	46.2
Fuel channel radius (cm)	3	7.5	—
Active core diameter/height (m)	8.6	4.6	3
Plena height (m)	0.09	0.3	0.75
Radial/axial reflector thickness (m)	0.5/1.3	0.5/1.3	0.5/1.3
Reflector material	Graphite	Graphite	Nickel-based alloy

2.2 Selection of Salt Compositions

The MA loading in the core is an important factor that determines the transmutation capability, which is a major restriction on achieving a high transmutation rate in solid-fueled reactors. In MSR, there is no fuel rod fabrication which extends the feasibility of MA mass loading. However, the solubility limit of TRU in the molten salt is an important restriction on the MA mass loading. Therefore, choice of the fuel salt is also one of the most important tasks for the MA transmutation since the solubility limit of TRU is tightly dependent on the molten salt compositions. In the past decades, the physico-chemical properties of various salt compositions were researched for selection of fuel and coolant compositions for MSR. Until now, there are three typical carrier salts proposed in different MSR designs, which are Flibe, Fli, and Flinak, respectively [19]. The Flibe salt with 99.995% of Li-7 enrichment has excellent neutron economy, and has been widely used in various MSRs. The PuF_3 solubility limit for the fuel compositions of 77% LiF–17% BeF_2–6% ThF_4 is 4.0%. The Fli carrier salt removes BeF_2 to accommodate a higher fraction of actinide tetrafluorides, which has been selected as the fuel compositions in the molten salt fast reactor (MSFR) (78% LiF–22% ThF_4) for Th breeding and MA transmutation. The 78% LiF–22% ThF_4 fuel salt has a worse neutron economy but has a higher solubility limit (5.2%) than the Flibe salt. Compared to the Flibe and Fli salts, the Flinak carrier salt has the highest solubility of actinides but the worst neutron economy due to the large absorptions of Na-23 and K-39. For the composition 46.5% LiF, 11.5%

NaF and 42% KF, the solubility limits for ThF_4, UF_4, PuF_3 and AmF_3 are as high as 37.5%, 45%, 30%, and 43%, respectively. In this work, the Flibe and Flinak carrier salts are selected for comparing the MA transmutation capability with different MA loadings.

The initial MA mole fraction for the Flibe salt is loaded with its limit value of 4.0% while 10% of MA is loaded for the Flinak salt in the thermal, epithermal and fast MSR cores. Th-232 and U-233 are used as the fertile and fissile fuels in the three MSR cores due to the very low TRU production. To enhance the MA transmutation rate and to ensure the stability of fuel salt simultaneously, MA including Np, Am and Cm are fed online into the core to keep the total inventory of MA and Pu constant during the entire operation. Furthermore, Th-232 and U-233 are also fed into the fuel salt continuously to maintain the criticality of reactor and keep the total heavy metal inventory in the fuel salt constant.

The MA compositions come from the spent fuel of current light water reactors (LWRs), which are mainly composed of Np-237, Am-241, Am-243, Cm-243, Cm-244 and Cm-245 in the proportions of 56.2%, 26.4%, 12%, 0.03%, 5.11% and 0.26%, respectively [4].

2.3 Calculation Tools

The SCALE6.1 code system, which has powerful functions for criticality, depletion and shielding calculations for critical reactors, was used to establish the simulation model of the MSR core [17]. Meanwhile, the molten salt reactor reprocessing system (MSR-RS) developed by our research group was applied to simulate the characteristics of online refueling and reprocessing in MSR. In this paper, the MSR-RS sequence is also used to calculate MA transmutation in MSR [18]. MA is fed continuously to enhance the transmutation rate and the feeding MA inventory is adjusted by keeping the total TRUs inventory constant to assure the TRU fraction in the fuel salt below the solubility limit. Meanwhile, the total heavy metal mass in the core is always kept constant by feeding Th-232 and U-233 during the entire operation time. The 238-group ENDF/B-VII cross section database is selected, and 388 nuclides are tracked in trace quantities in this simulation which contains most of the nuclides in the fuel cycle chain with very deep burnup.

3 Results and Discussion

To evaluate the MA transmutation capability of the thermal, epithermal and fast MSRs, the neutron spectra and fission/capture cross sections of MA varying with different MA loadings are first analyzed. Then the transmutation capability and the radiotoxicity of TRU are evaluated in detail. Finally, the related safety parameters will be discussed.

3.1 Neutron Spectra Variation with Addition of MA into the Fuel Salt

Neutron spectrum is a key core parameter which determines the MA transmutation performances mainly through the capture and fission reactions. When MA is refueled into the fuel salt, the neutron spectra in the three MSR cores will have a different shift due to the variation of cross sections of MA in different neutron energies. In further, the neutron spectrum change will have an influence on the fission/capture cross sections of MA. The neutron spectrum variations with MA addition at the beginning of life in the three MSR cores are displayed in Fig. 3. One can be seen that the neutron spectra in the thermal, epithermal and fast MSR cores harden significantly with addition of MA into the fuel salt. Furthermore, the variation in the epithermal core is more obvious than those in the other two cores as most of MA have strong resonance absorptions. In addition, the neutron spectrum with the Flinak salt shifts to a slightly fast region due to the parasitic absorption of Na-23 and K-39.

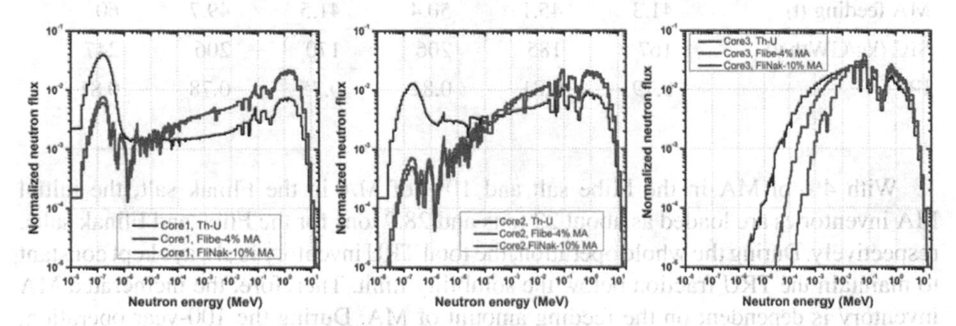

Fig. 3. Neutron spectrum variations with MA loading in the three cores

3.2 Transmutation Capability in the Three MSR Cores

To evaluate the transmutation capability of MA in an MSR, several parameters are introduced, namely the specific MA incineration consumption (SIC), the MA incineration ratio (TR) and the disappearance rate (DR) of each element in MA. The SIC is defined as

$$\mathrm{SIC} = \frac{MA(T=0) + MA(feeding) - MA(residue) - HN(residue)}{P \times T}$$

where $MA(T=0)$, $MA(feeding)$, $MA(residue)$ and $\mathrm{Pu}(residue)$ denote the initial MA loading, the online MA feeding and the MA residue, respectively. The HN residue in the fuel salt is the total TRU inventory except MA. P is the thermal power of MSR, while T is the entire operation time.

The TR is another important parameter for evaluating the MA incineration efficiency, which is defined as the ratio of the incinerated MA mass to the total loaded MA mass and calculated by

$$\mathrm{TR} = \frac{MA(T=0) + MA(feeding) - MA(residue) - HN(residue)}{MA(T=0) + MA(feeding)}$$

In addition, the disappearance rate (DR) of each element in MA is also applied to evaluate the transmutation capability of Np, Am and Cm in the thermal, epithermal and fast MSR cores, which is defined as

$$\mathrm{DR(i)} = \frac{M(i)(T=0) + M(i)(feeding) - M(i)(residue)}{(M(i)(T=0) + M(i)(feeding)) \times P \times T}$$

where $M(i)(T = 0)$, $M(i)(feeding)$ and $M(i)(residue)$ represent initial loading, the online feeding and the residue inventory of element i, respectively.

Table 2. SIC and TR for two salts in the three MSR cores

scenario	Flibe			Flinak		
	Core1	Core2	Core3	Core1	Core2	Core3
MA initial loading (t)	22.0	22.0	22.0	28.9	28.9	28.9
MA feeding (t)	41.3	45.1	50.4	41.5	49.7	60
SIC (kg/GWth.y)	167	185	206	170	206	247
IR	0.79	0.81	0.82	0.75	0.78	0.81

With 4% of MA in the Flibe salt and 10% of MA in the Flinak salt, the initial MA inventories are loaded as about 22 tons and 28.9 tons for the Flibe and Flinak salts, respectively. During the whole operation, the total TRU inventory is always kept constant to maintain the TRU fraction below the solubility limit. Therefore, the incinerated MA inventory is dependent on the feeding amount of MA. During the 100-year operation, the total MA feeding inventory with the Flibe salt ranges from 41.3 to 50.4 tons in the thermal, epithermal and fast MSR cores while it varies from 41.5 to 60 tons with the Flinak salt for the three cores. The STC and TR with Flibe and Flinak salts in the thermal, epithermal and fast MSR cores during 100-year operation are presented in Table 2. As most TRU nuclides have preferred fission ability in the fast neutron spectrum, which is an efficient way for MA incineration, a preferred MA transmutation capability is achieved in the fast MSR core for both of the two salts. One can be seen that the SICs with the Flibe salt for the three MSR cores at the end life of 100-year operation are 167, 185 and 206 kg/GWth.y, respectively. Due to the higher loading of MA for the Flinak salt, higher SICs are achieved correspondingly in the thermal, epithermal and fast MSR cores, which are 170, 206 and 247 kg/GWth.y, respectively. The discrepancy of SIC for the two salt compositions is more obvious in the fast spectrum due to the fact that the core with the Flinak salt hardens the spectrum more efficiently than that with the Flibe salt, which benefits the fission ability of most TRU nuclides and facilitates the MA incineration rate. The IR is an important parameter that evaluates the transmutation efficiency of MA. The IR variances of the SIC among the three MSR cores are similar for both of the two salts. However, the IR with the Flinak salt is lower than that with the Flibe salt, especially in the thermal MSR core. This is because that there is an inferior neutron economy in the Flinak salt with significant capture cross sections of Na-23 and K-39, especially in the thermal energy region. With the high MA loading and the feasibility of online refueling, MSR can attain an alluring transmutation efficiency of MA with the TR over 0.75, which means that more than 75% of MA loaded in the core is incinerated. The DR is an important indicator to evaluate the transmutation capability of a single element, which is displayed in Fig. 4. It can be seen that Np has a higher DR than the other two MA of Am and Cm as Np has a higher fraction in MA and a larger absorption cross section than those of Am and Cm, which ranges from 120.7 to 141.9 kg/GWh.y with the Flibe salt and from 122.1 to 163.3 kg/GWh.y with the Flinak salt in the above three cores. The DR of Am is inferior to that of Np due to a lower mass loading, a smaller absorption cross section and a significant production from Pu. In the transmutation chains, a significant amount of Cm will be accumulated by successive neutron captures and β decays from Np, Pu and Am, which impedes the DR of Cm significantly except its own low mass loading.

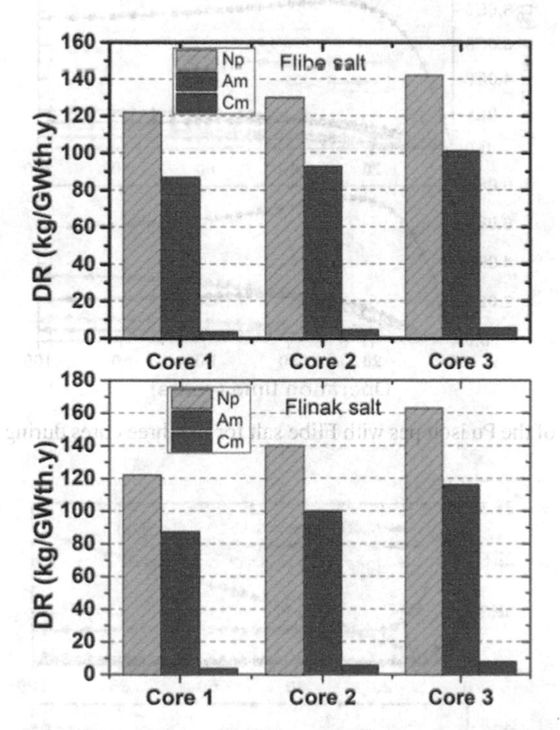

Fig. 4. DRs of Np, Am and Cm with Flibe and Flinak salts in the three MSR cores

3.3 TRU Evolution and Radiotoxicity

In the transmutation chains of MAs, the disappearance ways differ significantly in different neutron energies. Some MAs such as Np-237, Am-241 and Am-243 with larger capture cross sections than their fission ones in the thermal region are transmuted by capturing neutrons consecutively to form Pu-239, Am-242, Am-242m, and Cm-245 with very large fission cross sections. In the fast neutron spectrum, most MA can be transmuted by fissions with higher fission cross sections than the capture ones. Therefore, the TRU evolution varies significantly with the neutron spectrum, which imposes a variation on the radiotoxicity.

Figure 5 and Fig. 6 display the evolutions of the Pu isotopes and new created MA for the Flibe salt in the three MSR cores, respectively. One can find that the produced Pu isotopes are the majority of the new created nuclides, whose production rates are

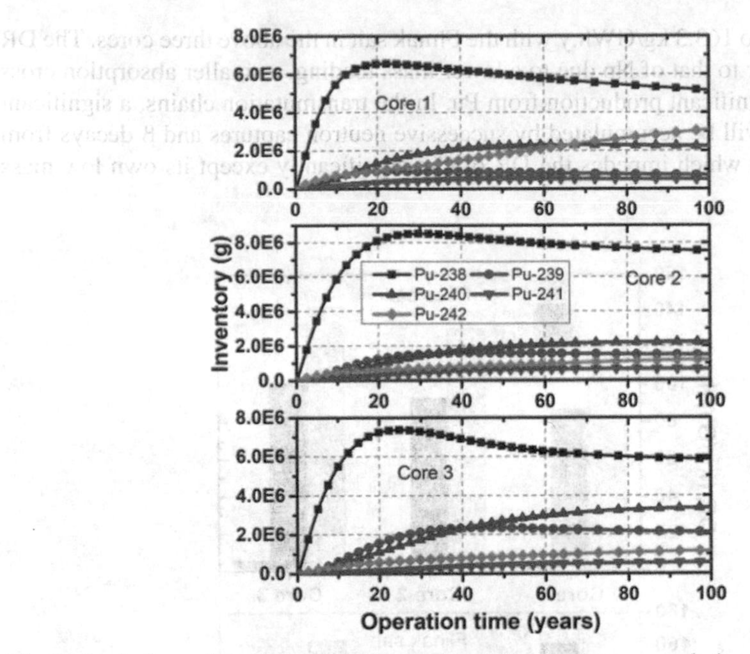

Fig. 5. Evolutions of the Pu isotopes with Flibe salt for the three cores during 100-year operation

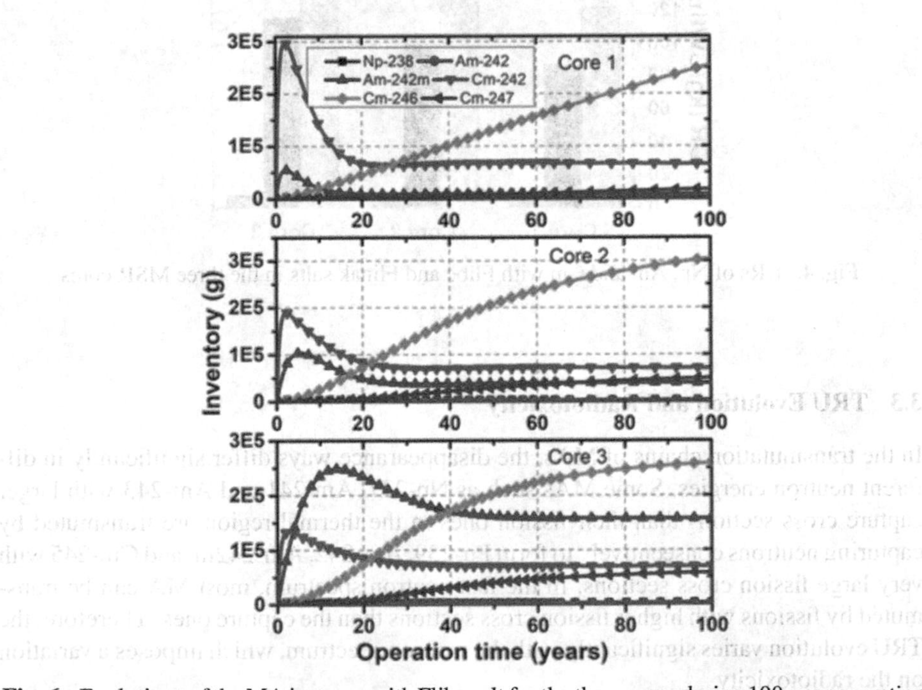

Fig. 6. Evolutions of the MA isotopes with Flibe salt for the three cores during 100-year operation

as high as about 116, 130 and 129 kg/y for the thermal, epithermal and fast cores, respectively. For the Pu isotopes, there are two main reaction chains to produce Pu-238 from Np-237 and Am-241 by successive (n, γ) and β/α decay reactions. In further, Pu-239 produced from the Pu-238 capture has a large fission cross section, which is the major disappearance way for Pu-239. Hence, the inventories of the higher Pu isotopes which are mainly produced by Pu-239 capture are much less than Pu-238. It can be seen in Fig. 5 that Pu-238 accounts for the majority of the transmuted products as the source nuclides of Np-237 and Am-241 are the main isotopes of the initial MA with the fraction of as high as about 86.2%. On the other hand, the accumulated inventory of Pu-238 in the epithermal core is higher than the other two cores due to the fact that most TRU have significant capture resonances in the epithermal region than in the fast region and the total loading of MA in the epithermal core is higher than that in the thermal core. The evolutions of the other Pu isotopes also reveal significant variations because of the different reaction cross sections and MA loadings. At the end of 100-year operation, the mass fraction of Pu-238 in the Pu isotopes is 44.4%, 57.6% and 44.9% in the thermal, epithermal and fast MSR cores, respectively, which is advantageous from the view point of non-proliferation. Similarly, the evolutions of new created MAs in the three cores reveal a significant variation which is tightly related with the MA loadings and the neutron spectrum. The production rate of the new created MA is 3.5, 4.7 and 5.5 kg/y for the thermal, epithermal and fast MSR cores, respectively, which is much lower than that of the Pu isotopes.

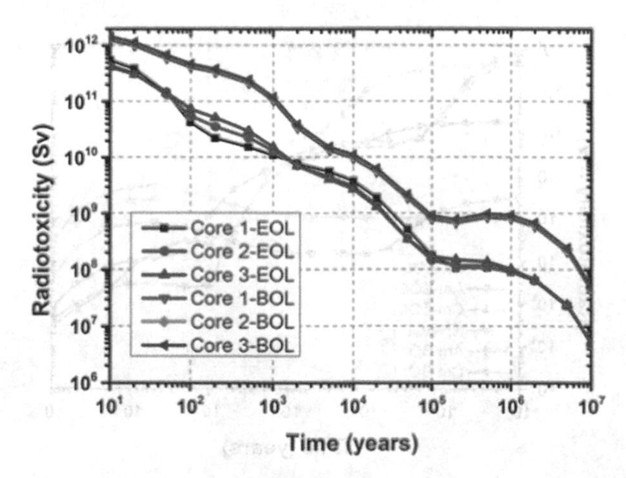

Fig. 7. Evolutions of the total radiotoxicity with Flibe and Flinak salts in the three MSR cores

The radiotoxicity is an important parameter to evaluate the effect of radionuclides to human health by ingestion or digestion, which is defined as:

$$R(t) = \sum_i R_i(t) = \sum_i r_i \lambda_i N_i(t)$$

where $R_i(t)$ represents the radiotoxicity of nuclide i at time t in unit of Sv; r_i refers to the effective dose coefficient in Sv/Bq by ingestion for the public [20], which depends on the decay mode and the emitted energy of particles; λ_i and $N_i(t)$ represent the decay constant and atoms of nuclide i, respectively. In the SCALE6.1 software system, the ORIGEN-S module is feasible to calculate the radioactivity of each nuclide and the radiotoxicity is calculated with multiplying the effective dose coefficient by the radioactivity of each nuclide.

To evaluate the radiotoxicity of MA with different MA loadings in the thermal, epithermal and fast MSR cores, the total radiotoxicities of MA and the radiotoxicities of Np, Am and Cm at the beginning and end of lifetime are chosen for discussion. One can see from Fig. 7 that the total radiotoxicitiy of MA for the Flibe salt in the three different MSR cores after 100-year operation is reduced significantly, about 45% lower than that at the beginning of lifetime. For the total radiotoxicity of MA, the highest contribution is from Cm because of the high dose coefficients of most Cm isotopes and their daughter products even though the total Cm inventory is extremely small (as shown in Fig. 8). For this reason, the total radiotoxicity of MA for the three different cores are very similar since the difference of the MA inventory can be negligible. Due to a significant accumulation of Np-238 with a very short half-life, the radiotoxicity of Np at the end of lifetime is increased slightly in the first 100 years, which is decreased rapidly in the following decay time and is about 80% lower than that at the beginning of lifetime. Am has a significant reduction on radiotoxicity, with more than an order of magnitude lower than that at the beginning of lifetime, which is the major factor to decrease the total radiotoxicity of MA.

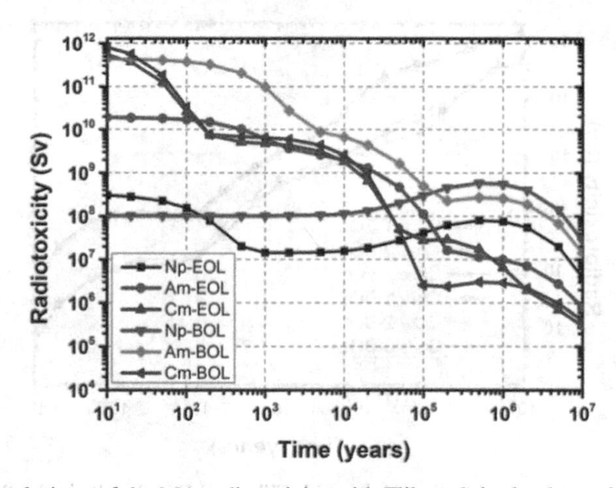

Fig. 8. Evolutions of the MA radiotoxicity with Flibe salt in the thermal MSR core

3.4 Evaluation on Safety Parameters

To evaluate the influence of MA loading on the reactor safety of the thermal, epithermal and fast MSR cores, two important parameters of the temperature feedback coefficient (TFC) and the effective delayed neutron fraction (β_{eff}) are analyzed in this work.

The TFC is an essential indicator for the inherent safety issues, which is required to be negative during the whole lifetime of the reactor. The total TFC in a reactor is calculated as the sum of the fuel TFC and the moderator TFC. In further, the fuel TFC can be broken down into the Doppler effect and the fuel salt expansion.

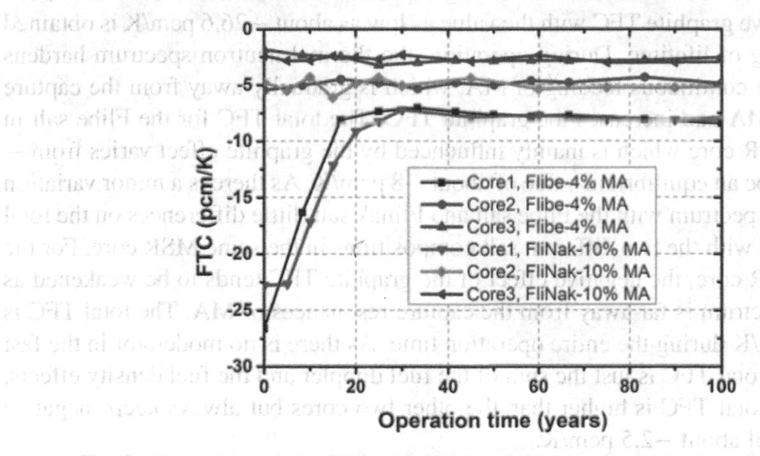

Fig. 9. Evolutions of the TFC with Flibe and Flinak salts in the three MSR cores

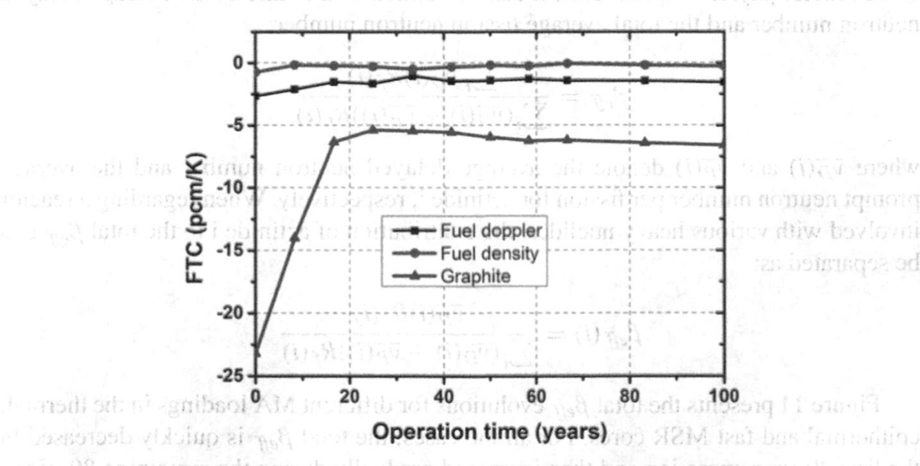

Fig. 10. Evolutions of the TFC with Flibe salt in the thermal MSR core

The evolutions of the TFC for different MA loadings in the three MSR cores are displayed in Fig. 9. One can be seen that the total TFC is significantly related with the neutron spectrum and slightly varied with the evolutions of the fuel compositions. To evaluate the contribution of TFC from fuel density, fuel Doppler and graphite, the three items of TFC for the Flibe salt in the thermal core are exampled for discussion, as shown in Fig. 10. For the MA from the spent fuel of PWR, the MA nuclides except Cm-245 (with the mole fraction in MA of 0.26%) are poisonous to reactivity, especially in a thermal region. Furthermore, Np-237 at 0.5 eV and Am-241 at 0.3 eV, 0.6 eV and 1.1 eV reveal significant capture resonance cross sections (>1000 barns), respectively, which is benefit for decreasing the TFC as the MA captures with a Doppler broadening can counteract the increased fissions of fissile nuclides with the fuel temperature increasing. In addition, when the graphite temperature is increased, the Maxwellian spectrum shifts to a higher energy region, where it is closer to the strong capture resonance cross-sections of MA. Hence, a negative graphite TFC with the value as low as about −26.6 pcm/K is obtained at the beginning of lifetime. During operation, the thermal neutron spectrum hardens gradually as the continuous feeding of MA, which is gradually away from the capture resonances of MA and increases the graphite TFC. The total TFC for the Flibe salt in the thermal MSR core which is mainly influenced by the graphite effect varies from −26.6 pcm/K to be an equilibrium value of about −8 pcm/K. As there is a minor variation on the neutron spectrum with the Flibe salt and Flinak salt, little differences on the total TFC are caused with the two different salt compositions in the same MSR core. For the epithermal MSR core, the negative effect of the graphite TFC tends to be weakened as the neutron spectrum is far away from the capture resonances of MA. The total TFC is around −5 pcm/K during the entire operation time. As there is no moderator in the fast MSR core, the total TFC is just the sum of the fuel doppler and the fuel density effects. Therefore, the total TFC is higher than the other two cores but always keeps negative with the value of about −2.5 pcm/K.

β_{eff} is another important parameter for both kinetics reactivity controlling safety and static reactor physics experiments. It can be defined as the ratio of the average delayed neutron number and the total average fission neutron number:

$$\beta_{eff} = \frac{\sum_i \overline{\nu_D}(i) R_f(i)}{\sum_i (\overline{\nu_D}(i) + \overline{\nu_P}(i)) R_f(i)}$$

where $\overline{\nu_D}(i)$ and $\overline{\nu_P}(i)$ denote the average delayed neutron number and the average prompt neutron number per fission for actinide i, respectively. When regarding a reactor involved with various heavy nuclides, the contribution of actinide i to the total β_{eff} can be separated as:

$$\beta_{eff}(i) = \frac{\overline{\nu_D}(i) R_f(i)}{\sum_i (\overline{\nu_D}(i) + \overline{\nu_P}(i)) R_f(i)}$$

Figure 11 presents the total β_{eff} evolutions for different MA loadings in the thermal, epithermal and fast MSR cores. For all the cases, the total β_{eff} is quickly decreased in the first 20-year operation and then increased gradually during the remaining 80 years. To explore the source of the variation of β_{eff}, the separate contributions of the main actinides in the case of 4% MA in the Flibe salt for the thermal MSR are evaluated, which is listed in Fig. 12. One can see that the β_{eff} of U-233 at the beginning of the lifetime is as high as 274 pcm, which accounts for the vast majority of the total β_{eff} (287 pcm). This is because that U-233 contributes the majority of fissions in the thermal core. When the neutron spectrum hardens, more U-233 is required which increases the total β_{eff} correspondingly. Furthermore, an inferior neutron economy for the Flinak salt also requires more fissile fuels for criticality. Therefore, when the neutron spectrum ranges from thermal to fast, the initial total β_{eff} rises from 287 pcm to 298 pcm for the Flibe salt while it increases from 289 pcm to 315 pcm for the Flinak salt. During the first

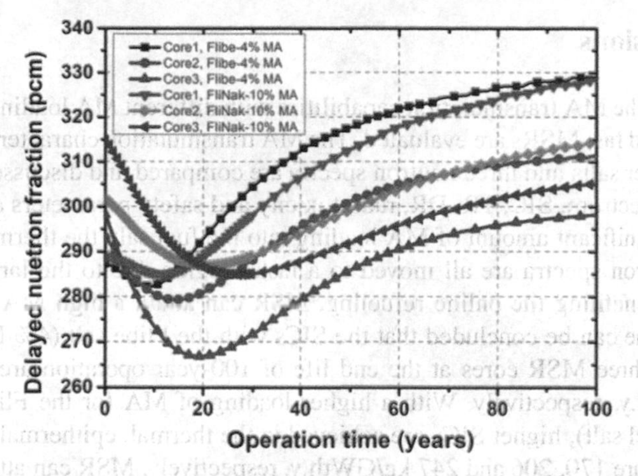

Fig. 11. Evolutions of the total β_{eff} with Flibe and Flinak salts in the three MSR cores

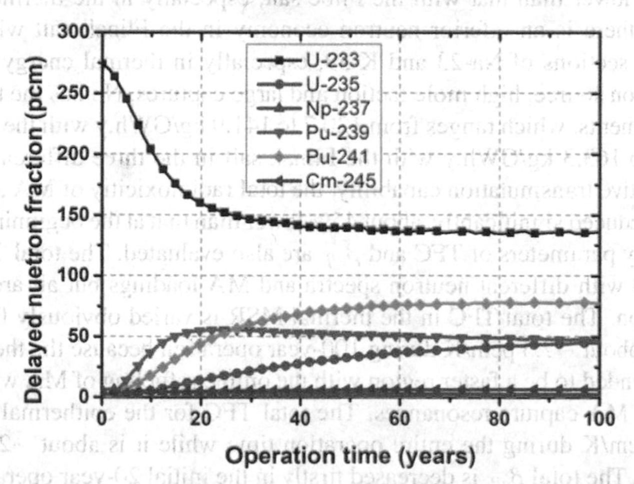

Fig. 12. Evolutions of the β_{eff} contribution of main nuclides with Flibe salt in the thermal MSR core

20-year operation, several fissile isotopes (Pu-239, Pu-241, Cm-245) transmuted from MA are quickly accumulated which weakens the requirement of U-233 significantly. As the single β_{eff} of Pu-239, Pu-241 and Cm-245 are much smaller than that of U-233, the total β_{eff} decreases correspondingly during the first decades of operation. Since most of the TRU nuclides have a higher ratio of fission to capture in a fast neutron spectrum, the requirement of U-233 is reduced in further. Hence, the total β_{eff} decreases more rapidly when the neutron spectrum hardens. The total β_{eff} increases gradually to be an equilibrium state in the remaining 80-year operation due to the accumulated fissile isotopes except U-233.

4 Conclusions

In this paper, the MA transmutation capabilities with different MA loadings for thermal, epithermal and fast MSRs are evaluated. The MA transmutation characteristics with two types of carrier salts and three neutron spectra are compared and discussed. The effects on neutron spectrum, SIC, TR, DR, radiotoxicity and safety parameters are analyzed.

With a significant amount of MA loading into the fuel salt, the thermal, epithermal and fast neutron spectra are all moved to a faster region due to the large captures of most MA. Benefiting the online refueling, MSR can attain a high MA transmutation capability. One can be concluded that the SICs with the Flibe salt (4% MA in the fuel salt) for the three MSR cores at the end life of 100-year operation are 167, 185 and 206 kg/GWth.y, respectively. With a higher loading of MA for the Flinak salt (10% MA in the fuel salt), higher SICs are achieved in the thermal, epithermal and fast MSR cores, which are 170, 206 and 247 kg/GWth.y, respectively. MSR can attain an alluring transmutation efficiency of MA with the TR over 0.75. In addition, the IR with the Flinak salt is lower than that with the Flibe salt, especially in the thermal MSR core, because that there is an inferior neutron economy in the Flinak salt with significant capture cross sections of Na-23 and K-39, especially in thermal energy region. With little production source, high mole faction and large captures, Np has the highest DR in three MA elements, which ranges from 120.7 to 141.9 kg/GWh.y with the Flibe salt and from 122.1 to 163.3 kg/GWh.y with the Flinak salt in the three different MSR cores. With an effective transmutation capability, the total radiotoxicitiy of MA after 100-year operation is reduced significantly, about 45% lower than that at the beginning of lifetime.

Two safety parameters of TFC and β_{eff} are also evaluated. The total TFC is significantly varied with different neutron spectra and MA loadings but all are located in a negative region. The total TFC in the thermal MSR is varied obviously from below $-$20 pcm/K to about $-$7.5 pcm/K during 100-year operation because the thermal neutron spectrum is tended to be a faster region with the online refueling of MA which weakens the effects of MA capture resonances. The total TFC for the epithermal MSR is kept around $-$5 pcm/K during the entire operation time while it is about $-$2.5 pcm/K for the fast MSR. The total β_{eff} is decreased firstly in the initial 20-year operation and then increased gradually to be an equilibrium state in the remaining 80-year operation. The total β_{eff} at the equilibrium state varies from about 298 pcm to 329 pcm for all cases.

In conclusion, it is feasible to transmute MA in an MSR and achieve the goals of reduction of MA long-term radioactive hazards. A higher SIC is obtained with a higher MA loading for the Flinak salt while a higher TR is achieved with a better neutron economy for the Flibe salt.

Acknowledgement. This work was supported by the National Natural Science Foundation of China (Grant Nos. 12005291 and 12175300), Youth Innovation Promotion Association CAS and the Chinese TMSR Strategic Pioneer Science and Technology Project (Grant No. XDA02010000).

References

1. Salvatores, M., Palmiotti, G.: Radioactive waste partitioning and transmutation within advanced fuel cycles: achievements and challenges. Prog. Part. Nucl. Phys. **66**(1), 144–166 (2011)
2. Bala, A., Namadi, S.: A review of the advantages and disadvantages of partitioning and transmutation. Int. J. Sci. Adv. Technol. **5**, 11–14 (2015)
3. OECD-Nuclear Energy Agency: Status and Assessment Report on Actinide and Fission Product Partitioning and Transmutation (1999)
4. Mukaiyama, T., Gunji, Y., Ogawa, T., et al.: Minor actinide transmutation in fission reactors and fuel cycle considerations. IAEA-TECDOC-693, Vienna, Austria: IAEA 86 (1993)
5. OECD, 2014. Minor Actinide Burning in Thermal Reactors: A Report by the Working Party on Scientific Issues of Reactor Systems, Nuclear Science, OECD Publishing, Paris
6. Liu, B., Han, J., Liu, F., et al.: Minor actinide transmutation in the lead-cooled fast reactor. Prog. Nucl. Energy **119**, 103148 (2020)
7. Takeda, T.: Minor actinides transmutation performance in a fast reactor. Ann. Nucl. Energy, **95**, 48–53 (2016)
8. Zhou, S., Wu, H., Zheng, Y.: Flexibility of ADS for minor actinides transmutation in different two-stage PWR-ADS fuel cycle scenarios. Ann. Nucl. Energy **111**, 271–279 (2018)
9. Jiang, M.H., Xu, H.J., Dai, Z.M.: Advanced fission energy program-TMSR nuclear energy system. Bull. Chin. Acad. Sci. **27**, 366 (2012)
10. Zhou, B., Xiaohan, Y., Zou, Y., et al.: Study on dynamic characteristics of fission products in 2 MW molten salt reactor. Nucl. Sci. Tech. **31**(17), 1–17 (2020)
11. Zou, C.Y., Cai, X.Z., Yu, C.G., et al.: Transition to thorium fuel cycle for TMSR. Nucl. Eng. Des. **330**, 420–428 (2018)
12. Zou, C.Y., Zhu, G.F., Yu, C.G., et al.: Preliminary study on TRUs utilization in a small modular Th-based molten salt reactor (smTMSR). Nucl. Eng. Des. **339**, 75–82 (2018)
13. Ignatiev, V., Feynberg, O., Merzlyakov, A., et al.: Progress in development of MOSART concept with Th support. In: Proceedings of ICAPP, 12394 (2012)
14. Heuer, D., Merle-Lucotte, E., Allibert, M., et al.: Towards the thorium fuel cycle with molten salt fast reactors. Ann. Nucl. Energy **64**, 421 (2014)
15. Zou, C., Yu, C., Wu, J., et al.: Parametric study on minor actinides transmutation in a graphite-moderated thorium-based molten salt reactors. Int. J. Energy Res. (12) (2021)
16. Ma, K.F., Yu, C.G., Cai, X.Z., et al.: Transmutation of ^{129}I in a single-fluid double-zone thorium molten salt reactor. Nucl. Sci. Tech. **31**(1) (2020)
17. ORNL: Scale: A Comprehensive Modeling and Simulation Suite for Nuclear Safety Analysis and Design. ORNL/TM-2005/39 (2011)
18. Yu, C., Zou, C., Wu, J., et al.: Development and verification of molten salt reactor refueling and reprocessing system analysis code based on SCALE. At. Energy Sci. Technol. **52**, 2136–2142 (2018)
19. Lizin, A.A., Tomilin, S.V., Gnevashov, O., et al.: PuF_3, Am $_F3$, Ce_F3, and Nd_F3 solubility in LiF-NaF-KF MELT. At. Energ. **115**, 11 (2013)
20. ICRP: Compendium of dose coefficients based on ICRP publication 60. In: ICRP Publication 119. Ann. ICRP **41**(Suppl.) (2012)

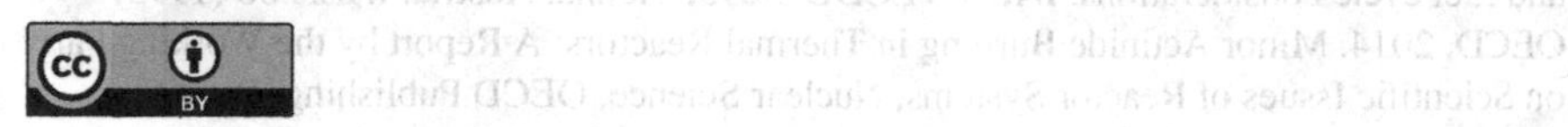

Development and Preliminary Verification of a Neutronics-Thermal Hydraulics Coupling Code for Research Reactors with Unstructured Meshes

Mingrui Yang[1], Chixu Luo[1], Dan Wang[2], Tianxiong Wang[2], Xiaojing Liu[1], and Tengfei Zhang[1](✉)

[1] Shanghai Jiao Tong University, Shanghai, China
{yangmr96,zhangtengfei}@sjtu.edu.cn

[2] Nuclear Power Institute of China, Chengdu, Sichuan, China

Abstract. To maximize their adaptability and versatility, research reactors are designed to adapt to various operational conditions. These requirements result in more complex configurations and irregular geometries for research reactors. Besides, there is usually a strong coupling of neutronics-thermal hydraulics (N-TH) fields inside the reactor. A three-dimensional N-TH coupling code has been developed named CENTUM (CodE for N-Th coupling with Unstructured Mesh). Steady-state and transient neutronic analyses are performed using a 3D triangular-z nodal transport solver with the stiffness confinement method (SCM). Meanwhile, thermal-hydraulics calculations adopt a multi-channel model. For a preliminary verification of the code, we examine CENTUM with benchmark problems including TWIGL, 3D-LMW, and NEACRP. CENTUM produces maximum power errors of −1.27% and −0.45% for the TWIGL A1 and A2 cases, respectively. For the 3D-LMW benchmark, the largest relative power error of 3.84% is observed at 10 s compared with the reference SPANDEX code. For the NEACRP N-TH coupling benchmark, CENTUM results in a 0.35 ppm error in critical boron concentration, a 2.16 °C discrepancy in the fuel average Doppler temperature, and a 0.63% overestimation in the maximum axial power. Moreover, transient results considering thermal-hydraulics feedback are in good agreement with the PARCS reference solutions, with the maximum relative power deviation being only 0.055%.

Keywords: Reactor Kinetics · Stiffness Confinement Method · Neutronics-Thermal Hydraulics Coupling · Safety Analysis · Research Reactor

1 Introduction

Research reactors have served as the workhorse for nuclear fuel and material irradiation testing, and they can also be used for secondary missions such as isotope production and electricity generation. They serve as research, development, and demonstration platforms for fuels, materials, and other critical components. To maximize their adaptability and

C. Liu (Ed.): PBNC 2022, SPPHY 283, pp. 673–689, 2023.
https://doi.org/10.1007/978-981-99-1023-6_58

versatility, research reactors are designed to adapt to a variety of operational conditions. These requirements result in more complex configurations and irregular geometries for research reactors than for conventional Pressurized Water Reactors (PWRs) and Boiling Water Reactors (BWRs). Direct application of conventional reactor analysis codes to research reactors is challenging due to their complex geometrical configurations and high neutron streaming caused by frequent control rod movement. It is noted that geometry complexity is a feature shared by almost all research reactors, which limits the feasibility of conventional methods based on rectangular or hexagonal meshes. As a result, the ability to accurately model research reactors require the ability to describe unstructured meshes. Additionally, the local neutron spectrum in a research reactor varies significantly with position, and frequent control rod operation results in significant neutron flux heterogeneity. Therefore, it is required to solve the neutron transport equation that can describe the angular anisotropy and to take into account the coupling of various physical fields in order to accurately simulate the behavior of the reactor. This will surely use a lot of computational resource during the core's transient analysis. However, due to the physical nature of the reactor, the use of conventional numerical methods, such as implicit Euler method [1] and Runge-Kutta methods [2], encounters serious problems due to the "rigidity" of the set of equations. Solving with these methods requires very small time-step sizes in order to ensure the stability of the method, and thus many unnecessary information will be computed while the computation continues for long transition times, leading to a huge waste of computational resources, and may also contain large cumulative errors. The precursor concentration equation introduced stiffness into the system, so we employ the stiffness confinement method (SCM) technique to decouple the neutron flux density equation and the delayed neutron density equation, eliminating the stiffness introduced by the precursor concentration equation [3].

Due to the strong coupling between reactor neutronics and thermal hydraulics, a multi-channel model is used to describe the coolant convection and the finite difference method to calculate the thermal conductivity of the fuel rods, so as to provide feedback on the cross-section used for the neutronics calculation. CENTUM's adopts the OSSI [4] (Operator Splitting Semi-Implicit) method for coupling. Figure 1 depicts the coupling flow of the OSSI method. Each time step begins with a neutronics calculation, and the power rate for each assembly is transferred to the thermal hydraulics module. Without iterating, the process enters directly to the next time step after the thermal hydraulics calculation is finished.

This paper is organized as follows. Section 2 illustrates the steady-state triangular-z node neutron transport model, the transient SCM method and the thermal hydraulics models embedded in CENTUM. As a preliminary verification, Sect. 3 presents numerical verification results using TWIGL, LMW and NEACRP benchmark cases. Specifically, the TWIGL benchmark and the LMW benchmark are neutronics transient problems, and the NEACRP is a problem to demonstrate CENTUM's capability of modeling the N-TH coupling effect.

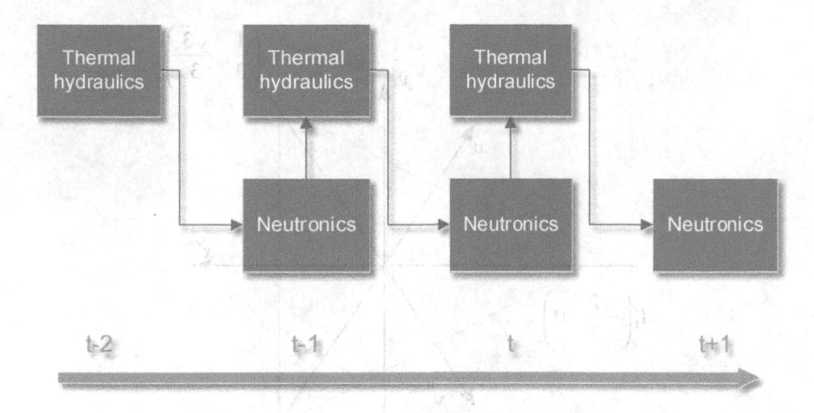

Fig. 1. The OSSI method in CENTUM

2 Methodology

2.1 Triangular-Z Node Neutron Transport Model

Considering isotropic scattering and using S_N method, the three-dimensional multi-group neutron transport equation in the triangular prism can be written as [5]:

$$\mu^m \frac{\partial \Psi_g^m(x, y, z)}{\partial x} + \eta^m \frac{\partial \Psi_g^m(x, y, z)}{\partial y} + \frac{\xi^m}{h_z} \frac{\partial \Psi_g^m(x, y, z)}{\partial z} + \Sigma_t^g \varphi_g^m(x, y, z) = \hat{Q}_g(x, y, z) \tag{2.1}$$

Here, μ^m, η^m, ξ^m are the components of the angular direction m on the x, y, z axes; m is a certain angular direction after using S_N discretization; $\hat{Q}_g(x, y, z)$ is The neutron source term includes fission sources and scattering sources ($cm^{-3} \cdot s^{-1}$); Ψ_g^m represents the neutron angular flux density of the g group in the m direction ($cm^{-2} \cdot s^{-1}$). Usually, the solved triangular node is arbitrary, and it needs to be transformed into a unified coordinate system (Fig. 2).

Using the coordinate transformation, we obtain the equation as:

$$\begin{aligned} &\mu_x^m \frac{\partial \psi_g^m(x', y', z)}{\partial x'} + \eta_x^m \frac{\partial \psi_g^m(x', y', z)}{\partial y'} + \frac{\xi^m}{h_z} \frac{\partial \psi_g^m(x', y', z)}{\partial z} + \Sigma_t^g \psi_g^m(x', y', z) \\ &= Q_g(x', y', z) \end{aligned} \tag{2.2}$$

where

$$\mu_x^m = \frac{(-y_n + y_p)\mu^m + (x_n - x_p)\eta^m}{2\Delta}$$

$$\eta_x^m = \frac{\left(-x_k + \frac{1}{2}x_n + \frac{1}{2}x_p\right)\eta^m + \left(y_k - \frac{1}{2}y_n - \frac{1}{2}y_p\right)\mu^m}{\sqrt{3}\Delta}$$

$$-2/3 \le x' \le 1/3, \quad -y_s(x') \le y' \le y_s(x'), \; y_s(x') = (x' + 2/3)/\sqrt{3}, \; -1/2 \le z \le 1/2$$

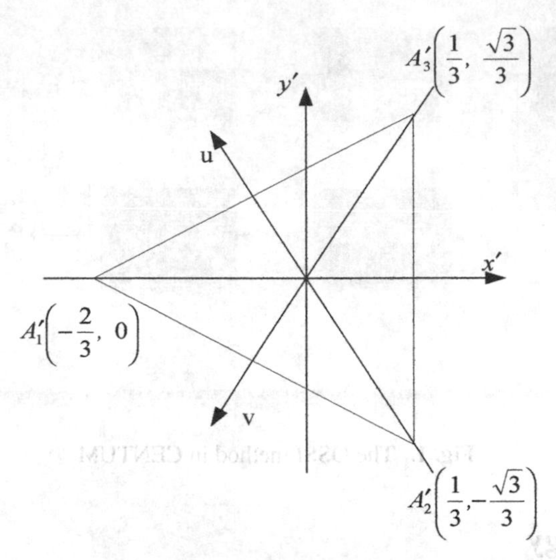

Fig. 2. Equilateral triangle in the calculated coordinate system

The nodal balance equation is given by:

$$2\mu_x^m\overline{\psi}_x^m + 2\mu_u^m\overline{\psi}_u^m + 2\mu_v^m\overline{\psi}_v^m + \frac{\xi^m}{h_z}\left(\overline{\psi}_{z+}^m - \overline{\psi}_{z-}^m\right) + \Sigma_t\overline{\psi}^m = \overline{Q} \tag{2.3}$$

$\overline{\psi}_i^m$ is the averaged outgoing surface flux in the x, u, v directions, as is shown in Fig. 1; $\overline{\psi}_{z\pm}^m$ are the outgoing surface averaged flux of the upper and lower sides of the node; $\overline{\psi}^m$ is the averaged flux in the node. The transport computation is performed using a specific sweeping and source iteration approach.

2.2 Transient Neutronic Model

For solving transient problems, CENTUM adopts the SCM [6]. By considering that the dynamic frequency of the angular flux is isotropic:

$$\omega_g(r,t) = \frac{\partial}{\partial t}\ln\varphi_g(r,t) \tag{2.4}$$

where $\omega_g(r,t)$ is the flux dynamic frequency of group g; $\varphi_g(r,t)$ is the scalar flux of group g.

The flux dynamic frequency can be further decomposed into the shape frequency $\omega_{s,g}(r,t)$ and the amplitude frequency $\omega_t(t)$.

$$\omega_g(r,t) = \omega_{s,g}(r,t) + \omega_t(t) \tag{2.5}$$

Similarly, the precursor frequency of group i is defined as:

$$\mu_i(r,t) = \frac{\partial}{\partial t}\ln C_i(r,t) \tag{2.6}$$

Inserting (2.4) and (2.6) into the 3-D multigroup time-space neutron transport equation within a triangular-z prism, the time-dependent neutron transport equation can be transformed into an equation for solving the eigenvalue problem:

$$\begin{aligned}&\mu^m \frac{\partial}{\partial x}\varphi_g^m(r,t) + \eta^m \frac{\partial}{\partial y}\varphi_g^m(r,t) + \frac{\xi^m}{hz}\frac{\partial}{\partial z}\varphi_g^m(r,t)\\&+\Sigma'_{t,g}(r,t)\varphi_g^m(r,t) = \frac{\chi'_g}{k_D}\sum_{g'=1}^{G}(\nu\Sigma_f)_{g'}(r,t)\varphi_g^m(r,t)\\&+\sum_{g'=1}^{G}\varphi_g^m(r,t)\Sigma_{g'\to g}(r,t)\end{aligned} \tag{2.7}$$

k_D is the dynamic eigenvalue; $\Sigma'_{t,g}$ is the dynamic total cross-section and χ'_g is the dynamic fission spectrum, which are respectively defined as:

$$\Sigma'_{t,g}(r,t) = \Sigma_{t,g}(r,t) + \frac{\omega_g(r,t)}{v_g} \tag{2.8}$$

$$\chi'_g = \chi_g(1-\beta) + \sum_{i=1}^{M}\chi_{ig}\lambda_i\frac{\beta_i}{\mu_i+\lambda_i} \tag{2.9}$$

Here, v_g represents the neutron velocity of the g group. The unknown quantities are the flux dynamic frequency and the precursor frequency. Solving for k_D iteratively by Eq. (2.7), combined with the secant method, we get amplitude frequency:

$$\omega_t^{(m+1)}(t_n) = \omega_t^{(m)}(t_n) + \left[\omega_t^{(m-1)}(t_n) - \omega_t^{(m)}(t_n)\right]\frac{1-k_D^{(m)}}{k_D^{(m-1)} - k_D^{(m)}} \tag{2.10}$$

Based on the isotropic approximation, the average scalar flux in the nodal ν is used to update the node wise shape frequency:

$$\overline{\omega}_{\nu,S,g}(t_n) = \frac{1}{\Delta t_n}\ln\left[\frac{\int_\nu drf^T(r)\widehat{\varphi}_{\nu,g}(t_n)}{\int_\nu drf^T(r)\varphi_{\nu,g}(t_{n-1})}\right] \tag{2.11}$$

in which $\widehat{\varphi}_{\nu,g}$ is the flux normalized according to the neutron density; $\overline{\omega}_{\nu,S,g}(t_n)$ is the average shape frequency within $[t, t+\Delta t]$. Meanwhile the actual flux can be written as:

$$\varphi_{\nu,g}(r,t) = \widehat{\varphi}_{\nu,g}(r,t)e^{\frac{\omega_T(t_n)+\omega_T(t_{n-1})}{2}\Delta t_n} \tag{2.12}$$

Assume that the precursor concentration is uniform within each node, the analytical solution for the precursor concentration is expressed as:

$$C_{\nu,i}(t_n) = C_{\nu,i}(t_{n-1})e^{-\lambda_i\Delta t_n} + \beta_i e^{-\lambda_i\Delta t_n}\int_{t_{n-1}}^{t_n} Q_\nu(t)e^{\lambda_i t}dt \tag{2.13}$$

where β_i is the share of group i precursor, λ_i is its decay constant; Q_ν is the the node-wise fission source. Combined with Eq. (2.13) the precursor frequency can be updated by:

$$\mu_{\nu,i}(t_n) = \begin{cases}\beta_i\frac{Q_\nu(t_n)}{C_{\nu,i}(t_n)} - \lambda_i & C_{\nu,i}(t_n) \neq 0\\ 0 & C_{\nu,i}(t_n) = 0\end{cases} \tag{2.14}$$

After obtaining the flux frequency and the precursor frequency, we can iteratively Solve k_D by adjusting dynamic total cross-section and the dynamic fission spectrum by using Eq. (2.8) and (2.9). The neutronics transient solving process is shown below (Fig. 3).

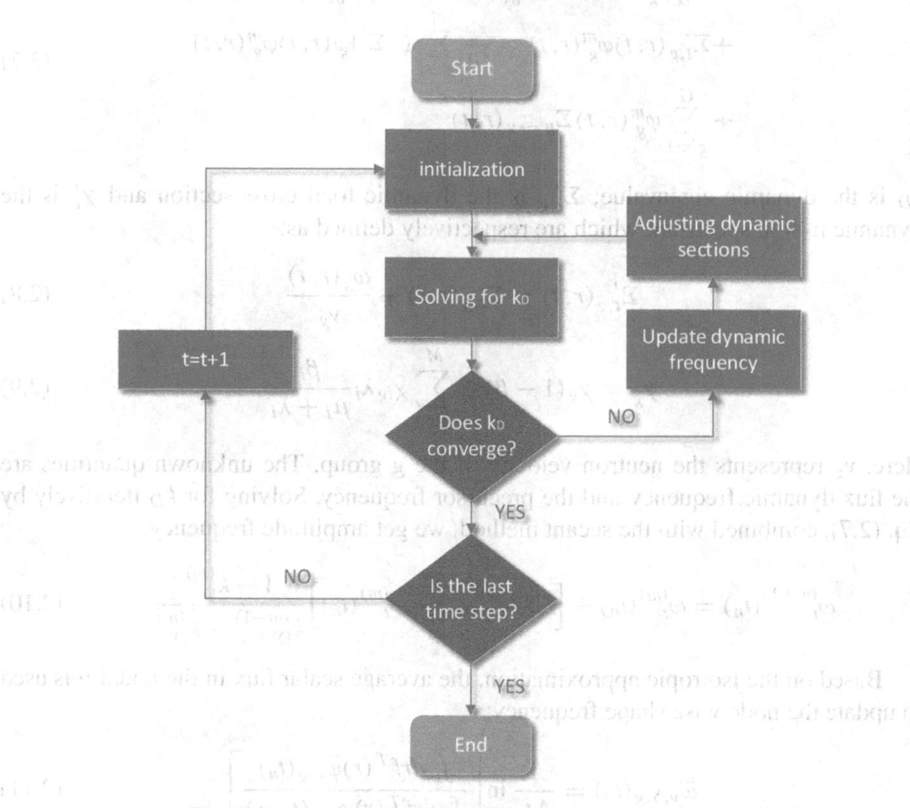

Fig. 3. Flowchart of the transient calculations with SCM

2.3 Thermal Hydraulics Model

In the N-Th coupling process, the thermal hydraulics calculation is used to obtain the thermal hydraulic parameters of the fuel rods and moderator. These parameters are then used to calculate the effect on the cross-sections used for the neutronics calculations.

2.3.1 Moderator Thermal Hydraulics Equation

CENTUM solves the thermal-hydraulics field using multi-channel models. The one-dimensional thermal hydraulics equations are expressed as follows [7]:

Moderator mass conservation equation:

$$\frac{\partial}{\partial t}(\rho)+\frac{\partial}{\partial z}(\rho u)=0 \tag{2.15}$$

Moderator energy conservation equation:

$$\frac{\partial}{\partial t}(\rho h)+\frac{\partial}{\partial z}(\rho uh)=Q^{f} \tag{2.16}$$

Moderator momentum conservation equation:

$$\frac{\partial}{\partial t}(\rho u)+\frac{\partial}{\partial z}(\rho u^{2})+\frac{\partial P_{fric}}{\partial z}+g\rho+\frac{\partial P}{\partial z}=0 \tag{2.17}$$

Moderator control equations are solved using a parallel multi channel model, the flow distribution is based on equal pressure drop in each parallel multi-channel where the inlet and outlet are at the same isobaric surface. The time derivative term is treated with the full-implicit backward difference method and the space variables are treated with the finite difference method.

2.3.2 Fuel Rod Heat Transfer Equation

In a one-dimensional cylindrical coordinate system, ignoring the effect of the axial direction, Fuel rod heat transfer equation as shown below [8]:

$$\rho c\frac{\partial T}{\partial t}+\frac{d^{2}T}{dr}+\frac{1}{r}\frac{dT}{dr}+\frac{Q}{\lambda}=0 \tag{2.18}$$

where T is the distribution of temperature; Q is the volumetric heat release rate of the fuel; λ, ρ and c are the thermal conductivity, density and heat capacity of the fuel material respectively. These properties change with temperature.

The spatial variable in the thermal conductivity equation is treated by the finite difference method. The cylindrical geometry from the inside to the outside are taken successively as the fuel zone, the air gap zone and the cladding zone. For the time discretization, the Crank-Nicholson implicit difference method is used, where θ equals 0.5 in Eq. (2.19).

$$\begin{gathered}\frac{\partial A(r,t)}{\partial t}=B(r,t)\\ \frac{A(r,t+\Delta t)-A(r,t)}{\Delta t}=(1-\theta)B(r,t+\Delta t)+\theta B(r,t)\end{gathered} \tag{2.19}$$

By discretization, the following tridiagonal matrix can be obtained:

$$\begin{pmatrix} b_1 & c_1 & 0 & 0 & \cdots & 0 & 0 \\ a_2 & b_2 & c_2 & 0 & \cdots & 0 & 0 \\ 0 & a_3 & b_3 & c_3 & \ddots & \vdots & \vdots \\ 0 & 0 & \ddots & \ddots & \ddots & 0 & 0 \\ \vdots & \vdots & \ddots & a_{n-2} & b_{n-2} & c_{n-3} & 0 \\ 0 & 0 & \cdots & 0 & a_{n-1} & b_{n-1} & c_{n-1} \\ 0 & 0 & \cdots & 0 & 0 & a_n & b_n \end{pmatrix}\begin{pmatrix} \overline{T_1}^{t+\Delta t} \\ \overline{T_2}^{t+\Delta t} \\ \overline{T_3}^{t+\Delta t} \\ \\ \overline{T_{n-2}}^{t+\Delta t} \\ \overline{T_{n-1}}^{t+\Delta t} \\ \overline{T_n}^{t+\Delta t} \end{pmatrix}=\begin{pmatrix} q_1 \\ q_2 \\ q_3 \\ \\ q_{n-2} \\ q_{n-1} \\ q_n \end{pmatrix} \tag{2.20}$$

in which n is the node number; a_n, b_n, c_n and q_n are constants related to the geometry, material and the temperature distribution at the previous time step.

3 Results and Discussion

The primary purpose of this section is to verify CENTUM's the transient function. TWIGL and LMW benchmark problem are mainly to verify the neutronics model of CENTUM, and NEACRP is to verify the N-Th coupling function. All results are evaluated with the angular discretization order of S_2. The triangular-z nodes are generated by Gmsh. All calculations are performed on a personal computer with a 2.90 GHz Intel i7-10700 CPU processor useing serial computation.

3.1 TWLGL Benchmark Problem

This section illustrates the preliminary verification of CENTUM using a simplified two-dimensional two-group dynamics benchmark problem with one set of precursor dynamics parameters [9, 10]. The core consists of three fuel zones forming a square core with a side length of 160 cm, and the fuel assembly size is 8 cm × 8 cm. The geometric arrangement of the core is shown in Fig. 4. The outer boundary condition of the original problem is a zero-flux density boundary, and since the transport calculations cannot handle this type of boundary, vacuum boundary conditions are used instead. The total number of triangular meshes used in the calculation is 400 as shown in Fig. 5, with one radial layer, and reflection boundary conditions are set on the upper and lower surfaces.

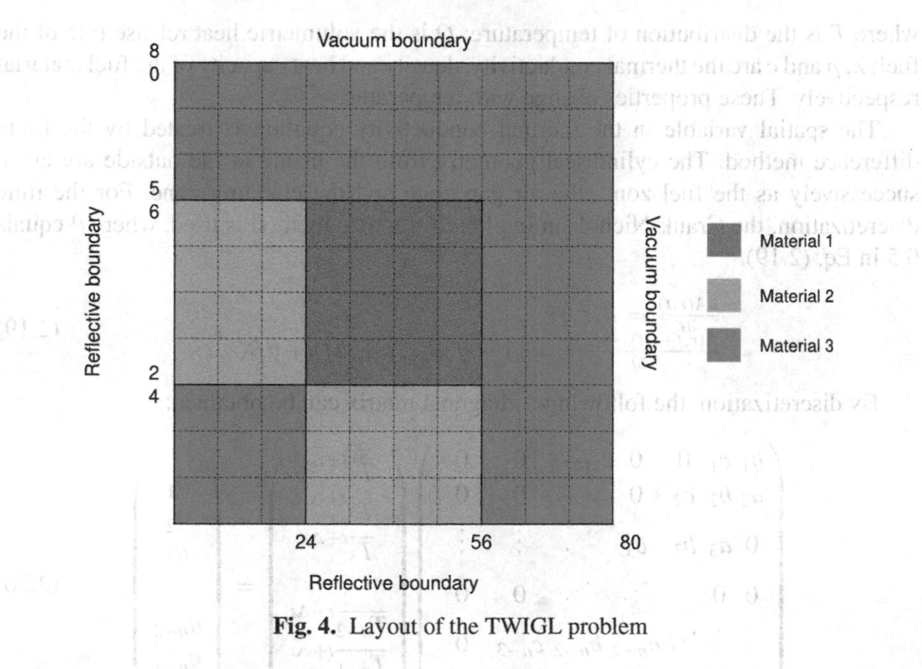

Fig. 4. Layout of the TWIGL problem

The calculation area is 1/4 of the core, and the transient process lasts for 0.5 s. The original problem includes two delayed supercritical problems, A1 and A2. The two problems introduce perturbation ramp and stepwise, respectively. For A1 and A2 problems, two sets of reference values are used, one is the transport calculation result

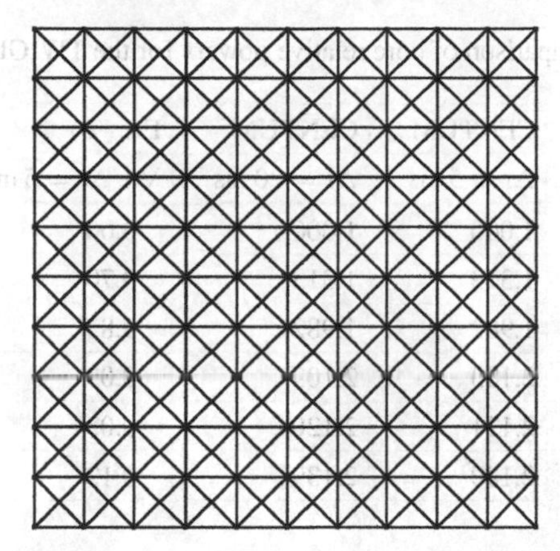

Fig. 5. Mesh of TWIGL benchmark in CENTUM

of UTK's improved quasi-static method code TDTORT; the other is the diffusion calculation result of the time direct discrete nodal code SPANDEX which adopts the θ method with a time-step sizes of 0.1 ms [9, 10]. As can be seen from the results in Fig. 6, the results of CENTUM (S_2) agree well with TDTORT (S_4), while the results of both transport codes are higher than the diffusion code. This discrepancy is mainly due to the zero-flux boundary condition used for the diffusion calculation, which leads to the core internal total power value being slightly smaller for the same perturbation case.

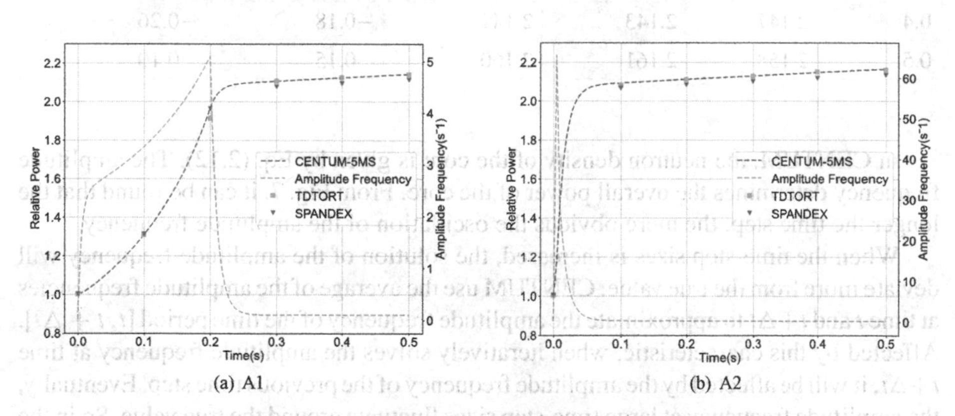

Fig. 6. Results of the TWIGL A1, A2 problem

To further compare the effect of time-step sizes on the calculation results, Table 1, Table 2 gives the normalized power values for the A1, A2 problems.

Compared with TDTORT, the results of CENTUM (20 ms) are in good agreement in both cases, the maximum deviation of the A1 case is 3.83%, and the maximum deviation of the A2 case is 0.41%. And the results obtained by CENTUM using the two time-step sizes are consistent, it show that CENTUM ensures acceptable accuracy even with large time-step sizes.

Table 1. Comparison of core relative powers for the TWIGL A1 problem

Time (s)	TDTORT	CENTUM	CENTUM	Err (%)	
		$\Delta t = 5$ ms	$\Delta t = 20$ ms	Vs. $\Delta t = 5$ ms	Vs. $\Delta t = 20$ ms
0.0	1.000	1.000	1.000	0.00	0.00
0.1	1.304	1.313	1.313	0.70	0.71
0.2	1.909	1.982	1.982	3.83	3.83
0.3	2.104	2.103	2.103	−0.03	−0.03
0.4	2.122	2.121	2.120	−0.03	−0.07
0.5	2.137	2.140	2.139	0.12	0.11

Table 2. Comparison of core relative powers for the TWIGL A2 problem

Time (s)	TDTORT	CENTUM	CENTUM	Err (%)	
		$\Delta t = 5$ ms	$\Delta t = 20$ ms	Vs.$\Delta t = 5$ ms	Vs.$\Delta t = 20$ ms
0.0	1.000	1.000	1.000	0.00	0.00
0.1	2.079	2.089	2.087	0.46	0.41
0.2	2.106	2.107	2.106	0.04	0.00
0.3	2.124	2.125	2.123	0.05	−0.03
0.4	2.147	2.143	2.141	−0.18	−0.26
0.5	2.158	2.161	2.160	0.15	0.10

In CENTUM, the neutron density of the core is given by Eq. (2.12). The amplitude frequency determines the overall power of the core. From Fig. 7, it can be found that the longer the time step, the more obvious the oscillation of the amplitude frequency.

When the time-step sizes is increased, the solution of the amplitude frequency will deviate more from the true value. CENTUM use the average of the amplitude frequencies at time t and $t+\Delta t$ to approximate the amplitude frequency of the time period $[t, t+\Delta t]$. Affected by this characteristic, when iteratively solves the amplitude frequency at time $t+\Delta t$, it will be affected by the amplitude frequency of the previous time step. Eventually, the amplitude frequency at large time-step sizes fluctuate around the true value. So in the end we can get accurate results as long as the average between the two time steps is close to the true value, this feature can effectively reduce the cumulative error. However, when the time-step sizes is too large and the reactivity changes sharply, the average amplitude frequency at two time points cannot reflect the real change very well.

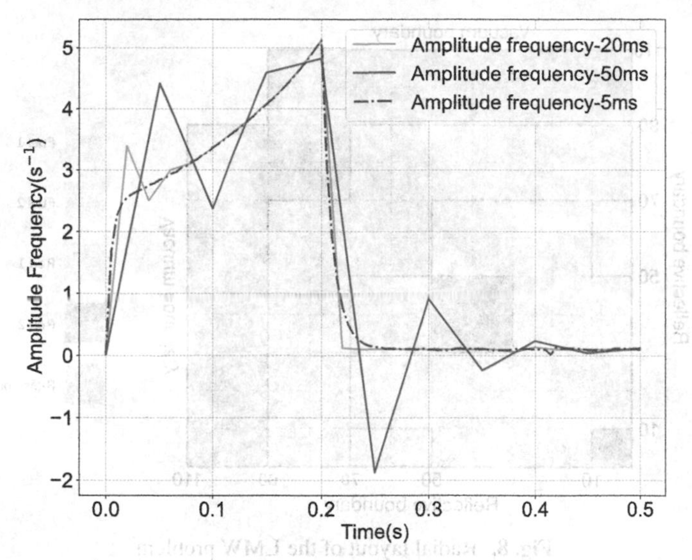

Fig. 7. Amplitude frequencies at different time-step sizes in TWIGL A1

3.2 3D-LMW Benchmark Problem

The 3D-LMW is a 3D diffusion benchmark [9, 10], which contains six sets of delayed neutron dynamics parameters, and the outer boundary is a vacuum boundary. In this paper, the calculation is performed with a quarter-core model. Figure 8 and 9 depict the problem's radial and axial geometrical arrangements, respectively. The motion of two groups of control rods is what causes the transient process: at the start of the transient, the first group of rods are inserted into the middle of the core at a height of 100 cm from the bottom, and the second group rods are withdraw from the active core; between 0.0 s and 26.666 s, the first group rods were lifted out of the active area of the core at a speed of 3.0 cm/s, and the second group rods were gradually inserted into the core from 7.5 s to 47.5 s. CENTUM uses spatial volume weights to deal with the cuspate effect of the control rods.

The size of the LMW assemblies is 20 cm $\times$ 20 cm. In CENTUM calculations, a total of 468 triangular meshes are divided radially, as shown in Fig. 10, and the axial 200 cm is divided into 10 layers. The entire transient process lasts for 60 s, and the reactivity changes during the process are slight, so a large time-step sizes of 0.5 s chosen for calculation.

The calculation results shown in Fig. 11. The two sets of reference solutions used for comparison are diffusion codes. The relative power trend of CENTUM is in good agreement with the reference solution, the largest relative power error of 3.84% occurs at 10s compared with SPANDEX [9, 10]. However, at this time step, the deviation between SPANDEX and SIMULATE is also quite significant. Likewise, this difference attributed to the discrepancy between the transport method and the diffusion method.

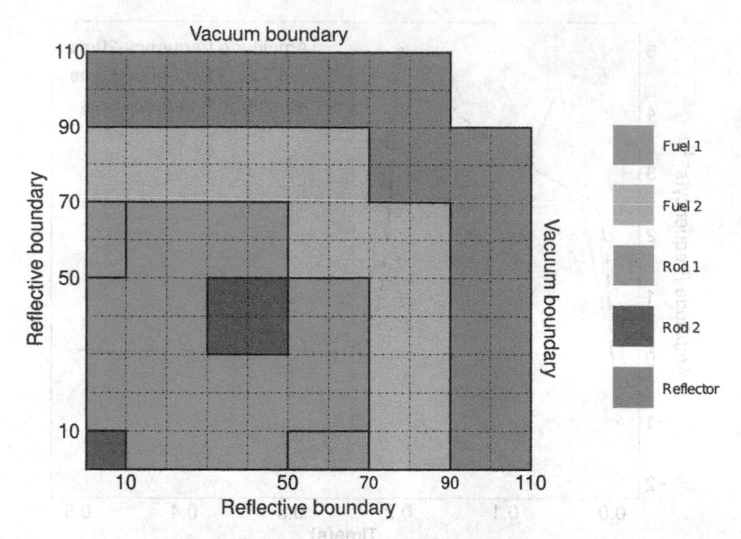

Fig. 8. Radial layout of the LMW problem

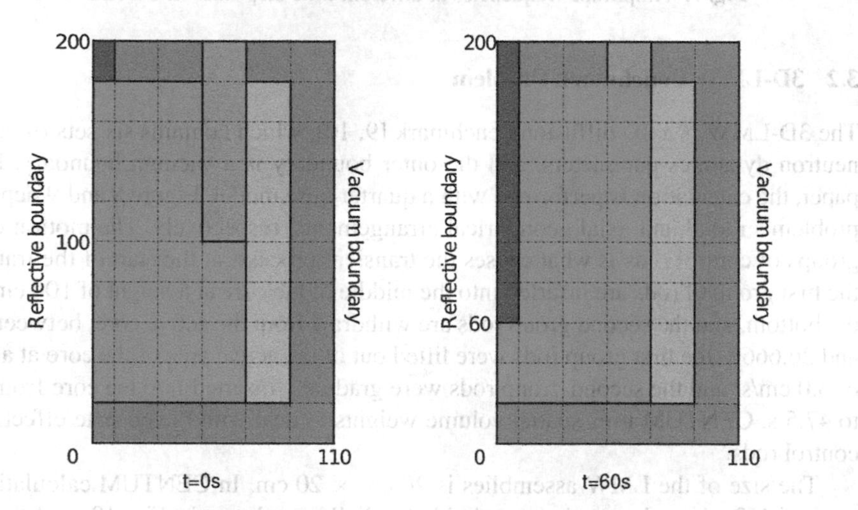

Fig. 9. Axial layout of the LMW problem

3.3 NEACRP Benchmark Problem

The NEACRP rod ejection benchmark includes two types of reactors, i. e. pressurized water reactor and boiling water reactor. It is mainly used for the verification of the neutronics-thermal hydraulics coupling codes of the light water reactor core [11]. The PWR benchmark refers to the geometric size and operating state of a typical PWR. The core consists of 157 assemblies, each measuring 21.606 cm in width. The fuel assemblies are made up of assemblies with various numbers of absorber rods and fuels with various levels of enrichment. Reflectors are set on the periphery region of the core. In the axial direction, the height of the active core is 367.3 cm. The control rod

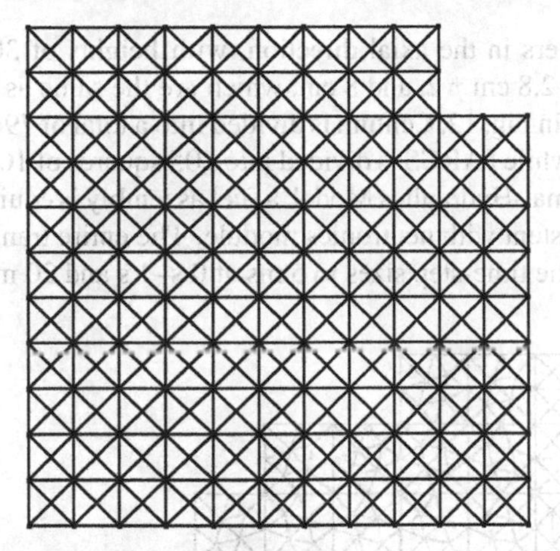

Fig. 10. Mesh of LMW benchmark in CENTUM

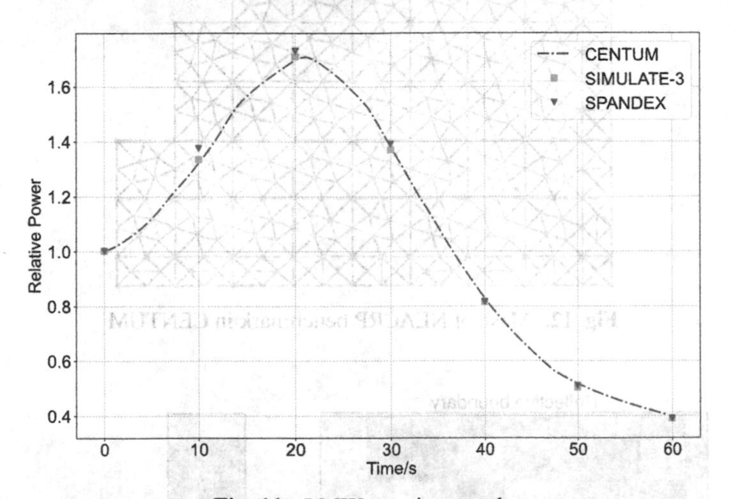

Fig. 11. LMW transient results

has a length of 363.195 cm, the height of the bottom of the absorbent rod when fully inserted is 37.7 cm, and the height when fully ejected is 401.183 cm. The cross section at a numerical node with control rod is determined by adding the cross section $\Delta\Sigma_{CR}$ contributed from the control rod to the cross sextion without control rod [11]:

$$\Sigma_{withCR} = \Sigma_{withoutCR} + p\Delta\Sigma_{CR} \tag{3.1}$$

where p is the relative insertion in the node, i.e. $0 \leqq p \leqq 1$.

The problem contains a total of 6 operating conditions. For simplicity, we select problem A2 as an example to demonstrate the accuracy of CENTUM. Case A2 represents the HFP (hot full power, 2775 MW) condition of the reactor, and the control rods in the center position (blue area in the Fig. 13) are ejected. In case A2, the central control rod eject to the top from a height of 196.12 cm from 0 ms to 100 ms.

There are 18 layers in the axial direction, with heights of 30 cm 7.7 cm, 11 cm, 15 cm, 30 cm * 10, 12.8 cm * 2, and 8 cm, which are the same as PARCS [12] used for reference. As shown in Fig. 12, Centum is divided into a total of 790 triangular meshes in the radial direction, while PARCS is divided into 205 squares of 10.803 cm * 10.803 cm. In CENTUM's Thermal-Hydraulics Model, a fuel assembly is equivalented as a channel, Axial meshing consistent with neutronics module. The entire transient process lasts for 5 s. CENTUM sets the time-step sizes to 5 ms at 0 s–1 s and 20 ms at 1 s–5 s.

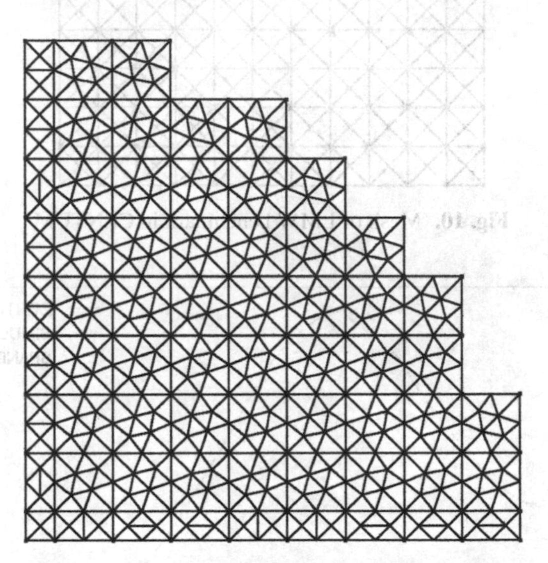

Fig. 12. Mesh of NEACRP benchmark in CENTUM

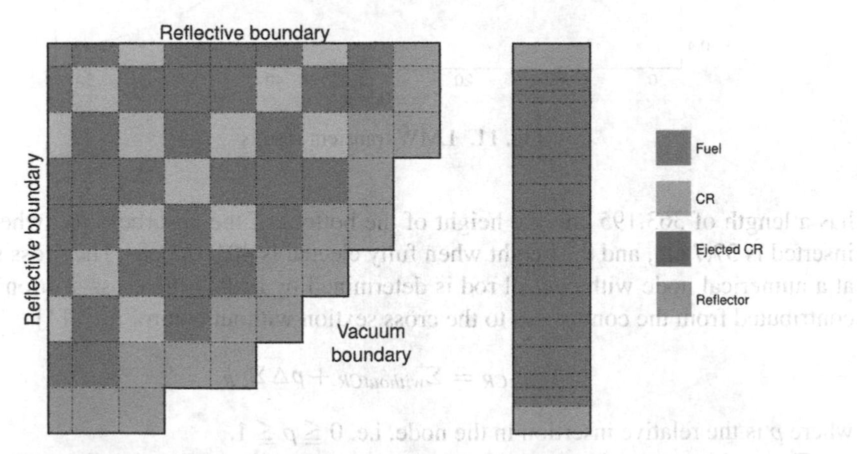

Fig. 13. The layout of case A2

For the steady-state coupling results, the critical boron concentration deviation of the two codes is 0.35 ppm, and the fuel average Doppler temperature deviation is 2.16 °C. Both of these are aggregate parameters of the core, so they match up well. The maximum fuel temperature difference is 30 °C. By comparing the axial power distribution in Fig. 14, it can be found that the maximum axial power of CENTUM is 0.63% higher. Figure 14 also shows the deviation of the radial power distribution, which is higher for the CENTUM outer assemblies compared to PARCS. The difference in power distribution is the most important reason for the difference in maximum fuel temperature. The transport method can better handle the various anisotropies of the angles, plus CENTUM uses a vacuum boundary condition, while PARCS uses a zero-flux boundary condition, all of which can lead to differences in the power distribution.

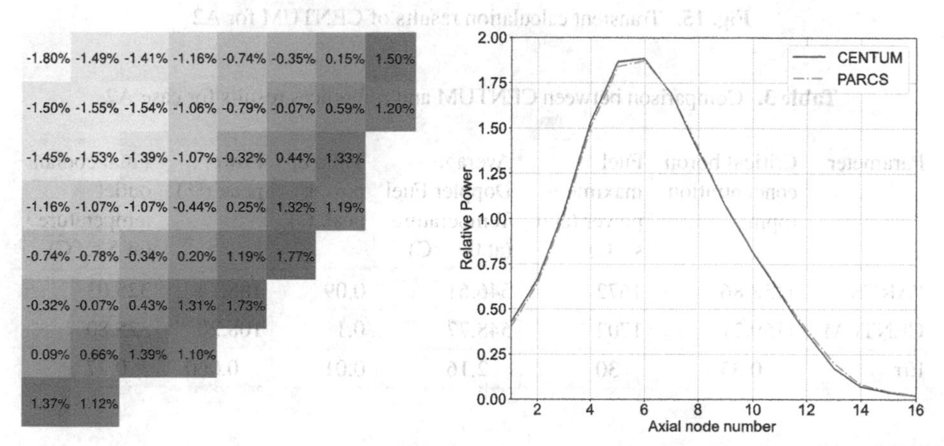

Fig. 14. Comparison of radial and axial power distribution between CENTUM and PARCS at steady-state

As can be seen in Fig. 15, the results of CENTUM are in good agreement with the PARCS reference solutions, with the maximum relative power deviation being 0.055%. Because CENTUM adopts the OSSI method, each neutronics calculation uses the thermal-hydraulic calculation results of the previous time step. This results in a delayed feedback to the neutronics calculation, which can also be observed in Fig. 15. The temperature rise curve of the coolant outlet is slightly deviated, which is mainly caused by the slightly different description of the coolant channel between the two codes (Table 3).

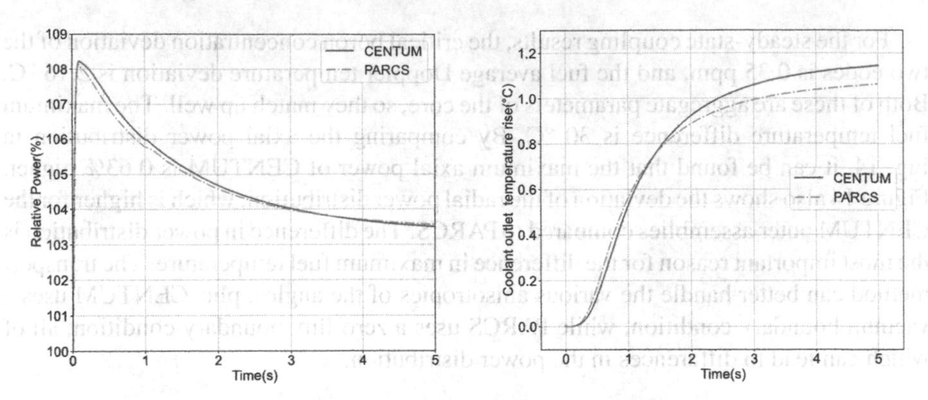

Fig. 15. Transient calculation results of CENTUM for A2

Table 3. Comparison between CENTUM and references results for case A2

Parameter	Critical boron concentration (ppm)	Fuel maximum power (at 0 s, °C)	Average Doppler Fuel Temperature (at 0 s, °C)	Time of power peak (s)	Power peak (%)	Final coolant outlet temperature (at 5 s, °C)
PARCS	1158.86	1672	546.61	0.09	108.14	325.03
CENTUM	1159.21	1702	548.77	0.1	108.20	325.80
Err	0.35	30	2.16	0.01	0.060	0.77

4 Conclusions

In this study, the CENTUM code is developed to analyses steady-state and transient neutronics problems using a 3D triangular-z nodal transport solver with the SCM. Meanwhile, the N-TH coupling calculation can be carried out using the multi-channel model. For neutron dynamics calculations, CENTUM agrees well with the references in the TWIGL and LMW benchmarks. The maximum errors with the TWIGL A1, A2 reference solution are −1.27% and −0.45%, respectively, when using a 20 ms time-step sizes. The maximum error with the LMW reference solution is 3.84%, and the time-step sizes set is 0.5 s. The results show that SCM can maintain good accuracy in longer time-step sizes, which can effectively reduce computing resources.

A preliminary comparison with PARCS shows that CENTUM can accurately simulate the core N-Th coupling process. It should be noted that all test cases are based on Cartesian geometry assemblies which cannot reflect the complex geometrical design of research reactors. We will preserve the verifications and applications of CENTUM on research reactors as an important research direction in the future work.

Acknowledgements. This research is supported by National Key R&D Program of China under grant number 2020YFB1901900, and National Natural Science Foundation of China (NSFC) [12175138].

References

1. Zimin, V.G., Ninokata, H.: Nodal neutron Kinetics model based on nonlinear iteration procedure for LWR analysis. Ann. Nucl. Energy **25**(8), 507–528 (1998)
2. Lu, D., Guo, C., Sui, D.: A three-dimensional nodal neutron kinetics code with a higher-accuracy algorithm for reactor core in hexagonal-z geometry. Ann. Nucl. Energy **101**, 250–261 (2017)
3. Chao, Y.-A., Attard, A.: A resolution of the stiffness problem of reactor kinetics. Nucl. Sci. Eng. **90**(1), 40–46 (1985)
4. Toth, A., et al.: Analysis of Anderson acceleration on a simplified neutronics/thermal hydraulics system. Oak Ridge National Lab.(ORNL), Oak Ridge, TN (United States). Consortium for Advanced Simulation of LWRs (CASL) (2015)
5. Zhang, T., et al.: A 3D transport-based core analysis code for research reactors with unstructured geometry. Nuclear Eng. Des. **265**, 599–610 (2013)
6. Xiao, W., et al.: Application of stiffness confinement method within variational nodal method for solving time-dependent neutron transport equation. Comput. Phys. Commun., 108450 (2022)
7. Křepel, J., et al.: DYN3D-MSR spatial dynamics code for molten salt reactors. Ann. Nuclear Energy **34**(6), 449–462 (2007)
8. Ghiaasiaan, S.M., et al.: Heat conduction in nuclear fuel rods. Nucl. Eng. Des. **85**(1), 89–96 (1985)
9. Goluoglu, S.: A deterministic method for transient, three-dimensional neutron transport. The University of Tennessee (1997)
10. Ban, Y., Endo, T., Yamamoto, A.: A unified approach for numerical calculation of space-dependent kinetic equation. J. Nucl. Sci. Technol. **49**(5), 496–515 (2012)
11. Finnemann, H., et al.: Results of LWR core transient benchmarks. No. NEA-NSC-DOC-93-25. Nuclear Energy Agency (1993)
12. Joo, H.G., et al.: PARCS: a multi-dimensional two-group reactor kinetics code based on the nonlinear analytic nodal method. PARCS Manual Version **2** (1998)

Preliminary Multi-physics Coupled Simulation of Small Helium-Xenon Cooled Mobile Nuclear Reactor

Xiangyue Li, Xiaojing Liu, Xiang Chai, and Tengfei Zhang(✉)

Shanghai Jiao Tong University, Shanghai, China
zhangtengfei@sjtu.edu.cn

Abstract. For the prediction of the internal physical process of SIMONS (Small Innovative helium-xenon cooled MObile Nuclear power Systems), this research created a coupled three-dimensional high-fidelity calculation platform of the neutronics/ thermo-elasticity analysis called FEMAS (FEM based Multi-physics Analysis Software for Nuclear Reactor). This platform allows for the multi-physics coupling calculations of neutron diffusion/ transport, thermal diffusion, and thermal elasticity. It is based on the open-source Monte Carlo code OpenMC and the open-source finite element codes Dealii and Fenics. In this paper, a simplified SIMONS reactor core is analyzed using the coupling platform. The results demonstrate that the coupling platform is capable of accurately predicting the effective multiplication factor change curve, power and temperature distribution, and thermal expansion phenomenon of SIMONS. With 240 kW of thermal power, the local temperature difference of the whole reactor is 390.1 K, and thermal stress-related deformation occurs at a rate of 2.4%. The reactivity feedback caused by the monolith's heating and thermal expansion is 30.5 pcm. Leveraging the high-precision computing hardware, this platform can assess the core performance to ensure that the core design satisfies the design criteria of ultra-long life and inherent safety.

Keywords: Multi-physics Coupling · Numerical Simulation · Mobile Nuclear Power Systems · Thermal Expansion · Finite Element Method

1 Introduction

In some circumstances, a micro transportable nuclear reactor power system can function as an independent micro-grid that can offer a long-term, high-power emergency power supply. It will lessen the loss of life and property due to power outages, helping in communication, transportation, the military, medicine, natural disaster rescue and many other industries, offering a wide range of application prospects in the future.

At present, the United States has carried out projects such as the Prometheus [1], eVinci [2], and Holos reactors [3]. Reactors have been designed using a wide variety of cutting-edge technologies aiming at output powers ranging from kilowatts to megawatts. In China, Tsinghua University has designed a 200-kW gas-cooled reactor with power called IGCR-200 [4].

C. Liu (Ed.): PBNC 2022, SPPHY 283, pp. 690–702, 2023.
https://doi.org/10.1007/978-981-99-1023-6_59

Designed to be packaged in an ISO container and transported by truck or train, the design of the core of the micro transportable nuclear reactor power system is particularly important. Compared with regular reactors, mobile reactors have strict restrictions for size, weight, power-density and lifetime. Besides, it has been demonstrated that the thermal expansion effect constitutes a significant portion of the reactivity feedback for micro nuclear reactors [5]. Conventional methods of research and analysis based on a single physical field and static geometry can hardly meet these needs.

In order to precisely forecast the internal physical processes of SIMONS, this research developed FEMAS (FEM-based Multi-physics Analysis Software for Nuclear Reactor), a coupled three-dimensional high-fidelity computation platform for neutronics/thermo-elasticity analysis. For neutronics, it employs a high-fidelity 3D multi-group neutron diffusion/ transport model with unstructured grids to estimate the neutron flux field and heat release distribution of the nuclear reactor core. For thermo-elasticity, it can establish a 3D thermal diffusion model and stress analysis model to accurately obtain the temperature field and deformation. This multi-physics platform enables us to carry out high-precision analysis of neutron physics, heat-conduction, deformation, providing technical assistance for the core design of the mobile nuclear reactor power system.

2 Methodology

2.1 Method

2.1.1 Neutronics

The neutronics model of FEMAS is based on the finite element method (FEM) [6] to support a high-fidelity geometrical modeling of SIMONS. The FEM is a basic strategy based on "discrete approximation", a numerical method for solving Partial Differential Equations (PDEs). As an illustration, consider the steady-state neutron diffusion equation of a two-group [7]:

$$\mathcal{L}\Phi + \mathcal{S}\Phi = \frac{1}{\lambda}\mathcal{F}\Phi \tag{1.1}$$

where

$$\mathcal{L} = \begin{pmatrix} -\vec{\nabla}\cdot\left(D_1\vec{\nabla}\right) + \Sigma_{a1} + \Sigma_{s12} & 0 \\ 0 & -\vec{\nabla}\cdot\left(D_2\vec{\nabla}\right) + \Sigma_{a2} \end{pmatrix},\ \mathcal{S} = \begin{pmatrix} 0 & 0 \\ -\Sigma_{s12} & 0 \end{pmatrix},$$

$$\mathcal{F} = \begin{pmatrix} \nu\Sigma_{f1} & \nu\Sigma_{f2} \\ 0 & 0 \end{pmatrix},\ \Phi = \begin{pmatrix} \phi_1 \\ \phi_2 \end{pmatrix},\ \chi = \begin{pmatrix} 1 \\ 0 \end{pmatrix}$$

where ϕ is the neutron flux, D is the diffusion factor, Σ_s is the scattering cross section, $\nu\Sigma_f$ is the neutron production cross section, and χ is the fission spectrum. The rest notations are conventional in the reactor physics field, thus will not be elaborated here.

By introducing the finite element shape function and the Galerkin method, the steady-state neutron diffusion equation of two-group in finite element form can be obtained as follows:

$$A^{\lambda}\tilde{\phi}^{\lambda} = \frac{1}{\lambda}B^{\lambda}\tilde{\phi}^{\lambda} \tag{1.2}$$

where

$$A^{\lambda} = \begin{pmatrix} L_{11} & 0 \\ S_{21} & L_{22} \end{pmatrix},\ B^{\lambda} = \begin{pmatrix} F_{11} & F_{12} \\ 0 & 0 \end{pmatrix},\ \tilde{\phi}^{\lambda} = \begin{pmatrix} \tilde{\phi}_1^{\lambda} \\ \tilde{\phi}_2^{\lambda} \end{pmatrix}$$

and the matrices elements (a, b) are given by

$$L_{11(ab)} = \sum_{c=1}^{N_c} D_1^c \int_{V_k} \vec{\nabla}\mathcal{N}_a\vec{\nabla}\mathcal{N}_b \mathrm{d}V - D_1^c \int_{\Gamma_k} \mathcal{N}_a\vec{\nabla}\mathcal{N}_b \mathrm{d}\vec{S} + (\Sigma_{a1}^c + \Sigma_{12}^c)\int_{V_c} \mathcal{N}_a\mathcal{N}_b \mathrm{d}V,$$

$$L_{22(ab)} = \sum_{c=1}^{N_c} D_2^c \int_{V_k} \vec{\nabla}\mathcal{N}_a\vec{\nabla}\mathcal{N}_b \mathrm{d}V - D_2^c \int_{\Gamma_k} \mathcal{N}_a\vec{\nabla}\mathcal{N}_b \mathrm{d}\vec{S} + \Sigma_{a2}^c \int_{V_c} \mathcal{N}_a\mathcal{N}_b \mathrm{d}V,$$

$$S_{21(ab)} = \sum_{c=1}^{N_c} -\Sigma_{12}^c \int_{V_c} \mathcal{N}_a\mathcal{N}_b \mathrm{d}V,$$

$$F_{11(ab)} = \sum_{c=1}^{N_c} \nu\Sigma_{f1}^c \int_{V_c} \mathcal{N}_a\mathcal{N}_b \mathrm{d}V,$$

$$F_{12(ab)} = \sum_{c=1}^{N_c} \nu\Sigma_{f2}^c \int_{V_c} \mathcal{N}_a\mathcal{N}_b \mathrm{d}V.$$

where $\mathcal{N}_i$ and $\mathcal{N}_j$ are the shape functions in the finite element. Similarly, this principle can be applied to the multigroup neutron diffusion equation.

2.1.2 Thermal Conduction

The differential equation of thermal conduction in Cartesian coordinates is:

$$\rho c\frac{\partial T}{\partial t} = \mathrm{div}(\lambda\ \mathrm{grad}(T)) + q_v \tag{1.3}$$

This equation can be cast into the variational form:

$$\int_{\Omega} \nabla T \cdot \nabla v \mathrm{d}x = \int_{\Omega} \frac{q_v}{\lambda} v \mathrm{d}x \tag{1.4}$$

Equation (1.4) can be easily solved with the open-source finite element platform Fenics, as will be illustrated in Sect. 2.2.

2.1.3 Thermal-Elasticity

The thermal-elasticity model in FEMAS is based on a perfectly elastic isotropic body assumption. The relationship between the deformation component (strain: ε) and the stress component (stress: σ) is [8]:

$$\begin{aligned} \sigma_x &= \lambda(\varepsilon_x + \varepsilon_y + \varepsilon_z) + 2\mu\varepsilon_x \\ \sigma_y &= \lambda(\varepsilon_x + \varepsilon_y + \varepsilon_z) + 2\mu\varepsilon_y \\ \sigma_z &= \lambda(\varepsilon_x + \varepsilon_y + \varepsilon_z) + 2\mu\varepsilon_z \\ \tau_{xy} &= 2\mu\varepsilon_{xy} \\ \tau_{yz} &= 2\mu\varepsilon_{yz} \\ \tau_{zx} &= 2\mu\varepsilon_{zx} \end{aligned} \tag{1.5}$$

where λ and μ are Lame constants. The transformation relationship between them and elastic modulus E, Poisson's ratio v is:

$$\begin{aligned} \lambda &= \frac{vE}{(1+v)(1-2v)} \\ \mu &= \frac{E}{2(1+v)} \end{aligned}$$

Therefore, the expression for stress can be written as:

$$\sigma = \lambda\,\mathrm{tr}(\varepsilon)I + 2\mu\varepsilon \tag{1.6}$$

where I denotes the identity tensor. The expression for strain can be written as:

$$\varepsilon_{ij} = \frac{1}{2\mu}\left(\sigma_{ij} - \frac{\lambda}{3\lambda + 2\mu}\sigma_{kk}\delta_{ij}\right) \tag{1.7}$$

The stress governing equation of the whole system is:

$$-\nabla \cdot \sigma = f, \text{ in } \Omega \tag{1.8}$$

where $\sigma = \lambda \mathrm{tr}(\varepsilon)I + 2\mu\varepsilon$, $\varepsilon = \frac{1}{2}\left(\nabla u + (\nabla u)^T\right)$, and f is the external force per unit volume in the entire system. Similar as Sect. 2.1.2, the variational form of the stress equation is:

$$\int_\Omega \sigma(u) : \in (v)\mathrm{d}x = \int_\Omega f \cdot v\mathrm{d}x \tag{1.9}$$

Temperature variations induce deformations in elastically unrestrained solids. Therefore, mechanical and thermal processes form the global strain field. In the context of the theory of linear small deformation, the total strain can be decomposed into the sum of mechanical and thermal components as:

$$\varepsilon_{ij} = \varepsilon_{ij}^{(M)} + \varepsilon_{ij}^{(T)} \tag{1.10}$$

For isotropic materials,

$$\varepsilon_{ij}^{(T)} = \alpha(T - T_o)\delta_{ij} \tag{1.11}$$

where α is the linear thermal expansion coefficient of the material.

Therefore, the expression for the total strain is:

$$\varepsilon_{ij} = \frac{1+\nu}{E}\sigma_{ij} - \frac{\nu}{E}\sigma_{kk}\delta_{ij} + \alpha(T - T_o)\delta_{ij} \tag{1.12}$$

Thus, the linearized thermoelastic constitutive equation can be given by:

$$\sigma_{ij} = \lambda\varepsilon_{kk}\delta_{ij} + 2\mu\varepsilon_{ij} - (3\lambda + 2\mu)\alpha(T - T_o)\delta_{ij} \tag{1.13}$$

Similar to the elasticity equation, the weak form of the thermal expansion equation can be obtained as:

$$\int_\Omega \sigma(u) : (v)\mathrm{d}x = \int_\Omega f \cdot v\mathrm{d}x \tag{1.14}$$

2.2 Coupling Framework

In order to achieve high-precision simulation of full reactor, this research developed the multi-physics calculation platform FEMAS, integrated open-source codes (OpenMC, Dealii and Fenics) based on the external coupling framework.

The iteration process goes as follows:

(1) Use OpenMC to perform neutron transport simulation and obtain a cross-section library.
(2) Based on the pre-set temperature distribution and geometric parameters in cold state, use Dealii to perform multi-group neutron diffusion calculation, thus getting the spatial distribution of power.
(3) Given the boundary conditions of the helium-xenon cooling channel and the spatial distribution of power, the temperature distribution was calculated by solving the thermal diffusion equation using Fenics.
(4) Combined with the temperature field and the mechanical boundary conditions, the structural displacement is obtained by solving the thermo-elasticity equation using Fenics.
(5) Update the geometry, density and cross sections of the model.
(6) Dealii then performs the neutron diffusion calculation again. The steps above are repeated until certain physical quantities meet the convergence criteria or the execution reaches the maximum number of iteration steps (Fig. 1).

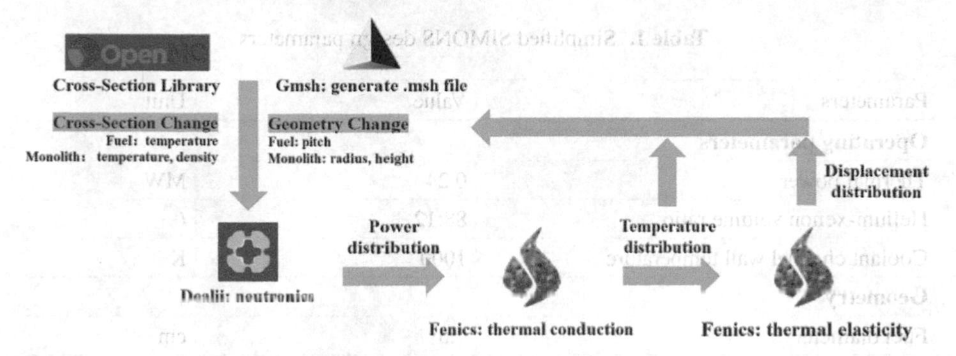

Fig. 1. Schematic view of the coupling framework

2.3 Model

2.3.1 Model

To meet the demand for terrestrial transportable nuclear reactor power supply, this research takes the overall conceptual design of SIMONS as the test case. SIMONS is designed to be an intelligent micro land-based transportable nuclear reactor featuring small size, high power density and inherent safety. Its thermal power is 20 MWe and it can operate continuously for 3300 EFPDs without refueling.

As a preliminary test of the coupling platform, a simplified scaling model of the SIMONS core is employed to reduce the computational cost. The scaled core exhibits a prismatic core design, using graphite as the monolith material. There are several holes in the monolith with a hexagonal arrangement to accommodate fuel rods and coolant. The overall height of the core is 30 cm, and the overall radius is 23 cm. The axial-radial schematic diagram of the core is shown in Fig. 2.

Fig. 2. Schematic diagram of the core

The operating parameters, geometric information and material selection of the simplified SIMONS modeling are shown in Table 1.

Table 1. Simplified SIMONS design parameters

Parameters	Value	Unit
Operating parameters		
Thermal power	0.24	MW
Helium-xenon volume ratio	88:12	/
Coolant channel wall temperature	1000	K
Geometry		
Fuel diameter	1.5	cm
Coolant channel diameter	0.9	cm
Fuel pitch	1.5	cm
Number of fuel rods	61	/
Number of Coolant channels	138	/
Fuel height	10	cm
Monolith radius	13	cm
Radial reflector width	10	cm
Axial reflector height	5	cm
Material		
Fuel	UC	/
Monolith	Graphite	/
Reflector	Be	/
Coolant	Helium-xenon	/

2.3.2 Neutronics

The coupling calculations employ a high-fidelity modelling approach by explicitly describing all the rods and holes using three-dimensional unstructured meshes. Homogenized cross sections are generated using two-dimensional OpenMC calculations for every material zone. This research uses a four-group structure which is divided as:

1) Group-1 (497.87 keV–20 meV)
2) Group-2 (9.11882 keV–497.87 keV)
3) Group-3 (0.625 eV–9.11882 meV)
4) Group-4 (0 eV–0.625 eV)

The cross sections are stored as data tables with considering different fuel and monolith temperatures. During the neutronics calculations, the cross sections for each material zone are updated using linear interpolation method. The cross sections of fuel are only related to the temperature, while those of monolith are dependent on both temperature and density.

Due to the simplification of the core design, setting vacuum boundary conditions will make the k_{eff} too small. Therefore, the reflective boundary condition is imposed on the outer boundaries in the calculation. Figure 3 shows the mesh used in the neutronics calculation.

Fig. 3. Meshes used in neutronics simulation

2.3.3 Thermo-Elasticity

Figure 4 depicts the mesh used in thermo-elasticity calculations. In the heat diffusion calculation, it is assumed that the thermal conductivities of the fuel and the monolith are constant. Currently, the coolant channels are treated as Dirichlet boundaries, and subsequent work will consider the heat transfer portion of the flow in the channel. Based on the temperature distribution obtained by solving the thermal diffusion equation, the thermal expansion calculation can be performed.

3 Results

3.1 Coupling Results

In the coupling calculation, the maximum number of iterations is set to 10. In fact, between the second and third steps, the k_{eff} error dropped below the predetermined convergence threshold (1e−5). However, the iteration step is set to 10 in order to study the characteristics of the coupled calculation results in greater detail. The results of various fields of neutronics/thermos-elasticity coupling calculations are shown in the Fig. 5.

The 3D four-group neutron flux distribution is shown in Fig. 6. It can be observed that the fast neutron flux are generated in the fuel rods, while thermal neutrons presents at the periphery of the active core due to the strong thermalization effect of the Be reflectors.

For the fuel region, based on the aforementioned neutron flux distribution and the combined total power, the heat release rate then can be calculated. Through thermal diffusion calculation, the temperature distribution of the simplified SIMONS can be obtained, shown in Fig. 7. Maximum temperature difference across the whole reactor is 390.1 K.

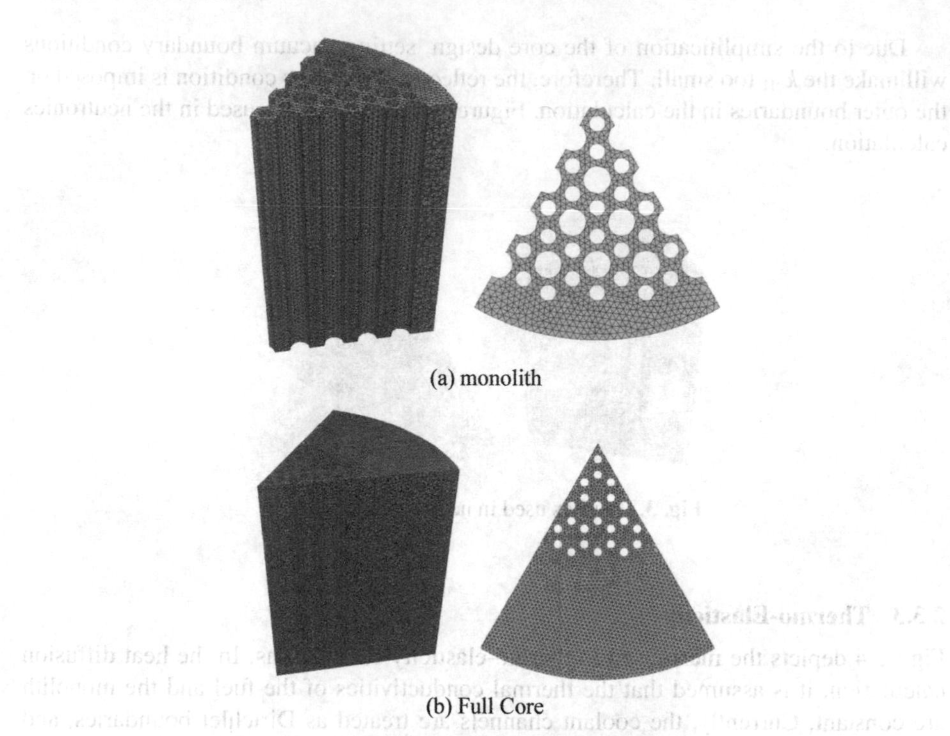

(a) monolith

(b) Full Core

Fig. 4. Meshes used in thermos-elasticity simulation

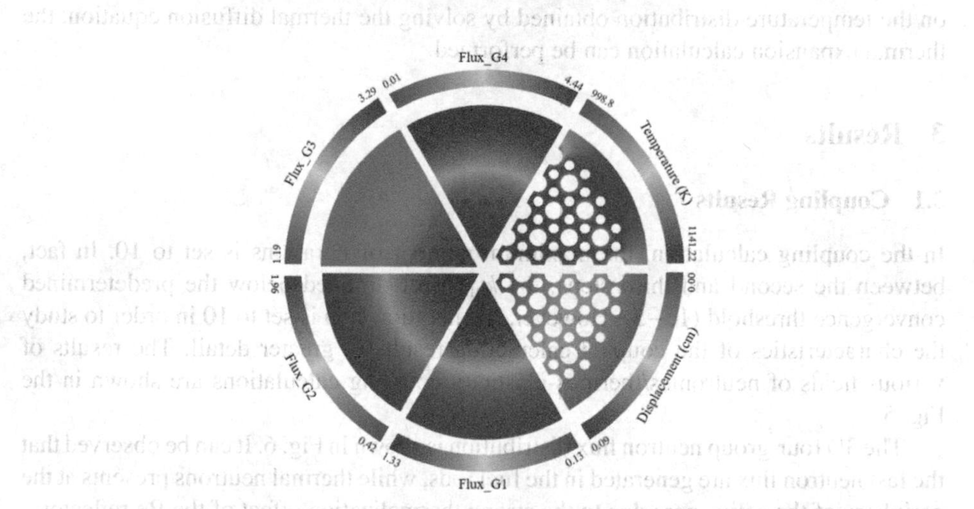

Fig. 5. 2D Results of neutronics/ thermos-elasticity coupling calculation (H = 15 cm)

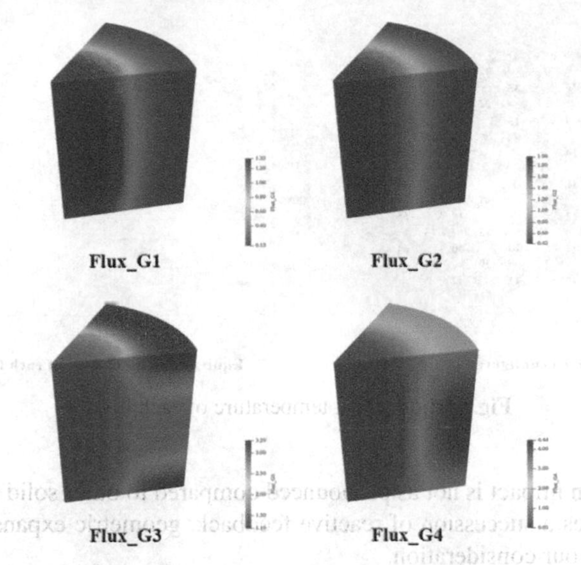

Fig. 6. Neutron flux distribution

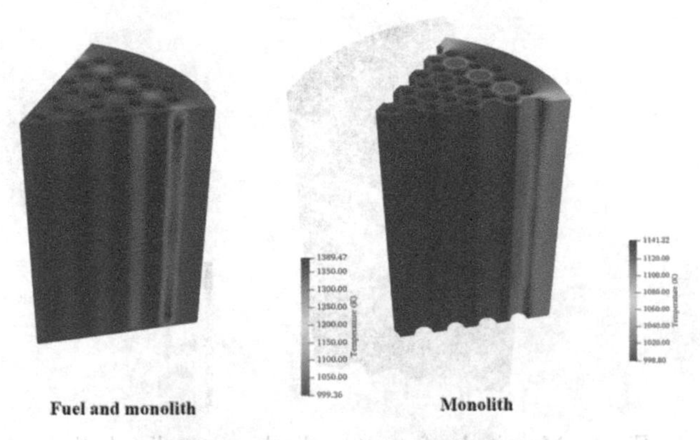

Fig. 7. Temperature distribution

According to the temperature field, the effective temperature of the fuel rod can be calculated, shown in Fig. 8. The results indicate that the fuel rods on the periphery have the highest temperatures due to the thermalization of neutrons in the reflector. Within the fuel rod, the middle of it has the highest temperature. This is because given the reflective boundary condition, the neutron fluxes of the third and fourth groups are predominantly spread outside the core, resulting in a higher power here. Moreover, The hot spot locations are also attributed to less numbers of surrounding coolant channels than other fuel locations.

Figure 9 depicts the thermal expansion calculation results. It reveals that the maximum displacement is about 0.09 cm, and the core expands radially from the center to the periphery. Graphite's linear thermal expansion coefficient is quite small, hence the

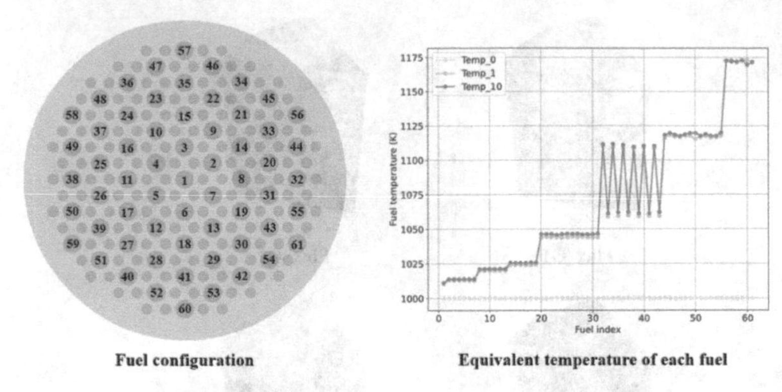

Fig. 8. Equivalent temperature of each fuel

thermal expansion impact is not as pronounced compared to other solid reactors. However, it still creates a succession of reactive feedback, geometric expansion, and other effects worthy of our consideration.

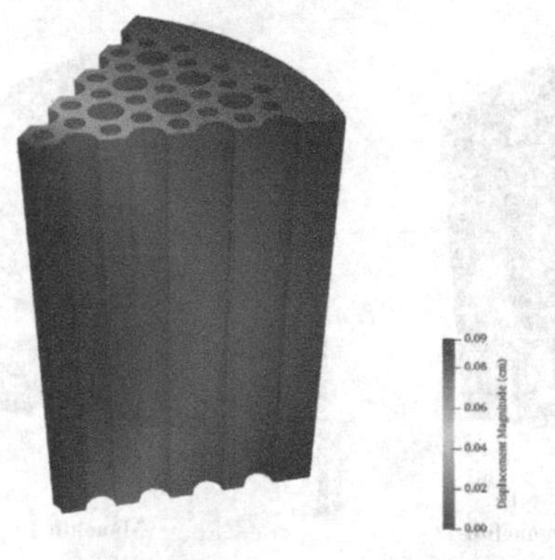

Fig. 9. Magnitude of structure displacement distribution

3.2 Thermal Results Analysis

Figure 10 illustrates the convergence diagram and monolith geometrical parameters for multi-physics calculations. At the first step, the k_{eff} is calculated by directly applying the OpenMC-generated cross section in the diffusion calculation, yielding a value of 1.65501. Meanwhile, at the last step, the value of k_{eff} is 1.65471. Comparing their k_{eff} values reveals that the adoption of thermo-elasticity calculation results in a 30.5 pcm decrease in reactivity because of the enhanced neutron leakage effect. Regarding the monolith's geometric specifications, its radius expanded from 13 cm to 13.08 cm, and its height increased from 10 cm to 10.06 cm. Consequently, its density is lowered to 97.6% of its original value.

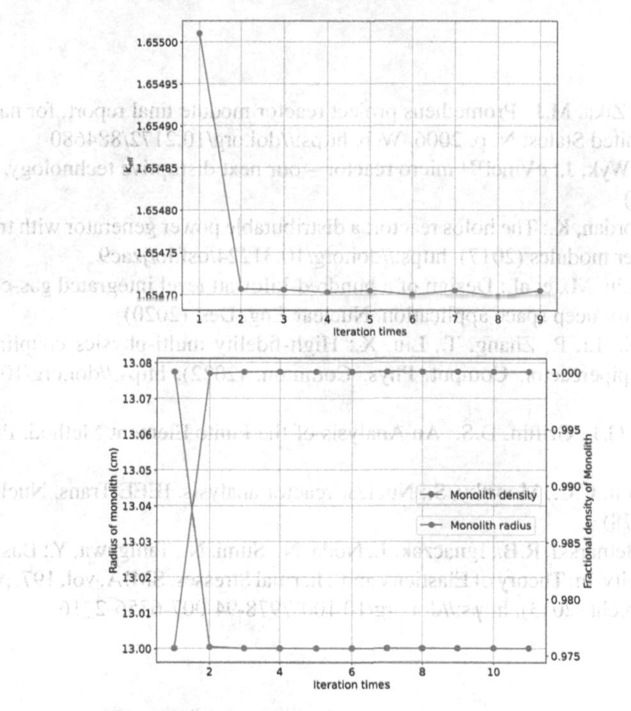

Fig. 10. Convergence diagram of k_{eff} and geometry

4 Conclusion

In response to the research and development requirements of innovative helium-xenon cooled mobile nuclear power systems, this research developed a three-dimensional high-fidelity multi-physics coupling platform FEMAS. This platform is built on the Picard iteration and incorporates the open source OpenMC, Dealii, and Fenics codes, which enables the multi-physics modeling of neutronics/ thermoelasticity.

Simultaneously, based on the scaled model of SIMONS, this research preliminarily concludes a series of simulations and analyses, including from the OpenMC cross-section generation, reactor modeling, and multi-physics simulations. The results demonstrate that the coupling platform can predict the power distribution, temperature distribution and thermal expansion of SIMONS. Given 240-kW of thermal power, the local temperature difference of the whole reactor is 390.1 K, and the deformation rate caused by thermal stress is 2.4%, and the reactivity feedback due to heat conduction and thermal expansion is 30.5 pcm.

In the era of high-precision computing, this platform can evaluate the core performance of the core to guarantee that the design of the core meets the standards for ultra-long life and inherent safety.

Acknowledgement. This research is supported by National Key R&D Program of China under grant number 2020YFB1901900, and National Natural Science Foundation of China (NSFC) [12175138].

References

1. Wollman, M.J., Zika, M.J.: Prometheus project reactor module final report, for naval reactors information. United States: N. p. 2006. Web. https://doi.org/10.2172/884680
2. Arafat, Y., Van Wyk, J.: eVinci™ micro reactor – our next disruptive technology. Nucl. Plant J., 34–37 (2019)
3. Filippone, C., Jordan, K.: The holos reactor: a distributable power generator with transportable subcritical power modules (2017). https://doi.org/10.31224/osf.io/jzac9
4. Li, Z., Sun, J., Liu, M., et al.: Design of a hundred-kilowatt level integrated gas-cooled space nuclear reactor for deep space application. Nuclear Eng. Des. (2020)
5. Xiao, W., Li, X., Li, P., Zhang, T., Liu, X.: High-fidelity multi-physics coupling study on advanced heat pipereactor. Comput. Phys. Commun. (2022). https://doi.org/10.1016/j.cpc.2021.108152
6. Strang, G., Fix, G.J., Griffin, D.S.: An Analysis of the Finite Element Method. Prentice-Hall (1973)
7. Henry, A.F., Scott, C.C., Moorthy, S.: Nuclear reactor analysis. IEEE Trans. Nucl. Sci. **24**(6), 2566–2567 (1978)
8. Eslami, M.R., Hetnarski, R.B., Ignaczak, J., Noda, N., Sumi, N., Tanigawa, Y.: Basic equations of thermoelasticity. In: Theory of Elasticity and Thermal Stresses. SMIA, vol. 197, pp. 391–422. Springer, Dordrecht (2013). https://doi.org/10.1007/978-94-007-6356-2_16

Study on the Steady-State Performance of the Fuel Rod in M^2LFR-1000 Using KMC-Fueltra

Guangliang Yang, Weixiang Wang, Wenpei Feng, Tao Ding, and Hongli Chen(✉)

University of Science and Technology of China, Hefei, Anhui, China
hlchen1@ustc.edu.cn

Abstract. Operating conditions in the liquid metal fast reactor, like higher power density, higher temperature gradient, and higher burnup, are more severe to the fuel rod comparing to the light water reactor. The integrity and safety of the fuel rod is very essential to the reactor safety. In this study, the fuel rod designed for M^2LFR-1000, which is a typical pool type lead cooled fast reactor, is evaluated using a fuel performance analysis code named KMC-Fueltra. Irradiation behaviors and material properties for the MOX fuel and T91 cladding are established and introduced into the code. The steady-state performance of the fuel rod is analyzed. Results concerning both fuel and cladding performance are discussed based on indicative design limits collected from the open literatures. This study is useful to improve the conceptual design of the M^2LFR-1000.

Keywords: Conceptual design · Fuel rod performance · Lead-cooled fast reactor · M^2LFR-1000 · KMC-Fueltra

1 Introduction

Ensuring the integrity and safety of the fuel rod, which is the core component in the reactor, is the most essential issue in the design of a nuclear reactor [1]. For the fast reactor, the power density, temperature gradient and burnup are higher than light water reactor. This brings a greater challenge for the fuel rod in the fast reactor.

Many materials have been developed and studied in order to meet the above characteristics of fast reactors. Material properties and irradiation behaviors are implemented into the fuel performance code to evaluate their performance in the specific reactor. Considering the transmutation and proliferation, MOX fuel has been widely used in fast reactors. The cladding material are mainly stainless steel. One is austenitic stainless steel; the other is ferritic martensitic stainless steel. Due to the better thermal creep resistance, the austenitic steel 15-15-Ti has been chosen to be the cladding material in the ALFRED, MYRRHA and ASTRID reactors [2]. Marcello extended the TRANSURANS code [3] and it is adopted for the simulation of the fuel and cladding performance covering the average and the hottest reactor conditions in the ALFRED reactor [2]. 316SS is also used

C. Liu (Ed.): PBNC 2022, SPPHY 283, pp. 703–716, 2023.
https://doi.org/10.1007/978-981-99-1023-6_60

in fast reactors like JOYO and EBR-II [4], but the it has poor compatibility with lead coolant. Corrosion with lead coolant should be considered at certain coolant velocity and temperature in the lead cooled fast reactor. Benefit from the good corrosion resistance to the lead coolant, T91 is selected as the cladding in 1000Mth Medium-size Modular Lead-cooled Fast Reactor (M^2LFR-1000) [5]. In order to ensuring the safety of the reactor, the integral performance of the fuel rod design should be investigated under the operational condition in M^2LFR-1000.

In this paper, material properties and irradiation behaviors of MOX fuel and T91 cladding are incorporated into the fuel performance code KMC-Fueltra. The widely used failure mechanism and design limits of some important parameters applicable for the fast reactors are collected from the public literatures to evaluate the fuel rod design. The steady-state performance of the fuel rod in M^2LFR-1000 is studied using KMC-Fueltra. Results are discussed and some improvements are suggested.

2 Material Properties

Material properties are of great importance to the fuel performance analysis. The thermal and mechanical properties of the MOX fuel and T91 cladding implemented in the KMC-Fueltra are presented in this part. The main point is put on the basic material properties. The unique irradiation behaviors did not list but can be found in the former work.

2.1 MOX Fuel

(1) **Thermal conductivity**

Many correlations have been proposed for MOX fuel concluded from the fresh or irradiation fuel data. A wide set of factors like fuel temperature, burnup, plutonium content, stoichiometry, and porosity have effects on the MOX fuel thermal conductivity. Magni et al. [6] assessed the most recent and reliable experimental data statistically and proposed a new correlation for the MOX fuel with above factors taking into consideration. The correlation of thermal conductivity for the fresh MOX fuel is:

$$k_0(T,x,[Pu],p) = \frac{\left(\dfrac{1}{A_0 + A_x \cdot x + A_{Pu} \cdot [Pu] + (B_0 + B_{Pu}[Pu])T} + \dfrac{D}{T^2}e^{-\frac{E}{T}}\right)}{(1-p)^{2.5}} \quad (1)$$

where T is the temperature, K; $[Pu]$ is the plutonium atomic fraction; p is the porosity fraction; Considering the thermal conductivity of degradation at certain burnup values, the correlation after the irradiation is:

$$k_{irr}(T,x,[Pu],p,bu) = k_{inf} + \left(k_0(T,x,[Pu],p) - k_{inf}\right) \cdot e^{-\frac{bu}{\varphi}} \quad (2)$$

where $k_0(T,x,[Pu],p)$ is the fresh MOX thermal conductivity; bu is the burnup, GWd/tHM; k_{inf} asymptotic thermal conductivity, $W/m/k$; φ is the fitted coefficient, GWd/tHM; Parameters used in this correlation is listed in the Table 1.

Table 1. Parameters in the correlation

Parameters	Value
A_0 $(m \cdot K/W)$	0.01926
A_x $(m \cdot K/W)$	1.06×10^{-6}
A_{Pu} $(m \cdot K/W)$	2.63×10^{-8}
B_0 (m/W)	2.39×10^{-4}
D_{Pu} (m/W)	1.37×10^{-13}
D $(W/m/k)$	5.27×10^{9}
E (k)	17109.5
k_{inf} $(W/m/k)$	1.755
φ (GWd/tHM)	128.75

(2) Young's modulus and Poisson's ratio

Young's modulus of MOX fuel is related to the Pu content and stoichiometric state [7]. The relation can be expressed as:

$$E = E(UO_2) \cdot (1 + 0.15w_{Pu})exp(-Bx) \tag{3}$$

where B is constant, 1.34 for the hyper-stoichiometric fuel and 1.75 for the hypo-stoichiometric fuel; x is the deviation from stoichiometry; w_{Pu} is the weight fraction of PuO_2; $E(UO_2)$ is the Young's modulus of UO_2 and can be denoted as:

$$E(UO_2) = 2.334 \times 10^{11}[1 - 2.752(1 - D)]\left(1 - 1.0915 \times 10^{-4}\right) \cdot T \tag{4}$$

where D is the theoretical density fraction; T is the temperature, K.

Possion's ratio of MOX fuel is given as a function of the weight fraction of PuO_2:

$$v(MOX) = w_{Pu} \cdot v(PuO_2) + (1 - w_{Pu}) \cdot v(UO_2) \tag{5}$$

$$v(PuO_2) = 0.276 + \frac{T - 300}{2800}(0.5 - 0.276) \tag{6}$$

$$v(UO_2) = 0.316 + \frac{T - 300}{2800}(0.5 - 0.316) \tag{7}$$

where $v(UO_2)$ and $v(PuO_2)$ is the Possion's ratio of UO_2 and PuO_2 respectively; T is the temperature, K; w_{Pu} is the weight factor of PuO_2.

(3) Thermal expansion

Thermal expansion of the MOX fuel is obtained by weight factor of different components [8].

$$\varepsilon^{th}(MOX) = (1 - w_{Pu}) \cdot \varepsilon^{th}(UO_2) + (w_{Pu}) \cdot \varepsilon^{th}(PuO_2) \tag{8}$$

where, ε^{th} is thermal expansion, %; w_{Pu} is the weight factor of PuO_2. Thermal expansion of UO_2 and PuO_2 is expressed by:

$$\varepsilon_i^{th} = K_1 \cdot T - K_2 + K_3 exp\left(-\frac{E_D}{kT}\right) \tag{9}$$

The subscript i takes 1 and 2, which represent UO_2 and PuO_2, respectively. Parameters used in this correlation is listed in the Table 2.

Table 2. Parameters used in thermal expansion model

Parameters	UO_2	PuO_2
$K_1(K^{-1})$	9.8×10^{-6}	9.0×10^{-6}
K_2	2.61×10^{-3}	2.7×10^{-3}
K_3	3.16×10^{-1}	7.0×10^{-2}
E_D (J)	1.32×10^{-19}	7.0×10^{-20}

(4) **Creep model**

Creep can effectively release the stress caused by irradiation behaviors like swelling and thermal expansion in the fuel, which can protect the fuel from reaching the safety limits [9]. Creep of the MOX fuel is divided into two parts: thermal creep and irradiation creep [4]. Thermal creep is composed of diffusional creep and dislocation creep and is expressed as follows:

$$\dot{\varepsilon_{th}} = 3.23 \times 10^9 \cdot \frac{\sigma_{eff}}{a^2} \exp\left(-\frac{Q_1}{RT}\right) + 3.24 \times 10^6 \cdot \sigma_{eff}^{4.4} \cdot exp\left(-\frac{Q_2}{RT}\right) \tag{10}$$

where $\dot{\varepsilon_{th}}$ is the fuel thermal creep rate, $1/h$; a is the grain size, μm; σ_{eff} is the equivalent stress, MPa; $Q_1 = -92500$ and $Q_2 = -136800$ are activation energy; R is the universal gas constant.

Irradiation creep is denoted as;

$$\dot{\varepsilon_{irr}} = 1.78 \times 10^{-26} \sigma_{eff} \cdot \varphi \tag{11}$$

where ε_{irr} is the fuel irradiation creep rate, $1/h$; φ is the fission rate, $fission/(m^3 \cdot s)$.

2.2 T91 Cladding

(1) **Thermal conductivity**

Thermal conductivity of the cladding is denoted as a function of temperature as follows [10]:

$$k = 21.712 + 0.011T - 9.5483 \times 10^{-6}T^2 + 3.627 \times 10^{-9}T^3 \tag{12}$$

where T is the temperature, K.

(2) **Young's modulus and Poisson's ratio**

Young's modulus of the cladding is a function of temperature and can be expressed as [11]:

$$E = 2.11458 \times 10^5 - 21.24 \cdot T - 7.94 \times 10^{-2} \cdot T^2 \quad (13)$$

where E is Young's modulus, Mpa; T is the temperature, °C, with the range of applicability in $20 < T < 760$ °C.

Possion's ratio of the cladding is set as a constant.

$$\nu = 0.3 \quad (14)$$

(3) **Thermal expansion**

Thermal expansion of the cladding is expressed as [10]:

$$\varepsilon^{th} = \frac{\Delta L}{L} = -3.0942 \times 10^{-3} + 1.1928 \times 10^{-5} \cdot T - 6.7979 \times 10^{-9} \cdot T^2 + 7.9606 \times 10^{-12} \cdot T^3 - 2.546 \times 10^{-15} \cdot T^4 \quad (15)$$

where ε^{th} is thermal expansion, %; T is the temperature, K.

(4) **Creep model**

Similar to the MOX fuel, creep model of the cladding is also divided into two parts: thermal creep and irradiation creep. Thermal creep is expressed as [11]:

$$\dot{\varepsilon}_{th} = A\sigma^n exp(-Q/RT) \quad (16)$$

where $Q = 728 \pm 35\,\mathrm{kJ/mo\,l}$ is the activation energy; $n = 5$.

Irradiation creep is expressed as:

$$\dot{\varepsilon_{irr}} = 1.8 \times 10^{-22} \varphi_v(t) \cdot \sigma_{eq} \quad (17)$$

where $\varphi_v(t)$ is the neutron flux rate, $n/(cm^2 \cdot s)$; σ_{eq} is the equivalent stress, Mpa; t is the time, h.

(5) **Plasticity model**

Plasticity is an important aspect in the fuel rod mechanical performance, which is related to the fuel failure highly. Plasticity model of the cladding is denoted as a function of temperature [11].

$$\sigma_y = 536.1 - 0.4878 \cdot T + 1.6 \times 10^{-3} \cdot T^2 - 3 \times 10^{-6} \cdot T^3 + 8 \times 10^{-10} \cdot T^4 \quad (18)$$

where σ_y is the yield stress, Mpa; T is the temperature, °C, with the range of applicability in $20 < T < 700$ °C.

(6) **Cumulative damage function**

Fuel rod failure time is a critical parameter to the reactor safety. Predicting this parameter as accurately as possible is one of the main tasks of fuel performance analysis. The traditional cumulative damage function (CDF) is used to evaluate the fuel rod failure.

$$CDF = \sum_{i=1}^{n} \frac{\Delta t}{t_r(\sigma_i, T_i)} \tag{19}$$

where Δt is the time step, s; $t_r(\sigma_i, T_i)$ represents time to failure at certain stress and temperature. It can be calculated by the Larson-Miller parameter (LMP) [12], defined as:

$$LMP(\sigma) = T\left(C + \log_{10} t_r\right) \tag{20}$$

For the T91 cladding, the LMP is fitted by the polynomial function.

$$LMP(\sigma) = 38387.008 - 84.880\sigma + 0.403\sigma^2 - 1.15 \times 10^{-3}\sigma^3 + 1.254 \times 10^{-6}\sigma^4 \tag{21}$$

where σ is the equivalent stress, *Mpa*; C is set as 33.

3 Model Implementation

3.1 Description of M²LFR-1000

The 1000MWth Medium-size Modular Lead-cooled Fast Reactor (M²LFR-1000) is a typical pool-type fast reactor developed by USTC [5]. It adopts a rod-shaped fuel element design. The pellet is MOX with Pu enrichment about 20%, the cladding is T91 stainless steel and the coolant is lead. The operating temperature is 400–480 °C. The main design parameters of the fuel rod are outlined in Table 3.

Table 3. Fuel specifications for M²LFR-1000

Property	Value
Fuel type	MOX
Cladding type	T91
Coolant type	Pb
Fuel pellet inner diameter (mm)	1.9

(continued)

Table 3. *(continued)*

Property	Value
Fuel pellet outer diameter (mm)	8.6
Cladding inner diameter (mm)	8.9
Cladding outer diameter (mm)	10.0
Fuel pellet density (% of TD)	95
Pitch (mm)	14.0
Active fuel height (mm)	1000
Average linear power (W/mm)	20.5
Average neutron flux rate (n/cm^2/s)	1.55×10^{15}
Initial gas pressure (MPa)	0.5
Plenum to fuel ratio	1.15
Oxygen-to-Metal ratio	1.97
Coolant inlet temperature (°C)	400
Coolant inlet velocity (m/s)	1.5
Coolant pressure (MPa)	0.1

3.2 Introduction of KMC-Fueltra

KMC-Fueltra is a 1.5D fuel performance analysis code designed for the liquid metal fast reactor. It is applicable for the steady-state and transient operating conditions and covers typical materials used in LMFRs [13, 14]. Figure 1 shows the flow diagram of KMC-Fueltra. It can perform the thermal, thermal migration, fission gas release, and mechanical analysis of the fuel rod. Based on the above calculation results, the fuel rod failure analysis is also performed. Irradiation behaviors considered in this code contain thermal expansion, swelling, densification, relocation, cracking-healing, restructuring, creep, and plasticity for the pellet as well as thermal expansion, swelling, creep, and plasticity for the cladding. In this article, KMC-Fueltra is used to evaluate the steady-state performance of the fuel rod in the M^{2}LFR-1000.

3.3 Indicative Design Limits

Design limits are important for the conceptual design of the fuel rod. Luzzi et al. [2] has concluded some conservative design limits for many important parameters that influence thermal and mechanical performance from the open literatures. These limits are adopted for the conceptual design of fuel rods in M^{2}LFR-1000 as well. Some limits are concluded from the view of the corrosion and erosion problem of the lead environment, others are from the material properties or irradiation experiments. Table 4 lists the conservative design limits of some important parameters.

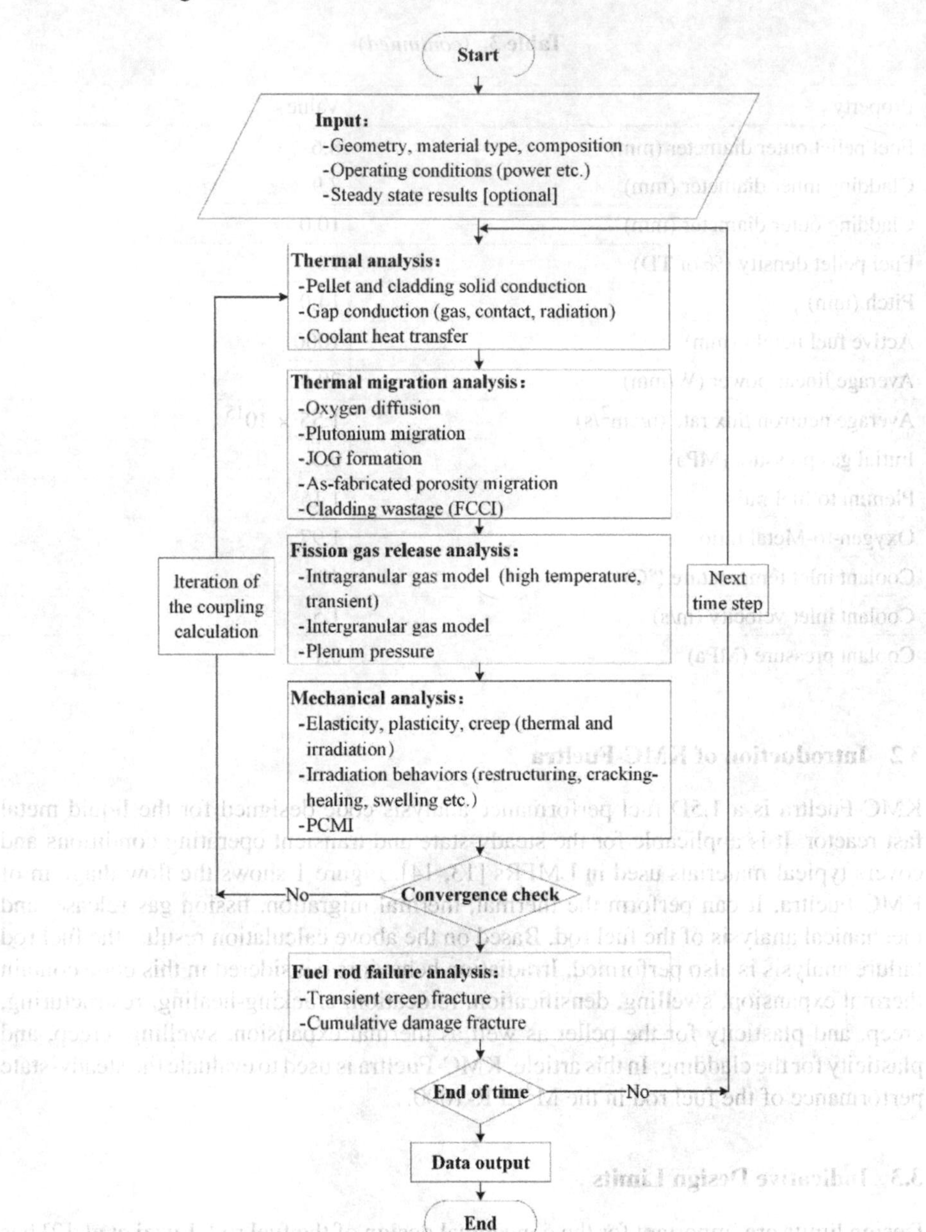

Fig. 1. Flow diagram of KMC-Fueltra.

Table 4. Indicative design limits for M^2LFR-1000

Quantity	Value
Peak fuel temperature	<2300 °C
Peak cladding temperature	<550 °C
Maximum coolant velocity	<2 m/s
Plenum pressure	<5 MPa
Total creep strain	<3%
Instantaneous plastic strain	<0.5%
Maximum cladding stress	<128 MPa
Cumulative damage function	<0.2/0.3

4 Results and Discussions

4.1 Thermal Performance

The change of fuel rod temperature with time is shown in Fig. 2. The maximum fuel pellet temperature is about 1314.4 °C, which appears at the half height of the fuel rod and is consistent with the axial power distribution. The maximum cladding temperature is about 529.6 °C, which appears at the top of the fuel rod. The margin to the safety limits is abundant. The maximum temperature of the fuel pellet becomes larger with time while the cladding did not show too much change. Burnup effect on the fuel rod can account for this. Thermal conductivity of the fuel pellet in Eq. 2 gets smaller with burnup and the gap conductance also decreases at a certain burnup as can be seen in Fig. 3. These factors lead to the deterioration of heat conduction in fuel rod. Gap conductance is affected by many factors including the gap with, gas temperature, plenum pressure, and gas content. Its change depends on which factor is dominant in the operational life. Due to the swelling in the fuel pellet, the gap width decreases with time, which improves the gap conduction at the initial stage. However, the gap conductance decreases later because of the increasing released fission gases.

4.2 Fission Gas Release

Figure 4 gives the change of fission gas release fraction in MOX fuel. At the initial stage, fission gases diffuse and agglomerate in the fuel matrix with the fission process, so the release fraction is almost zero. When the fission gases atoms accumulate to a certain amount in the gas bubbles, the migration process starts. These gas bubbles migrate to the intergranular and then to the gap under the temperature gradient. The released fission gases cause the plenum pressure getting larger as shown in Fig. 5. The initial gas pressure is 0.5 MPa and the maximum plenum pressure is about 1.8 MPa at the end of burnup.

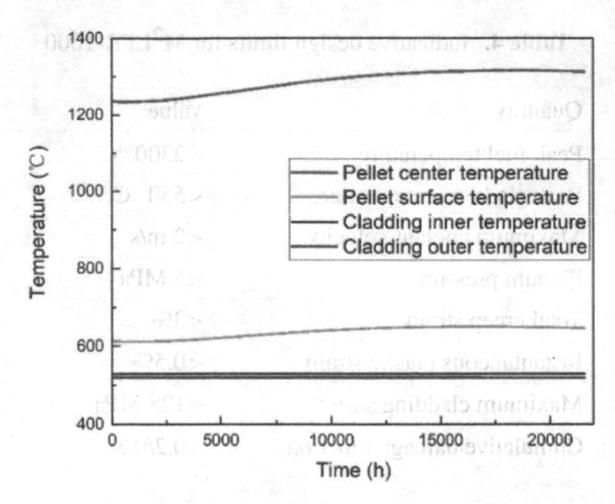

Fig. 2. Evolution of the fuel rod maximum temperature.

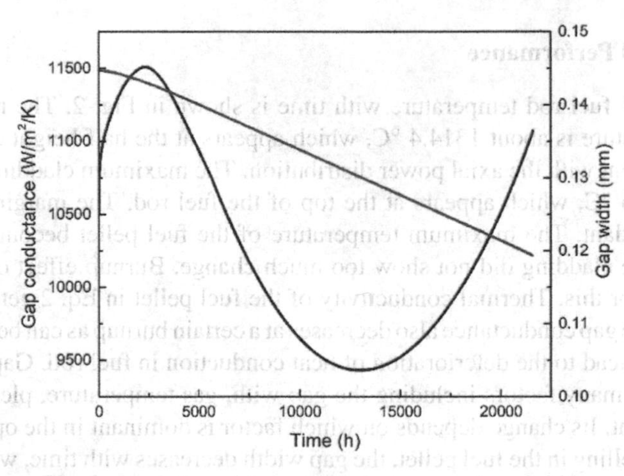

Fig. 3. Evolution of the gap conductance and gap width.

4.3 Mechanical Performance

In addition to the temperature field, the effect of irradiation behaviors on the fuel performance is also shown in the deformation of the fuel rod. Figure 6 illustrates the evolution of fuel rod size. The biggest change is the fuel pellet outer diameter. Irradiation behaviors that occur within the fuel pellet is more intense where the neutron flux is the largest. Thermal expansion and swelling contribute a lot to the expansion of the fuel pellet. Although densification can decrease the pellet deformation, its contribution is small and it ends at the initial stage quickly. The size of the cladding does not show too much change during the entire operational life. Table 5 gives the comparison results of the parameters with the design limits. The safety margin of fuel pellet temperature is very large while the cladding is only 20 °C. This should be paid attention in the later optimization of the fuel rod design. Some burnup related parameters like plenum pressure, total creep strain,

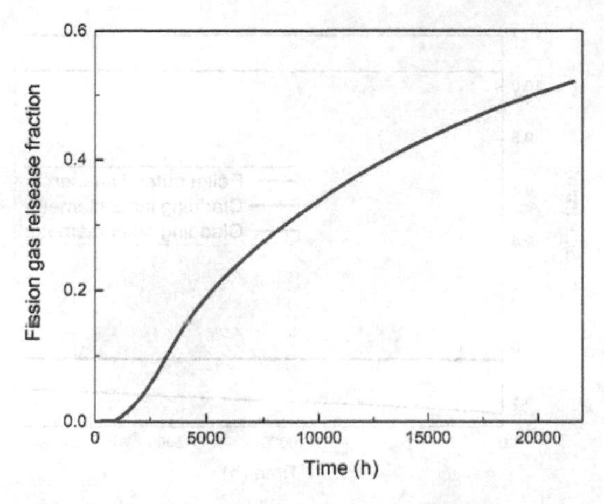

Fig. 4. Evolution of the fission gas release fraction.

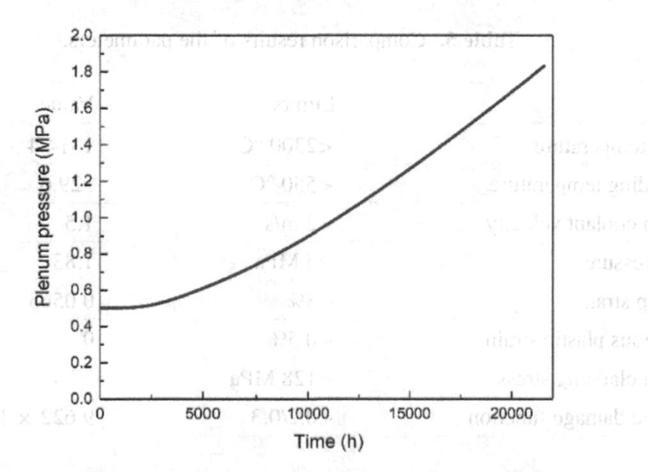

Fig. 5. Evolution of the plenum pressure.

maximum cladding stress and CDF is far from the design limits. Meanwhile, due to the low stress state, the plastic strain does not appear in the fuel rod. A major reason is that burnup of the fuel rod, as shown in Fig. 7, is not very large compared to the other fast reactors of the same type. The maximum burnup is about 4.2 at%. For better economy, it can enlarge by modifying the reactor design if needed.

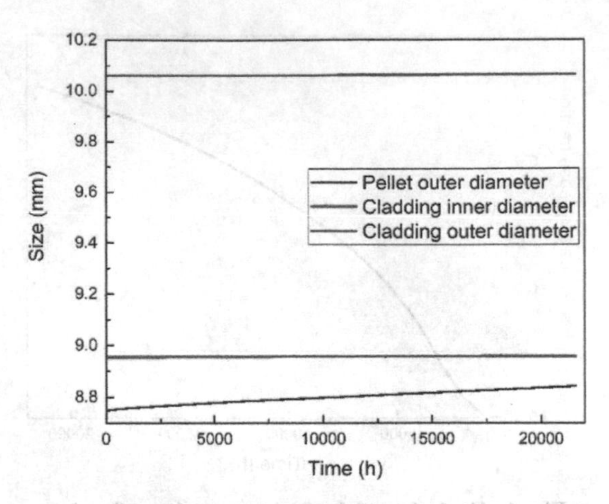

Fig. 6. Evolution of the fuel rod size.

Table 5. Comparison results of the parameters.

Quantity	Limits	Value
Peak fuel temperature	<2300 °C	1314.4
Peak cladding temperature	<550 °C	529.6
Maximum coolant velocity	<2 m/s	1.5
Plenum pressure	<5 MPa	1.83
Total creep strain	<3%	0.0565
Instantaneous plastic strain	<0.5%	0
Maximum cladding stress	<128 MPa	27.4
Cumulative damage function	<0.2/0.3	9.622×10^{-11}

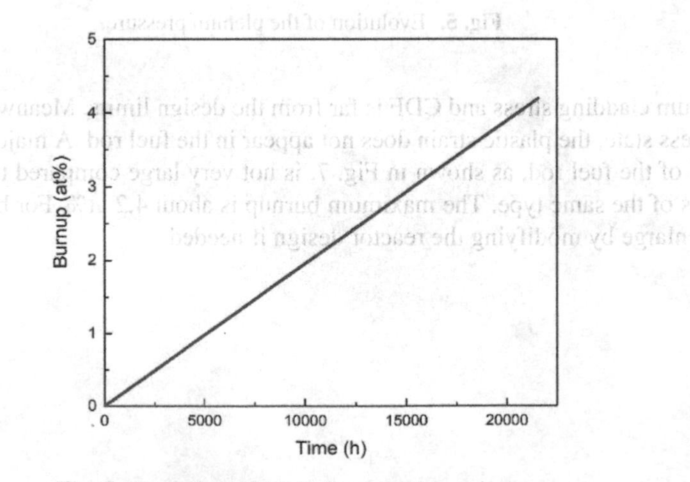

Fig. 7. Evolution of the fuel rod maximum burnup.

5 Conclusions

In this paper, the steady-state fuel rod performance in M^{2}LFR-1000 has been studied using the fuel performance code KMC-Fueltra. Material properties for the MOX fuel and T91 cladding have been incorporated into the code. An important evaluation model and some indicative design limits are collected from the open literatures to help find out the shortcoming of the fuel rod design. The corresponding thermal and mechanical parameters of the fuel rod are evaluated and discussed. Some important parameters concerning the thermal and mechanical performance of the fuel pellet and cladding do not exceed the design limits, proving the safety of the fuel rod design. Even some parameters are far from the design limits, leaving a lot of room for optimization. Burnup of the fuel rod in M^{2}LFR-1000 is about 4.2 at%, which is not very deep compared to fast reactors of the same type. It can also be optimized in the later design process.

References

1. Van Uffelen, P., Hales, J., Li, W., et al.: A review of fuel performance modelling. J. Nucl. Mater. **516**(INL/JOU-18-45934-Rev000) (2018)
2. Luzzi, L., Cammi, A., Di Marcello, V., et al.: Application of the TRANSURANUS code for the fuel pin design process of the ALFRED reactor. Nucl. Eng. Des. **277**, 173–187 (2014)
3. Di Marcello, V., Botazzoli, P., Schubert, A., et al.: Improvements of the TRANSURANUS code for FBR fuel performance analysis (2011)
4. Karahan, A., Buongiorno, J.: Modeling of thermo-mechanical and irradiation behavior of mixed oxide fuel for sodium fast reactors. J. Nucl. Mater. **396**(2–3), 272–282 (2010)
5. Chen, H., Zhang, X., Zhao, Y., et al.: Preliminary design of a medium-power modular lead-cooled fast reactor with the application of optimization methods. Int. J. Energy Res. **42**(11), 3643–3657 (2018)
6. Magni, A., Barani, T., Del Nevo, A., et al.: Modelling and assessment of thermal conductivity and melting behaviour of MOX fuel for fast reactor applications. J. Nucl. Mater. **541**, 152410 (2020)
7. Siefken, L.J., Coryell, E.W., Harvego, E.A., et al.: SCDAP/RELAP5/MOD 3.3 code manual MATPRO - a library of materials properties for light-water-reactor accident analysis. NUREG/CR-6150; INEL-96/0422, Idaho Falls (2000)
8. Luscher, W., Geelhood, K., Porter, I.: Material property correlations: comparisons between FRAPCON-4.0, FRAPTRAN-2.0, and MATPRO[R]. PNNL-19417 Rev.2, Richland (2015)
9. Yang, G., Liao, H., Ding, T., et al.: Preliminary study on the thermal-mechanical performance of the U3Si2/Al dispersion fuel plate under normal conditions. Nucl. Eng. Technol. **53**(11), 3723–3740 (2021)
10. Angelo, A.D., Casaccia, E., Torino, P., et al.: Benchmark on beam interruptions in an accelerator-driven system final report on phase II calculations. Nuclear Energy Agency, Organisation for Economic Cooperation and Development (2003). (7)
11. Agosti, F., Botazzoli, P., Marcello, V.D., et al.: Extension of the TRANSURANUS code to the analysis of cladding materials for liquid metal cooled fast reactors: a preliminary approach (2013)
12. Larson, F.R.: A time-temperature relationship for rupture and creep stresses. Trans. ASME **74**, 765–775 (1952)

13. Yang, G., Guo, Z., Wang, P., et al.: Research and validation on the numerical algorithm of mechanical module in a transient fuel rod performance code for fast reactor. Ann. Nucl. Energy **171**, 108991 (2022)
14. Yang, G.L., Liao, H.L., Ding, T., et al.: Development and validation of a new oxide fuel rod performance analysis code for the liquid metal fast reactor. Nucl. Sci. Tech. **33**, 66 (2022)

Heterogeneous Reactivity Effect Analysis of Pu Spots Considering Grain Size Distribution Based on MOC

Masato Ohara(✉), Akifumi Ogawa, Takanori Kitada, and Satoshi Takeda

Osaka University, Suita, Osaka, Japan
m-ohara@ne.see.eng.osaka-u.ac.jp

Abstract. In the process of Mixed Oxide (MOX) fuel fabrication, plutonium grains, called plutonium spots (Pu spots), occur in MOX fuel because uranium and plutonium cannot be mixed completely. The previous study showed that the prediction accuracy of criticality is improved by considering the heterogeneity of Pu spots in MOX fuel in the analysis of some critical experiments. The analysis of the heterogeneity has been performed by Monte Carlo calculations, to ensure that core geometry and self-shielding effects are accurately considered in nuclear calculations. However, the results of Monte Carlo calculations are obtained with the statistical errors, thus the small reactivity worth is difficult to be analyzed. Therefore, in the previous study, the heterogeneous model by using a method devised by R. Sanchez was introduced into OpenMOC that is deterministic calculation code. Although there are numerous sizes of Pu spots in MOX fuel, an only single size of the grain could be considered in the modified OpenMOC. Therefore, in this study, the modified OpenMOC code was additionally modified to consider the grain size distribution of Pu spots.

The calculation result obtained by the additionally modified OpenMOC shows that the heterogeneous reactivity effect caused by numerous grain sizes of Pu spots is evaluated at $-0.190\%\Delta k/kk'$ at highly enriched fuels. On the other hand, a heterogeneous reactivity effect by single grain size is evaluated at $-0.225\%\Delta k/kk'$. The additionally modified OpenMOC makes it possible to study the effect of considering the grain size distribution of Pu spots in MOX fuel.

Keywords: MOX fuel · Pu spots · OpenMOC · GMVP · Heterogeneous model · Grain size distribution · Heterogeneous reactivity effect

1 Introduction

MOX fuel contains plutonium oxide grains (Pu spots) within the pellet. Pu spots burn locally and cause a problem with local power distribution in the fuel. Currently, the size of Pu spots is measured and controlled to ensure that local power distribution does not occur. The previous studies showed that the accuracy of criticality predictions can be improved by considering the heterogeneity of Pu spots [1]. The analysis of considering the heterogeneity of Pu spots is performed by Monte Carlo calculations to consider

C. Liu (Ed.): PBNC 2022, SPPHY 283, pp. 717–723, 2023.
https://doi.org/10.1007/978-981-99-1023-6_61

the exact core geometry and self-shielding effects. However, the results of Monte Carlo calculations are obtained with statistical errors, thus the small reactivity worth is difficult to be analyzed.

The deterministic calculation is suitable for the analysis of small reactivity worth. One deterministic approach to treating the self-shielding effect is introducing the reduced order-modeling [3–6]. Another approach is introducing the heterogeneous model by using a method devised by R. Sanchez [7–10]. In the previous study, the heterogeneous model was introduced into the MOC code OpenMOC [11]. Although there are numerous sizes of Pu spots in MOX fuel, the heterogeneous model in the modified OpenMOC could consider only a single grain size. Therefore, in this study, the modified OpenMOC code was additionally modified to consider the grain size distribution of Pu spots, and the effect of grain size distribution on criticality was investigated.

2 Calculation Method and Condition

2.1 Calculation Flow

The OpenMOC code was modified so that numerous grain sizes can be calculated to treat the grain size distribution. The calculation flow of the modified OpenMOC code is shown in Fig. 1. In Fig. 1, the input module was modified to read the numerous sizes of the grain, and the angular flux calculation function was modified to calculate the flux in each grain size. At the beginning of the calculation, the escape probabilities and transmission probabilities required to introduce the heterogeneous model are calculated.

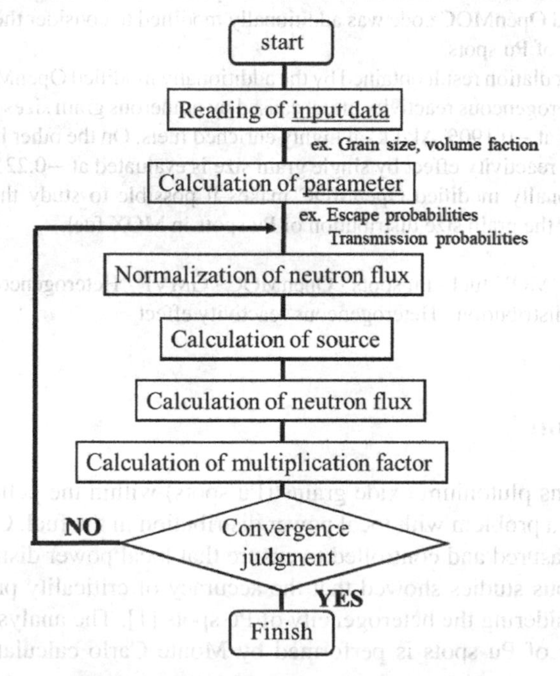

Fig. 1. Calculation flow of modified code

2.2 Calculation Condition

This section shows the calculation condition of the modified OpenMOC and additionally modified OpenMOC. The effective cross section is obtained by the collision probability method (PIJ) of SRAC2006 [12]. 107 energy group structure are used. The fast neutron energy region is divided into 61 groups, and the thermal neutron energy region is divided into 46 groups. The modified OpenMOC calculations was performed in the pin-cell model. The geometry of pin-cell model is shown in Fig. 2. The fuel temperature is 900K and moderator temperature is 600K. JENDL-4.0 is used [13]. U-235 enrichment of matrix is 0.2wt%. Pu enrichment of each region is shown in Table 1. Isotopic ratio of Pu is shown in Table 2. The calculations of the modified OpenMOC are performed under the conditions shown in Table 3.

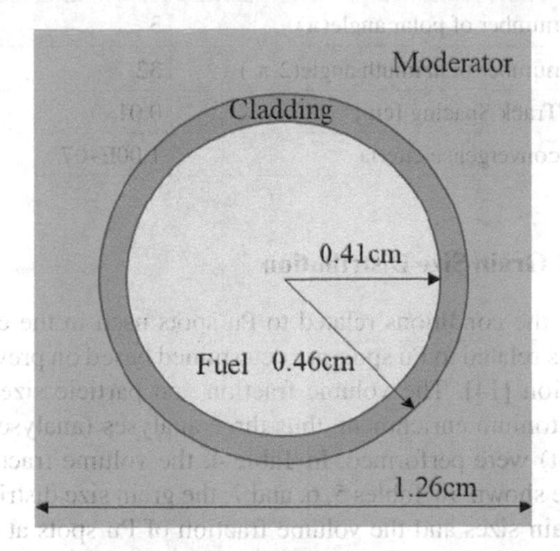

Fig. 2. Geometry of Pin-cell model

Table 1. Pu enrichment of each region

Fuel type	Entire region	Grain region	Matrix region
High enrichment	10.6%	25.0%	3.0%
Middle enrichment	6.2%		3.7%
Low enrichment	4.5%		3.0%

In this study, the calculations were performed for both the homogeneous model, in which uranium and plutonium are homogeneously mixed, and the heterogeneous model, in which Pu spots are included. The effect of Pu spots is expressed as a heterogeneity reactivity effect. The heterogeneity reactivity effect is defined as the reactivity which is obtained by changing from the homogeneous model to heterogeneous model. The effect is defined as:

$$heterogeneity\ reactivity\ effect = \frac{1}{k(homogenenous\ model)} = \frac{1}{k'(homogenenous\ model)} \tag{1}$$

Table 2. Isotopic Ratio of Pu (wt%)

Pu-238	5
Pu-239	55
Pu-240	20
Pu-241	15
Pu-242	5

Table 3. Calculation Conditions on the modified OpenMOC

number of polar angle(π)	3
number of azimuth angle(2 π)	32
Track Spacing [cm]	0.01
convergence citeria	1.00E–07

2.3 Estimation of Grain Size Distribution

This section shows the conditions related to Pu spots used in the calculations. In this study, the conditions related to Pu spots are determined based on previously investigated grain size distribution [14]. The volume fraction and particle size distribution of Pu spots vary with plutonium enrichment, thus three analyses (analyses for high, middle, and low enrichment) were performed. In Table 4, the volume fractions of Pu spots at each enrichment are shown. In Tables 5, 6, and 7, the grain size distribution is expressed in terms of four grain sizes and the volume fraction of Pu spots at each grain size are shown.

Table 4. Volume fraction of Pu spots at each enrichment

Fuel type	Volume fraction of Pu
High enrichment	30.3%
Middle enrichment	10.1%
Low enrichment	5.8%

Table 5. Volume fraction of each grain size in high enrichment fuel

Grain Size	125 μm	45 μm	25 μm	10 μm
Volume fraction	23.80%	1.49%	2.59%	2.42%

Table 6. Volume fraction of each grain size in middle enrichment fuel

Grain Size	75 μm	45 μm	25 μm	10 μm
Volume fraction	6.35%	0.95%	0.97%	1.86%

Table 7. Volume fraction of each grain size in low enrichment fuel

Grain Size	65 μm	35 μm	25 μm	10 μm
Volume fraction	4.50%	0.36%	0.30%	0.68%

In addition, calculations that simulate single grain size are also carried out to investigate the effect of considering the grain size distribution. In the high, middle, and low enrichment fuels, most of Pu spots are 125, 75, and 65 μm in diameter respectively. Therefore, these grain sizes are used in the calculation of the single grain size. The single grain sizes used in the calculations are shown in Table 8.

Table 8. The single grain sizes used in the calculations at each enrichment

	single grain size
High enrichment	125 μm
middle enrichment	75 μm
low enrichment	65 μm

3 Calculation Result

The calculation results of single grain size and four grain sizes are shown in Fig. 3.

In all conditions, the results of the heterogeneity reactivity effect are negative because the ratio of Pu-239 is high and the ratio of Pu-240 is low in the Pu isotope ratio. Pu 239 has a fission resonance at 0.3 eV and the fission reaction is reduced by the self-shielding effect of Pu spots. The reason for the negative heterogeneity reactivity effect is the high proportion of Pu-239. On the other hand, the isotope ratio with the high proportion of Pu-240 is used, the heterogeneity reactivity effect shifts to the positive side. The reason is that Pu-240 has a neutron absorption resonance at 1.0eV and neutrons escaping this resonance can cause fission.

At all enrichments, the heterogeneity reactivity effect of the calculation of single grain size is larger than four grain sizes. Because the larger the grain size, the larger the self-shielding effect. In this study, the calculations of single grain size are carried out by replacing smaller grain sizes with the largest one. Therefore, the calculations of single grain size have a large proportion of Pu spots with a larger grain size than the

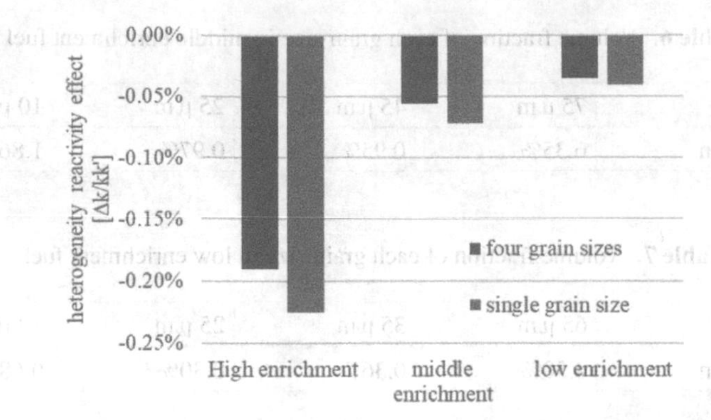

Fig. 3. Comparison of calculation results between the case of single grain size and four grain sizes

calculations of four grain sizes. As a result, the heterogeneity reactivity effect is larger in the calculation of single grain size.

At the high enrichment fuel, the difference between the calculation results of single grain size and four grain sizes was the largest. The difference is 0.035%k/kk'. When calculating a slight difference in reactivity effect by Monte Carlo calculation, a huge calculation cost is required to suppress statistical error. This study makes it possible to carry out calculations that consider the grain size distribution of Pu spots in MOX fuel, using a deterministic calculation method that does not have statistical errors.

4 Conclusions

The heterogeneous model introduced in previous studies could only consider a single grain size. In practice, there is a distribution of Pu spots sizes in MOX fuel. Therefore, the purpose of this study is to investigate the effect of the Pu spots size distribution on criticality. Initially, the modified OpenMOC was additionally modified to consider the grain size distribution, thus the effect of the grain size distribution on criticality can be evaluated without statistical errors. At high enrichment fuel, the difference between the calculation results of single grain size and four grain sizes was found to be 0.035Δk/kk'. The additionally modified OpenMOC makes it possible to study the effect of considering the grain size distribution of Pu spots in MOX fuel. Further improvement of the criticality prediction accuracy in MOX fuel is expected by the additionally modified OpenMOC.

References

1. Reactor Integral Test Working Group, JENDL Committee "Compilation of the Data Book on Light Water Reactor Benchmark to Develop the Next Version of JENDL" Japan Atomic Energy Agency (2017)
2. William, B., Shaner, S., Li, L., et al.: The open MOC method of characteristics neutral particle transport code. Ann. Nucl. Energy **68**, 43–52 (2014)
3. Yamamoto, A., Endo, T., Takeda, S., et al.: A resonance calculation method using energy expansion bases based on a reduced order model. In: M&C 2019, 25–29 August 2019 (2019)

4. Yamamoto, A., Kondo, R., Endo, T., et al.: Resonance calculation using energy spectral expansion based on reduced order model: application to heterogeneous geometry. In: 2019 Transactions of the American Nuclear Society, 7–21 November 2019 (2019)
5. Kondo, R., Endo, T., Yamamoto, A., et al.: A resonance calculation method using energy expansion based on a reduced order model: Use of ultra-fine group spectrum calculation and application to heterogeneous geometry. In: PHYSOR 2020, 28 March–2 April 2020 (2020)
6. Kondo, R., Endo, T., Yamamoto, A., et al.: A new resonance calculation method using energy expansion based on a reduced order model. Nucl. Sci. Eng. **195**(7), 694–716 (2021)
7. Sanchez, R., Pomraning, G.C.: A statistical analysis of the double heterogeneity problem. Ann. Nucl. Energy **18**(7), 371–395 (1991)
8. Sanchez, R., Masiello, E.: Treatment of the double heterogeneity with the method of characteristics. In: PHYSOR 2002, 7–10 October 2002 (2002)
9. Sanchez, R.: Renormalized treatment of the double heterogeneity with the method of characteristics. In: PHYSOR 2004, 25–29 April 2004 (2004)
10. Joo, H.G., Park, T.K.: Derivation of Analytic Solution and MOC Calculation Procedure for Double Heterogeneity Treatment. Seoul National University. SNURPL-SR-001 (2007)
11. Ogawa, A.: Modification of MOC for considering double heterogeneity due to pu spots in MOX fuel. In: Proceedings of the Reactor Physics Asia 2019 Conference, pp. 86–89(2019)
12. Okumura, K., Kugo, T., Kaneko, K., Tsuchihashi, K.: SRAC2006: a comprehensive neutronics calculation code system. Japan Atomic Energy Research Institute (2007)
13. Shibata, K., et al.: JENDL4.0: a new library for nuclear science and engineering. J. Nucl. Sci. Technol. **48**(l), 1(2011)
14. Documentation of the 4th Safety Review Meeting handout 4–3. Miyagi Prefecture. Accessed 27 June 2022. [In Japanese]. https://www.pref.miyagi.jp/documents/10438/4042.pdf

Preparation and Properties of Graphite Surface Vitrification Y_2O_3 Coating

Zhen Lei(✉), Hongya Li, Bingzhai Yu, and Binghan Geng

The Fourth Filial Company of 404 Company Limited, CNNC, Lanzhou 732850, Gansu, China
576442476@qq.com

Abstract. The Mo-Y_2O_3 composite coating was prefabricated on the surface of the graphite crucible by plasma spraying, and then the coating surface was remelted by high-power laser. The coating was subjected to SEM analysis, X-ray diffraction, bond strength testing, and molten terbium metal corrosion test. The results show, After laser cladding treatment, the problems of pores and unmelted particles on the surface of the Mo-Y_2O_3 coating have been significantly improved, and the porosity has been reduced to less than 1.5%, Vitrified Y_2O_3 with a thickness of about 80–120 μm and columnar crystal growth was formed on the surface of the coating, and the bonding strength was increased from 2–3 MPa before treatment to 7–11 MPa, In the corrosion assessment test of molten terbium metal, the coating was intact and did not fall off, the surface of the ingot was not adhered, and there was no contamination by impurity elements, which could be easily demolded.

Keyword: Surface treatment · Laser micro-nano sintering · Performance optimization · Composite materials

1 Introduction

In recent years, industries such as nuclear energy technology, aerospace, new energy, new materials, marine ships and medical equipment have been changing rapidly. The progress in these fields is inseparable from the rapid development of high-performance special alloy materials. Such alloy materials are often active metals with a high melting point, the raw materials are precious, and are easily reacted with various gases and crucible materials at high temperatures. Therefore, solving the problem of "sticking and contamination" of melting crucibles has become a hotspot of surface engineering research. Graphite material has excellent thermal shock resistance, thermal shock resistance and ease of processing, and is the most commonly used crucible material in high melting point active metal smelting. However, at high temperature, the CO gas and free carbon vapor generated in the furnace can easily react with the smelted metal to form carbides, resulting in melt carbon pollution, which affects the properties of metal materials [1]. Xiao [2] studied the thermal shock resistance of Nb/ZrO_2(Y_2O_3), Nb/$CaZrO_3$, Mo/ZrO_2(Y_2O_3) composite coatings, and the results showed that the coatings basically cracked and peeled off at 1500 °C for 30 min. Koger J W [3] prepared a Nb/ZrO_2/Y_2O_3 composite coating, which can withstand the test of metal smelting many times, but

C. Liu (Ed.): PBNC 2022, SPPHY 283, pp. 724–732, 2023.
https://doi.org/10.1007/978-981-99-1023-6_62

the process technology is harsh. Kim [4] prepared a Y_2O_3 coating that can be used for ternary alloy casting, but it is difficult to prepare. Petitbon [5] added Al_2O_3 powder when remelting the plasma sprayed zirconia coating by laser to obtain the Al_2O_3/ZrO_2 composite coating. The coating's strength, wear resistance and high temperature corrosion resistance are significantly increased Experimental materials and methods. Ahmaniemi [6] used the laser cladding method to seal the ZrO_2 coating, the lattice distortion of the ZrO_2 crystal occurred, the coating became dense, and the microhardness increased.

At present, most of the researches focus on the optimization of the plasma spraying process parameters and the repair of the coating on the surface of the crucible, and the process is complicated and tedious, while the research and application of the coating required for the smelting of special materials are very few. In order to solve the above shortcomings, this paper takes graphite/Mo/Y_2O_3 as the research system, and uses the laser micro-nano sintering method to clad the plasma sprayed Mo/ Y_2O_3 coating, and obtain the vitrified Y_2O_3 coating. The coating can withstand multiple corrosion tests of molten terbium metal. The preparation method and formation mechanism of vitrified Y_2O_3 coating on graphite surface were studied in order to provide some theoretical guidance for production practice.

2 Test Materials and Methods

High-purity graphite has the advantages of light weight, high strength and good thermal shock resistance. It is a commonly used crucible material in high-temperature smelting and high-temperature smelting of rare earth metals, niobium silicide-based superalloys and some actinide metals. However, carbon as an impurity element needs to be strictly controlled in the metallurgical processing of metal materials, and the increase of carbon content in the smelting process should be avoided as much as possible. Therefore, it is necessary to spray a layer of oxide on the surface of the crucible to prevent carbon increase. As the intermediate transition layer of the composite coating, the thermal expansion coefficient of Mo is between high-purity graphite and yttrium oxide, which can effectively prevent the coating from peeling, cracking and falling off due to high temperature melt erosion. Y2O3 is stable and does not react or wet with high temperature melts, which can ensure the purity of the melts.

The materials selected in this study and their properties are shown in Table 1.

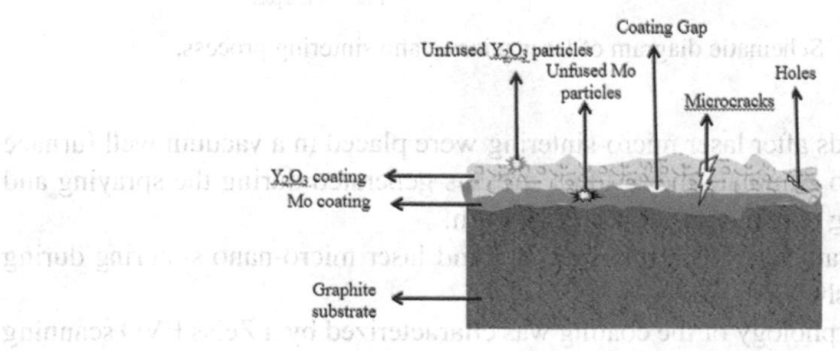

Fig. 1. Schematic diagram of plasma spray coating structure.

Firstly, the Mo-Y_2O_3 composite gradient coating was prefabricated on the surface of the graphite mold crucible by plasma spraying method, and the coating structure is shown in Fig. 1.

Table 1. Material types and their properties.

High purity graphite	Ash content (ppm)	Density (g/cm^3)	Graininess (mm)	Thermal conductivity w/(m·k)	Coefficient of thermal expansion (°C^{-1})
	≤40	≥1.8	≤0.02	≥110	4.0–4.8 × 10^{-6}
Molybdenum powder	Principal content (%)		Particle size distribution (μm)		5.2 × 10^{-6}
	≥99.90		45–96		
Yttrium powder	Principal content (%)		Particle size distribution (μm)		8.0 × 10^{-6}
	≥99.98		11–53		

Next, the graphite crucible was dried in a blast drying oven at a temperature of 50–60 °C for 30 min, again, the crucible was fixed on a special fixture after the dust on the surface was blown off by compressed air, and the equipment was turned on to perform the laser fusion treatment of the coating. The micro-nano sintering process is shown in Fig. 2.

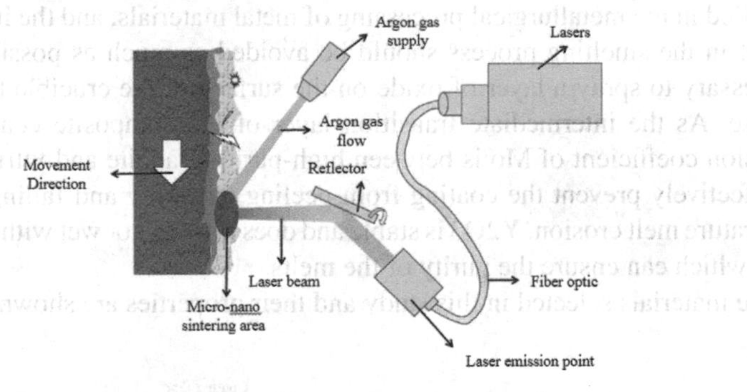

Fig. 2. Schematic diagram of laser micro-nano sintering process.

Finally, the molds after laser micro-sintering were placed in a vacuum well furnace for heat treatment to eliminate the residual stresses generated during the spraying and laser micro-sintering experiments at 400 °C for 4 h.

The process parameters of plasma spraying and laser micro-nano sintering during the experiment are shown in Table 2.

The surface morphology of the coating was characterized by a Zeiss EVO scanning electron microscope. And use Image J image processing software to calculate apparent porosity. The phase structure and composition of the coating were determined by X-ray

Table 2. Plasma spraying and laser micro-nano sintering process parameters.

Plasma spray	Voltage (V)	Current (A)	Spraying distance (mm)	Main air flow (Ar/MPa)	Auxiliary gas flow (He/MPa)
	45–55	600–750	90	2.00–3.00	0.20–0.40
Laser Micro-Nano Sintering	Overlap rate (%)	Scan speed (°/min)	Laser power (w)	Energy Density (w/mm^2)	Laser spot diameter (mm)
	50	2600–2800	105–120	145	2–3

diffractometer XRD-6100. The bond strength of the coating was tested by the pull-off method.

3 Experimental Results and Analysis

3.1 Coating SEM, XDR Analysis

The SEM micro-morphologies of the coatings before and after laser micro-nano sintering are shown in Fig. 3.

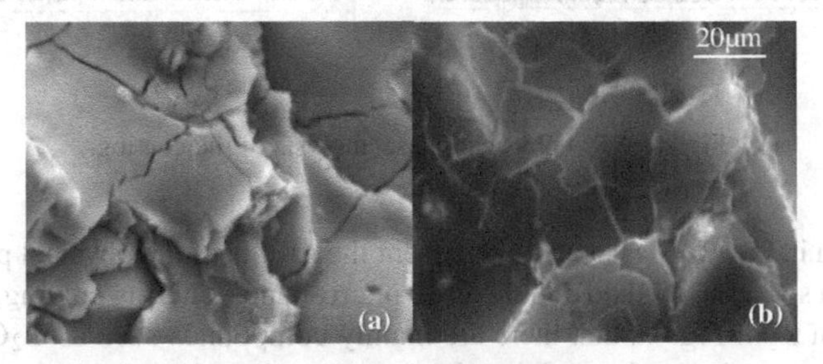

Fig. 3. Microscopic surface morphology of the coating after plasma spraying and laser micro-nano sintering.

As shown in Fig. 3, a and b indicate the microscopic morphology of the coating by plasma spraying and laser micro-nano sintering, respectively. The surface of the plasma sprayed coating is uneven, loose and porous, and has a lamellar structure. The connection between the coating and the substrate belongs to mechanical bonding. Because in the spraying process, the molten powder is not shot to the surface of the substrate at a high speed, and a porous layered coating is formed after deformation and spreading. Some powder particles that are not fully melted reach the surface of the substrate to form spherical particles of different sizes, and through layer-by-layer superposition, many holes are formed in the coating, causing defects. In addition, during the spraying process, the air in the surrounding environment absorbed by the molten particles cannot be completely eliminated during cooling and solidification, which also forms a defect.

After laser micro-nano sintering, the surface of the coating is flat and smooth, and there are no obvious holes. This shows that the coating performance is optimized, and the bonding between the coatings is closer to metallurgical bonding. After micro-nano sintering, micro-cracks can still be seen on the surface (the bright part in Fig. 3b), which is caused by the combined effect of tensile stress and compressive stress on the coating surface during the rapid heating and cooling of the coating. After laser remelting, the surface lamellar structure almost disappeared. After micro-nano sintering, the coating formed a vitrified Y_2O_3 film with a thickness of about 80–120 μm and columnar crystal growth. The defects such as pores and inclusions of the coating are greatly reduced, and the density has been greatly improved.

The XRD analysis of the plasma sprayed coating and the laser micro-nano sintered coating was carried out using an X-ray diffractometer XRD-6100, and the results are shown in Fig. 4.

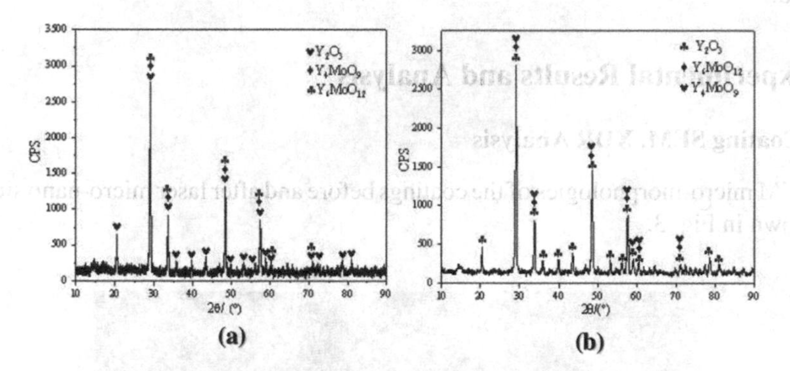

Fig. 4. XRD physical phase analysis of the coating.

As shown in Fig. 4, Figs. a and b represent the results of the physical phase analysis of the plasma spraying method and the laser micro-nano-sintering coating, respectively. It can be seen from Fig. 4a that there are mainly compounds such as Y_2O_3, Y_4MoO_9, and Y_6MoO_{12} on the surface of the plasma sprayed yttrium oxide coating. At high temperature, Y_2O_3 and MoO_3 react as follows, which is consistent with Fig. 4b.

$$2Y_2O_3 + MoO_3 = Y_4MoO_9 \tag{1}$$

$$3Y_2O_3 + MoO_3 = Y_6MoO_{12} \tag{2}$$

Because the plasma temperature is much higher than the sublimation temperature of MoO_2 (700 °C) during the spraying process, the coating material does not contain MoO_2. The X-ray diffraction peaks corresponding to MoO_3 are mainly at $2\theta = 13°, 26°$ and 39°, so it can be considered that the coating surfaces are all Y_2O_3. It can be seen that the laser micro-nano sintering process does not change the phase composition of the coating surface, nor does it introduce other impurities.

3.2 Coating Bonding Strength Score

The bonding strength of the coating was tested by the pull-off method. Three samples were prepared with different process parameters. The test results are shown in Table 3. Plasma Binding force test.

Control tests 1 and 2 show that the overall bonding force of the plasma sprayed layer is low, and the fractured part is on the surface of the substrate. Because the surface temperature of the substrate is very high after spraying, a large internal stress will be generated inside the coating, which is directly quenched in the air, and defects such as microcracks will be generated during the stress release process, which will reduce the bonding force of the coating. From experiments 1 and 3, it can be seen that laser micro-nano sintering can increase the bonding force of the coating from 2–3 MPa to 5–10 MPa, because the coating is converted from mechanical bonding to metallurgical bonding after laser micro-nano sintering. It can be seen from experiments 3 and 4 that heat treatment can further optimize the laser cladding coating, eliminate the internal stress, and further improve the bonding force of the coating.

Table 3. Plasma Binding force test.

No	Laser power/W	heat treatment	Experimental results and binding force (MPa)
1	0	-	Fractured from graphite substrate after stretching, bonding force 2–3
2	0	4 h at 700 °C	Fractured from graphite substrate after stretching, bonding force 2.5–4
3	110		Fractured from graphite substrate after stretching, bonding force 7–9
4	110	4 h at 700 °C	Fractured from yttrium oxide coating after stretching, bonding force 8–11

3.3 Terbium Metal Melting Test Coating Test

At 1550 °C, under 8.5×10^{-2} Pa, the metal terbium in the distilled state is smelted by an intermediate frequency induction heating furnace. Then, the liquid terbium metal was cast into Mo/Y_2O_3 graphite crucibles with two coatings (plasma spray coating and laser cladding coating) respectively, and the temperature was kept for 20 min. Finally, the bottom of the crucible is cooled, filled with inert gas, and after cooling to room temperature, it is released from the furnace and demolded.

As shown in Fig. 5 a, b, and c represent the inner wall of the crucible treated by plasma spraying, the local morphology of the coating and the appearance of the terbium ingot after demolding. d, e, and f represent the inner wall of the crucible, the partial morphology of the vitrified coating after laser micro-nano sintering, and the appearance of the terbium ingot after demolding. Terbium casting test results show that the plasma sprayed Mo/Y_2O_3 composite coating falls off after contacting with terbium melt. During

Fig. 5. Terbium metal casting test.

the solidification and shrinkage process of the terbium melt, the melt penetrates into the interior of the coating through pores and cracks, and forms a mosaic bonding structure with the coating surface. Causes the molybdenum and yttrium oxide coatings to peel off in a large area, and black and white spots appear (as shown in Figure c).

It can be seen from this: the bonding force of the plasma sprayed coating is low, and this phenomenon is consistent with the results of Test 1 in Table 3. After the Mo/Y_2O_3 composite coating is sintered by laser micro-nano, the surface of the terbium ingot is smooth and clean, and there is no "sticking to the ingot" phenomenon, as shown in Fig. f. Because the heating and cooling speeds are extremely fast during the laser cladding process, and the solidification speed is about 10^4 °C/s, the surface of the Y_2O_3 coating will be vitrified, and the voids and cracks on the coating surface will be repaired. The rapid melting makes the surface produce fine-grain strengthening, and the interfacial bonding force increases. This phenomenon is consistent with the results of Test 4 in Table 3.

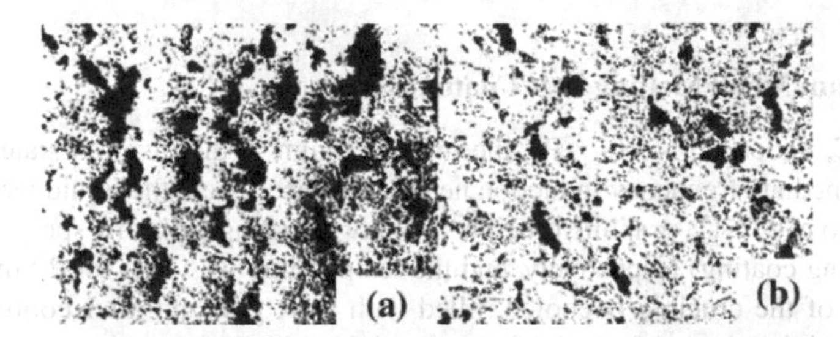

Fig. 6. Pore space on the surface of the coatin.

Apparent porosity was calculated using Image J image processing software. Eight SEM photos of the coating were randomly selected in different areas and at the same magnification, and then the apparent porosity of each photo was calculated by the gray-scale method, and the arithmetic mean was taken as the apparent porosity of the coating.

As shown in Fig. 6,a and b represent the micro-morphologies of the plasma spray coating and the laser micro-nano sintering coating after grayscale processing of the Image image, respectively. After calculation, the apparent porosity of the coating after laser micro-nano sintering decreased from 5–6% to less than 1.5%. It shows that laser micro-nano sintering can significantly reduce the surface porosity of the coating, effectively reduce the infiltration path of the melt, and alleviate the peeling of the coating.

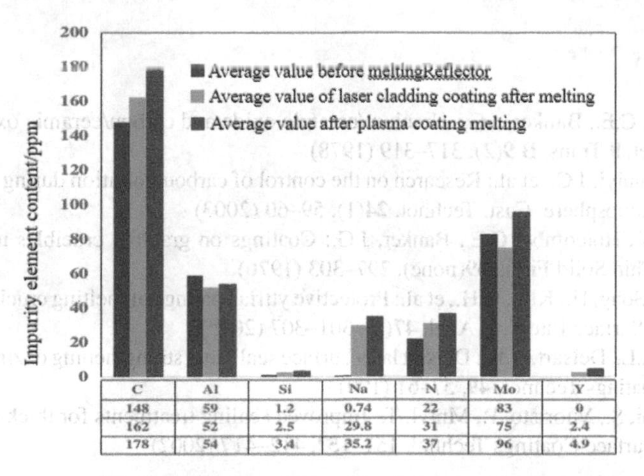

C	Al	Si	Na	N	Mo	Y
148	59	1.2	0.74	22	83	0
162	52	2.5	29.8	31	75	2.4
178	54	3.4	35.2	37	96	4.9

Fig. 7. Variation of impurity element content.

After After demoulding, the content of impurity elements on the surface of the ingot was analyzed. The impurity content on the surface of the ingot is shown in Fig. 7.

Compared with the terbium metal raw material, the content of each impurity element in the ingot after the plasma spray coating is smelted has increased, especially the C content exceeds the standard, which does not meet the smelting requirements. After laser cladding coating smelting, the C content of the ingot increased by 4 ppm on average, the Al, Na, Mo, N content decreased by 7 ppm, 29 ppm, 8 ppm and 9 ppm respectively, the content of Y element met the requirements. The average impurity content is within the control range, and there is no obvious carbon increase, indicating that the Mo/Y_2O_3 coating after laser micro-nano sintering has the effect of preventing carbon completely. The surface of the terbium metal ingot has no white spots attached (Fig. 5f), the interface between the coating and the terbium metal melt does not infiltrate, the coating does not adhere to the surface of the ingot, the ingot is easily demolded, and the melting target is achieved.

4 Conclusion

(1) The surface of the coating after laser micro-nano sintering is smooth and flat, the porosity is reduced to less than 1.5%, and a vitrified Y_2O_3 coating with a thickness of about 80–120 μm and columnar crystal growth is formed. It shows that laser micro-nano sintering can repair coating cracks, and can make the bonding between coatings change from mechanical bonding to metallurgical bonding.

(2) After the plasma sprayed layer is sintered by laser micro-nano, the surface of the coating has a fine-grain strengthening effect due to rapid melting, and the bonding strength of the coating is increased from 2–3 MPa before treatment to 7–11 MPa.
(3) The laser micro-nano sintering coating has the function of blocking carbon without introducing impurities and meeting the smelting requirements.

References

1. Holcombe, C.E., Banker, J.G.: Uranium/ceramic oxide and carbon/ceramic oxide interaction studies. Metall Trans. B **9**(2), 317–319 (1978)
2. Xiao, Y., Shuiyi, J.C., et al.: Research on the control of carbou pollution during metal smelting in carbon atmosphere. Cast. Technol. **24**(1), 59–60 (2003)
3. Koger, J.W., Holcombe, C.E., Banker, J.G.: Coatings on graphite crucibles used in melting uranium. Thin Solid Films **39**(none), 297–303 (1976)
4. Kim, J.H., Song, H., Kim, K.H., et al.: Protective yttria coatings of melting crucible for metallic fuel slugs. Surface Interface Anal. **47**(3), 301–307 (2015)
5. Petitbon, A.L., Delsart, B.D.: Delsart laser surface sealinand strengthening of zirconia coatings. Surface Coatings Technol. **49**, 57–61 (1991)
6. Ahmaniemi, S., Vuoristo, P., Mntyl, T.: Improved sealing treatments for thick thermal barrier coatings. Surface Coatings Technol. **151–152**, 412–417 (2002)

Numerical Study on the Mechanism of Oxygen Diffusion During Oxygen Control Process in Heavy Liquid Metals

Ying Li, Liang Guo, Chao Zhang(✉), Hongbo You, and Yuanfeng Zan

CNNC Key Laboratory on Nuclear Reactor ThermoHydraulics Technology, Nuclear Power Institute of China, Chengdu, Sichuan, China

zhangchao_turbo@163.com

Abstract. In order to investigate the mechanisms and characteristics of oxygen diffusion during oxygen control in Heavy Liquid Metals (HLM). Turbulence model coupling species transport model are used to predict the oxygen transport and distribution in liquid lead-bismuth eutectic (LBE) during gas oxygen control process. The oxygen mass flow coefficients are calculated under diffident LBE temperatures, inlet LBE flow rates, inlet oxygen concentrations and gas/liquid interface oxygen concentrations. A mass transfer correlation for oxygen transport is finally acquired. The simulated results indicate that oxygen transport is mainly influenced by the combined effect of convection and diffusion. Specifically, the mass flow coefficient increases with the increase of mass flow coefficient and LBE temperature. The gas/liquid interface oxygen concentration has little effect on the oxygen mass transfer coefficient in the oxygen transfer device of this study. Comparatively, the smaller inlet oxygen concentration leads to the larger average oxygen mass transfer coefficient and the effects of inlet oxygen concentration become weaker with the increase of flow rate. The numerical data presented in this study represents an essential step to reveal the mechanism of oxygen transport in flowing LBE and provides the theoretical basis for guiding the design of oxygen transfer device in HLM-cooled nuclear reactors.

Keywords: Heavy liquid metal · Oxygen diffusion · Gas oxygen control · Oxygen concentration

1 Introduction

Heavy Liquid Metals (HLM) such as liquid lead (Pb) and lead-bismuth eutectic (LBE) are considered as candidate coolants in Generation IV Nuclear Reactors thanks to their suitable thermo-physical and chemical properties. However, the oxidation of HLM and corrosion of structural steels have become major issue in the development of HLM-cooled nuclear reactors. One of the most viable methods for protection of HLM-cooled nuclear reactors is to control the oxygen content dissolved in the HLM among an appropriate range, which could guarantee the structural steels and avoid the HLM oxidation at the same time.

C. Liu (Ed.): PBNC 2022, SPPHY 283, pp. 733–741, 2023.
https://doi.org/10.1007/978-981-99-1023-6_63

Several researchers have studied the characteristics of oxygen control technics based on different facilities [1–5]. Schroer et al. [1] investigated the oxygen-transfer to flowing lead alloys inside a gas/liquid transfer device in the CORRIDA loop. Marino et al. [2] acquired a mass transfer correlation for the lead oxide mass exchanger and the model was validated using experimental data from the CRAFT loop. Chen et al. [3] explored the LBE flow and oxygen transport in a simplified container under the gas oxygen control using a lattice Boltzmann simulation. However, the mechanisms and characteristics of oxygen diffusion during oxygen control are still not well understood.

In this study, a CFD model in the specific oxygen transfer device has been developed to determine the oxygen transport and distribution in flowing LBE. A mass transfer correlation for oxygen transport was obtained in terms of the Sherwood with simulation results. The numerical data presented in this study represents an essential step to reveal the mechanism of oxygen transport in flowing LBE and provides the theoretical basis for guiding the design of oxygen transfer device in HLM-cooled nuclear reactors.

2 Numerical Modeling

2.1 Geometry and Mesh

The simulated oxygen transfer device has a similar geometry with that in the CORRIDA loop [6], which has a specific inner diameter of 400 mm and a length of 1300 mm. As shown in Fig. 1, the LBE flows through vessels with a liquid level of 1/3 diameter. The control gas is pumped into the oxygen transfer device in the opposite direction through top vessels. The computational domain is simplified and depicted in Fig. 2. The minimum mesh size is set as $5 \cdot 10^{-3}$ m and the total number of the elements is about 2150977 after the mesh independence analysis [7].

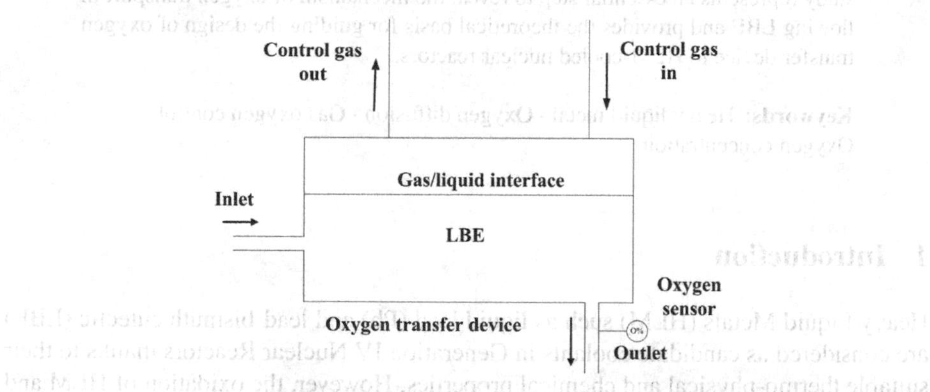

Fig. 1. Schematic diagram of the gas control apparatus

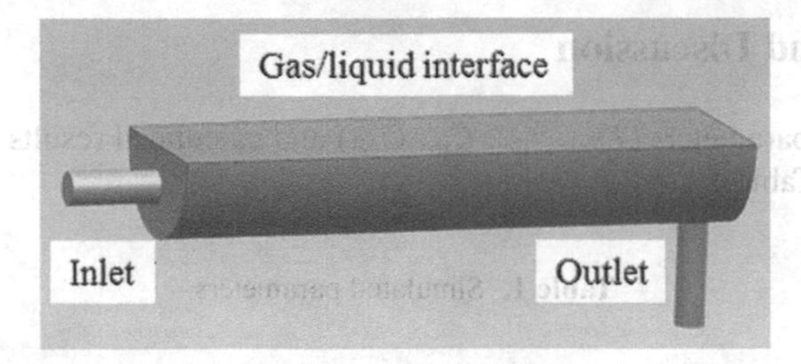

Fig 2. Geometry of the simulated oxygen transfer device

2.2 Governing Equations and Boundary Conditions

The transport of oxygen obeys the Reynolds-averaged transport equation for turbulent flow:

$$\frac{\partial C_o}{\partial t} + \vec{u} \cdot \nabla C_o = \nabla \cdot (D + \frac{\mu_t}{Sc_t})\nabla C_o + q_o \tag{1}$$

where *Co* is the oxygen concentration, *u* is the velocity of LBE, μ_t is the eddy viscosity, *Sct* is the turbulent Schmidt number, and q_o is the oxygen source, *D* is the diffusion coefficient of oxygen. *D* ($cm^2 \cdot s^{-1}$) is given by the following equation [8]:

$$D = 0.0239e^{-\frac{43073}{RT}} \quad 473K < T < 1273K \tag{2}$$

where *R* is the molar gas constant.

Based on the research in [9], the gas/liquid interface is simulated as a free-surface boundary with constant oxygen concentration.

The average dissolution rate *q* ($kg \cdot s^{-1}$) of oxygen in the flowing LBE is described by:

$$q = m(C_{out} - C_{in}) \tag{3}$$

where C_{out} and C_{in} are the oxygen concentration at the outlet and inlet of oxygen transfer device (wt%), respectively. m is the mass flow rate.

The average mass transfer coefficient *k* ($kg \cdot m^{-2} \cdot s^{-1}$) of oxygen in the flowing LBE is calculated by:

$$k = \frac{q}{A(C_{g/l} - C_{ave})} \tag{4}$$

where *A* is the area of gas/liquid interface (m^2), $C_{g/l}$ is the oxygen concentration at the gas/liquid interface (wt%), C_{ave} is the average oxygen concentration of inlet and outlet (wt%).

The Sherwood number (*Sh*) representing the dimensionless form of the average mass transfer coefficient is defined by:

$$Sh = \frac{k \cdot l}{\rho \cdot D} \tag{5}$$

where *l* is a characteristic length and ρ is the density of LBE.

3 Results and Discussion

The simulation parameters (T_{lbe}, u_{lbe}, C_{in}, $C_{g/l}$) and calculated results (C_{out}, k, Sh) are summarized in Table 1.

Table 1. Simulated parameters

T_{lbe}/°C	u_{lbe}/m·s−1	C_{in}/wt%	$C_{g/l}$/wt%	C_{out}/wt%	k/kg·m^{-2}·s^{-1}	Sh
550	2	8×10^{-7}	1×10^{-3}	1.30×10^{-5}	1.79	0.26
550	1	8×10^{-7}	1×10^{-3}	1.54×10^{-5}	1.07	0.16
550	1.5	8×10^{-7}	1×10^{-3}	1.32×10^{-5}	1.36	0.20
550	0.5	8×10^{-7}	1×10^{-3}	2.57×10^{-5}	0.92	0.13
550	2	8×10^{-7}	1×10^{-4}	2.00×10^{-6}	1.77	0.26
550	1	8×10^{-7}	1×10^{-4}	2.24×10^{-6}	1.06	0.16
550	1.5	8×10^{-7}	1×10^{-4}	2.02×10^{-6}	1.36	0.20
550	0.5	8×10^{-7}	1×10^{-4}	3.29×10^{-6}	0.93	0.14
550	2	8×10^{-7}	1×10^{-5}	9.05×10^{-7}	1.67	0.24
550	1	8×10^{-7}	1×10^{-5}	9.21×10^{-7}	0.96	0.14
550	1.5	8×10^{-7}	1×10^{-5}	9.07×10^{-7}	1.28	0.19
550	0.5	8×10^{-7}	1×10^{-5}	1.02×10^{-6}	0.86	0.13
550	2	1×10^{-8}	1×10^{-5}	1.22×10^{-7}	1.65	0.24
550	1	1×10^{-8}	1×10^{-5}	1.57×10^{-7}	1.08	0.16
550	1.5	1×10^{-8}	1×10^{-5}	1.34×10^{-7}	1.36	0.20
550	0.5	1×10^{-8}	1×10^{-5}	2.90×10^{-7}	1.04	0.15
500	2	8×10^{-7}	1×10^{-5}	1.72×10^{-6}	1.24	0.27
500	1	8×10^{-7}	1×10^{-5}	1.91×10^{-6}	0.72	0.16
500	0.5	8×10^{-7}	1×10^{-5}	9.63×10^{-7}	0.66	0.14
450	2	8×10^{-7}	1×10^{-5}	1.38×10^{-6}	0.89	0.30
450	1	8×10^{-7}	1×10^{-5}	1.67×10^{-6}	0.52	0.18
450	0.5	8×10^{-7}	1×10^{-5}	9.19×10^{-7}	0.48	0.16
400	2	8×10^{-7}	1×10^{-5}	1.19×10^{-6}	0.61	0.35
400	1	8×10^{-7}	1×10^{-5}	1.48×10^{-6}	0.36	0.21
400	0.5	8×10^{-7}	1×10^{-5}	8.84×10^{-7}	0.34	0.20
350	2	8×10^{-7}	1×10^{-5}	8.24×10^{-7}	0.38	0.41
350	1	8×10^{-7}	1×10^{-5}	8.29×10^{-7}	0.36	0.21
350	0.5	8×10^{-7}	1×10^{-5}	8.56×10^{-7}	0.23	0.24

Figure 3 depicts the distribution of oxygen concentration and velocity under the typical working condition ($T_{lbe} = 550$ °C, $u_{lbe} = 0.5$ m/s, $C_{in} = 8 \times 10^{-7}$ wt%, $C_{g/l} = 1 \times 10^{-5}$ wt%). In the simulated model, the oxygen concentration at gas/liquid interface is considered as a constant 1×10^{-5} wt%. The oxygen concentration inside oxygen transfer device is set as 8×10^{-7} wt% in the initial condition. It can be seen from Fig. 3(a), after the transport of oxygen, the oxygen concentration inside the device increases from inlet to outlet with the flow of LBE. Especially, the oxygen distributed near inlet and outlet is obviously higher as a result of the reversed flow of LBE, as shown in Fig. 3(b).

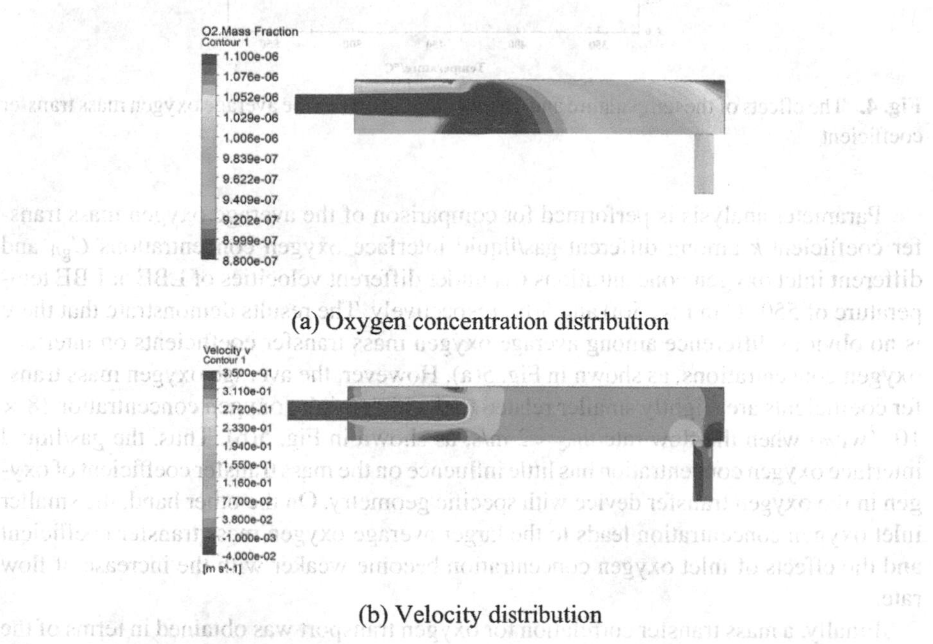

(a) Oxygen concentration distribution

(b) Velocity distribution

Fig. 3. The distribution of oxygen concentration and velocity under the typical working condition ($T_{lbe} = 550$ °C, $u_{lbe} = 0.5$ m/s, $C_{in} = 8 \times 10^{-7}$ wt%, $C_{g/l} = 1 \times 10^{-5}$ wt%)

Figure 4 shows the effects of the temperature and the velocity of LBE on the average oxygen mass transfer coefficient. It can be obtained that the average oxygen mass transfer coefficient increases with the increase of temperature and velocity of LBE. The reason is that the thermophysical properties such as viscosity and diffusion coefficient change with the LBE temperature and the mass transfer of oxygen is improved with the combined effect of convection and diffusion.

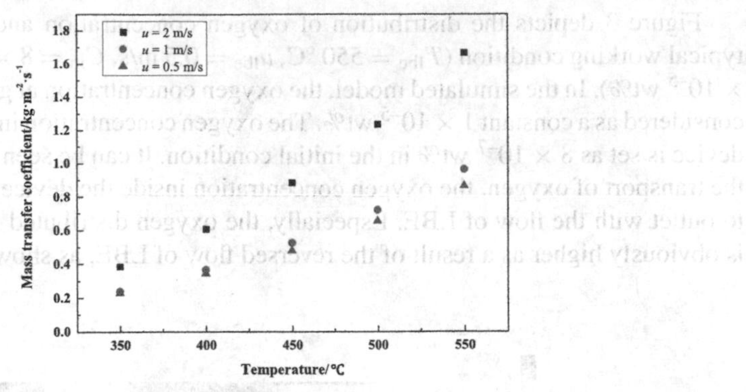

Fig. 4. The effects of the temperature and the velocity of LBE on the average oxygen mass transfer coefficient

Parameter analysis is performed for comparison of the average oxygen mass transfer coefficient k among different gas/liquid interface oxygen concentrations $C_{g/l}$ and different inlet oxygen concentrations C_{in} under different velocities of LBE at LBE temperature of 550 °C in Fig. 5(a) and 5(b), respectively. The results demonstrate that there is no obvious difference among average oxygen mass transfer coefficients on interface oxygen concentrations, as shown in Fig. 5(a). However, the average oxygen mass transfer coefficients are slightly smaller related to the higher inlet oxygen concentration (8×10^{-7}wt%) when the flow rate u_{lbe} <2 m/s, as shown in Fig. 5(b). Thus, the gas/liquid interface oxygen concentration has little influence on the mass transfer coefficient of oxygen in the oxygen transfer device with specific geometry. On the other hand, the smaller inlet oxygen concentration leads to the larger average oxygen mass transfer coefficient and the effects of inlet oxygen concentration become weaker with the increase of flow rate.

Finally, a mass transfer correlation for oxygen transport was obtained in terms of the Sherwood with simulation results, as described by

$$Sh = A \cdot Pe^B \tag{6}$$

The corresponding coefficients A and B are listed in Table 2.

The deviations between simulated results and calculated results by (6) are shown in Fig. 6. It can be noted that the predictions of the correlation equation are all in agreement with the simulated Sh numbers in general with the maximum deviation of ± 26%.

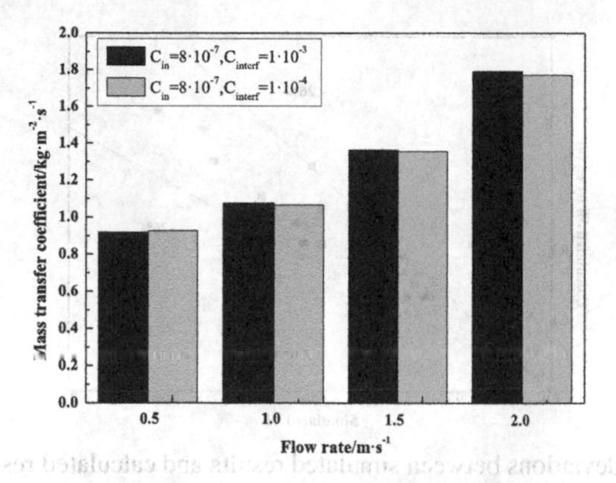

(a) The effects of the gas/liquid interface oxygen concentration on the average oxygen mass transfer coefficient

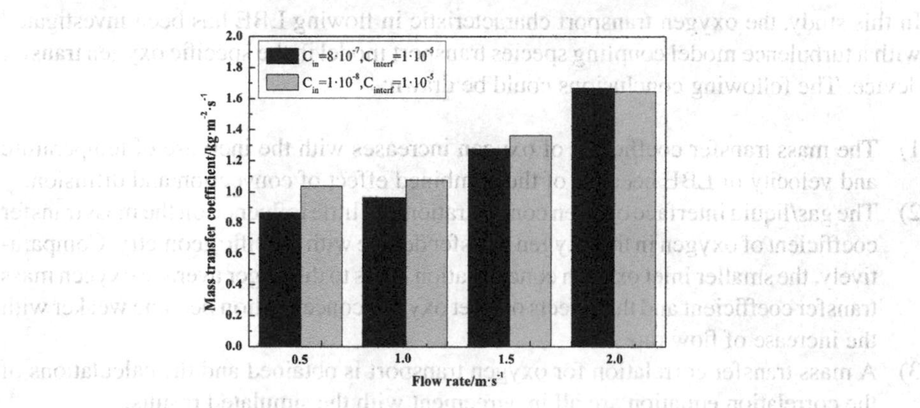

(b) The effects of the inlet oxygen concentration on the average oxygen mass transfer coefficient

Fig. 5. Comparison of the average oxygen mass transfer coefficient k among different gas/liquid interface oxygen concentrations $C_{g/l}$ and different inlet oxygen concentrations C_{in} under different velocities of LBE at temperature of 550 °C

Table 2. Coefficients in (6)

Coefficient	Value
A	0.0169
B	0.321

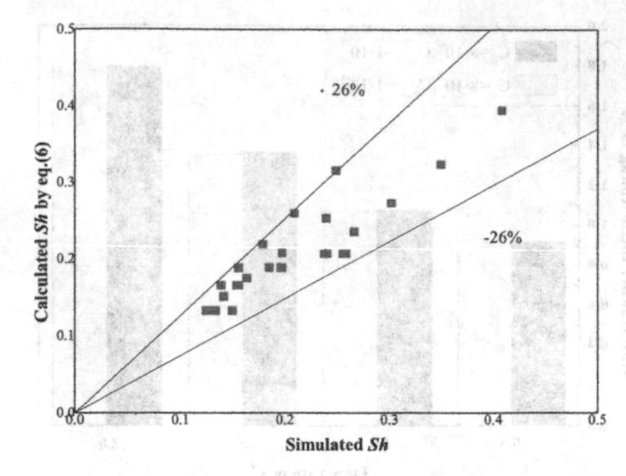

Fig. 6. The deviations between simulated results and calculated results by (6)

4 Conclusions

In this study, the oxygen transport characteristic in flowing LBE has been investigated with a turbulence model coupling species transport model in the specific oxygen transfer device. The following conclusions could be drawn:

1) The mass transfer coefficient of oxygen increases with the increase of temperature and velocity of LBE because of the combined effect of convection and diffusion.
2) The gas/liquid interface oxygen concentration has little influence on the mass transfer coefficient of oxygen in the oxygen transfer device with specific geometry. Comparatively, the smaller inlet oxygen concentration leads to the larger average oxygen mass transfer coefficient and the effects of inlet oxygen concentration become weaker with the increase of flow rate.
3) A mass transfer correlation for oxygen transport is obtained and the calculations of the correlation equation are all in agreement with the simulated results.

References

1. Schroer, C., Wedemeyer, O., Konys, J.: Gas/liquid oxygen-transfer to flowing lead alloys. Nucl. Eng. Des. **241**(5), 1310–1318 (2011)
2. Marino, A., Lim, J., Keijers, S., et al.: A mass transfer correlation for packed bed of lead oxide spheres in flowing lead–bismuth eutectic at high Péclet numbers. Int. J. Heat Mass Transf. **80**, 737–747 (2015)
3. Chen, Y., Chen, H., Zhang, J.: Numerical investigation on enhancement of oxygen transfer by forced convection in liquid lead–bismuth eutectic system. Int. J. Heat Mass Transf. **50**(11–12), 2139–2147 (2007)
4. Müller, G., Heinzel, A., Schumacher, G., et al.: Control of oxygen concentration in liquid lead and lead-bismuth. J. Nucl. Mater. **321**(2–3), 256–262 (2003)
5. Aerts, A., Gavrilov, S., Manfredi, G., et al.: Oxygen-iron interaction in liquid lead-bismuth eutectic alloy. Phys. Chem. Chem. Phys. **18**(29), 19526 (2016)
6. Brissonneau, L., Beauchamp, F., Morier, O., et al.: Oxygen control systems and impurity purification in LBE: learning from DEMETRA project. J. Nucl. Mater. **415**, 348–360 (2011)

7. ANSYS. ANSYS fluent theory guide (2016)
8. OECD, Handbook on Lead-bismuth Eutectic Alloy and Lead Properties, Materials Compatibility, Thermalhydraulics and Technologies. Organisation for Economic Co-Operation and Development (2015)
9. Wang, C., Zhang, Y., Zhang, D., et al.: Numerical study of oxygen transport characteristics in lead-bismuth eutectic for gas-phase oxygen control. Nucl. Eng. Technol. **53**, 2221–2228 (2021)

Research of Steady-State Heat Transfer Performance of Heat Pipe Inside Mobile Heat Pipe Reactor

Huaichang Lu[1,2,3], Tao Zhou[1,2,3](✉), Wenbin Liu[1,2,3], Shang Mao[1,2,3], Dong Wei[1,2,3], Yao Yao[1,2,3], and Tianyu Gao[1,2,3]

[1] Department of Nuclear Science and Technology, School of Energy and Environment, Southeast University, Nanjing 210096, China
101012636@seu.edu.cn

[2] Institute of Nuclear Thermal-Hydraulic Safety and Standardization, Beijing, China

[3] National Engineering Research Center of Power Generation Control and Safety, Nanjing 210096, China

Abstract. High-temperature heat pipe is an important element in a mobile heat pipe reactor, and the study of the steady-state heat transfer performance of the heat pipe is of great value to the design and safe application of the heat pipe. Based on COMSOL software, a three-dimensional heat pipe model is established to study the effects of the input power of the evaporative section of the heat pipe and the horizontal acceleration of the heat pipe due to its movement on the heat transfer performance of the potassium heat pipe in steady-state operation. The results show that the overall temperature of the outer wall surface of the heat pipe and the axial temperature variation of the center of the heat pipe are less affected by the horizontal acceleration and more affected by the input thermal power within the study range; the thermal resistance of the heat pipe decreases with the increase of the input power of the evaporation section of the heat pipe, and shows a trend of decreasing and then increasing with the increase of the horizontal acceleration.

Keywords: Potassium heat pipe · Heating power · Horizontal acceleration · Thermal resistance

1 Introduction

China's total energy consumption is growing, and although the rate of energy consumption growth is decreasing year by year, the overall total energy consumption [1] is still high. China plans to achieve carbon peaking by 2030 and carbon neutrality by 2060. According to Lin Boqiang [2], China's current carbon emissions are mainly concentrated in the power generation, industrial and transportation sectors, and deep decarbonization of the power sector is the key to achieving China's carbon neutrality goal. Nuclear power, as the second largest source of low-carbon electricity in the world after hydropower, will be the optimal choice to solve the problem of fossil energy depletion in the future. Heat

C. Liu (Ed.): PBNC 2022, SPPHY 283, pp. 742–752, 2023.
https://doi.org/10.1007/978-981-99-1023-6_64

pipe reactors are an emerging type of reactor with the advantages of compact structure, high inherent safety, and modularity. Heat pipe reactors were initially proposed for application in space nuclear reactors [3], and heat pipe cooled reactors [4] use a solid state reactor design concept that directly exports heat from the core through high temperature heat pipes and are more suitable as a technology option for small nuclear power sources. Heat pipe reactors do not contain conventional reactor equipment such as main cooling circuits, circulation pumps, and various valves. The system design is extremely simplified compared to conventional reactors, and these features make heat pipe reactors suitable as a technology option for small mobile reactors. The mobile small reactor focus is located in islands, plateaus, polar regions and other special environments or isolated island power supply applications under major disasters, and has a strong national defense and civilian significance.

The change of conditions such as horizontal acceleration due to the movement of the heat pipe small stack has a certain effect on the heat transfer performance of the heat pipe. Tian Z X [5] et al. used a thermal resistance network model and wrote a program to study the heat transfer characteristics of a high-temperature potassium heat pipe with a liquid-absorbing core in steady-state operation, and obtained the effect of the inclination angle on the heat transfer performance of the heat pipe at different heating powers. Liu S Y [6] et al. experimentally investigated the comparison of the steady-state operating performance of loop heat pipes in an accelerated environment to provide some guidance on the design of loop heat pipes for cooling airborne electronic equipment. Xiao Lv [7] et al. studied the temperature oscillations of a double-compensated cavity loop heat pipe under acceleration conditions, and analyzed and discussed the effects of different loading modes, thermal loads, acceleration directions and amplitudes, and other control parameters on the loop temperature oscillations. Hao S [8] et al. experimentally studied the thermal behavior of a horizontal high temperature heat pipe under motion conditions. The current studies focus on the heat transfer performance during conventional heat pipe start-up and heat pipe steady-state operation, and there are still relatively few studies on the heat transfer performance of heat pipes in mobile heat pipe reactors. Therefore, the study of heat transfer performance of heat pipes in mobile heat pipe reactors under different operating conditions can provide a certain theoretical basis for the design and application of mobile heat pipe reactors.

2 Research Object

2.1 Geometric Model

The heat pipe used is shown in Fig. 1.

It can be seen from the Fig. 1 that the heat pipe from below to above is the evaporation section, adiabatic section and condensing section.

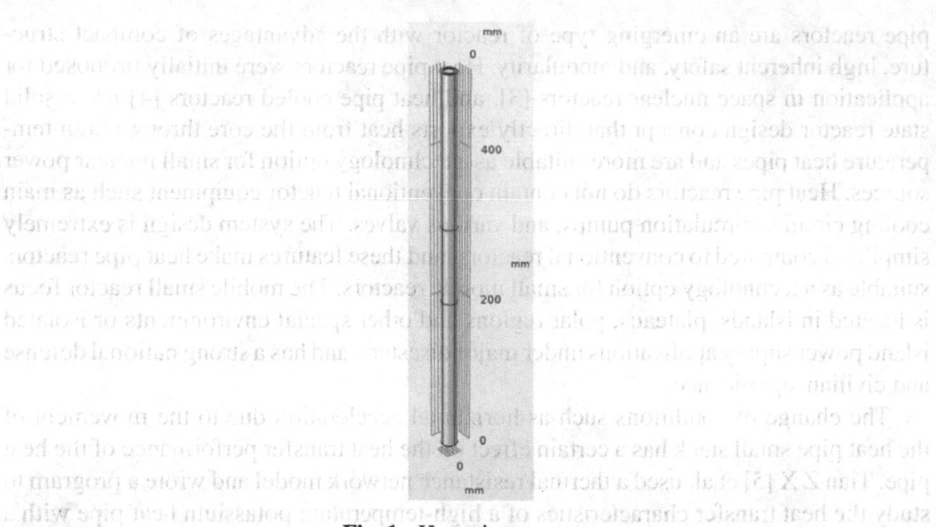

Fig. 1. Heat pipe structure

2.2 Calculation Parameters

Heat pipe structure parameters and calculation parameters are shown in Table 1.

Table 1. Calculated parameters of heat pipe

Name	Unit	Quantity
Working medium in heat pipe	-	Potassium
Evaporation section length	mm	200
Condensation section length	mm	200
Adiabatic section length	mm	100
Heat pipe outer diameter	mm	20
Pipe wall thickness	mm	2
Wick thickness	mm	2
Convective heat transfer coefficient	W/(m^2·K)	300
Environmental temperature	K	300
Heating power of evaporation section	W	150/250/350/450
Pipe wall material	-	Haynes 230

2.3 Grid Irrelevance Verification

In the simulation process, in order to ensure the accuracy of the model and at the same time improve the efficiency of the calculation, the mesh will be verified for irrelevance.

Under the premise of ensuring the mesh quality, four different cell size meshes are used to divide the model, and the mesh numbers are 151107, 222363, 277810, and 378839, respectively. the results of the axial temperature distribution at the center of the heat pipe in steady state with the four mesh numbers are shown in Fig. 2.

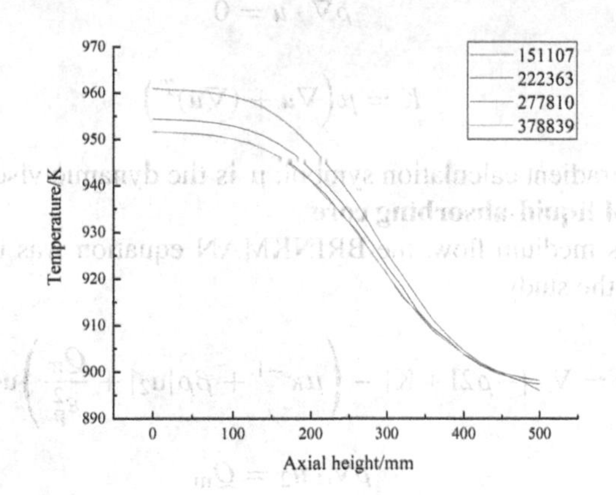

Fig. 2. Axial temperature distribution in the center of the heat pipe at steady state

From Fig. 2, it can be seen that the change of axial temperature distribution in the center of heat pipe tends to be stable with the increase of the number of grids, and the temperature difference of evaporating section of heat pipe is larger when the number of grids is 151107 and 222363, while the trend of axial temperature distribution in the center of heat pipe is almost the same when the number of grids is 277810 and 378839. Within the margin of error, 277810 meshes are appropriate considering that too many meshes will reduce the calculation efficiency.

3 Calculation Formula

(1) **Thermal conductivity model of pipe wall**

$$\rho C_{\mathrm{p}} \mathrm{u} \cdot \nabla T + \nabla \cdot q = Q \tag{1}$$

$$q = -k \nabla T \tag{2}$$

ρ is the density, kg/m^3; C_{p} is the constant pressure heat capacity, J/(kg·K); K is the effective thermal conductivity, W/(m·K). q_{v} is the heat source, W/m^3.

(2) **Thermal conductivity equation of liquid-absorbing core**

$$\rho_{\mathrm{f}} C_{\mathrm{p,f}} \mathrm{u} \cdot \nabla T + \nabla \cdot q = Q \tag{3}$$

$$q = -k_{\mathrm{eff}} \nabla T \tag{4}$$

$(\rho C_{\mathrm{p}})_{\mathrm{eff}}$ is the effective volumetric heat capacity at constant pressure, J/(m^3·K); k_{eff} is the effective thermal conductivity, W/(m-K).

(3) **Vapor flow model**

$$\rho(u \cdot \nabla)u = \nabla \cdot [-\rho \mathrm{I} + \mathrm{K}] + F \tag{5}$$

$$\rho \nabla \cdot u = 0 \tag{6}$$

$$K = \mu\left(\nabla u + (\nabla u)^T\right) \tag{7}$$

$\bigtriangledown$ is the gradient calculation symbol; μ is the dynamic viscosity, Pa-s.

(4) **Flow model of liquid-absorbing core**

For porous medium flow, the BRINKMAN equation was used as a coupling calculation in the study

$$0 = \nabla \cdot [-\rho 2\mathrm{I} + \mathrm{K}] - \left(\mu\kappa^{-1} + \beta\rho|\mathrm{u}_2| + \frac{Q_\mathrm{m}}{\varepsilon_\mathrm{p}^2}\right)\mathrm{u}_2 + F \tag{8}$$

$$\rho \nabla \cdot u_2 = Q_\mathrm{m} \tag{9}$$

where, ρ is the liquid density, kg/m^3; Q_m is the flow rate, kg/s; g is the acceleration of gravity, m/s^2; $\bigtriangledown$ is the Laplace operator.

(5) **Equivalent thermal resistance of heat pipe**

The heat pipe heat transfer characteristics are studied using the form of calculating the overall thermal resistance of the heat pipe with the following expressions.

$$R_\mathrm{sum} = \frac{T_\mathrm{e,ave} - T_\mathrm{c,ave}}{Q_\mathrm{in}} \tag{10}$$

R_sum is the overall thermal resistance of the gravity heat pipe, W/K; $T_\mathrm{e,ave}$ is the average temperature of the wall of the evaporating section of the heat pipe, K; $T_\mathrm{c,ave}$ is the average temperature of the wall of the condensing section of the heat pipe, K; Q_in is the thermal power input to the evaporating section of the heat pipe, W.

4 Results and Discussion

4.1 Effect of Horizontal Acceleration on Heat Pipe Temperature

(1) **Surface temperature of outer wall**

The temperature distribution on the outer wall surface of the heat pipe when the horizontal acceleration is changed under the heat pipe evaporation section input thermal power of 450 W is shown in Fig. 3.

From Fig. 3, it can be seen that the temperature distribution of the outer wall surface of the heat pipe under different horizontal acceleration conditions has basically the same trend, and the heat pipe is in a stable working condition. The temperature change of the

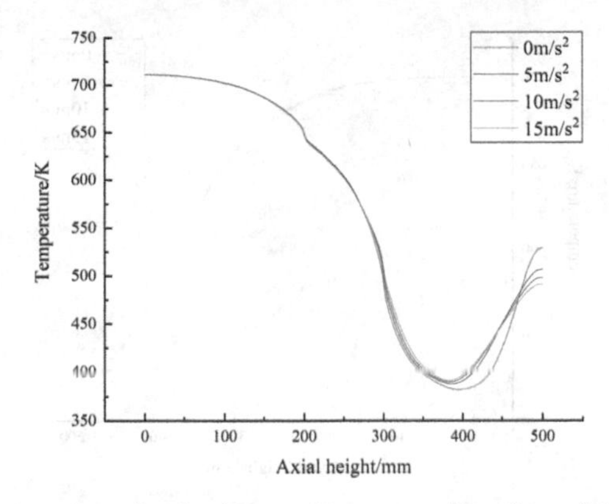

Fig. 3. Temperature distribution on the outer wall surface of the heat pipe

outer wall surface of the heat pipe shows a trend of decreasing first and then increasing. In the evaporation and adiabatic sections of the heat pipe, the change of horizontal acceleration almost does not affect the temperature of the outer wall surface, and the temperature of the outer wall surface gradually decreases along the axial direction. At the end of the condensing section of the heat pipe, the temperature of the outer wall surface increases, while the temperature increase at the end of the condensing section decreases with the increase of the horizontal acceleration, and the uniformity of the heat pipe increases. This is because with the operation of the heat pipe, the evaporating section of the heat pipe is constantly fed with thermal power, and a large amount of superheated steam gathers at the top of the steam chamber of the heat pipe, and the steam entering the suction core condenses into small liquid beads or liquid film adsorbed on the wall of the top of the heat pipe and exothermic. In addition, because the liquid beads or liquid film by gravity decline, the thickness of the liquid film from the top of the heat pipe to the bottom gradually increase, the upper part is thinner, the lower part is thicker, there is a difference in thermal resistance, making the temperature of the outer wall surface appears to rise. And with the increase of horizontal acceleration, it is equivalent to the heat pipe working under a certain inclination angle, which makes the instability of liquid phase workpiece flow inside the suction core reduced, thus offsetting part of the temperature increase of the outer wall surface.

(2) **Central axial temperature**

The axial temperature distribution at the center of the heat pipe when the horizontal acceleration is changed under the input thermal power of 450 W in the evaporation section of the heat pipe is shown in Fig. 4.

From Fig. 4, it can be seen that the trend of axial temperature distribution in the center of the heat pipe under different horizontal acceleration conditions is basically the same, and the heat pipe is in steady state. At this time, the heat pipe shows good homogeneity in the evaporation section, the adiabatic section and the beginning of the condensation section, and the temperature decreases faster at the end of the condensation section of the heat pipe, and finally stabilizes. With the increase of horizontal acceleration, the

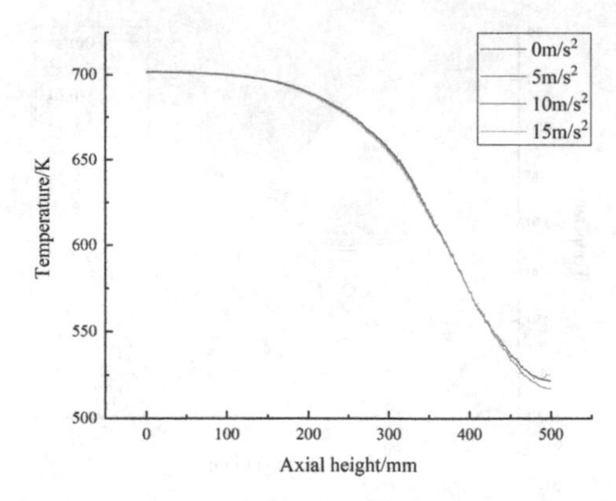

Fig. 4. Axial temperature distribution in the center of the heat pipe

temperature at the end of the condensing section increases, which indicates that the heat transfer performance of the heat pipe is improved.

4.2 Effect of Horizontal Acceleration on Heat Transfer Performance of Heat Pipes

In order to compare the heat transfer capability of potassium heat pipe easily, the equivalent thermal resistance is used to describe the heat transfer performance of the heat pipe, which can be given by Eq. (10). The trend of the equivalent thermal resistance of the heat pipe with increasing horizontal acceleration is given in Fig. 5.

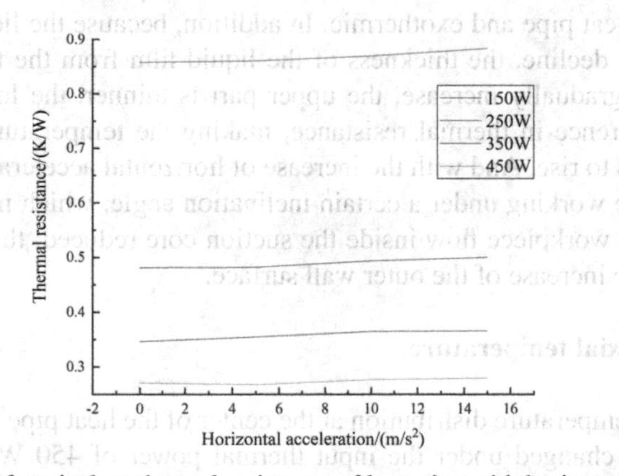

Fig. 5. Variation of equivalent thermal resistance of heat pipe with horizontal acceleration

From Fig. 5, it can be seen that the equivalent thermal resistance of the heat pipe under different input thermal power has basically the same trend, and the equivalent thermal resistance shows a trend of first decreasing and then slightly increasing with the increase of horizontal acceleration. Under the same horizontal acceleration, the equivalent thermal resistance of the heat pipe decreases with the increase of the input thermal power.

The minimum thermal resistance of the heat pipe occurs between 5 m/s^2 and 10 m/s^2 acceleration. When the heat pipe is in a smaller horizontal acceleration condition, the heat pipe moves with the heat pipe reactor, the effect of gravitational acceleration is greater than the effect of horizontal acceleration, and the heat pipe as a whole shows the form of vertical heat transfer; when the horizontal acceleration exceeds the gravitational acceleration, the effect of gravitational acceleration on the heat transfer of the mass in the heat pipe decreases, and the heat pipe gradually changes from vertical heat transfer to horizontal heat transfer, the equivalent thermal resistance increases, and the heat transfer performance of the heat pipe decreases.

4.3 Influence of Input Thermal Power on Heat Pipe Temperature

(1) **Surface temperature of outer wall**

The distribution of the temperature on the outer wall surface of the heat pipe when the input thermal power of the evaporating section of the heat pipe is changed under no horizontal acceleration is shown in Fig. 6.

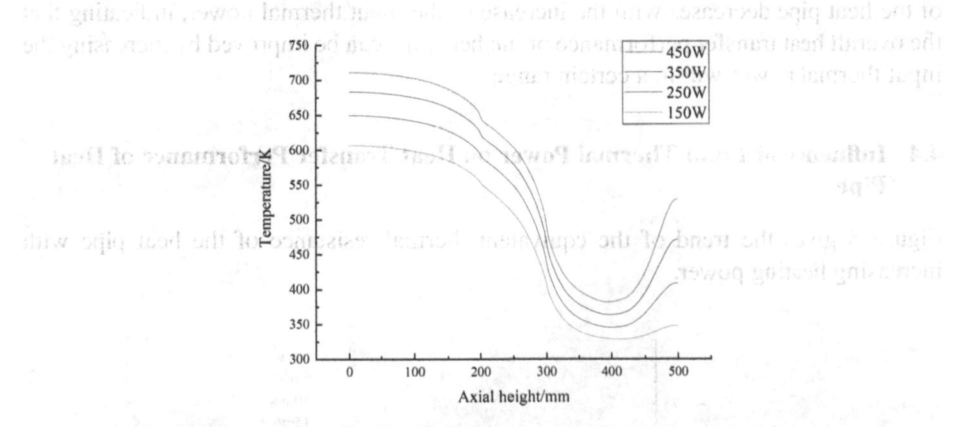

Fig. 6. Temperature distribution on the outer wall surface of the heat pipe

From Fig. 6, it can be seen that the temperature distribution of the outer wall surface of the heat pipe under different input thermal power conditions is basically the same, showing a trend of first decreasing and then increasing. The evaporation and adiabatic sections of the heat pipe show good temperature homogeneity, and the end of the condensing section also shows the phenomenon of temperature increase, and the degree of temperature increase increases with the increase of the input thermal power.

(2) **Central axial temperature**

The distribution of the axial temperature at the center of the heat pipe when the input thermal power of the evaporating section of the heat pipe is changed without horizontal acceleration is shown in Fig. 7.

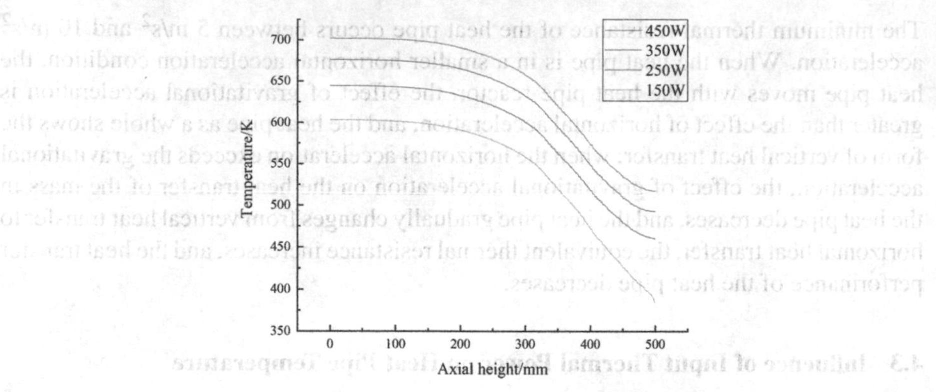

Fig. 7. Axial temperature distribution in the center of the heat pipe

From Fig. 7, it can be seen that the trend of axial temperature change in the center of the heat pipe under different input thermal power conditions is basically the same. The heat pipe is in steady-state operation, and the axial temperature difference at the center of the heat pipe decreases with the increase of the input thermal power, indicating that the overall heat transfer performance of the heat pipe can be improved by increasing the input thermal power within a certain range.

4.4 Influence of Input Thermal Power on Heat Transfer Performance of Heat Pipe

Figure 8 gives the trend of the equivalent thermal resistance of the heat pipe with increasing heating power.

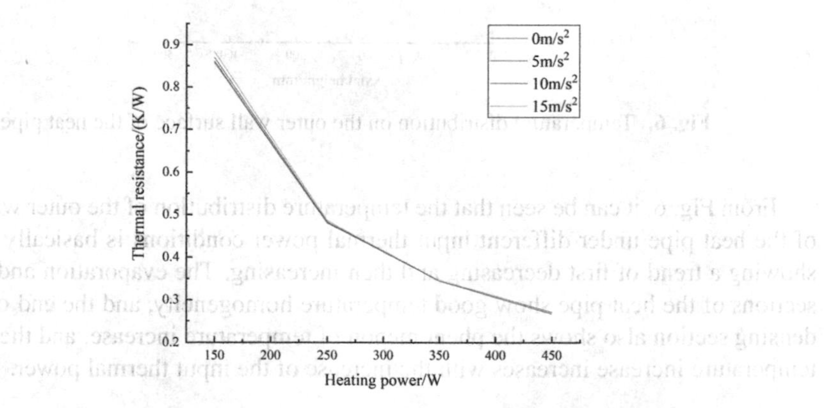

Fig. 8. Variation of equivalent thermal resistance of heat pipe with heating power

From Fig 8, it can be seen that the trend of equivalent thermal resistance of heat pipe under different horizontal acceleration conditions is basically the same. The equivalent thermal resistance shows a decreasing trend with the increase of heating power. In the studied heating power range, when the heating power increases, the evaporation in the heat pipe evaporation section intensifies, the thermal resistance decreases, and the overall heat transfer performance of the heat pipe is improved. With the increase of heating power, the degree of thermal resistance affected by heating power gradually decreases.

5 Conclusion

Based on the multi-physics field simulation software COMSOL Multiphysics, the heat transfer characteristics of the steady-state operation of heat pipes in a mobile heat pipe stack are investigated, and the effects of horizontal acceleration and variation of input thermal power on the heat pipe center temperature, wall temperature and equivalent thermal resistance are obtained.

(1) The overall temperature variation of the outer wall surface of the heat pipe is less influenced by horizontal acceleration and more influenced by the input thermal power in the study range. The wall surface temperature increases with the increase of the input thermal power. The temperature increase existing at the end of the condensing section is suppressed by the increase in acceleration and the decrease in input thermal power.
(2) The axial temperature variation at the center of the heat pipe is less affected by the horizontal acceleration and more affected by the input thermal power in the studied range. The wall surface temperature increases with the increase of input thermal power.
(3) The equivalent thermal resistance of the heat pipe shows a trend of decreasing and then increasing with the increase of horizontal acceleration in the study range, and decreases with the increase of input thermal power, and the effect of the change of acceleration on the thermal resistance becomes smaller with the increase of input thermal power.

Acknowledgements. This project was supported by "Inherently Safe Modular Metal Dispersion Heat Pipe Reactor Design Technology", 2020YFB1901701.The authors also thank the support of "All-in-one mobile nuclear power supply overall system design technology", 2020YFB1901703 and "Jiangsu Province Dual Innovation Talent Funding", JSSCRC2021500.

References

1. Jiang, H.L., Liu, Y.Q., Feng, Y.M., Zhou, B.Z., Li, Y.X.: Analysis of power generation technology trends in the context of carbon peaking and carbon neutrality in the 14th five-year plan period. Power Gener. Technol. **43**(01), 54–64 (2022). (in Chinese)
2. Lin, B.Q.: China's difficulties and the way forward towards carbon neutrality. New Finan. **07**, 26–29 (2021). (in Chinese)
3. Yan, B.H., Wang, C., Li, L.G.: The technology of micro heat pipe cooled reactor: a review. Ann. Nucl. Energy **135**, 106948 (2020)

4. Grover, G.M., Cotter, T.P., Erickson, G.F.: Structures of very high thermal conductance. J. Appl. Phys. **35**(10), 3072–3072 (1964)
5. Tian, Z.X., Liu, Y., Wang, C.L., Su, G.H., Tian, W.X., Qiu, S.Z.: Study on heat transfer characteristics of high-temperature potassium heat pipe in steady-state operation. At. Energy Sci. Technol. **54**(10), 1771–1778 (2020). (in Chinese)
6. Liu, S.Y., Xie, Y.Q., Su, J., Zhang, H.X., Li, G.G.: Comparative study of steady-state operating performance of loop heat pipe in accelerated environment. J. Aeronaut. 1–12, 28 June 2022. (in Chinese)
7. Lv, X., Xie, Y., Zhang, H., et al.: Temperature oscillation of a dual compensation chamber loop heat pipe under acceleration conditions. Appl. Therm. Eng. **198**(3), 117450 (2021)
8. Sun, H., Liu, X., Liao, H., et al.: Experiment study on thermal behavior of a horizontal high-temperature heat pipe under motion conditions. Ann. Nucl. Energy **165**, 108760 (2022)

Direct Contact Condensation Characteristics of Steam Injection into Cold-Water Pipe Under Rolling Condition

Zhiwei Wang[1], Yanping He[1(✉)], Zhongdi Duan[1(✉)], Chao Huang[1], and Shiwen Liu[2]

[1] State Key Laboratory of Ocean Engineering, Shanghai Jiao Tong University, Shanghai, China
{hyp110,duanzhongdi}@sjtu.edu.cn

[2] Science and Technology on Reactor System Design Technology Laboratory, Nuclear Power Institute of China, Chengdu, China

Abstract. Direct contact condensation (DCC) is widely occurred in nuclear power systems and leads to undesired phenomena such as condensation-induced water hammer. For ocean nuclear power ships, DCC is inevitable in the passive heat removal system and influenced by sea conditions. In this paper, the characteristics of DCC under rolling conditions are analyzed. The numerical model of DCC is established based on computational fluid dynamics approach. The VOF model, SST k–ω turbulence model and the additional inertia force model are incorporated to describe the liquid-gas two-phase flow under the rolling motion. The condensation model based on surface renewal theory (SRT) is used to simulate steam-water DCC phenomenon. The simulation results are compared with the experimental data and show reasonable agreement. The effects of rolling motion on DCC for steam injection into a horizontal pipe filled with cold water are numerically investigated. The results show that the additional inertial forces and the average condensation rate increase with the increase of the rolling angle and frequency. The reverse flow of the seawater induced by rolling motion leads to the accumulation of the steam at the lower part of the pipe, resulting in a large pressure pulse. With the increase of rolling angle and frequency, the pressure pulse induced by DCC increases.

Keywords: Ocean nuclear power ships · Direct contact condensation · Rolling motion · Average condensation rate · Pressure pulse

1 Introduction

For the ocean nuclear power ships, the direct contact condensation (DCC) phenomenon is inevitably occurred due to the efficient heat transfer characteristics [1, 2]. In addition, when the nuclear power system is working in the marine environment, it will inevitably be affected by the ocean environment and produce a series of motions, such as heaving, pitching and rolling [3, 4]. These motions will affect the two-phase flow and heat transfer characteristics in the cooling systems [5]. Therefore, the research of the transient

C. Liu (Ed.): PBNC 2022, SPPHY 283, pp. 753–763, 2023.
https://doi.org/10.1007/978-981-99-1023-6_65

behaviors of the DCC phenomenon under ocean conditions is of great significance for the application of the ocean nuclear power ships.

At present, many experimental and numerical simulation researches have been conducted to focus on the DCC under static conditions [6]. Prasser et al. (2008) experimentally observed the DCC events induced by steam-water counterflow in the horizontal pipe, and focused on the number of pressure pulse and formation of steam slug [7]. Chong et al. (2020) experimentally studied the DCC phenomenon induced by steam discharged into a horizontal pipe and focused on the condensation induced water hammer induced phenomenon by the DCC [8]. Sun et al. (2020) experimentally observed the DCC phenomenon in the passive heat removal system for offshore application and analyzed the formation mechanism [9]. Wang et al. (2022) simulated the DCC phenomenon in the horizontal pipe by the ANSYS FLUENT software, and captured the formation process of DCC phenomenon by the VOF method [10].

The typical motion of the floating structures under ocean conditions, such as rolling motion, will generate various additional forces and further affect the coolant flow and heat and mass transfer. Peng et al. (2020) numerically studied the effect of rolling motion on the void distribution of subcooled flow boiling. The results indicated that the void fraction distribution is sensitive to the rolling period [11]. Wang et al. (2022) numerically investigated the flow characteristics of gas-liquid two-phase flow under rolling condition. They found that the void fraction presents periodical variation and will induce complex secondary-flow phenomenon [12]. Chen et al. (2022) numerically studied the dominant oscillation frequency of the unstable steam jet under rolling condition and indicated that the Coriolis force has a great influence on the dominant oscillation frequency [13]. Therefore, the effects of rolling motion on boiling two-phase flow and gas-liquid two-phase flow have attracted enough attention from relevant researchers.

Based on the literature review, the present study aims to study the transient behaviors of the direct contact condensation (DCC) phenomenon under rolling conditions. The DCC numerical model frame was established based on the volume of fluid (VOF) model, the condensation model and the rolling motion model. The additional inertial forces, formation process, average condensation rate, and pressure behaviors were obtained to study the effects of rolling motion on the DCC phenomenon.

2 Mathematic Model

2.1 Two-Phase Flow Mode

The direct contact condensation usually occurs at the steam-subcooled water interface, so obtaining a clear steam-subcooled water interface is the key to the numerical simulation. Previous studies indicated that the VOF model has significant advantages in tracking the gas-liquid interface [14]. Hence, the VOF model is used to describe the steam-subcooled water two-phase flow in the present study. The tracking of the steam-subcooled water interface is by solving the continuity equation for the volume fraction of each phase. The conservation of mass as follow:

$$\frac{\partial(\alpha_v \rho_v)}{\partial t} + \nabla \cdot \left(\alpha_v \rho_v \vec{u}\right) = S_m \quad (1)$$

There exists a closed equation for the volume fraction of gas phase and liquid phase:

$$\alpha_v + \alpha_l = 1 \tag{2}$$

In the VOF model, the gas and liquid phase share a set of momentum equations. The surface tension and the additional inertial forces are added to the momentum equation in the form of source terms. The momentum equation as follow:

$$\frac{\partial(\rho \vec{u})}{\partial t} + \nabla \cdot (\rho \vec{u} \vec{u}) = -\nabla P + \nabla\left[\mu\left(\nabla \vec{u} + (\nabla \vec{u})^T\right)\right] + \rho \vec{g} + \vec{F}_{st} + \vec{F}_{roll} \tag{3}$$

Similarly, the gas and liquid phase share a set of momentum equation, as follow:

$$\frac{\partial}{\partial t}(\rho E) + \nabla \cdot (\vec{v}(\rho E + P)) = \nabla\left(k_{eff} \nabla T - \sum_j h_j \vec{J}_j + \left(\overline{\overline{\tau}}_{eff} \cdot \vec{v}\right)\right) + Q_m \tag{4}$$

where: the Q_m represents the energy source term and is modeled by the UDF, the specific form refers to the Sect. 2.2.

In the present study, the SST k − ω turbulence model is used to compute the turbulence characteristics induced by the rolling motion [15]. The governing equations as follow:

The equation of turbulent kinetic energy:

$$\frac{\partial(\rho k)}{\partial t} + \frac{\partial}{\partial x_j}(\rho k u_j) = \frac{\partial}{\partial x_j}\left[\Gamma_k \frac{\partial k}{\partial x_j}\right] + G_k - Y_k + S_k \tag{5}$$

The equation of specific dissipation rate:

$$\frac{\partial(\rho \omega)}{\partial t} + \frac{\partial}{\partial x_j}(\rho \omega u_j) = \frac{\partial}{\partial x_j}\left[\Gamma_\omega \frac{\partial \omega}{\partial x_j}\right] + G_\omega - Y_\omega + D_\omega + S_\omega \tag{6}$$

where: the G_k is the turbulent kinetic energy term due to velocity gradient, Γ_k and Γ_ω represent the convective term of k and ω, Y_k and Y_ω represent the effective diffusion term of k and ω induced by turbulence, Y_k represents the cross convective term. S_k and S_ω represent the user-defined source terms.

2.2 Condensation Phase Change Model

During the process of condensation induced water hammer, only the condensation phenomenon occurs. It is assumed that condensation occurs only at the steam-subcooled water interface. The heat and mass transfer during the condensation is calculated by the UDF, and then added to the momentum and energy conservation equation [16]. The energy source term represents the heat transfer from the steam phase to the water phase at the interface, which is expressed as:

$$Q_m = HTC \cdot a_{int} \cdot (T_{sat} - T_l) \tag{7}$$

where: the *HTC* is the heat transfer coefficient, a_{int} is the interfacial area density, T_{sat} is the saturated temperature, T_l is the subcooled water temperature.

For the VOF model, the absolute value of volume fraction gradient is defined as the interfacial area density [17], which is expressed as:

$$a_{int} = |\nabla \alpha_l| = \frac{\partial \alpha_l}{\partial n} \tag{8}$$

where: α_l is the liquid volume fraction, n is the unit vectors.

Mass transfer source can be obtained on the basis of the energy transfer source, which is expressed as:

$$S_m = \frac{Q_m}{h_{lv}} \tag{9}$$

where: h_{lv} is the latent heat with a constant value of 1026 kJ/ kg

2.3 Rolling Motion Model

In the present study, the additional inertial force method is utilized to calculate the effects of rolling motion. Figure 1 shows the relationship between the non-inertial frame and inertial frame [18]. The additional inertial forces mainly include the centripetal force (F_{ce}), tangential force (F_{ta}), and Coriolis force (F_{co}).

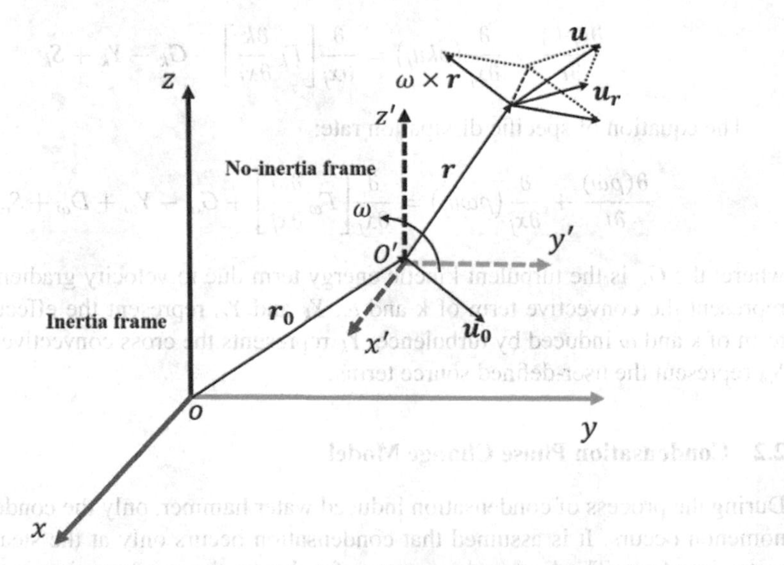

Fig. 1. Relationship between the non-inertial frame and the inertial frame.

The additional inertial force method involves two reference frames, an inertial frame and a non-inertial frame. The geometric model moves relative to the inertial frame, and we only considered the rolling motion in the YOZ plane. Hence, the equations of rolling motion are as follows [30]:

$$\theta(t) = \theta_m \sin\left(\frac{2\pi t}{T}\right)\vec{i} \tag{10}$$

$$\omega(t) = \frac{2\pi}{T}\theta_m \cos\left(\frac{2\pi t}{T}\right)\vec{i} \tag{11}$$

$$\varepsilon(t) = -\frac{4\pi^2}{T^2}\theta_m \sin\left(\frac{2\pi t}{T}\right)\vec{i} \tag{12}$$

In conclusion, the additional inertial force can be calculated as follows:

$$\vec{F}_{roll} = -\rho(\vec{g} + \vec{\varepsilon}(t) \times \vec{r} + \vec{\omega}(t) \times (\vec{\omega}(t) \times \vec{r}) + 2\vec{\omega}(t) \times \vec{u}_r) \tag{13}$$

2.4 Defined Geometry and Mesh Generation

For marine floating structures, the nuclear reactor is usually arranged in the reactor room. The DCC events mainly occurs in the upper horizontal pipe, as depicted in Fig. 2 (a). In addition, the length and inner diameter of the geometric model is 2.0 m and 0.1 m respectively [19]. In order to ensure the calculation convergence, the structured grid was generated in the ANSYS ICEM software, and the boundary layer was densified, as shown in Fig. 2 (b).

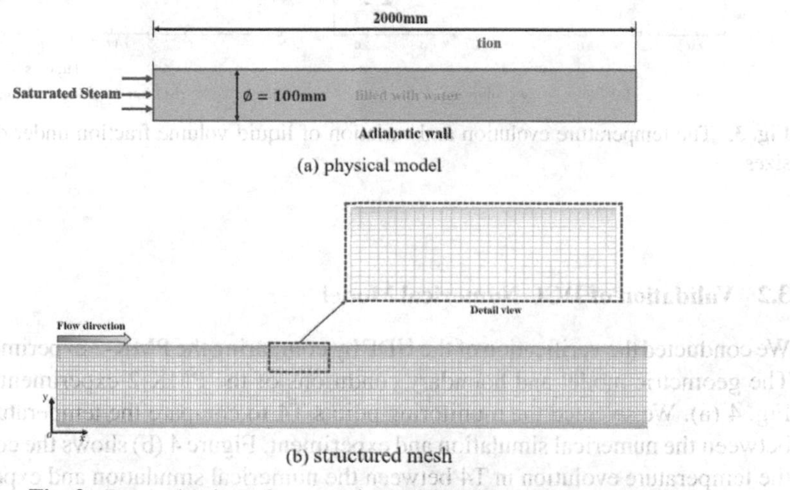

Fig. 2. Determination of geometric model and mesh generated

In this paper, the commercial CFD software Ansys Fluent 2020 has been utilized to conduct the numerical simulation. The numerical simulations adopt the pressure-based solver. The pressure- velocity coupling scheme adopts the pressure-implicit with splitting of operators (PISO). The QUICK discretization scheme is applied to the momentum and energy equations [20]. Geo-Reconstruct scheme is applied to the spatial discretization for volume fraction.

3 Results and Discussion

3.1 Mesh Independence Test

The mesh independence test was conducted with three different mesh sizes in the present study. The minimum size of the three meshes were selected 1 mm, 0.5 mm and 0.25 mm, respectively. Then, the temperature evolution and the variation of liquid volume fraction under different mesh sizes were analyzed. As depicted in Fig. 3 (a), with the decrease of the minimum size, the trend of the temperature evolution is gradually consistent. The mesh sizes also affect the variation of the liquid volume fraction. With the mesh size decreased from 0.5 mm to 0.25 mm, the variation trend of the liquid volume fraction is gradually consistent, as shown in Fig. 3 (b). Therefore, the subsequent numerical simulations are based on the mesh with the minimum size 0.5 mm.

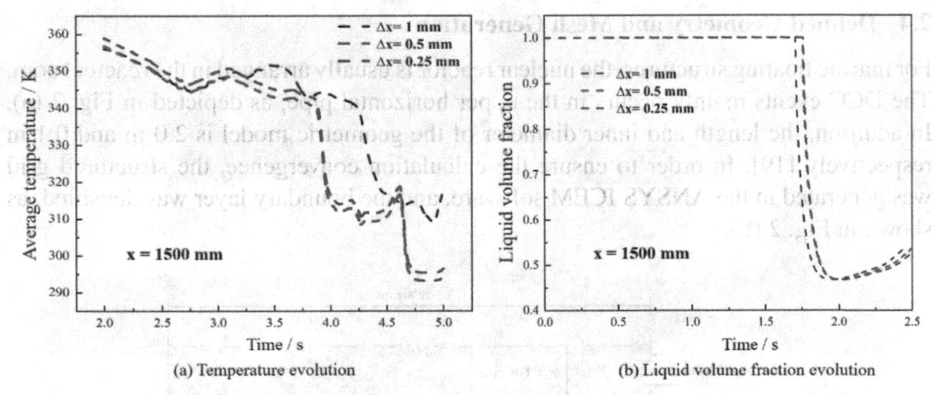

Fig. 3. The temperature evolution and variation of liquid volume fraction under different mesh sizes

3.2 Validation of DCC Numerical Model

We conducted the verification of the UDF by comparing the PMK-2 experimental results. The geometric model and boundary conditions of the PMK-2 experiment is shown in Fig. 4 (a). We selected the monitoring points T4 to compare the temperature evolution between the numerical simulation and experiment. Figure 4 (b) shows the comparison of the temperature evolution in T4 between the numerical simulation and experiment. The results indicated that the temperature step time obtained by the numerical simulation is basically consistent with the experiment. In summary, the present condensation model can effectively simulate the condensation behaviors and provide a guarantee for the subsequent numerical simulation of the DCC events.

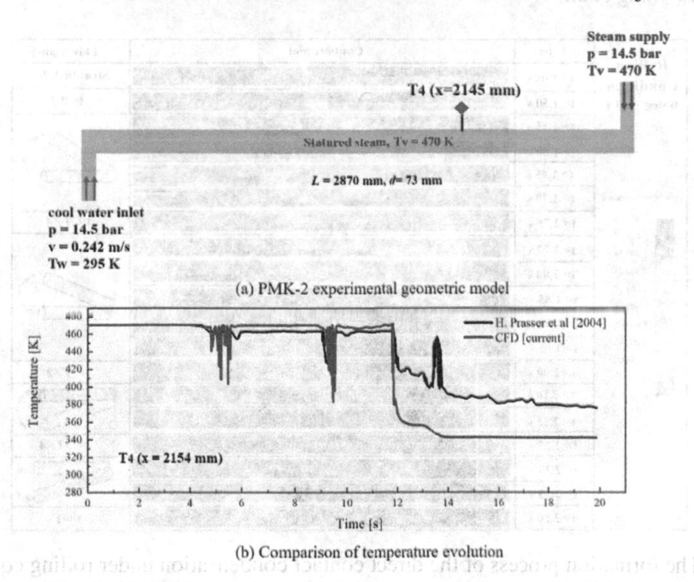

(a) PMK-2 experimental geometric model

(b) Comparison of temperature evolution

Fig. 4. PMK-2 experimental facility and temperature validation

3.3 DCC Process Under Rolling Conditions

Figure 5 shows the formation process of the CIWH phenomenon under rolling condition. Similarly, the stratified-wave flow is formed in the horizontal pipe at the early stage. From 0 to T/4, the additional inertial force accelerates the reverse flow of the subcooled water. During the process of the reverse flow, the saturated steam is continuously squeezed down the pipe until the isolated steam bubble is formed. From T/4 to T/2, the isolated steam bubble is surrounded by the subcooled water, and the steam bubble collapse quickly due to the DCC. At T/2, the condensation induced water hammer occurs. After the CIWH event, the horizontal pipe is filled with water and has a larger subcooling. Hence, from T/2 to T, the saturated steam flowing into the pipe is rapidly condensed, and the formation mechanism of the CIWH disappears. With the temperature field in the horizontal pipe increases, the saturated steam can flow into again, which in turn triggers a new CIWH event. In summary, the formation mechanism of the CIWH under static and rolling condition is basically the same. While, it is worth noting that the rolling motion complicates the reverse flow of the subcooled water and the DCC phenomenon.

3.4 Effect of Rolling Motion on Condensation Rate

The condensation rate is a key parameter to describe the DCC phenomenon. Thus, the effects of rolling motion on the condensation rate are numerically studied. The condensation rate is calculated by the UDF and recorded by the UDM function in the present study. As depicted in Fig. 6 (a) and (b), compared to the static condition, the average condensation rate under rolling conditions has a remarkable increase. The average condensation rate has a tendency to decrease as the rolling period increased, and has a tendency to increase with the rolling angle increased. This is mainly due to the larger

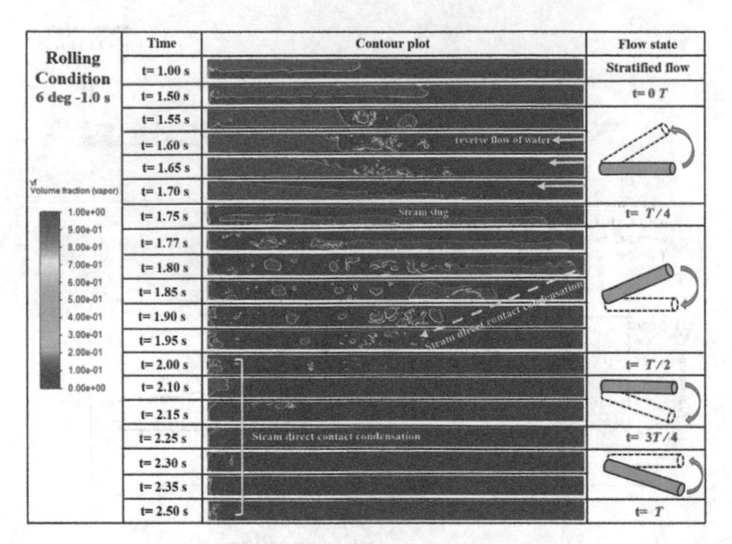

Fig. 5. The formation process of the direct contact condensation under rolling conditions.

rolling angle and high rolling frequency enhanced the contact area between the steam and subcooled water, which in turn improved the capacity of heat and mass transfer at the steam- subcooled water interface. In addition, the average condensation rate is proportional to the steam inlet velocity, as depicted in Fig. 6 (c) and (d). This is mainly due to the increase of the steam inlet velocity increased the amount of steam in the pipe, and then increased the average condensation rate.

3.5 Effect of Rolling Motion on Pressure Behaviors

Figure 7 shows the comparison of the pressure behaviors induced by the DCC events under static condition and rolling conditions. The main pressure behaviors observed in the time domain signals are as flows: (1) the pressure peak under rolling conditions is obviously greater than that under static condition. This is mainly due to the larger condensation rate and reverse flow of subcooled water under rolling condition aggravated the pressure peak. (2) the pressure behaviors under rolling conditions are more complex, which present periodic fluctuations after generating the violent pressure peak. This is mainly due to the rapid formation and collapse of the isolated steam bubbles in the early stage of the rolling motion. Then the subcooled water filled with the pipe, and the saturated steam is completely condensed at the pipe inlet due to the larger subcooling. Therefore, the periodic fluctuation is mainly induced by the subcooled water flow under the rolling motion. (3) the pressure peak decreased with the rolling period increase and increased with the rolling angle increase. This is mainly due to the high-frequency and large-angle rolling motions have aggravated the reverse flow of the subcooled water and increased the average condensation rate. In addition, we also found that as the steam inlet velocity is constant, the occurrence positions of the pressure peaks under different rolling parameters are consistent.

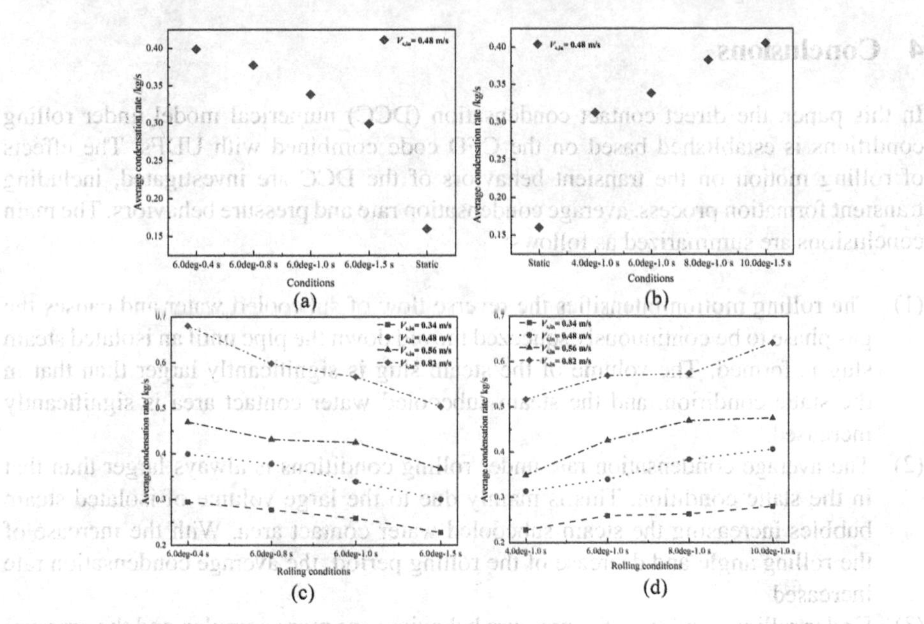

Fig. 6. Effects of rolling motion and steam inlet velocity on the average condensation rate

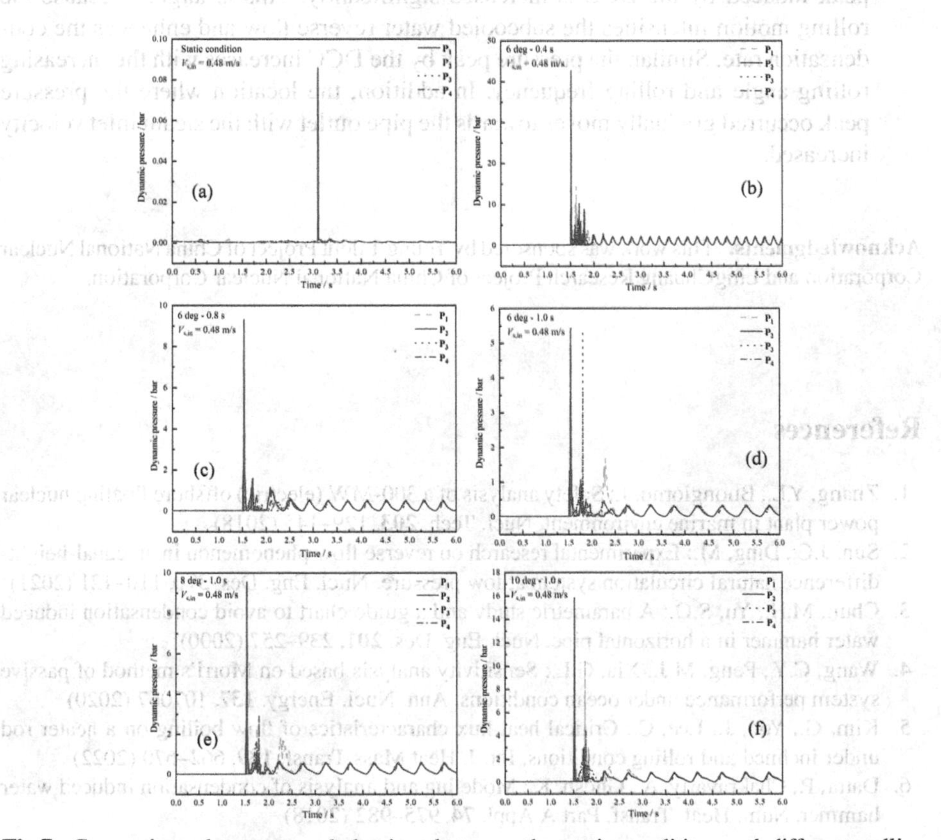

Fig.7. Comparison the pressure behaviors between the static condition and different rolling conditions

4 Conclusions

In this paper, the direct contact condensation (DCC) numerical model under rolling conditions is established based on the CFD code combined with UDFs. The effects of rolling motion on the transient behaviors of the DCC are investigated, including transient formation process, average condensation rate and pressure behaviors. The main conclusions are summarized as follows:

(1) The rolling motion intensifies the reverse flow of subcooled water and causes the gas phase to be continuously squeezed moved down the pipe until an isolated steam slug is formed. The volume of the steam slug is significantly larger than that in the static condition, and the steam-subcooled water contact area is significantly increased.
(2) The average condensation rate under rolling conditions is always larger than that in the static condition. This is mainly due to the large volume of isolated steam bubbles increasing the steam-subcooled water contact area. With the increase of the rolling angle and decrease of the rolling period, the average condensation rate increased.
(3) Under rolling condition, the pressure behaviors are more complex and the pressure peak induced by the DCC is increased significantly. This is largely because the rolling motion intensifies the subcooled water reverse flow and enhances the condensation rate. Similar, the pressure peak by the DCC increases with the increasing rolling angle and rolling frequency. In addition, the location where the pressure peak occurred gradually moves towards the pipe outlet with the steam inlet velocity increased.

Acknowledgments. This work was sponsored by Young Talent Project of China National Nuclear Corporation and LingChuang Research Project of China National Nuclear Corporation.

References

1. Zhang, Y.L., Buongiorno, J.: Safety analysis of a 300-MW (electric) offshore floating nuclear power plant in marine environment. Nucl. Tech. **203**, 129–145 (2018)
2. Sun, J.C., Ding, M.: Experimental research on reverse flow phenomenon in an equal-height-difference natural circulation system at low pressure. Nucl. Eng. Des. **370**, 110–121 (2021)
3. Chun, M.H., Yu, S.O.: A parametric study and a guide chart to avoid condensation induced water hammer in a horizontal pipe. Nucl. Eng. Des. **201**, 239–257 (2000)
4. Wang, C.Y., Peng, M.J., Xia, G.L.: Sensitivity analysis based on Morri's method of passive system performance under ocean conditions. Ann. Nucl. Energy. **137**, 107067 (2020)
5. Kim, G., Yoo, J., Lee, C.: Critical heat flux characteristics of flow boiling on a heater rod under inclined and rolling conditions. Int. J. Heat Mass Transf. **189**, 662–670 (2022)
6. Datta, P., Chakravarty, A., Ghosh, K.: Modeling and analysis of condensation induced water hammer. Num. Heat. Transf. Part A Appl. **74**, 975–982 (2018)

7. Prasser, H., Baranyai, G.: Water Hammer Tests, Condensation caused by Cold water Injection into Main Steam-Line of VVER-440-type PWR-Quick-Look Report (QLR), Technical report. 236–248 (2004)
8. Chong, D.T., Wang, L.T.: Characteristics of entrapped bubbles of periodic condensation-induced water hammer in a horizontal pipe. Int. J. Heat Mass Transf. **152**, 119–134 (2020)
9. Sun, J.C., Ding, M.: Experimental research on characteristics of condensation induced water hammer in natural circulation systems. J. Int. Commun. Heat Mass Transf. **114**, 104–112 (2020)
10. Wang, Z.W., He, Y.P., Duan, Z.D.: Numerical investigation on direct contact condensation-induced water hammer in passive natural circulation system for offshore applications. Num. Heat. Transf. Part A Appl. **63**(3), 974–985 (2022)
11. Peng, J., Chen, D.Q., Xu, J.J.: CFD simulation focusing on void distribution of subcooled flow boiling in circular tube under rolling condition. Int. J. Heat Mass Transf. **156**, 380–390 (2020)
12. Wang, Z.W., He, Y.P., Duan, Z.D.: Effects of rolling motion on transient flow behaviors of gas-liquid two-phase flow in horizontal pipes. Ocean Eng. **255**, 111–122 (2022)
13. Chen, W.X., Mo, Y.L., Wei, P.B.: Numerical study on dominant oscillation frequency of unstable steam jet under rolling condition. Ann. Nucl. Energy. **171**, 109–116 (2022)
14. Li, S.Q., Wang, P.: CFD based approach for modeling steam-water direct contact condensation in subcooled water flow in a tee junction. Prog. Nucl. Energy. **85**, 729–746 (2015)
15. Tan, H.B., Wen, N., Ding, Z.: Numerical study on heat and mass transfer characteristics in a randomly packed air-cooling tower for large-scale air separation systems. Int. J. Heat Mass Transf. **178**, 121–129 (2021)
16. Szijártó, R., Badillo, A., Prasser, H.: Condensation models for the water–steam interface and the volume of fluid method. Int. J. Multiphase Flow. **93**, 63–70 (2017)
17. Antham, R.: Condensation induced water hammer, Ph. D thesis, Chalmers University of Technology Gothenburg, Sweden (2016)
18. Liu, Z.P., Huang, D.S., Wang, C.L.: Flow and heat transfer analysis of lead–bismuth eutectic flowing in a tube under rolling conditions. Nucl. Eng. Des. **382**, 111–128 (2021)
19. Adumene, S., Islam, R., Amin, M.: Advances in nuclear power system design and fault-based condition monitoring towards safety of nuclear-powered ships. Ocean Eng. **251**, 111–126 (2022)
20. Li, M.Z., He, Y.P., Liu, Y.D.: Hydrodynamic simulation of multi-sized high concentration slurry transport in pipelines. Ocean. Eng. **163**, 691–705 (2018)

The Dynamic Response Analysis Method of Steel Containment in Floating Nuclear Power Plant

Weizhe Ren[1], Shuyou Zhang[1], Danrong Song[2], Meng Zhang[1](✉), and Wei Wang[2]

[1] Yantai Research Institute of Harbin Engineering University, Yantai, Shandong, China
{renweizhe,zhangmeng}@hrbeu.edu.cn
[2] Nuclear Power Institute of China, Chengdu, Sichuan, China

Abstract. During the operation of the floating nuclear power plant, if the floating nuclear power plant is anchored near the shore for power generation, it may be impacted by runaway ships. If the impact causes damage to the tank of the floating nuclear power plant and water inflow, and even damage to the containment structure, it will seriously threaten the structural safety of the floating nuclear power plant and even threaten nuclear safety. This paper uses the explicit dynamics method to establish two models of dry/wet impact. It assumes that the 5000T container ship will laterally impact the reactor compartment of the floating nuclear power plant at a speed of 2 m/s, and the effects of the fluid domain (water and air domain) are ignored/counted in the calculation process. Then, comparing the effects of the above two models of dry/wet impact on the calculation duration, the damage and water ingress state of the floating nuclear power plant, the stress of the steel containment (including support). Studies have shown that if the relevant personnel do not need to understand the effect of water intrusion into the damaged cabin, the influence of water and air domain on the impact simulation results can be ignored, and the dry model can be directly used for calculation.

Keywords: Floating nuclear power plant · dry/wet impact · fluid-structure interactions

1 Introduction

Ship impact refers to the accident that ships contact and cause damage at sea or in navigable waters connected with the sea. It often leads to disastrous consequences, such as casualties, sinking of ships, and even environmental pollution. There are many reasons for ship impact accidents, and researchers in many countries are trying to find ways to avoid ship impact accidents [1]. At present, due to the influence of human factors, ship impact accidents cannot be completely eliminated. For floating nuclear power plants, it is more important to study the dynamic response of impact. Once a ship collides with a floating nuclear power plant, it will not only cause serious economic losses, but also cause more terrible harm to the environment that is difficult to evaluate [2, 3].

There are three main methods for impact analysis of ship structures: experimental method, simplified analytical method and numerical simulation method. The test

C. Liu (Ed.): PBNC 2022, SPPHY 283, pp. 764–775, 2023.
https://doi.org/10.1007/978-981-99-1023-6_66

method includes actual accident investigation and real ship or ship model impact test. The results obtained by it are very intuitive and have self-evident guiding significance for the theoretical method [4, 5]. However, it often costs a lot of money. The accuracy of the simplified analytical method is relatively low. It simplifies a lot of ships in impact. Using the analytical method and some empirical data, it establishes the semi analytical/semi empirical mechanical equations of the hull or local ship structures to evaluate the impact characteristics of the hull. The advantage of numerical simulation method is that it can reproduce the real ship impact scene virtually. With the help of some finite element analysis software, various physical quantities in the process of ship impact can be output as results [6].

With the continuous upgrading and development of computer software and hardware technology and the increasing progress and maturity of finite element technology, explicit finite element numerical simulation technology has gradually received attention in the research of ship impact, and the research of ship structure impact based on full coupling technology has become gradually feasible. This paper mainly uses ANSYS-APDL software to establish the impact model, and uses the nonlinear explicit dynamics analysis software ANSYS workbench LS-DYNA to solve the analysis. Based on the comparison of dry/wet calculation results, the applicability of the two impact models is studied.

2 Modeling and Theoretical Basis

2.1 ANSYS APDL Model Parameter Settings

Container ships and floating nuclear power plants are modeled by 4-node shell181 shell element and beam188 beam element in classic ANSYS (after importing work bench LS-DYNA, it is automatically converted to LS-DYNA applicable element types: shell163 and beam161). The plate thickness is divided into 15, 18, 20, 22, 24, 26, 28, 30, etc. The grids at the bulbous bow of the container ship and the impact part on the port side of the floating nuclear power plant are densified (Figs. 1 and 2).

After the model is processed in classical ANSYS, three CBD files of stern containment, floating nuclear power plant (excluding stern containment) and container ship are written respectively, and then imported into workbench LS-DYNA module for pre impact model processing.

2.2 Contact Algorithms and Contact Types

The contact impact algorithm in LS-DYNA usually has the following methods:

(1) Dynamic constraint method

As the contact algorithm first used in dyna program, the basic principle of dynamic constraint method is: before calculating and running each time step ΔT, retrieve those slave nodes that currently have no penetration with the master surface, and retrieve whether these slave nodes have penetration with the master surface at this time step.

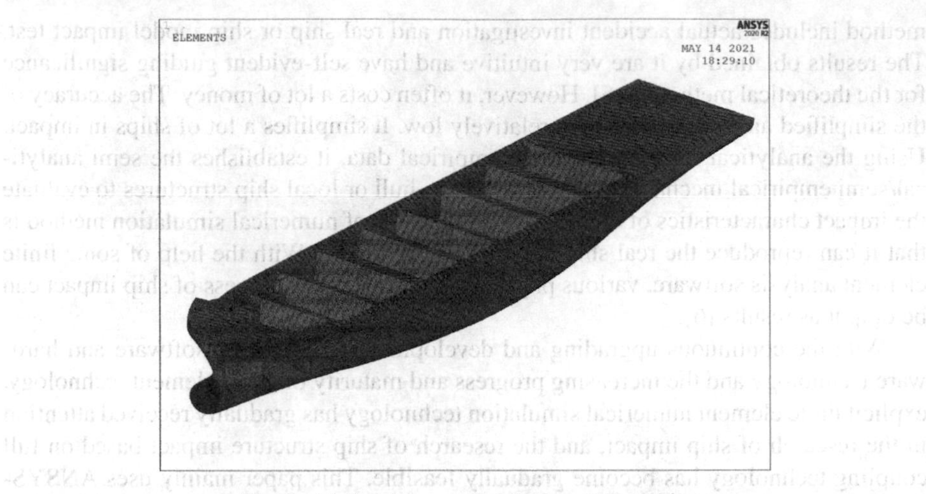

Fig. 1. Finite element model of container ship.

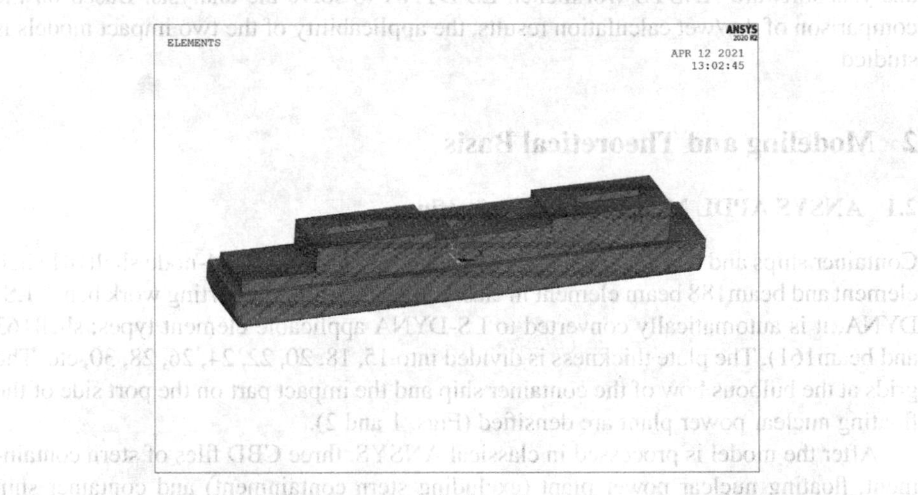

Fig. 2. Finite element model of floating nuclear power plant.

If there is penetration, reduce the time step ΔT. under the reduced time step, the slave node just contacts the main surface but does not penetrate the main surface. However, when the mesh division of the master surface is fine, some nodes on the master surface can penetrate the slave surface without constraints, which has a great impact on the accuracy of the calculation results. This algorithm will not be applicable. To sum up, this algorithm has certain limitations and is relatively complex, so it is only used for fixed connection and fixed connection disconnection contact at present.

(2) Distributive parameter method

The basic principle is: distribute half of the mass of each slave unit being contacted to the main surface being contacted, and distribute positive pressure at the same time.

Then, the principal surface acceleration is corrected. Finally, the acceleration and velocity constraints are imposed on the slave node to ensure that the slave node does not slide and avoid rebound.

(3) Penalty function method

This method is widely used in numerical calculation. The basic principle is: in each time step, check whether the slave node passes through the main surface. If not, do not deal with it. On the contrary, a larger surface is introduced between the slave node and the penetrated surface. The size of this contact surface is related to the penetration depth and the stiffness of the main surface, also known as the penalty function value. Its physical meaning is that a spring is placed between the two to limit the penetration, as shown in the following figure. The so-called symmetric penalty function method means that the program processes all master nodes according to the above steps, using the same algorithm as the slave nodes (Fig. 3).

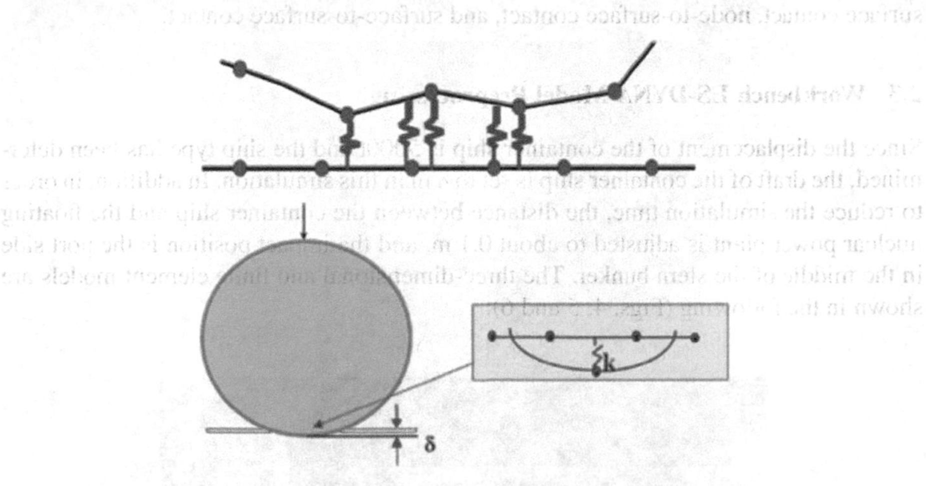

Fig. 3. Penalty function method.

The magnitude of contact force is expressed by this formula: In the above formula: k is contact interface stiffness, is penetration.

Because the symmetrical penalty function method is adopted, the calculation of this method is simple, the hourglass phenomenon in the impact process is not obvious, and there is no noise impact. The energy conservation in the system is accurate, symmetrical and the momentum conservation is accurate, and the impact conditions and release conditions are not required. If obvious penetration occurs in the calculation process, it can be adjusted by enlarging the penalty function value or reducing the time step.

Among the three algorithms, the dynamic constraint method is mainly applicable to the fixed interface, the distributed parameter method is mainly used for the sliding interface, and the symmetric penalty function method is the most commonly used method. The following problems in contact analysis should be paid attention to: first, in terms of data, try to make the material data accurate, because the accuracy of most nonlinear dynamic problems is related to the mass density of the input data; Second, in terms of units, the definition of material properties is to coordinate units, which will cause abnormal response and even calculation problems; Third, in terms of contact, the calculation efficiency varies greatly with different contact types. It is best to use automatic contact types and be particularly cautious about complex problems; Fourth, multiple contacts should be defined in the definition process.

Contact problem is state nonlinear, and is a highly nonlinear behavior, which requires more computer resources. There are two major difficulties in simulating contact problems with finite elements: the user usually does not know the contact area until the problem is solved. Moreover, it is difficult to estimate the time when their contact surfaces are separated. Changes in the two contact surface materials, loads, boundaries, and other conditions will also change the analysis of the contact problem. In LS-DYNA, contact is not simulated by elements that are in contact with each other, but by contact surfaces that may be in contact. By setting the contact type and parameters, the contact-impact interface algorithm is used to solve the problem. Common contact types are single-surface contact, node-to-surface contact, and surface-to-surface contact.

2.3 Workbench LS-DYNA Model Preprocessing

Since the displacement of the container ship is 5000t and the ship type has been determined, the draft of the container ship is set to 8 m in this simulation. In addition, in order to reduce the simulation time, the distance between the container ship and the floating nuclear power plant is adjusted to about 0.1 m, and the impact position is the port side in the middle of the stern bunker. The three-dimensional and finite element models are shown in the following (Figs. 4, 5 and 6):

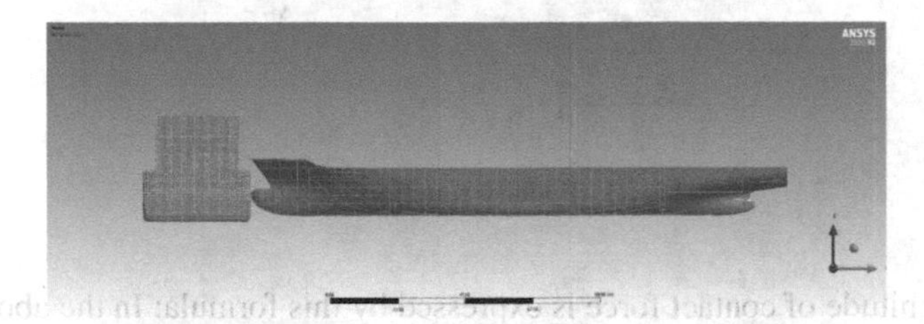

Fig. 4. Three dimensional model of impact between two ships.

Fig. 5. Finite element model of two ship impact.

Fig. 6. Local finite element mesh refinement model of two ship impact.

Define the material properties. The material of the floating nuclear power plant (except the stern containment) is defined as nonlinear structural steel, Poisson's ratio is set to 0.3, elastic modulus is set to 2.06×10^{11}Pa, tangent modulus is set to 1.45×10^{9}Pa, and yield strength is set to 3.55×10^{8}Pa. In order to make the floating nuclear power plant have the possibility of damage, it is necessary to add a plastic strain failure criterion, and its maximum equivalent plastic failure criterion is set to 0.0035, according to the young's modulus.

Connection relationship settings. The single surface only type suitable for the beam shell hybrid model in the body interaction is adopted, and the constraint for forming contact algorithm with high accuracy is adopted for the flexible constraint algorithm. In addition, the contact and connection between the stern containment and the floating nuclear power plant are realized by using body to body fixed, and the coupling point is set as flexible coupling (Fig. 7).

Fig. 7. Connection between containment and floating nuclear power plant.

Loads, boundary conditions and analysis settings of workbench LS-DYNA. Fix the bow and stern surfaces of the floating nuclear power plant rigidly, as shown in Figs. 8 and 9. Set the end time to 3.5 s, and the calculation accuracy is double precision. Set the speed of the container ship to 2 m/s and the direction to −y, and conduct body tracking for the container ship, the floating nuclear power plant and its containment, so as to check the contact force later. In addition, it can provide data for subsequent fatigue analysis and ultimate bearing capacity analysis.

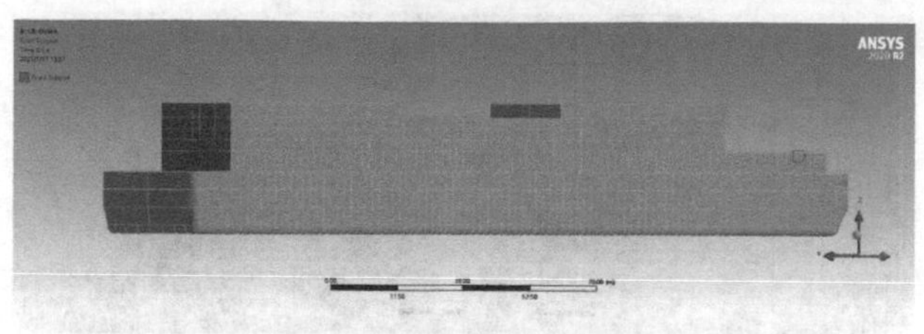

Fig. 8. Rigid fixed surface 1.

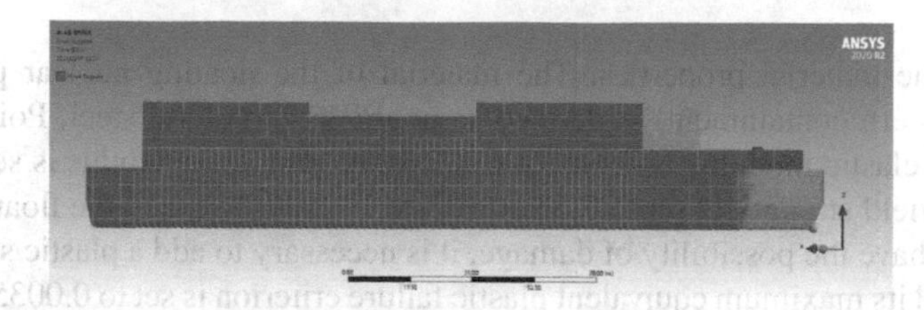

Fig. 9. Rigid fixed surface 2.

3 Comparison of Simulation Results

3.1 Equivalent Stress During Impact

For dry impact, the time history curve of the maximum equivalent stress of the whole ship is shown in Fig. 10. The maximum equivalent stress is 360 MPa, which occurs in 0.471 s, as shown in Fig. 11. The degree of damage is shown in Fig. 12. The simulation results show that the damage range of the hull outer plate is small, and the deformation of the bulkheads is small and undamaged.

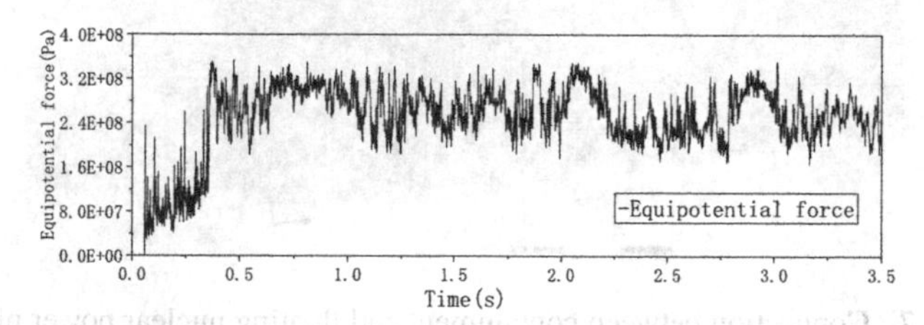

Fig. 10. The time history curve of the maximum equivalent stress of the whole ship

For dry impact, the maximum equivalent stress of the containment is 331.7 MPa, which occurs at 0.319 s as shown in Fig. 13.

For wet impact, the maximum equivalent stress of the whole ship is about 360 MPa, which occurs at 0.48 s, as shown in Fig. 14. The degree of damage is shown in Fig. 15. Figure 16 shows the details of the water ingress of the damaged tank. The maximum equivalent stress of the containment is 304.3MPa, which occurs at 0.320 s, as shown in Fig. 17.

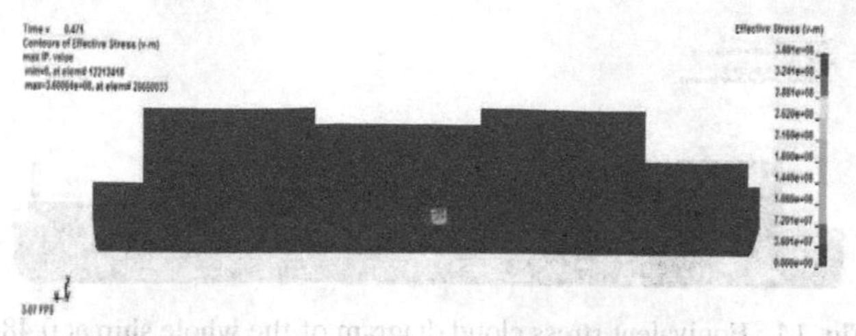

Fig. 11. Equivalent stress cloud diagram of the whole ship at 0.471 s

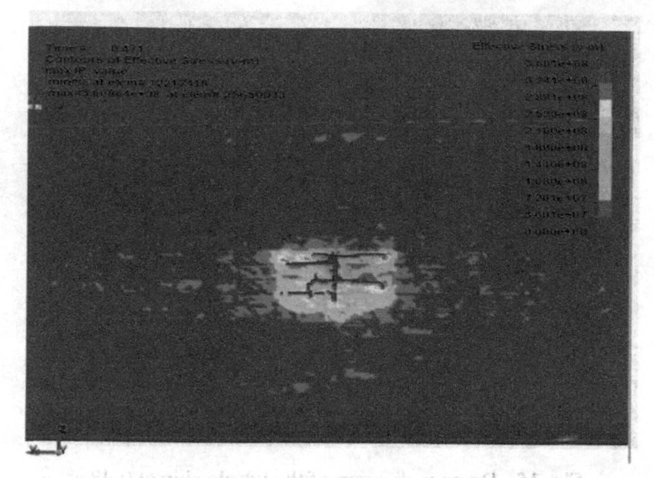

Fig. 12. Damage diagram of the whole ship at 0.471 s

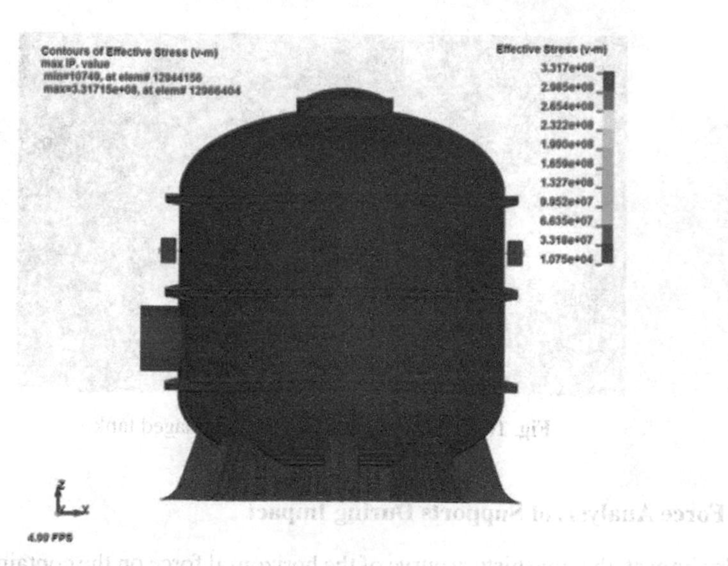

Fig. 13. Equivalent stress cloud diagram of containment at 0.319 s

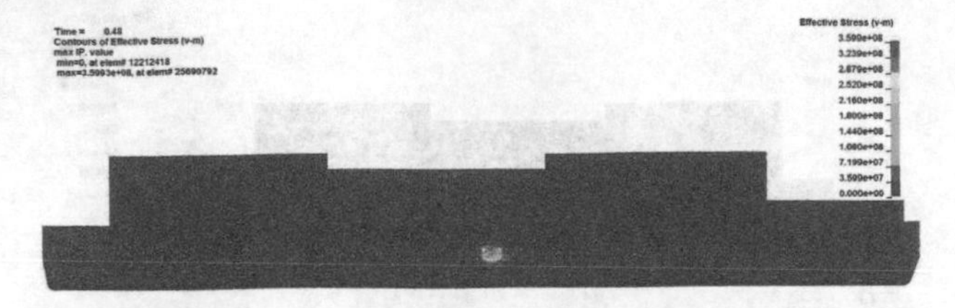

Fig. 14. Equivalent stress cloud diagram of the whole ship at 0.48 s

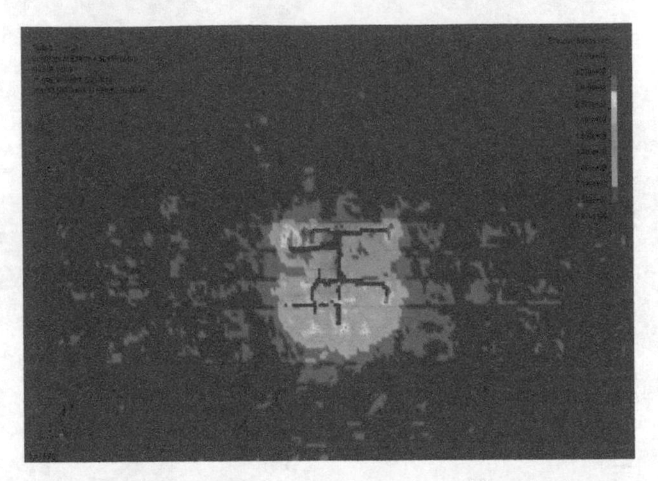

Fig. 15. Damage diagram of the whole ship at 0.48 s

Fig. 16. The water ingress of the damaged tank.

3.2 Force Analysis of Supports During Impact

For dry impact, the time history curve of the horizontal force on the containment support is shown in Fig. 18–20. The maximum translational force of the containment support in the X direction is $1.04 \times 10^6 N$, which occurs in 0.655 s; The maximum translational force

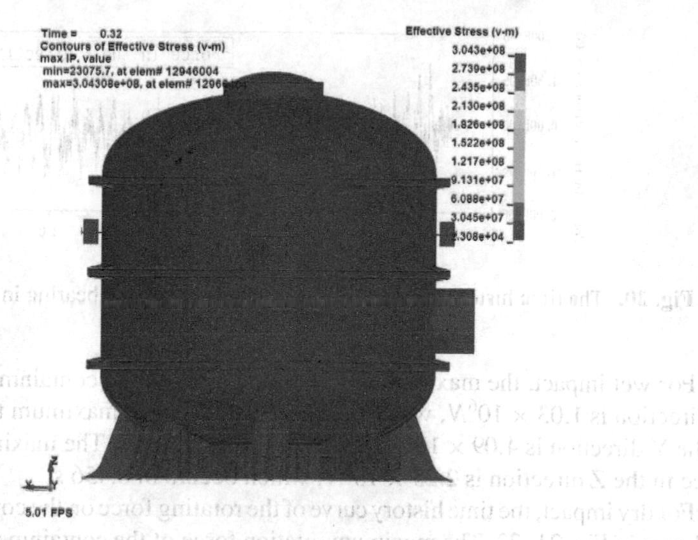

Fig. 17. Equivalent stress cloud diagram of containment at 0.320 s

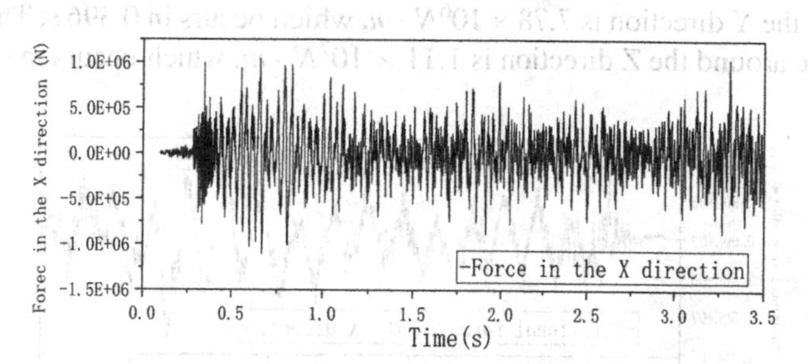

Fig. 18. The time history curve of the horizontal force on the bearing in the X direction

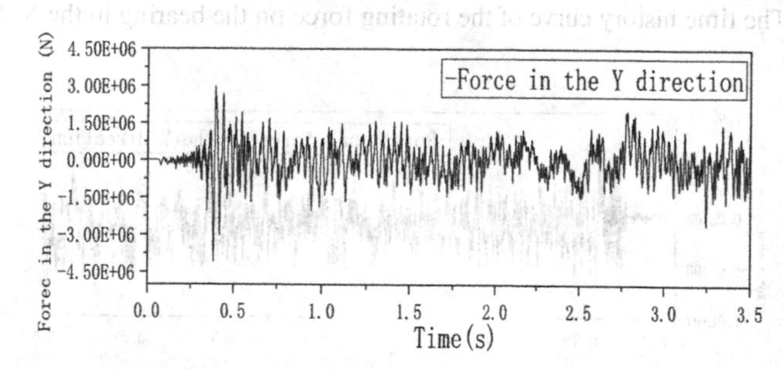

Fig. 19. The time history curve of the horizontal force on the bearing in the X direction

in the Y direction is $4.09 \times 10^6 N$, which occurs in 0.384 s; The maximum translational force in the Z direction is $2.25 \times 10^6 N$, which occurs in 0.453 s.

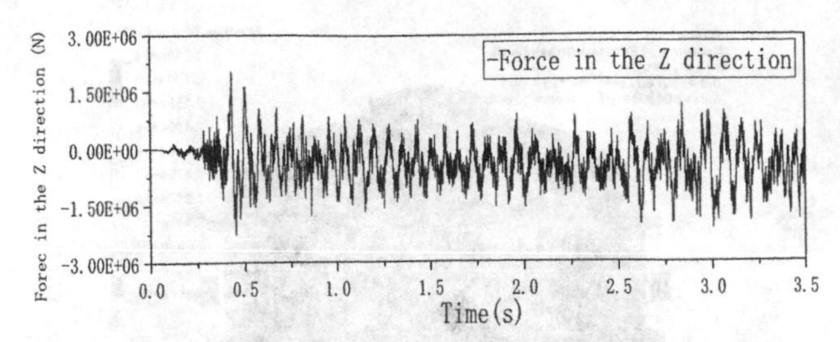

Fig. 20. The time history curve of the horizontal force on the bearing in the Z direction

For wet impact, the maximum translational force of the containment support in the X direction is $1.03 \times 10^6 N$, which occurs in 0.659 s; The maximum translational force in the Y direction is $4.09 \times 10^6 N$, which occurs in 0.388 s; The maximum translational force in the Z direction is $2.25 \times 10^6 N$, which occurs in 0.456 s.

For dry impact, the time history curve of the rotating force on the containment support is shown in Fig. 21–23. The maximum rotation force of the containment support around the X direction is $2.10 \times 10^7 N \cdot m$, which occurs in 2.753 s; The maximum rotation force around the Y direction is $7.78 \times 10^6 N \cdot m$, which occurs in 0.396 s; The maximum rotation force around the Z direction is $1.11 \times 10^7 N \cdot m$, which occurs in 0.384 s.

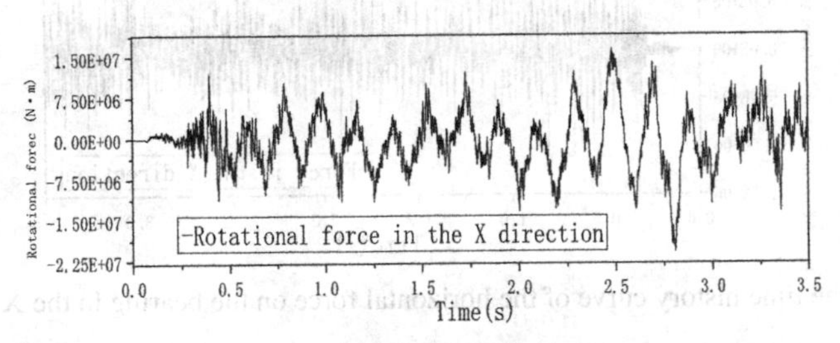

Fig. 21. The time history curve of the rotating force on the bearing in the X direction

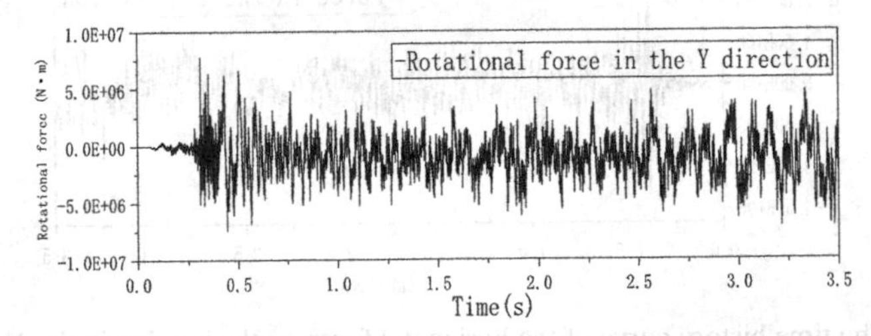

Fig. 22. The time history curve of the rotating force on the bearing in the Y direction

For wet impact, the maximum rotation force of the containment support around the X direction is $2.09 \times 10^7 N \cdot m$, which occurs in 2.807 s; The maximum rotation force around the Y direction is $7.78 \times 10^6 N \cdot m$, which occurs in 0.409 s; The maximum rotation force around the Z direction is $1.11 \times 10^7 N \cdot m$, which occurs in 0.384 s.

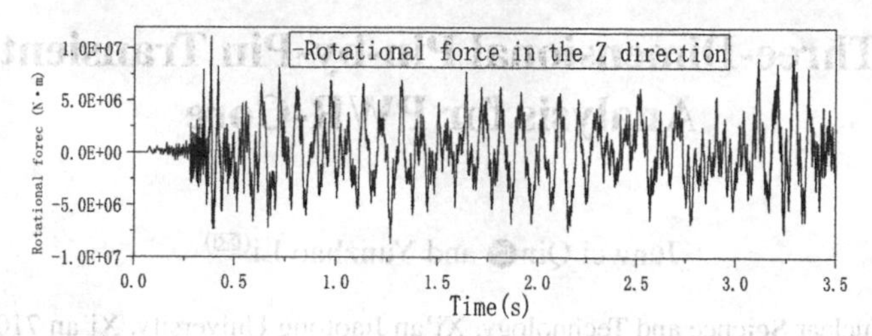

Fig. 23. The time history curve of the rotating force on the bearing in the Y direction

4 Conclusion

The simulation time of wet impact is 2.5 times that of dry impact, and the equivalent stress of the whole ship, the equivalent stress of the containment, the damage degree of the hull and the force of the support are not much different. So, if the relevant personnel do not need to understand the effect of water intrusion into the damaged cabin, the influence of air and water on the impact simulation results can be ignored, and the dry mode can be directly used for calculation.

References

1. Li, C., Wang, J.: Correlation analysis of ship collision accidents. J. Shanghai Marit. Univ. **42**(02), 70–74 (2021)
2. Wen, X., W., Lu, J., Cui, Z.: Numerical simulation and experiment research of ship collision process. J. Zhejiang Ocean Univ. Nat. Sci. **30**(1), 71–76 (2011)
3. Liu, K., L., Zhang, Y., Wang, Z.: Study on the influence of bow shape to the side structure during ship collision. Ship Eng. **32**(2), 12–14 (2010)
4. Zhang, W.: Numerical simulation research of ship collision analogue rule. Chin. J. Ship Res. **4**(3), 38–41 (2009)
5. Travanca, J., Hao, H.: Dynamics of steel offshore platforms under ship impact. Appl. Ocean Res. **47**, 352–372 (2014)
6. Oshieo, R.E., Alves, M.: Scaling of cylindrical shells under axial impact. Int. J. Impact Eng **34**, 89–103 (2007)

Three-Dimensional Pin-by-Pin Transient Analysis for PWR-Core

Junwei Qin and Yunzhao Li(✉)

School of Nuclear Science and Technology, Xi'an Jiaotong University, Xi'an 710049, China
yunzhao@xjtu.edu.cn

Abstract. To ensure the safety of PWR-core operation, three-dimensional whole-core transient analysis needs to be carried out for the sake of the pin-power distribution. For this purpose, this paper presents "Bamboo-Transient 2.0", a three-dimensional pin-by-pin transient analysis program. The program adopts a fully-implicit backward method with finite difference for time variable discretization, a method of exponential function expansion nodal (EFEN) SP_3 for the neutron transport calculation, and a multi-channel model for the thermal feedback calculation. In addition, Picard iteration is used to couple the neutronics with thermal-feedback, which is intended to guarantee the convergence of coupling iteration at each time step. Moreover, the program can perform parallel computing based on Message Passing Interface (MPI) for the whole-core pin-by-pin transient analysis. This developed program has been applied to two commercial PWRs, viz. AP1000 and CNP1000. Numerical results of this application demonstrate that Bamboo-Transient 2.0 can yield much more refined results than the traditional legacy coarse-mesh neutron-diffusion programs based on assembly homogenization. Its pin-wise distributions of state parameters are reliable and thus can satisfactorily meet the requirements and purpose of safety analysis.

Keywords: PWR · NECP-Bamboo · Pin-by-pin · Transient analysis

1 Introduction

Compared with the steady-state operation process, the reactor core is more dangerous in transient processes. For PWR, only by determining the hot spots and heat pipes at the core can its safe operation be ensured and the ultra-high temperature-caused core melting be prevented. To find the hot spots and heat pipes at the core, the transient analysis is supposed to be accurate in the fuel rod scale and be capable of providing various state parameters. Thus, it is necessary to track and predict the changes in the key parameters of the core to prevent accidents and reduce the harm after accidents if there is any. As the traditional transient analysis method is based on the diffusion calculation of assembly homogenization, only information in the assembly scale can be retained whereas other relevant information in the pin scale is ignored. Therefore, in order to obtain the power distribution of fuel rod scale, it is necessary to carry out the power reconstruction based on a series of approximations and assumptions. Consequently, the calculation accuracy of the traditional method is far from desirable [1].

C. Liu (Ed.): PBNC 2022, SPPHY 283, pp. 776–788, 2023.
https://doi.org/10.1007/978-981-99-1023-6_67

It is against such a background that fast and accurate transient analysis methods for nuclear reactors are becoming increasingly important and should be designed and developed. In recent years, a pin-by-pin transient analysis method based on pin homogenization has attracted extensive attention. With the pin transport calculation accomplished, this method can directly homogenize the calculation area and retain all kinds of information in the pin scale. As a result, in the core calculation, the core information in the pin scale can be obtained directly without introducing the error caused by the power reconstruction [2]. In addition, the channel of thermal feedback should be accurate in the rod scale so as to match with neutronics. Additionally, the transient process of PWR is a process of coupling neutronics with thermal-hydraulics [3]. This paper presents how the coupling of neutronics with thermal-feedback is calculated, which is to be suitable for the pin-by-pin transient analysis. In addition, due to the significantly increased number of computational meshes, it is urgent to improve the efficiency of whole-core pin-by-pin transient analysis. In this paper, the parallel technology is also discussed.

The rest of the paper is organized as follows. Section 2 introduces each method in detail. Section 3 introduces the numerical verification and analysis. Section 4, the last part of this paper, sums up the study and concludes the paper.

2 Theoretical Models

The fully implicit method and EFEN method are specially employed to solve the neutron kinetics equation, whereas a multi-channel model in the pin scale is employed to treat the heat transfer and flow process of coolant, and a 1D cylindrical heat conduction model is employed to treat the heat conduction process in fuel rods. Picard iteration is utilized at each time step to guarantee the convergence between neutronics and thermal-feedback. Using MPI of distributed memory, the same spatial domain decomposition is performed for both neutronics and thermal-feedback calculation for parallel computing, which can significantly shorten the computing time needed by transient analysis.

2.1 Calculation of Neutron Dynamics

In the transient process of PWR, considering the influence of delayed neutrons and adopting multi-group approximation, the neutron flux distribution at the core meets the spatiotemporal neutron transport equations, which is shown in Eq. (1). Where, v_g is the neutron velocity of group g/cm·s^{-1}, r the spatial location, Φ_g the neutron angular flux of group g/(cm3·s)$^{-1}$, $\boldsymbol{\Omega}$ the neutron motion direction, t the time/s, $\Sigma_{t,g}$, $\Sigma_{f,g}$ the total, fission cross sections of group g/cm^{-1}, $\Sigma_{s,g'\to g}$ the scattering cross section from group g' to group g/cm^{-1}, $\chi_{p,g}$ and $\chi_{d,g,i}$ the prompt neutron fission spectrum of group g and the delayed neutron fission spectrum of group g, delayed group i, v the number of neutrons per fission, C_i the precursor concentration of delayed group i, λ_i the decay constant of precursor delayed group i/s^{-1}, β_i the delayed neutron fractions of group i/pcm, $g = 1, 2, \ldots, G$; $i = 1, 2, \ldots, Nd$ the neutron energy group index and the delayed neutron precursor group index.

$$\begin{cases}\frac{1}{v_g(\mathbf{r})}\frac{\partial \mathbf{\Phi}_g(\mathbf{r},\mathbf{\Omega},t)}{\partial t}=-\mathbf{\Omega}\cdot\nabla\mathbf{\Phi}_g(\mathbf{r},\mathbf{\Omega},t)-\Sigma_{t,g}(\mathbf{r},t)\mathbf{\Phi}_g(\mathbf{r},\mathbf{\Omega},t)\\ \quad +\int\Sigma_{s,g'\to g}(\mathbf{r},\mathbf{\Omega}'\to\mathbf{\Omega},t)\mathbf{\Phi}_g(\mathbf{r},\mathbf{\Omega}',t)d\mathbf{\Omega}'\\ \quad +(1-\beta(\mathbf{r}))\frac{\chi_{p,g}(\mathbf{r})}{4\pi}\sum_{g'=1}^{G}\int\nu(\mathbf{r})\Sigma_{f,g'}(\mathbf{r},t)\mathbf{\Phi}_{g'}(\mathbf{r},\mathbf{\Omega}',t)d\mathbf{\Omega}'\\ \quad +\frac{1}{4\pi}\sum_{i=1}^{Nd}\chi_{d,g,i}(\mathbf{r})\lambda_i C_i(\mathbf{r},t)\\ \frac{\partial C_i(\mathbf{r},t)}{\partial t}=\beta_i(\mathbf{r})\sum_{g'=1}^{G}\int\nu(\mathbf{r})\Sigma_{f,g'}(\mathbf{r},t)\mathbf{\Phi}_{g'}(\mathbf{r},\mathbf{\Omega}',t)d\mathbf{\Omega}'-\lambda_i C_i(\mathbf{r},t)\end{cases} \tag{1}$$

As Eq. (1) cannot be solved straightforwardly, it has to be discretized. First, the fully implicit backward difference method is used for the time term [4] to obtain the equation of the angular flux at t_{n+1}, which is shown in Eq. (2):

$$\begin{aligned}&\mathbf{\Omega}\cdot\nabla\mathbf{\Phi}_g(\mathbf{r},\mathbf{\Omega},t_{n+1})+\overline{\Sigma}_{t,g}(\mathbf{r},t_{n+1})\mathbf{\Phi}_g(\mathbf{r},\mathbf{\Omega},t_{n+1})=\\ &\sum_{g'=1}^{N}\int\Sigma_{s,g'\to g}(\mathbf{r},\mathbf{\Omega}'\to\mathbf{\Omega},t_{n+1})\mathbf{\Phi}_g(\mathbf{r},\mathbf{\Omega}',t_{n+1})d\mathbf{\Omega}'\\ &+\frac{1}{4\pi}\overline{\chi}_g(\mathbf{r})\sum_{g'=1}^{N}\int\nu(\mathbf{r})\Sigma_{f,g'}(\mathbf{r},t_{n+1})\mathbf{\Phi}_{g'}(\mathbf{r},\mathbf{\Omega}',t_{n+1})d\mathbf{\Omega}'+Q_g(\mathbf{r},\mathbf{\Omega})\end{aligned} \tag{2}$$

where,

$$\overline{\Sigma}_{t,g}(\mathbf{r},t_{n+1})=\Sigma_{t,g}(\mathbf{r},t_{n+1})+\frac{1}{v_g(\mathbf{r})\Delta t}$$

$$\overline{\chi}_g(\mathbf{r})=(1-\beta(\mathbf{r}))\chi_{p,g}(\mathbf{r})+\sum_{i=1}^{Nd}\chi_{d,g,i}(\mathbf{r})\beta_i(\mathbf{r})-\sum_{i=1}^{Nd}\chi_{d,g,i}(\mathbf{r})\frac{\beta_i(\mathbf{r})}{1+\lambda_i\Delta t}$$

$$Q_g(\mathbf{r},\mathbf{\Omega})=\frac{1}{4\pi}\sum_{i=1}^{Nd}\chi_{d,g,i}(\mathbf{r})\lambda_i\frac{C_i(\mathbf{r},t_n)}{1+\lambda_i\Delta t}+\frac{1}{v_g(\mathbf{r})}\frac{\mathbf{\Phi}_g(\mathbf{r},\mathbf{\Omega},t_n)}{\Delta t}$$

Next, the P_1 approximation is used for the angle to obtain the diffusion fixed source equation as shown in Eq. (3):

$$-D\nabla^2\psi_g^0(x)+\overline{\Sigma}_{t,g}\psi_g^0(x)=S_g \tag{3}$$

The SP_3 approximation is used for the angle to obtain the SP_3 fixed source equations as shown in Eq. (4):

$$\begin{cases}-D\nabla^2[\psi_g^0(x)+2\psi_g^2(x)]+\overline{\Sigma}_{r,g}[\psi_g^0(x)+2\psi_g^2(x)]=S_g+2\overline{\Sigma}_{r,g}\psi_g^2(x)\\ -\frac{27}{35}D\nabla^2\psi_g^2(x)+\overline{\Sigma}_{t,g}\psi_g^2(x)=S_2+\frac{2}{5}(\overline{\Sigma}_{r,g}\psi_g^0(x)-S_g)\end{cases} \tag{4}$$

The derivation process is described in detail in Reference [4]. Equation (3) and (4) can be solved by using EFEN method, whose process is described in detail in Reference [5]. By solving the diffusion or SP_3 fixed source equation at each time step, the neutron dynamics is calculated.

2.2 Calculation of Thermal Feedback

The transient analysis of PWR entails that the coupling of neutron dynamics with transient thermal feedback is calculated. The solution to the neutron dynamics equation requires the cross-sections of all materials in the core. To obtain these cross-sections, the state parameters at the core are needed. Thus, thermal feedback calculation should provide the distributions of these state parameters. Calculation of the thermal feedback of the reactor core consists of two parts, viz. Fluid calculation of the coolant and heat conduction calculation of the fuel rod, in which the fluid calculation is 1D calculation in the axial direction of the flow area and the heat conduction calculation is 1D calculation in the radial direction of the cylindrical rod.

Fluid Model

Transient mass and energy conservation equations of the coolant are shown in Eq. (5) and (6). Where, ρ is the density of coolant/kg·cm^{-3}, h the enthalpy of coolant/J, u the velocity of coolant/cm·s^{-1}, A_c the circulation area/ cm^{-2}, q_c the heat release in fuel and q_w the fuel surface heat flux.

$$\frac{\partial \rho(z,t)}{\partial t}+\frac{\partial(\rho(z,t)u(z,t))}{\partial z}=0 \tag{5}$$

$$\frac{\partial(\rho(z,t)h(z,t))}{\partial t}+\frac{\partial(\rho(z,t)u(z,t)h(z,t))}{\partial z}=P_h\frac{q_w(z,t)}{A_c}+q_c(z,t) \tag{6}$$

The two equations are discretized by θ difference in time to obtain the equation of the nodal enthalpy rise. Given the parameters of the coolant at the channel inlet, the heat flux can be obtained via the heat conduction calculation. Hence, the temperature of the coolant can be solved in the axial direction from the inlet to the outlet.

Heat Conduction Model

1D transient heat conduction model of cylinder is shown in Eq. (7). Where, c_p is the specific heat capacity at constant pressure/(J/kg·K), k the thermal conductivity/(W/m·K), and T the temperature/K.

$$\rho(r,t)c_p(r,t)\frac{\partial T(r,t)}{\partial t}=\frac{1}{r}\frac{\partial}{\partial r}\left(k(r,t)\cdot r\cdot\frac{\partial T(r,t)}{\partial r}\right)+q(r,t) \tag{7}$$

In the radial direction, the fuel pellet is divided into n meshes, the gas gap divided into 1 mesh, and the clad divided into 3 meshes. Equation (7) is discretized by finite difference in space and θ difference in time. The radial temperature distribution of the fuel is obtained by Gauss-Seidel iterative calculation.

2.3 Coupling of Neutronics with Thermal-Hydraulics

As the mesh size of the whole-core pin-by-pin neutron dynamics calculation can be either that of a fuel rod, or a control rod or a water tunnel, the corresponding thermal feedback adopts multi-channels in pin scale division to simulate the coolant flow between the rods. The channel is the coolant area between fuel rods and guide tubes. The channels along the axis can be divided into several meshes, which can exchange mass and energy with each other, ignoring the exchange of momentums and the exchange between the radial channels. In addition, different physical fields adopt various meshing methods while the neutron and thermal fields uses rod-centered meshing methods and the flow field uses channel-centered meshing method. The various methods are shown in Fig. 1.

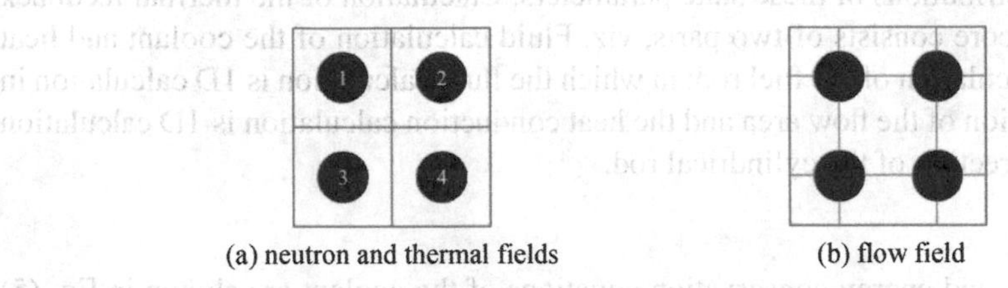

Fig. 1. Various meshing methods of physics fields

As the mapping relationships across neutron field, thermal field and flow field are complicated, five types of conversion relationships are considered for different physical fields as follows, which is shown in Fig. 2 as follows:

1) conversion from the nodal power of neutron field to the channel power of flow field;2) conversion from the coolant temperature of flow field to the cladding surface temperature of thermal fields;3) conversion from the nodal power of neutron field to the mesh power of thermal field;4) conversion from the coolant temperature of flow field to the nodal average coolant temperature of the neutron field;5) conversion from the fuel temperature of thermal field to the nodal effective fuel temperature of neutron field.

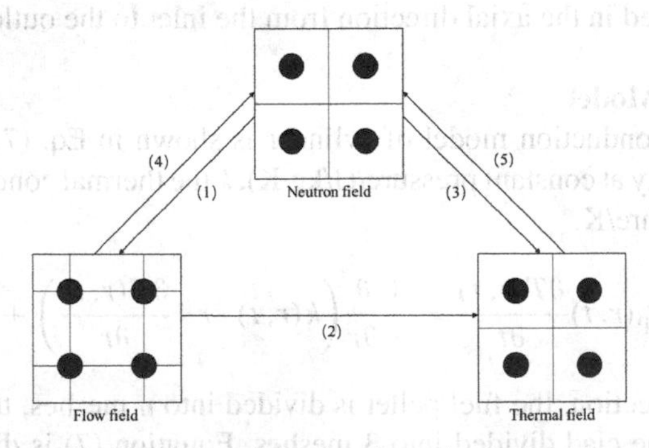

Fig. 2. Conversion of various physical fields

Traditional transient analysis calculation programs mostly use explicit coupling method. This method can solve the neutronics equation and thermal feedback calculation separately, but cannot perform the iteration between them. Thus, the convergence of the coupling cannot be guaranteed. To ensure the convergence, a shorter time step is needed for this method. However, due to lack of iteration, the cost of calculation at each time step is also low. However, implicit coupling, or rather, Picard iteration, solves the neutronics and thermal-feedback separately. In essence, Picard iteration is intended to solve the various physical fields in different ways through operator splitting, and then iterates between the two physical fields to converge the coupled parameters. However, it takes a longer time step and more convergence than the explicit coupling. In view of this, the current study employs Picard iteration to calculate the coupling of neutronics with thermal-feedback to accurately calculate the coupling. The neutron dynamics and thermal feedback are calculated at each time step. Only when the neutronics and thermal-hydraulics coupling iteration are converged can the next time step be calculated. The iterative process is shown in Fig. 3.

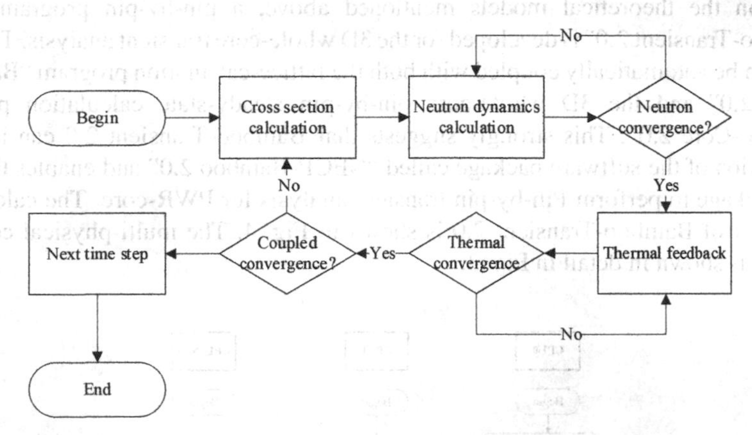

Fig. 3. Neutronics and thermal-hydraulics coupling iteration.

2.4 Parallel Calculation

The parallel efficiency is mainly affected by the following factors. 1) Communication overhead. As different threads need to communicate, it takes longer time. Thus, the longer communication lasts, the lower the parallel efficiency. 2) Parallel computing may degrade the iterative format, and as a result, increase the amount of computation. 3) Multithreading may cause worthless waiting to the processes caused by load imbalance. 4) Redundant computation can be caused by parallel algorithm per se.

Given the factors above, the parallelization of whole-core pin-by-pin transient calculation consists of the following aspects:

1) Broadcast of input parameters. To prevent multiple CPUs from using the same input channel at the same time, a designated CPU reads the input file and then uses the broadcast function of sending multiple data of the same type in batch at one time to send the input information to all CPUs.

2) Domain decomposition. The whole area on average is divided according to the number of CPUs and the scale of calculation problems to keep the load balance across different CPUs. Each CPU stores only the area information responsible for calculation.
3) Node sweep and communication. To improve the parallel efficiency, the Red-Black Gauss-Seidel node sweep method [6], which is suitable for parallel computing, is selected to avoid the degradation in iterative format caused by parallel computing. In addition, the fractional neutron currents only need to communicate once after the red and black node sweep to reduce the communication overhead.
4) Post-processing of calculation results. To prevent multiple CPUs from using the same output channel at the same time, the calculation results of all CPUs are exported uniformly by the designated CPU.

2.5 Interim Summary

Based on the theoretical models mentioned above, a pin-by-pin program called "Bamboo-Transient 2.0" is developed for the 3D whole-core transient analysis. This program can be automatically coupled with both the lattice-calculation program "Bamboo-Lattice 2.0" and the 3D whole-core pin-by-pin steady-state calculation program "Bamboo-Core 2.0". This strongly suggests that Bamboo-Transient 2.0 can improve the function of the software package called "NECP-Bamboo 2.0" and enables the software package to perform Pin-by-pin transient analysis for PWR-core. The calculation flow chart of Bamboo-Transient 2.0 is shown in Fig. 4. The multi-physical coupling iteration is shown in detail in Fig. 3.

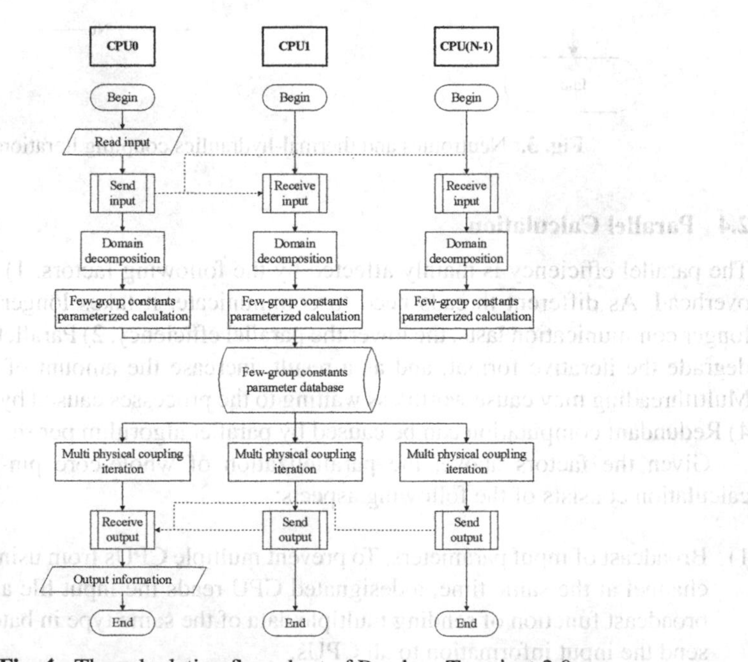

Fig. 4. The calculation flow chart of Bamboo-Transient 2.0.

3 Numerical Verification and Analysis

To verify the accuracy and the analytical ability of Bamboo-Transient 2.0, the program is applied to the transient analysis of two commercial PWRs. This section will show the numerical verification and analysis of CNP1000 and AP1000 reactor core by using Bamboo-Transient 2.0.

3.1 CNP1000

The rated power and the rated operating pressure of CNP1000 are 2895MWt and 15.5 MPa, respectively. There are 157 boxes of fuel assemblies at the reactor core and 57 control rod assemblies in the first cycle. The reactor core is divided into 9 groups according to the needs of the problem. The control rods are 362.49 cm in length, which is divided into 225 steps. The grouping of the control rods is shown in Fig. 5. The program calculates the rod ejection in the following case: the initial power level of the core is 1%. The control rods of the ninth group are fully inserted while the rest are all lifted. The control rods of the ninth group are all ejected in 0–0.1 s. It takes 0.4 s to accomplish the transient process, and the time step is divided into 0.001 s.

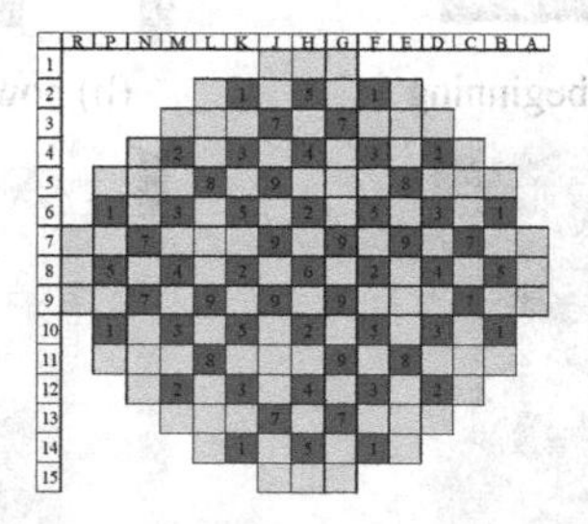

Fig. 5. The grouping of the control rods in CNP1000

The normalized power is shown in Fig. 6. As the control rods are gradually ejected from the core, the normalized power of the core increases rapidly. When all the control rods are ejected from the core, the core enters a stable state. The 3D distribution of power, the effective temperature of fuel and the temperature of coolant are shown in Fig. 7. At the beginning, due to the insertion of control rods, the power is distributed precipitously, and the distribution is high outside but low inside the core. With the control rods ejected, the power distribution is flattened. While the temperature distributions show the same regularity, the effective temperature of the fuel and the temperature of the coolant are more uniform than at the beginning in the radial direction after the rod ejection.

3.2 AP1000

AP1000 is the third-generation advanced passive PWR designed by Westinghouse. Compared with the traditional PWR nuclear reactor, AP1000 adopts a passive safety system, which further simplifies the structure of the power station and improves the safety of the

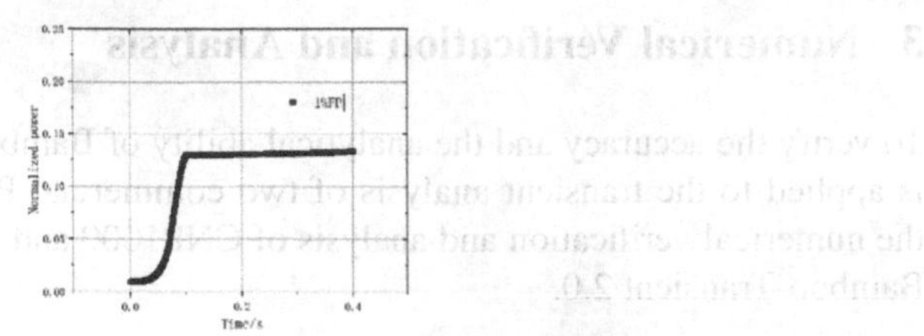

Fig. 6. Normalized power of CNP1000

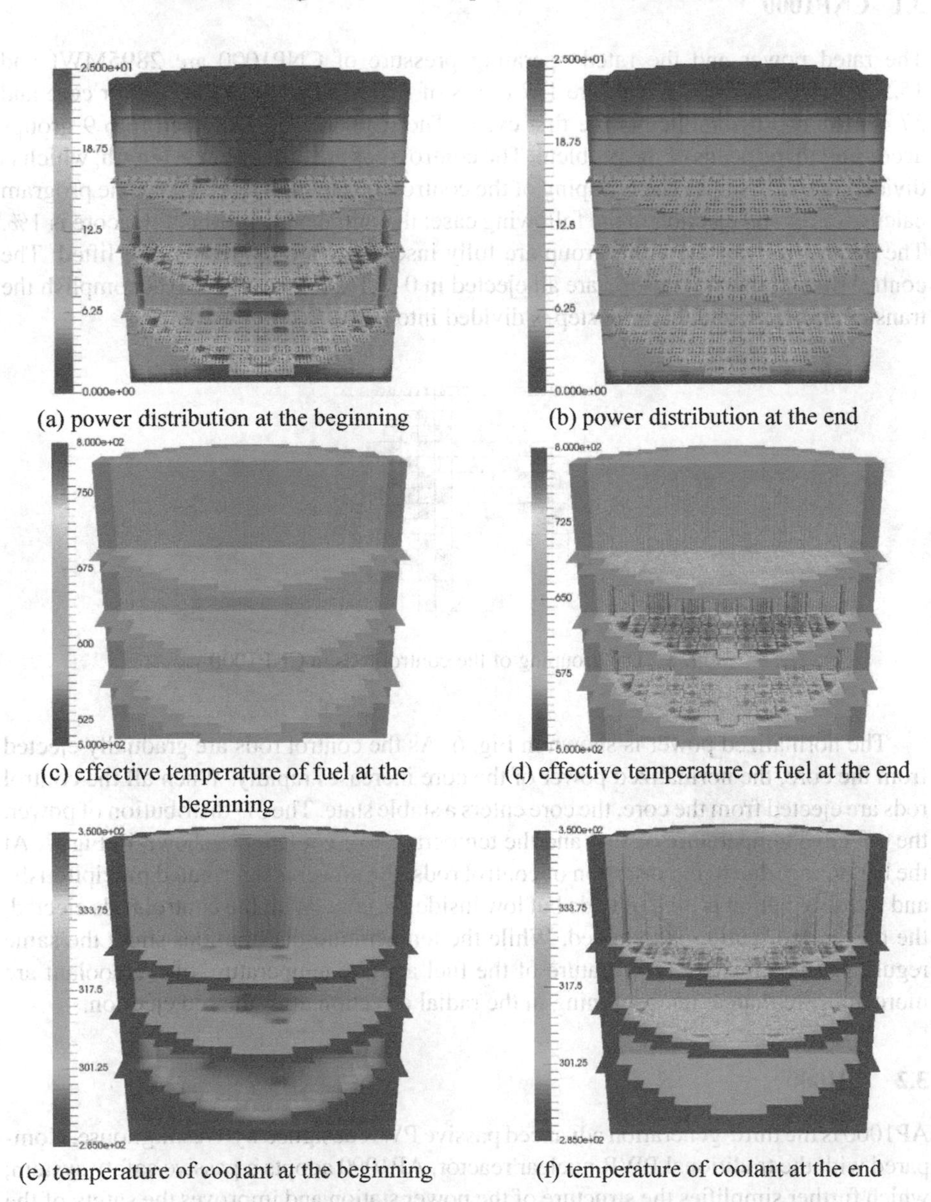

(a) power distribution at the beginning

(b) power distribution at the end

(c) effective temperature of fuel at the beginning

(d) effective temperature of fuel at the end

(e) temperature of coolant at the beginning

(f) temperature of coolant at the end

Fig. 7. Three-dimensional distribution at different time

reactor. There are 69 groups of control rods in AP1000, including 53 groups of black rods and 16 groups of gray rods. Some of the grey rods and black rods are employed to form the MSHIM system of the core, which controls the reactivity and power distribution of the core with the boron containing coolant, while the remaining black rods are employed to form the shutdown rod group [7]. The pattern of the control rods is shown in Fig. 8.

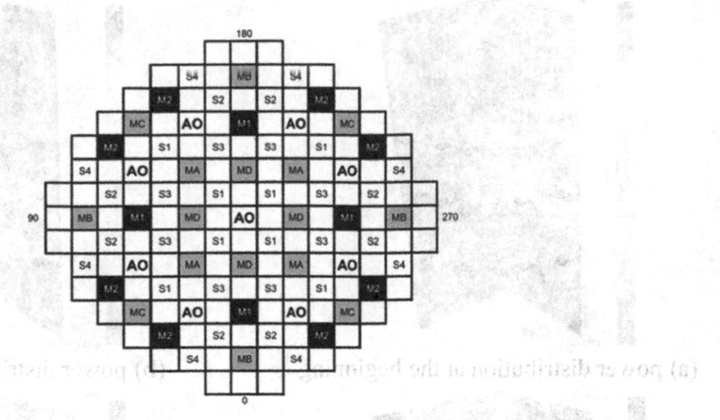

Fig. 8. The pattern of the control rods of AP1000

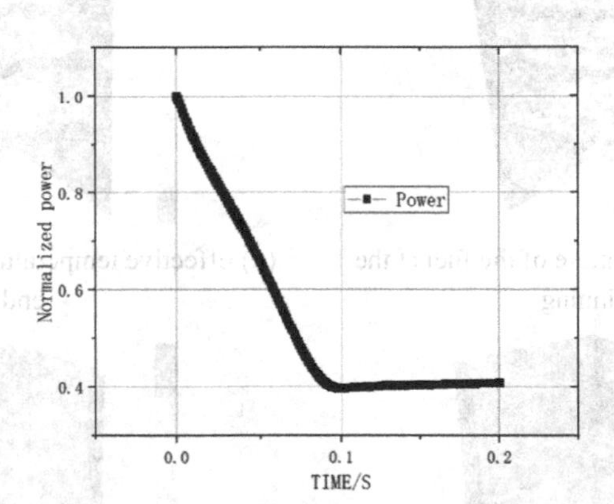

Fig. 9. Normalized power of AP1000

AP1000 has designed a rapid power reduction system (RPR system). To analyze the system, this study calculates the simulation of rod drop working condition of M1 and S2 rod groups under full power. M1 and S2 rod groups simultaneously fall into the core within 0.1 s while the program simulates the transient process within 0.2 s. The normalized power of the core is shown in Fig. 9. It can be seen that the core power decreased rapidly due to the insertion of control rods into the core. At 0.1 s, the normalized power is reduced to 40% of that at the initial time. Then, the core power increases slowly as the fuel temperature decreases. The 3D distribution at the beginning and at the end is shown in Fig. 10. The figure shows that the power distribution is distorted by the insertion of the control rod and that the power around the control rod is significantly decreased. Besides, the effective fuel temperature also shows this regularity. At the same time, the axial distribution of power and the fuel temperature become increasingly uneven while the temperature of the coolant changes little and decreases slightly.

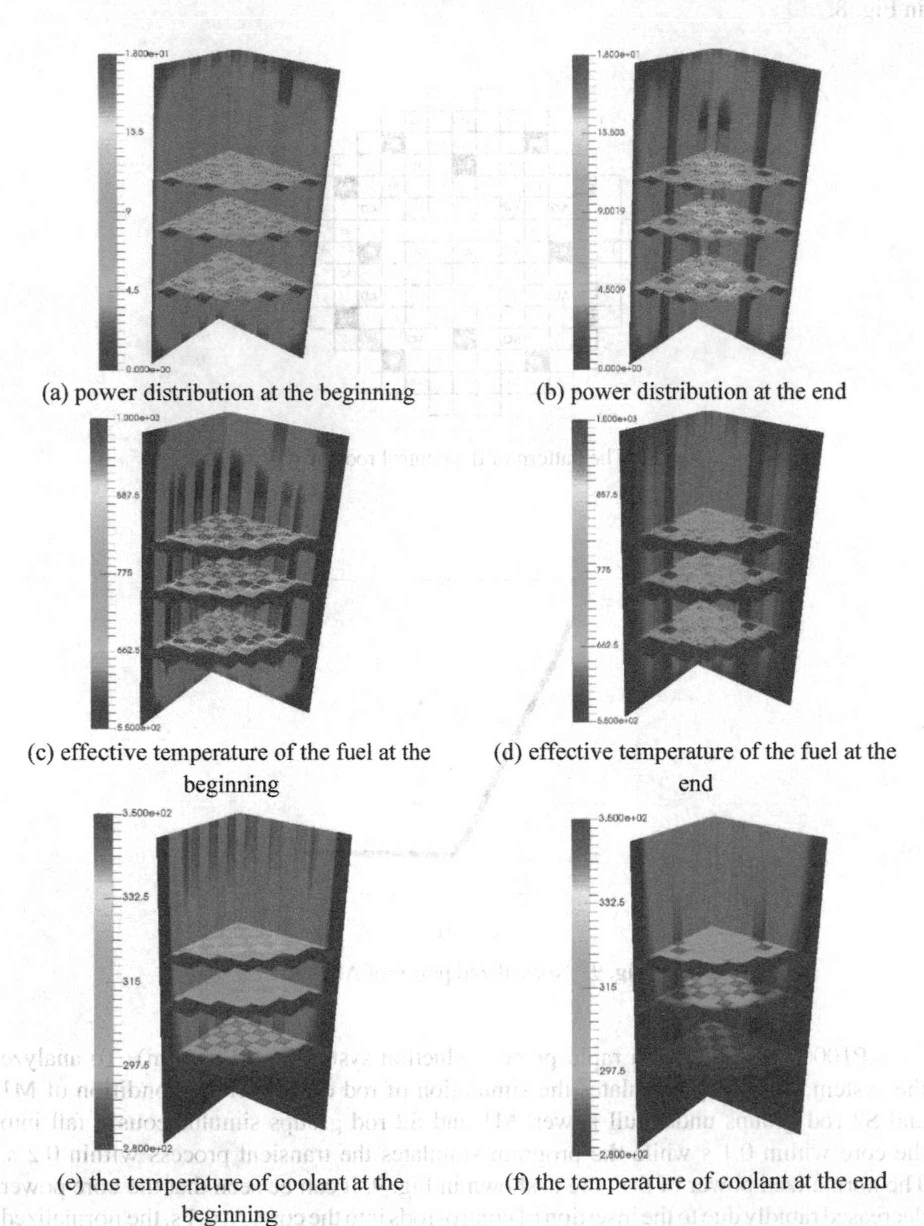

(a) power distribution at the beginning (b) power distribution at the end

(c) effective temperature of the fuel at the beginning (d) effective temperature of the fuel at the end

(e) the temperature of coolant at the beginning (f) the temperature of coolant at the end

Fig. 10. Three-dimensional distribution at different time

4 Conclusion

In order to perform the 3D whole-core pin-by-pin transient analysis, the theoretical model of the transient analysis is established in this study as follows:

1) The pin-by-pin neutron dynamics is calculated by using the fully implicit method and the exponential function expansion nodal method;
2) The model of 3D whole-core pin-by-pin transient thermal feedback is established by using multi-channel in the pin scale to simulate the heat transfer and flow process of coolant, and by using a 1D cylindrical heat conduction model to simulate the heat conduction process in fuel rods;
3) In each time step, Picard iteration is used to perform the iterative calculation of the coupling of neutronics with thermal-feedback;
4) Based on MPI, the parallel calculation of the 3D whole-core pin-by-pin transient analysis is accomplished, which significantly shortens the calculation time.

To sum up, in this study, the program system that couples the pin-by-pin transient neutronics with thermal-feedback is applied to the PWR-core transient analysis of the second-generation nuclear power CNP1000 and the third-generation nuclear power AP1000. Both the condition for rapid rod drop power reduction and that for rod ejection accident are analyzed. The rod power distribution and fuel temperature distribution are more refined than the traditional diffusion program of assembly homogenization. Bamboo-transient 2.0 makes full use of the ability of pin-by-pin transient analysis and calculation, which is more accurate than the traditional program. The working conditions of the reactor core under transient conditions are simulated and speculated finely, which provides a guarantee for the safe operation of PWR.

References

1. Li, Y., He, T., Liang, B., et al.: Development and verification of PWR-core nuclear design code system NECP-Bamboo: Part III: Bamboo-Transient. Nuclear Eng. Design **359** (2018)
2. Yang, W., Wu, H., Li, Y., et al.: Development and verification of PWR-core fuel management calculation code system NECP-Bamboo: Part II Bamboo-Core[J]. Nucl. Eng. Des. **337**, 279–290 (2018)
3. Grahn, A., Kliem, S., Rohde, U.: Coupling of the 3D neutron kinetic core model DYN3D with the CFD software ANSYS-CFX. Annals Nuclear Energy. **84**, 197–203 (2015)
4. Xie, W., Cao, L., Li, Y.: Study on Pin-by-pin Neutron Kinetics Calculation based on EFEN-SP_3 Method. Atomic Energy Sci. Technol. **53**(2), 7 (2019)
5. Yang, W., Wu, H., Li, Y., et al.: Acceleration of the exponential function expansion nodal SP 3 method by multi-group GMRES algorithm for PWR pin-by-pin calculation. Annals Nuclear Energy, **120**, 869–879 (2018)
6. Palmiotti, G., Carrico, C.B., Lewis, E.E.: VARIANT: Variational anisotropic nodal transport for multidimensional Cartesian and hexagonal geometry calculation. Technical Report Archive & Image Library (1995)
7. Li. X., Du, C.: Study on selection of rapid power reduction system banks for load rejection for AP1000 NPP. Nuclear Power Eng. (S02), 4 (2019)

S3R Advanced Training Simulator Core Model: Implementation and Validation

Jeffrey Borkowski[1], Lotfi Belblidia[2], and Oliver Tsaoi[3](✉)

[1] Studsvik Scandpower, Idaho Falls, Idaho, USA
[2] Studsvik Scandpower, Gaithersburg, MD, USA
[3] Studsvik Scandpower, Beijing, China
oliver.tsaoi@studsvik.com

Abstract. Modern training simulators core models are required to replicate plant data for neutronic response. Replication is required such that reactivity manipulation on the simulator properly trains the operator for reactivity manipulation at the plant. This paper discusses advanced models which perform this function in real-time using S3R, the real-time, time-dependent core model of the Studsvik Core Management System (CMS). This paper also discusses the coupled multi-physics of the Reactor Coolant System (RCS) model, using RELAP5 as a prototype. Finally, this paper discusses the implementation of S3R under the control of a server-based executive environment and instructor station, essential for training simulator applications.

Keywords: Operational Training · Real-Time Simulation · Cycle Specific · Just-Time Training · Reactivity Management

1 Introduction

The Studsvik nuclear reactor analysis code, SIMULATE-3, has been extended to transient applications for both engineering analysis and real-time operator training. The physics models used in S3R are much the same as those used for steady-state core design/safety analysis, except that no core design or depletion calculation are done, and some simplifications are introduced to run in real-time under the control of real-time executive.

S3R has become the standard in 3D real-time core models for training simulator. It has been installed in more than 90 sites worldwide. The neutronics model of S3R has been coupled to several real-time thermal hydraulic models, including RELAP5, used in training simulators.

C. Liu (Ed.): PBNC 2022, SPPHY 283, pp. 789–799, 2023.
https://doi.org/10.1007/978-981-99-1023-6_68

2 Neutronic Model Description

2.1 Features

The neutronics model used in S3R solves the 3-D, two-energy group, neutron diffusion equation with one radial node to represent each fuel assembly. In the axial direction, 24–25 nodes are typically used to represent the active portion of each fuel assembly, and one node is used to represent the upper and lower reflectors.

The S3R core model uses a fourth-order flux expansion to represent the neutron flux distribution within each node (in each of the three directions), and the spatial gradient of the flux can then be taken analytically (a third-order function, rather than the traditional first-order function). This results in a much more accurate representation for the flux than that of simpler methods.

The S3R core model uses explicit nodes (both radially and axially) to model the PWR baffle/reflectors. This permits direct solutions for the fluxes and leakages into the reflectors, without need for introduction of albedos (which are often used in simpler models) to treat the leakage out of the core. The baffle/reflector nodes are treated like any other node in the S3R core model.

2.2 Nuclear Data

Accuracy of the S3R core model depends not only on detailed 3-D neutronic and thermal-hydraulic modeling, but also on accurate representation of feedback parameters. These parameters include:

- Two-Group Macroscopic Cross Sections
- Fission Product Yields and Microscopic Cross Sections
- Assembly Discontinuity Factors (ADFs)
- Kinetics Data (Betas, Lambdas, Velocities)
- Spontaneous Fission/Alpha-n Neutron Sources
- Decay Heat Data (Fission Fractions by Isotope)
- Pin Power Distributions
- Detector Data

The functional dependence of the nuclear parameters is expressed as "base cross-section" and several "delta cross-sections" in the form:

$$
\begin{aligned}
&NP^{S3R}\left(\rho, \sqrt{TF}, N_{xe}, N_{sm}, N_{Bo}, wfct_2, wfct_3\right) = \\
&= NP_{Base}(\rho) \\
&+ \frac{\partial NP}{\partial\sqrt{TF}}(\rho)\cdot\left(\sqrt{TF}-\sqrt{TF^{REF}}\right)+ \\
&+ \frac{\partial NP}{\partial N_{Bo}}(\rho)\cdot\left(N_{Bo}-N_{Bo}^{REF}\right)+ \\
&+ \frac{\partial NP}{\partial CR2}(\rho)\cdot wfct_2 + \frac{\partial NP}{\partial CR3}\cdot wfct_3 + \\
&+ \delta NP(Xe) + NP(Sm)
\end{aligned}
$$

For a given core life, all the history effects are frozen and only the instantaneous effects are input to S3R. The only instantaneous dependence is due to the moderator density, fuel temperature, control fractions, and boron. By freezing all history dependence, the calculation of the base cross-section is reduced to a set of 1D interpolations in density.

2.3 Decay Heat

Following a fission event in the fuel, about 93% of the heat of fission is immediately released, and the remaining 7% is released slowly over time. Modeling of this decay heat is very important in transients which have large changes in power level (e.g., Reactivity excursions, SCRAMs, and LOCAs). The fission product heat generation in S3R is modeled by using the ASNI/ANS-5.1, 23-group data. The decay heat sources are initialized as part of the steady-state solution in S3R by assuming infinite-time operation at constant power. It can be reset at any time after shutdown (see example in Sect. 5.1).

The predominant isotopes that contribute to decay heat are U-235, U-238, Pu-239, and Pu-241. The split among these isotopes varies from node to node with exposure. Effect of neutron capture in fission products and contributions from heavy elements (U-239 and Np-239) are also included.

2.4 Detectors

For in-core detectors, the detector responses are predicted as the power average from the surrounding nodes. The geometrical weighting factors account for the axial position of the detector. The detector constants are specified individually for each detector string and its surrounding channels and are obtained from the data library. Their radial locations as well as the number of axial strings and axial locations at each radial location are provided in the S3R input file. Flux data at these locations will be accessible from the instructor station and process computer.

In the case of ex-core detectors, top and bottom detector signals are constructed based on weighted sums of the flux at the core boundary and reflect accurately power imbalances and flux tilts. Detector response is based on flux value at the location of the detectors. A weighting is used to relate the ex-core detector response to the powers of the bundles contributing to it. Changes in the downcomer density cause changes in the attenuation of neutron escaping from the core and reaching the detectors and its effect is accounted for by using an empirical function of downcomer density.

3 Coupling to System Code

This section describes the algorithm and software used to expand thermal-hydraulic data from an RCS model, such as RELAP5, to S3R and collapse nodal powers from S3R to RELAP5-3D. RELAP5-3D is used as an example herein; however, the coupling algorithm may be applied to any system code which can model core channels. The RELAP5-3D model uses a coarser nodalization in the active core regions than S3R and a method for expanding the RELAP5-3D data from this coarse nodalization to the fuel assembly wise data needed for S3R is needed. The axial nodalization may also be different between the models and this is addressed by the algorithm.

For a 3-loop PWR, typically RELAP5 groups the assemblies into 4 active channels (each representing 39¼ assemblies). The channels assignment is illustrated in the figure below. S3R uses 5 radial power zones. The fifth zone receives the average properties from the 4 RELAP5 channels. This is illustrated below.

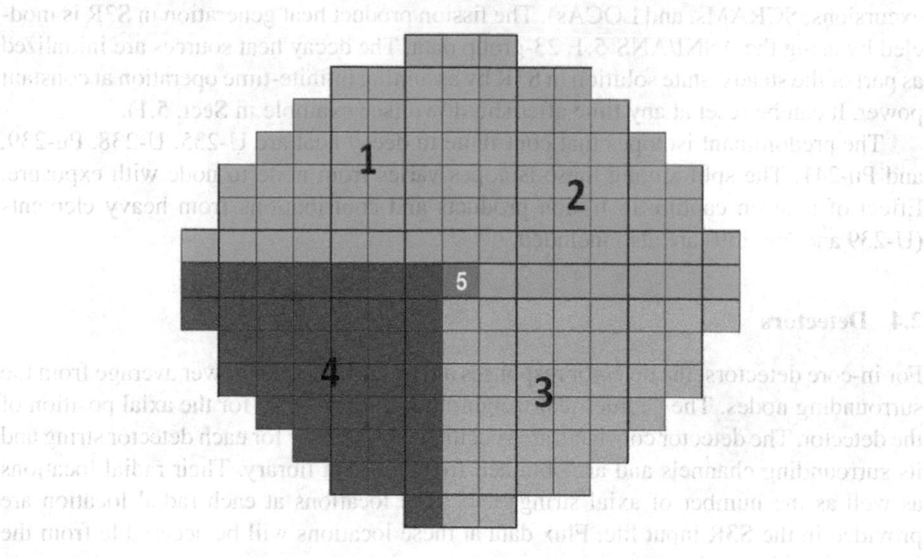

In the axial direction, a possible nodalization used in RELAP5 is:

– 6 hydraulic cells per channel
– 24 heat structures per channel

S3R uses 24 axial levels per assembly (see illustration below).

A mapping scheme that takes the RELAP5 data and expands it from the RELAP5 geometry to the S3R nodalization and collapses the S3R data for use in RELAP5 has been implemented in S3R.

The algorithm is comprised of three parts. The first part takes the thermal-hydraulic data and expands it to S3R axial nodalization utilizing an axial power weighting scheme in each thermal-hydraulic channel. The second part uses a power-weighting scheme to expand the fuel temperatures in the radial direction using the last time step 3D power distribution. The third part uses a simple enthalpy rise model and last time step 3D power distribution to calculate the nodal moderator densities.

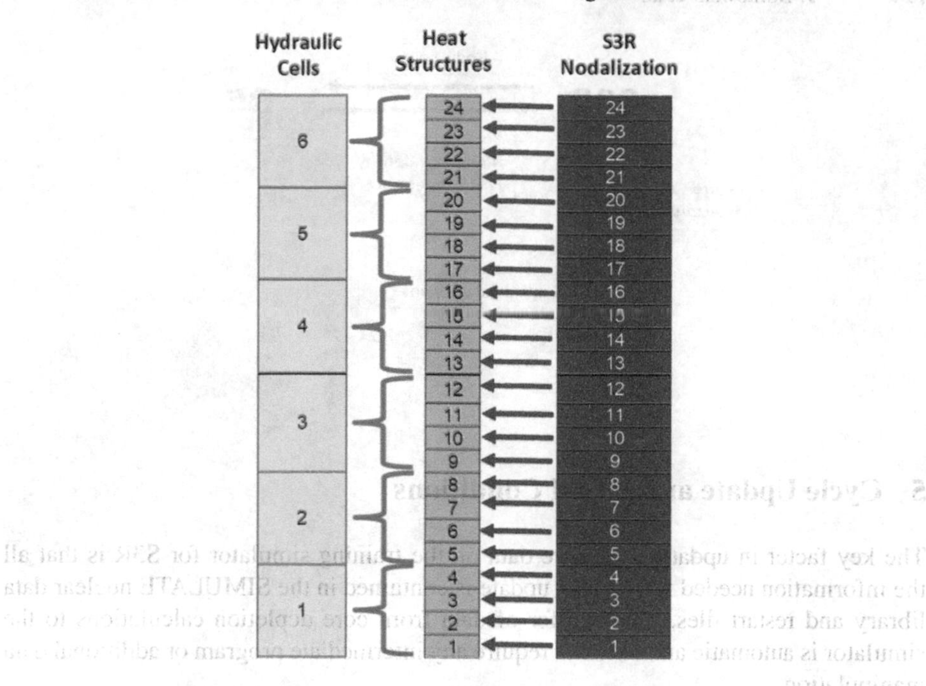

A few assumptions are made:

- The number of axial subdivisions in the RELAP5 volumes and/or heat structures for the active fuel shall be such that an integer multiple of S3R axial nodes are bound by a volume or heat structure height.
- The number of axial subdivisions in the RELAP5 volumes and/or heat structures shall not be greater than the S3R axial nodalization.
- The number radial flow paths used in the RELAP5 model shall only include full assemblies, assemblies may not be subdivided.
- Thermal-hydraulic data shall be provided for the lower and upper plenum regions of the RELAP5 model for use in the reflector cross-section calculations.

4 Interfface with Executive

S3R is used as a library used during the generation of the load regardless of the real-time executive used on the simulator. There are basically two interface routines used to transfer data back-and-forth between S3R and the rest of the simulator. An illustration of what is exchanged at each time step is shown here.

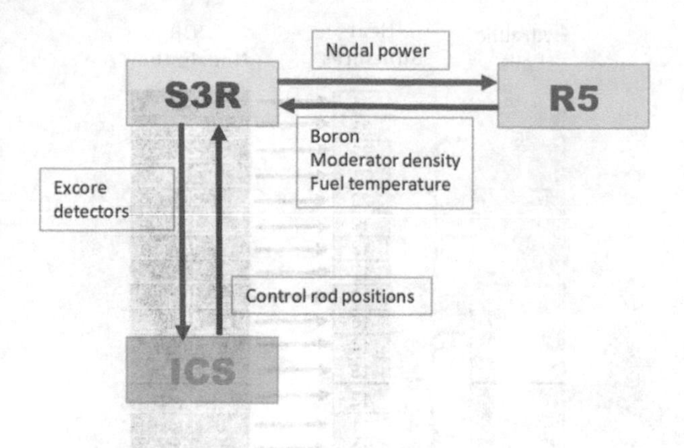

5 Cycle Update and Initial Conditions

The key factor in updating the core data on the training simulator for S3R is that all the information needed for such an update is contained in the SIMULATE nuclear data library and restart files. The transfer of data from core depletion calculations to the simulator is automatic and does not require any intermediate program or additional data manipulation.

To update the core model with data for a new cycle, the following information needs to be supplied by the organization that maintains the CASMO/SIMULATE model:

a) Nuclear data library for the new cycle from Studsvik's CMS system.
b) Restart file(s) for the new cycle with enough exposure points to be able to model all core lives of interest (e.g., BOC, MOC, EOC, etc.).
c) Input and output files from the S3 core depletion calculations. This is needed to conveniently determine the conditions used at each depletion point and the exposure points saved on the restart file.
d) The boron concentration used during the core depletion calculations for the core lives of interest.
e) Updated S3R input files (essentially only the 'RES' and 'LIB' cards) to provide names of the new data files and point to the exposure of interest.

The process of updating ICs with new core data is straightforward and includes the following steps:

- Acquire the nuclear data files (S3 restart and library files) for the new cycle
- Edit the S3R input file to point to the new nuclear data files
- Reset to an existing IC
- Set the boron concentration
- Snap to a temporary location
- Reset to snapped IC to reinitialize (this step accesses the new core data)

- Run until steady state has been established at the desired power level, while using the fast xenon flag to force xenon to equilibrium.
- Snap IC. Process completed.

Depending on the real-time executive, it may be possible to automatize this process using scripts. An example from one site includes the different parts in which the conditions of the existing ICs are used to generate an IC-specific restart file to be used in updating the ICs with the new cycle core data. This is illustrated below.

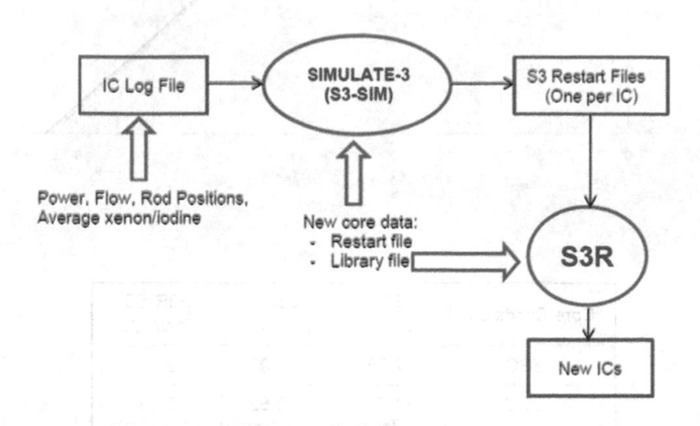

6 Physics Testing

The S3R core model has been validated against CMS results, vendor codes, and plant data when available. As part of the implementation of S3R on a training simulator, standalone physics testing is conducted to compare, in the case of PWRs, power distributions, critical boron concentrations, boron coefficients, temperature coefficients, bank worths, and xenon and samarium worths.

This is done by running S3R (with its own internal TH model) and compare it to the design code (SIMULATE-3 or SIMULATE5). Results from the vendor's code, typically available in the Nuclear Design Report, are also used to validate the S3R model.

Finally, S3R predictions are validated against plant data, such as data collected during the Low-Power Physics Tests conducted during the plant startup and flux maps when these are generated.

Examples of comparison results are show below for boron letdown, temperature coefficient, and bank worths.

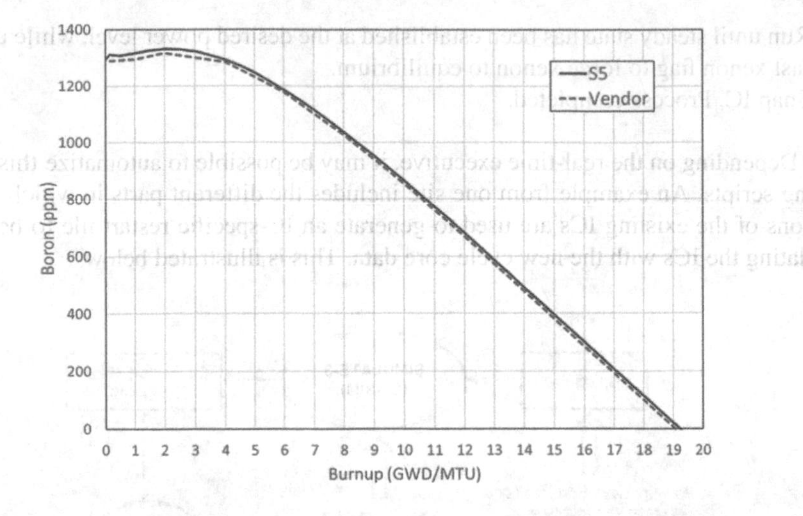

Core Condition	S5 (pcm/C)	S3R (pcm/C)	S3R-S5 (pcm/C)
ZBC	-2.38	0.00	2.38
BOC	-1.71	0.52	2.23
MOC	-11.84	-11.03	0.81
EOC	-30.47	-29.81	0.67

Configuration	S5	S3R	Measured
CA	1268	1309	1131
CB	270	286	294
CC	1180	1224	1241
CD	731	762	678
SA	789	819	775
SB	1320	1361	1252

7 Core Monitoring

Many LWRS use a core monitoring system in the plant control room. These core monitor systems combine measured data and physics calculations to provide operations assistance information. For a variety of reasons, these systems are frequently not available in the simulated control room or are available only via a simplified emulation.

Although conceptually simple, there have been obstacles to implementing core monitoring in the simulator control room. Besides cost and hardware, one important issue from a training point of view is accuracy. The simulator core models take the place of measured data in the plant. The core monitoring system takes measured data and performs calculations to predict things that are not measured. If the simulator is generating inaccurate "plant data," the simulator core monitoring system will generate an inconsistent plant state, and the predicted results will be unusable.

Since S3R is an engineering-grade core model and replicates closely design calculations, it can be used to provide data to a core monitoring system in lieu of the process computer in the plant. This has been demonstrated in several sites which use the simulator version of the Studsvik Core Monitoring product GARDEL. This version, called GARDEL-SIM, runs on its own server (PC, Linux, or Unix) and responds consistently to executive commands such a run, init, freeze, etc. It also responds to numbered Initial Conditions (ICs) or backtracks and gets "plant data" directly from the simulator database.

The data requirement for GARDEL-SIM is the same data required by S3R. No additional data is required.

8 Additional Items

8.1 Decay Heat Reset

S3R includes several fast flags to advance the fission products (Xe and Sm) solution or the decay heat solution faster than real time. One option for decay heat is to be able to reinitialize to a representative decay heat at a given time after shutdown. An example is shown below. This figure shows three curves:

- The base case in red with the expected decrease of decay heat after shutdown
- The case with multiple reinitialization in blue with:

  (1) Reinitialization to 10 days after shutdown
  (2) Reset to about 52 s after shutdown
  (3) Reset to about 17 s after shutdown
  (4) Reinitialization to 1 day after shutdown

- Verification that time behavior after reset to 52 s after shutdown is preserved (broken green line)

This code feature makes the control of the amount of decay heat after shutdown and its impact on the response of the system very straightforward alleviating any need for guessing or tuning.

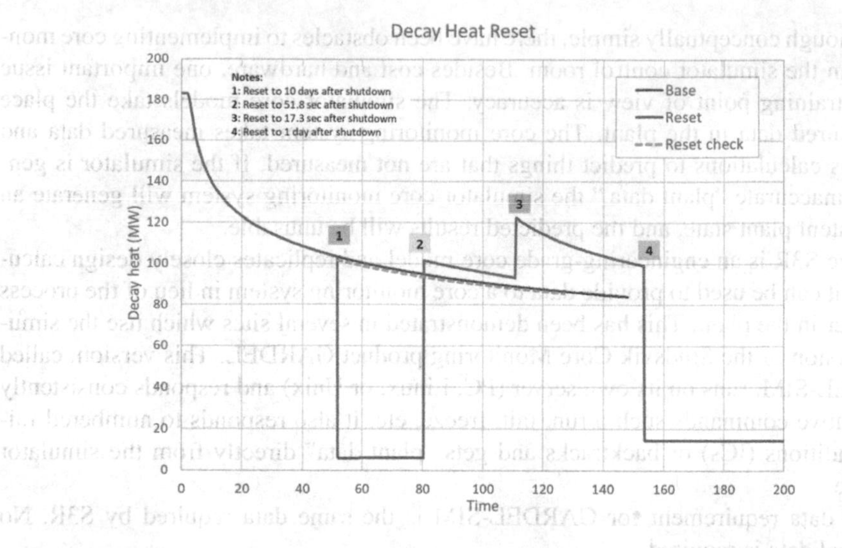

8.2 Xenon Worth

In most cases, the calculation of xenon worth edit is assumed to be proportional to the average xenon concentration. This is a valid assumption in typical LWRs and is one of the methods used in S3R. Not this is only used for providing a xenon.

One case where this assumption shows its limitation is when the core uses mixed UO2 and MOX assemblies. The MOX fuel assemblies have higher initial (equilibrium) Xenon number density due to larger yield and smaller absorption cross section, which dominates the total xenon in the core. On the other hand, the Xenon worth is larger in UO2 fuel (due to softer spectrum) than MOX, and this dominates the core reactivity. Therefore, the time to peak differs between MOX and UO2.

The figure below show change in xenon concentration (red) and the change in xenon worth (blue) following a scram. The largest xenon worth change occurs about one hour later than the peak xenon.

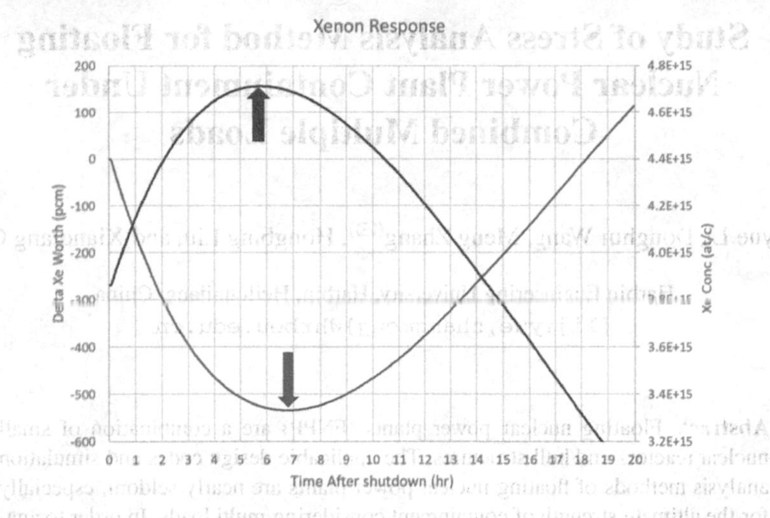

9 Conclusions

S3R and its connection to the design methods of Studsvik has been demonstrated as the go-to tool for training simulators. Since it can directly take data from the depletion calculations, it makes the cycle update automatic and the process of generating new ICs fast and easy. With S3R, training anywhere in the cycle or performing just-in-time training, is readily available.

References

1. Borkowski, J.: SIMULATE-3K Models and Methodology. SOA-98/13 Rev 0 (1998)
2. Borkowski, J.: Core monitoring applications in the simulator control room. In: Power Plant Simulation Conference, San Diego, California (2007)

Study of Stress Analysis Method for Floating Nuclear Power Plant Containment Under Combined Multiple Loads

Jiyue Li, Donghui Wang, Meng Zhang(✉), Hongbing Liu, and Xianqiang Qu

Harbin Engineering University, Harbin, Heilongjiang, China
{lijiyue,zhangmeng}@hrbeu.edu.cn

Abstract. Floating nuclear power plants (FNPP) are a combination of small nuclear reactors and hull structures. The applicable design codes and simulation analysis methods of floating nuclear power plants are nearly seldom, especially for the ultimate strength of containment considering multi loads. In order to analyze the structural strength of the steel containment of a floating nuclear power plant under the combined action of multiple loads, the structural response is analyzed in ANSYS considering the external hull loads and internal containment loads such as wave loads, wind loads, current loads, hull impact loads, internal pressure and temperature of the containment. The structural response result from wind, wave, current, internal pressure and temperature loads are calculated, separately, to obtain the stress field of the containment. Finally, the stress fields of the containment generated by each load are superimposed to obtain the stress distribution characteristics of the containment, and then strength assessment and stress analysis are performed.

Keywords: Floating nuclear power plant · combined multi-load action · quasi-static equivalence · containment · stress analysis

1 Introduction

As small reactor technology continues to develop, the advantages of using nuclear reactors on offshore floating platforms to provide energy such as electricity for areas such as offshore oil extraction or remote areas are emerging. In addition, offshore nuclear power plants can also meet a variety of needs such as heat supply and desalination. As a combination of small nuclear reactors and offshore vessels, offshore floating nuclear power plants are becoming a hot spot for engineering research applications. How to ensure the safety of reactors and floating platforms under various accidental operating conditions and extreme loads has become the focus of technical research on offshore floating nuclear power plants [1].

At present, the most mature floating nuclear power plant is the Russian "Lomonosov", which is a large unpowered barge carrying two "KLT-40" type nuclear reactors [2]. In the USA, MIT has designed and developed a new offshore cylindrical floating nuclear power plant that combines an advanced light water reactor with a floating platform,

C. Liu (Ed.): PBNC 2022, SPPHY 283, pp. 800–811, 2023.
https://doi.org/10.1007/978-981-99-1023-6_69

similar in structure to an offshore oil and gas plant [3]. South Korea and France have each developed concepts for the safe design of floating nuclear power plants by submerging them underwater [4].

The safety of floating nuclear power plants is a central factor in their structural design. Because floating nuclear power plants have a nuclear reactor compartment that houses the containment and support structures that encase the reactor, the safety requirements for the containment structure during the design phase are higher than those for conventional nuclear reactors and conventional marine platforms. Under the unique accident conditions of a nuclear power plant and the superposition of multiple accident loads in the sea, both the internal steel structure of the containment and the external support elements are subject to material yielding or structural failure. Therefore, we need to take into account the external hull loads and internal containment loads in various typical marine environments, such as wave loads, wind loads, current loads, hull impact loads, internal pressure in the containment and temperature.

In this paper, the structural response to the combined effects of multiple loads is analyzed in ANSYS, and the structural response to the wave, hydrostatic, internal pressure and temperature loads are calculated separately to obtain the stress field of the containment under the individual effects of the above loads.

2 Materials and Methods

2.1 Study Subjects

The finite element analysis software ANSYS is used to build the overall analysis model of the floating nuclear power plant including the hull and the containment (including support) structure, in which the hull part uses shell cells, the bone part uses beam or shell cells, and the steel containment (including support) uses solid cells.

The hull structure finite element model is constructed using shell181 and beam188 cells, the mesh size of the containment and support part is 0.1 m, the mesh size of the support and hull transition part is 0.2 m, and the rest of the mesh size is 0.8 m. For the pressure vessel structure which is the focus of analysis, the shell cell is used to simplify the analysis. Steel containment shells and internal components such as multi-layer platforms, core shells and pressure suppression pools were established. The geometric model is shown in Fig. 1.

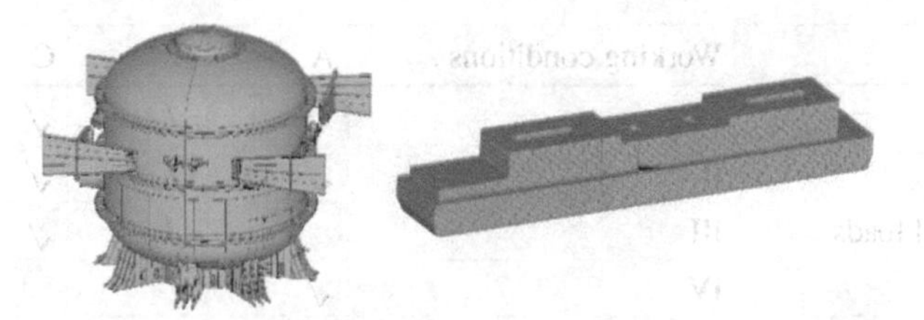

Fig. 1. Geometric model of vessel structure and FNPP

Because of the special characteristics of the nuclear reactor in the containment compartment, the steel plate material used for the containment structure cannot be traditional marine steel, but needs to be high strength steel. In this paper, SA738GrB steel is chosen as the material strength standard. The material properties are shown in the table below (Table 1).

Table 1. SA738GrB steel material properties

Material properties	SA738GrB
Young's modulus(MPa)	2.06E5
Poisson's ratio	0.3
Density(kg/m^3)	7850
Yield limit(MPa)	415
Tensile limit(MPa)	585

2.2 Combination of Working Conditions

The containment structure needs to meet the strength requirements of the load combinations under different categories of operating conditions. Combined with the design provisions of the MC class components in the ASME Code [5], the containment loads are divided into several categories of operating conditions such as A, B, C and D.

Referring to the onshore nuclear power plant containment analysis model, the floating reactor will focus on the marine environmental loads, and the design wave is selected as the ultimate wave load in this paper, replacing the onshore nuclear power plant seismic load. In addition, there are external loads caused by man-made events, such as ship collisions, as accidental load conditions for floating reactor containment. Therefore, the design conditions and load combinations for the containment are shown in Table 2. In the table, I to VIII denote gravity, hydrostatic pressure, ultimate wave load, operating pressure, abnormal pressure, accident pressure, thermal load under accident conditions, and collision load respectively.

Table 2. Combination of loads and working conditions

Load	Working conditions	A	B	C	D
Fixed load	I	√	√	√	√
	II	√	√	√	√
Environmental loads	III		√	√	√
Working load	IV	√	√		
Design accident load	V			√	
	VI				√
	VII			√	√
Artificial event load	VIII				√

2.3 Internal Pressure Loads and Temperature Loads

Nuclear power platforms with a range of design basis accidents based on regulatory or accident analysis assumptions are a fundamental requirement for system design to be met. For reactors, the design basis accident for floating nuclear power plant containment can be modelled on that for land-based nuclear power plants. It is generally accepted that the most serious baseline accident is the Loss of Coolant Accident (LOCA), which causes an increase in containment pressure and temperature [6], and the containment design should meet the structural integrity under this condition.

In the event of a LOCA accident, the pressure inside the containment rises rapidly from 0.1 MPa to 0.8 MPa, and under containment pressure suppression measures the pressure drops to about 0.6 MPa and the temperature drops to about 150 °C.

Combined with the operating conditions of a floating nuclear power plant, the maximum vessel temperature of the containment for accidental operation is 300 °F (148.89 °C) according to the US onshore AP1000 nuclear power plant containment design control document. In the workbench steady state thermal module, the temperature field of the containment and the floating nuclear power plant as a whole can be simulated, and the thermal stresses in the containment structure can then be calculated (Fig. 2).

Fig. 2. Temperature field of part of the structure

It is therefore considered that the containment abnormal pressure is 0.6 MPa and the accident pressure is 0.8 MPa. The accidental service heat load can be calculated indirectly by importing the.rth whole ship temperature field in ANSYS to calculate the whole ship temperature stress.

2.4 Wave Load

The principle of design waves is to replace randomly distributed waves with a series of regular wave equivalents. In order to be able to reflect the maximum force on the platform, the equivalent design waves should put the platform in the most dangerous wave loading condition. The design wave method can simplify the calculation process of wave loads and is widely used by the engineering community. At present, it is mainly divided into deterministic methods, stochastic methods and long-term forecasting methods.

The stochastic design wave method is a short-term forecast of the platform through the wave spectrum, which is more informative than the deterministic design wave method as it reflects the random nature of waves.

Firstly, the overall hydrodynamic characteristic response and load transfer function RAO are determined, then the appropriate wave spectrum $Sw(\omega)$ is selected based on the meaningful wave height in the short-term sea state, and the response spectrum SR(ω) of the wave load is calculated, i.e. $SR(\omega) = [RAO(\omega)]^2 Sw(\omega)$.

Finally, the maximum load response value R and the corresponding load transfer function are selected to calculate the design wave amplitude. The design wave amplitude, A, for stochastic design wave forecasting can be calculated by the following equation: $A = (Rmax/RAO)$. The process is shown in the diagram below (Fig. 3).

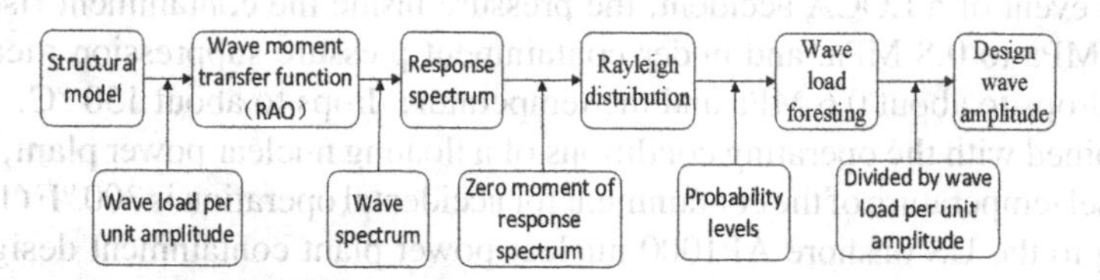

Fig. 3. Design wave calculation flow chart

2.4.1 Frequency Domain Wave Transfer Loads

Regarding the analysis of the load-bearing characteristics of the structural system of floating nuclear power plants under wave action, firstly, based on the overall analysis model of floating nuclear power plants established by ANSYS, the overall hydrodynamic analysis model of floating nuclear power plants can be established using ANSYS-AQWA in combination with the weight of the reactor of the floating nuclear power plant, the center of gravity and other parameters, and further carry out the overall hydrodynamic analysis of floating nuclear power plants under all-round wave incidence angle to obtain the relevant The wave load distribution characteristics of the overall structure of the floating nuclear power plant in the relevant sea area were obtained.

In this project, the waves are divided into 13 directions for analysis, with wave directions set at 15° intervals from 0° to 180°. Load them into the hull to calculate the structural response, the structural response is the corresponding structural under the action of unit wave amplitude, extract the combined force acting on top of the floating nuclear power plant containment, the transfer function is the modulus square of the extracted combined force, thus the transfer function obtained.

2.4.2 Wave and Response Spectra

Due to the shallow depth of the marine structures analyzed in this project and the combined distribution of effective wave height and spectral peak period, the improved JONSWAP spectrum is recommended for the prediction of wave loads in different wave directions. The improved JONSWAP spectrum is expressed in the following form.

$$S(f) = \beta_J H_{\frac{1}{3}}^2 T_P^{-4} f^{-5} \exp[-\frac{5}{4}(T_P f)^{-4}] \gamma^{\exp[-(\frac{f}{f_P}-1)^2 / 2\sigma^2]} \tag{1}$$

Wave loads on ships are forecast using spectral analysis. Short-term forecasts are generally within a few hours, so short-term waves can be considered as a smooth normal stochastic process with zero mean, where the response of the ship to the waves can be considered as a linear relationship, so the wave load response is also a smooth normal stochastic process with zero mean. The response spectrum of a ship's wave load can be expressed as the product of the wave spectrum and the transfer function squared, i.e.

$$S_w(\omega) = |H_w(\omega)|^2 \cdot S_\xi(\omega) \tag{2}$$

Considering the diffusion effect of the wave, a diffusion function can be introduced, which is the angle between the combined wave and the main wave direction. The diffusion function can generally be taken as:

$$fs(\theta) = \frac{2}{\pi}\cos^2\theta \tag{3}$$

Calculating the nth order spectral moment of the response spectrum yields a range of response values.

$$m_n = \int_0^{\infty}\int_{-\frac{\pi}{2}}^{\frac{\pi}{2}} \omega_n \cdot f_s(\theta) \cdot S_w(\omega) d\theta d\omega \tag{4}$$

where is the spectral moment of order of the response spectrum; is the angle between the combined wave and the main wave direction.

It has been shown that the short-term response of the ship's wave load amplitude follows a Rayleigh distribution with a probability distribution function F(x) of

$$F(x) = 1 - \exp(\frac{-x^2}{2m_0}) \tag{5}$$

m0 is the 0th order moment of the response spectrum. The amplitude of the wave load response of the ship corresponding to any exceeded probability level in a certain sea state can be obtained. The design value of wave load R for a ship at a certain number of fluctuations N is:

$$R = \sqrt{2m_0 \cdot \ln N} \tag{6}$$

2.4.3 Determination of Design Wave Parameters

The maximum meaningful wave height for a floating nuclear power plant in the vicinity of the operating sea is 5.2 m with a spectral peak period of 7.8 s. The wave load design calculates the wave bending moment at a probability level of 10^{-8} for a 500 years event. The design wave amplitude is calculated by taking the maximum load response value R and the corresponding load transfer function per unit wave amplitude. The design wave amplitude, A, for a stochastic design wave forecast can be calculated by the following equation:$A = (Rmax/RAO)$.

From this the design wave height corresponding to the wave bending moment at the 10^{-8} probability level can be calculated, as shown in the table below (Table 3).

Table 3. Design waves at the 10^{-8} probability level

Wave (°)	Wave moment forecasting(N*m)	RAO(N*m/m)	Design wave amplitude(m)
0	6.83E + 08	6.19E + 08	1.1
15	7.21E + 08	6.00E + 08	1.2
30	8.90E + 08	5.42E + 08	1.7
45	1.46E + 09	4.84E + 08	3.0
60	2.30E + 09	4.32E + 08	5.3
75	1.39E + 09	2.02E + 08	6.9
90	3.24E + 08	5.63E + 07	5.8
105	1.26E + 09	1.87E + 08	6.7
120	2.16E + 09	4.13E + 08	5.3
135	1.44E + 09	4.75E + 08	3.1
150	8.67E + 08	5.43E + 08	1.6
165	6.80E + 08	6.02E + 08	1.2
180	6.38E + 08	6.21E + 08	1.1

2.5 Crash Load

Collision simulation and quasi-static equivalence in the workbench LS-DYNA module. A container ship with a displacement of 5000 T was selected for this artificial accident load. Lateral collision stresses were calculated to simulate the most dangerous external accident. The impact site is the middle port side of the transom hold.

The dynamic forces of the collision need to be converted into equivalent static forces so as to superimpose with other physical field stresses, there are relevant international codes to simplify the calculation of ship collision forces, among which the European code and the American code are more commonly used and accurate. 1999 European code gives the ship collision force calculation formula to consider the ship navigation waters. The formula given in the 1999 Eurocode takes into account the impact of the ship's speed and tonnage on the collision force and is more applicable to this case.

The simplified Eurocode formula for ship collision force is shown below.

$$P = v\sqrt{Km} \tag{7}$$

where P is the collision force (N), v is the ship impact velocity, K is the ship equivalent stiffness, for sea area ships can be taken as 15 × 106 N/m, m is the collision ship mass (kg).

Through the formula calculation, 2m/s impact of 5000 tons ship equivalent collision static force is 1.73x107N.

2.6 Stress Analysis Methods

Based on the ASME Code, we use the Analytical method to carry out stress intensity checks. We classify the stresses at each check point along the thickness direction to obtain the corresponding membrane stress, bending stress and peak stress.

The membrane stress is the average of the target stresses integrated in the thickness direction and the bending stress is the total stress at the point of integration in the thickness direction minus the membrane stress, i.e. the sum of the membrane stress and bending stress is the stress on the surface of the shell unit. According to the stress intensity calculation method specified in the ASME Code, the stress intensity value of the target node is the Tresca stress value.

For a node connected to different units, the magnitude of the stress value extrapolated to the node by each unit is not necessarily the same. The FEA software usually uses the arithmetic mean of the above extrapolated stress values as the equivalent of the stress value at the node. However, in the case of a discontinuous structure or a local structure with a non-uniform distribution of stress levels, the stress values extrapolated to the node vary considerably and the use of arithmetic averaging to estimate the stress level at the node is not desirable. Therefore, when calibrating the strength of the containment, the extrapolated stress value of the unit in the region with the higher stress level should be used as the stress value at the node.

3 Results

3.1 Stress Evaluation

Different levels of service have different stress limits in the elastic analysis, for the design condition and the limits specified for the operational condition, as shown in the table below (Table 4).

Table 4. Limits and stress limits for different operating conditions during elastic analysis of pressure vessels

Restrictions	P_m(MPa)	P_L(MPa)	$P_m + P_b$(MPa)
Design constraints	$\leq S_m$	$\leq 1.5S_m$	$\leq 1.5S_m$
A	–	–	–
B	$\leq 1.1S_m$	$\leq 1.65S_m$	$\leq 1.65S_m$
C	$\leq \mathrm{Min}(S_y, 2/3S_n)$	$\leq \mathrm{Min}(1.5S_y, S_n)$	$\leq \mathrm{Min}(1.5S_y, S_n)$
D	$\leq 0.7S_n$	$\leq S_n$	$\leq S_n$

where S_y, S_m, S_n are the yield strength, design strength and tensile strength respectively.

3.2 Individual Load Conditions

The stress response of the whole vessel under the above loads acting separately and independently was calculated in ANSYS and the maximum stresses in the pressure vessel structure under each load were extracted. The results are shown in the Table 5 and Fig. 4.

Table 5. Stress strength assessment and calibration for each individual load condition

Load	P_L Max (MPa)	Stress limits (MPa)	$P_L + P_b$ Max (MPa)	Stress limits (MPa)
Design wave	96.2	390	99.53	585
0.8MPa	244.16	409.5	246.65	585
0.6MPa	170.95	409.5	171.64	585
0.1MPa	28.78	202.3	33.5	303.4
Hydro pressure	143.44	183.9	177.72	275.9
Collision	21.02	390	247.8	585
Thermal stress	415.45	409.5	441.7	585

After analysis, the film and bending stresses of all load components except thermal stresses meet the stress limits.

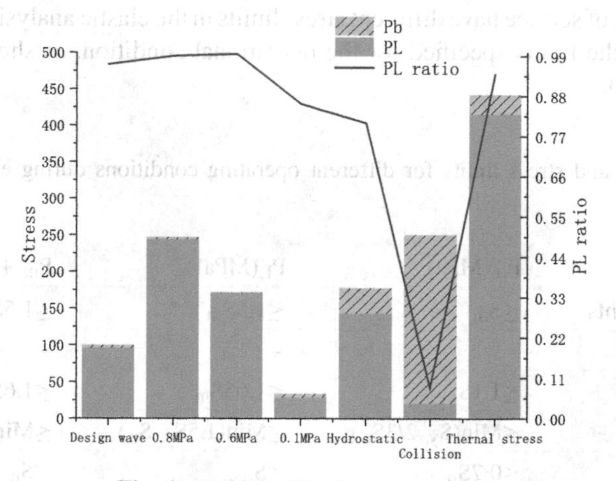

Fig. 4. Individual load stress analysis

Only one node exceeds the stress limit under thermal stress, the maximum stress occurs at the bottom support of the containment shell, the film stress at the node at the maximum stress location is 415.45MPa, the maximum film stress plus bending stress is 441.70MPa, the maximum stress occurs locally. The other locations of the bottom support are all less than 400MPa, the maximum stress at the containment shell is 85.63MPa and the maximum stress at the pressure suppression pool is 274.15MPa, both of which meet the stress limits.

3.3 Combined Load Conditions

The stress field results are calculated individually and obtained as rst results files, after which the stress field results are superimposed according to the combination of conditions to obtain the superimposed stress field and then analyze the structural stress distribution. The superimposed stress field for the whole ship and the containment stress field (Fig. 5).

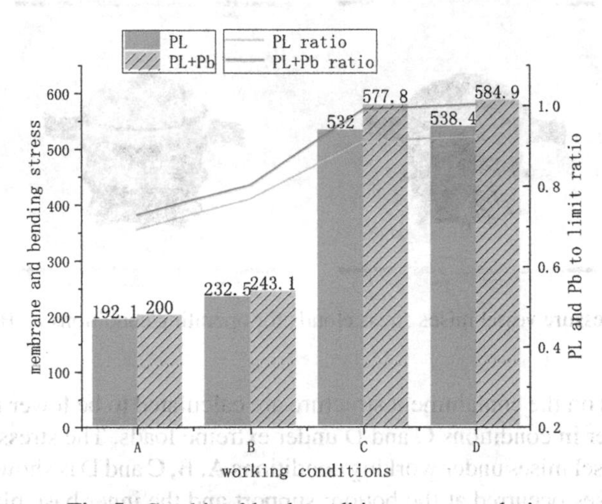

Fig. 5. Combined load conditions stress analysis

4 Discussion

Comparing the structural response of the hull under each condition individually, it can be seen that hydrostatic pressure, temperature load and internal pressure have a greater influence on the stresses of the whole ship and the containment structure.

Comparing the elastic stress composition of each condition, the stresses caused by internal pressure are primary stresses; the temperature load, design wave and other conditions belong to the stresses generated by the containment structure to meet the displacement restraint conditions or the continuous requirements of its own deformation after being subjected to displacement loads, which are secondary stresses. The internal pressure, hydrostatic pressure, thermal stress and design wave conditions are mainly composed of film stress, and bending stress is small; while the bending stress accounts for a large proportion of the stress in the collision condition.

For the composition of the combined working condition stress: for the maximum stress point at the bottom support, mainly the high temperature thermal stress and hydrostatic pressure have a large force on it. For the internal structure of the containment, especially the relatively high stress parts of the ground floor platform and the outer shell, are mainly affected by internal pressure and thermal stresses.

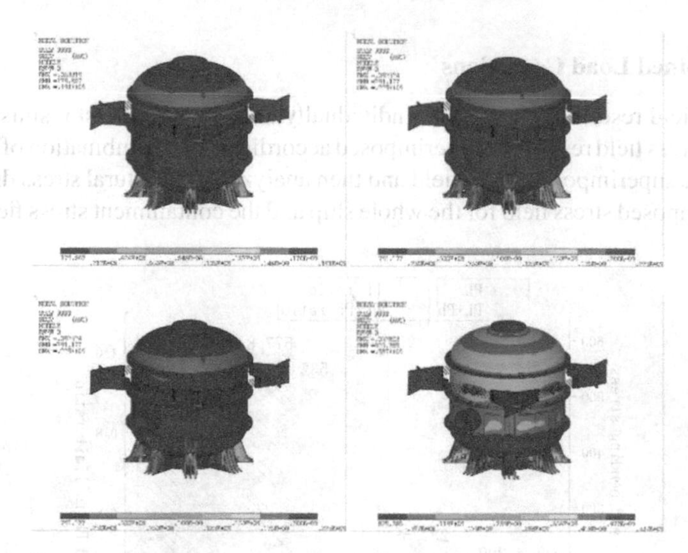

Fig. 6. Pressure vessel mises stress clouds for operating conditions A, B, C and D

The stresses on the containment structure are calculated to be lower in conditions A and B and higher in conditions C and D under extreme loads. The stress distribution of the pressure vessel mises under working conditions A, B, C and D is shown in Fig. 6. The maximum stresses occurred at the bottom support and the inner base plate weld, while the maximum film stress in accident condition D was 538 MPa, reaching 92% of the stress limit, and the sum of the maximum film stress and bending stress was 584.9 MPa, reaching 99.6% of the limit. These are local stress and the overall maximum film stress in the bottom support is approximately 254MPa, which is below the ASME Nuclear Code and Standard Stress Strength Assessment Value, and below the Offshore Mobile Platform Entry Code Allowable Strength and Material Yield Value. This indicates that the design of the structure satisfies the containment and floating nuclear power plant limits for self-sustainability.

References

1. Qiu, C.F., Deng, J., Chen, H.H., Li, F., Yu, N., Wu, P., et al.: Modular small reactor overpressure risk and design optimization study. Atomic Energy Sci. Technol. **50**(1), 5 (2016)
2. Dowdall, M., Standring, W.: Floating nuclear power plants and associated technologies in the northern areas. Norwegian Radiation Protection Authority (2008)
3. Kindfuller, V., Todreas, N., Buongiorno, J., et al.: Overview of security plan for offshore floating nuclear plant. In: International Conference on Nuclear Engineering, V005T15A076 (2016)
4. IAEA. Nuclear Technology Review. Section B, Advanced Fission And Fusion (2014)

5. American Society of Mechanical Engineers. ASME BPVC-III rules for construction of nuclear facility components. Volume 1 NE Sub-volume: MC-class components. 2004 Edition. Shanghai Science and Technology Press, Shanghai (2007)
6. Tan, M,. Li, P.F., Guo, J., et al.: Design of floating nuclear power plant containment under marine environment conditions. Chin. J. Ship Res. **15**(1), 107–112, 144 (2020)

Analytical Method to Study the Ultimate Bearing Capacity of Containment for Floating Nuclear Power Plants Considering Wave Loads

Shuo Mu[1,2], Lijuan Li[3], Meng Zhang[1](✉), Hongbing Liu[1,2], and Xianqiang Qu[1,2]

[1] Yantai Research Institute and Graduate School, Harbin Engineering University, Yantai, China
{mushuo,zhangmeng}@hrbeu.edu.cn

[2] College of Shipbuilding Engineering, Harbin Engineering University, Harbin, China

[3] Nuclear Power Institute of China, Chengdu, China

Abstract. In floating nuclear power plant (FNPP), important equipment such as nuclear pressure vessel, pressure pipeline and pressurizer are installed in the containment, which is the last safety barrier of the reactor primary circuit of FNPP. The ultimate bearing capacity of containment is one of the important safety indexes of FNPP. In this study, considering the wave load, internal pressure load and their corresponding combination conditions in the sea area of FNPP, the overall finite element model of FNPP and the local finite element model of containment are established by using ANSYS software. The ultimate bearing capacity of containment structure is analyzed by nonlinear method. The ultimate internal pressure that the containment can bear is analyzed based on the wave load. Two methods are used to analyze the ultimate internal pressure that the containment can withstand. One is direct analysis method to directly carry out nonlinear analysis of the whole ship model. The other, called indirect analysis method, first makes a linear analysis of the whole ship model to calculate the input load transmitted from the marine environment to the containment through the containment support, and then uses the nonlinear method to analyze the ultimate internal pressure that the containment can withstand. The analysis results of the above two methods show that the ultimate bearing capacity of the containment is basically consistent, while the efficiency of the indirect analysis method is higher. In addition, the ultimate bearing capacity of the containment is far beyond the design requirements considering the wave load.

Keywords: floating nuclear power plant · containment · ultimate bearing capacity · direct analysis method · indirect analysis method

1 Introduction

At present, the research on the ultimate strength of offshore platform structures has become a hot topic in the field of international offshore engineering. For the structural design of offshore platforms, the limit state based design method has been introduced [1]. Through the calculation of the ultimate strength of the platform, the dangerous

C. Liu (Ed.): PBNC 2022, SPPHY 283, pp. 812–826, 2023.
https://doi.org/10.1007/978-981-99-1023-6_70

component parts can be predicted, and the local strengthening of the components can timely control the degree of damage and reduce the potential crisis. Up to now, scholars have put forward different research methods for the ultimate strength of ship structures, and several of them have been widely recognized and gradually improved. At present, the ultimate strength analysis methods of hull structures mainly include: direct calculation method, nonlinear finite element method, progressive collapse method, ideal structural element method and model test method.

The direct calculation method is also called Caldwell method. In 1965, Caldwell [2] first proposed the plastic behavior of the total longitudinal strength of surface ships, and derived the analytical formula of the total longitudinal ultimate strength. In view of the limitations of Caldwell method, Smith [3] proposed a relatively simple method based on Caldwell method, taking into account the reduction of strength after the member reaches the ultimate strength, and thus the progressive collapse method is also called Smith method. Another simple method for progressive failure analysis of hull beams is the ideal structural element method (ISUM) proposed by Japanese scholar Ueda et al. [4] in the 1980s. With the general improvement of computer hardware level, the finite element method (FEM) has been applied more often to engineering problems. Compared with the above analysis methods, the physical test can more directly reflect the collapse process of the ship structure and measure the ultimate strength of the structure. Dow [5] established a hull beam model of a frigate at a scale of 1/3 and conducted ultimate strength tests. Nishihara [6] conducted a series of collapse tests under pure bending load on a box beam model similar to the corresponding ship structure. The phenomena and results observed in the experiment also have high reference value for theoretical calculation and numerical simulation.

2 Nonlinear Analysis Method

Chen [7] et al. proposed a finite element method for hull ultimate bearing analysis, which is applicable to any loading type and structural model. This method introduces three element types: beam element, plate element and orthotropic plate element, which can analyze the limit state of the structure under static or dynamic loads, and It can also analyze the overall response of a single member (deck, side, bulkhead, longitudinal girder, etc.) and consider the response of the hull under the combined action of bending moment, torque and shear force. Kutt [8] et al. used nonlinear analysis method to analyze the longitudinal ultimate strength of four hulls, reflecting the interaction between various local and overall failure modes and the interaction between the elastic−plastic, buckling and post buckling effects of materials.

With the development of calculation technology and nonlinear finite element, many large general finite element programs, such as NASTRAN, ANSYS, ABAQUS, Marc, have been applied to the ultimate strength calculation of ship structures. Based on the nonlinear finite element method, this chapter will use the finite element software ANSYS to study the solution method.

2.1 Arc length method

The arc length method is a static analysis method based on the Newton Raphson iterative solution of the nonlinear static equilibrium equation of the structure [9]. Its basic idea is to set a parameter (arc length l) to control the incremental iteration and convergence of the equilibrium equation.

$$\{P\} - \{I\} = 0 \tag{1}$$

$$[K_T]\{\Delta u\} = \{\Delta P\} - \{R\} \tag{2}$$

where $[K_T]$ is tangent stiffness matrix, $\{\Delta u\}$ is the displacement increment, $\{\Delta P\}$ is load increment, $\{R\}$ is residual force.

When using the arc length method, let the $i-1$ load step converge to (x_{i-1}, P_{i-1}) , for the i load step, j iterations are required to reach the new convergence point (x_i, P_i). Step i iterates the true load size $\{\Delta P\}$ acting on the structure, which is determined by the load increment factor $\Delta\lambda_i$ and the input external load $\{P_{ref}\}$,

$$\{\Delta P\}_i = \Delta\lambda_i\{P_{ref}\} \tag{3}$$

By introducing Eq. (3) into Eq. (2), the incremental form of iteration i of arc length method can be obtained:

$$[K_T]\{\Delta u\}_i = \Delta\lambda_i\{P_{ref}\} - \{R\}_i \tag{4}$$

The position of the equilibrium point in step i is obtained by Newton Raphson iteration with the equilibrium point calculated in step $i-1$ as the center of the circle and the arc length increment Δi as the radius, as shown in Fig. 1. The arc length increment Δl_i, load increment factor $\Delta\lambda_i$ and displacement increment $\{\Delta u\}_i$ of each step satisfy the governing equation of Eq. (5).

$$\Delta l_i^2 = |\{\Delta u\}_i|^2 + \left|\Delta\lambda_i\{P_{ref}\}\right|^2 \tag{5}$$

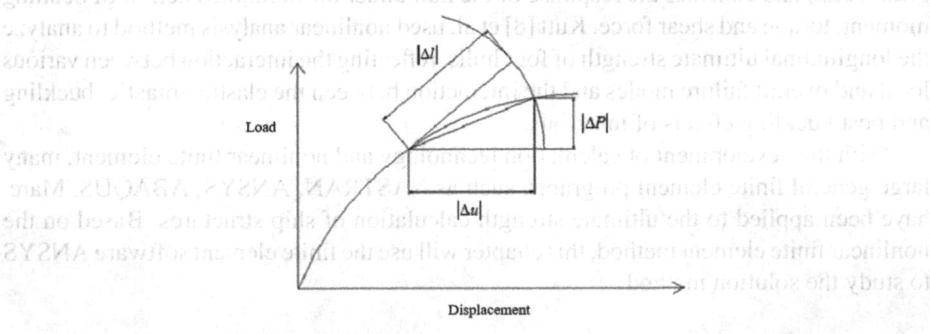

Fig. 1. Load displacement curve

Through iteration, until the residual force is within the tolerance {R}, when the iteration of step i is completed, there are:

$$\text{Load array } \{P\}i = \sum_i \{\Delta P\}_i = \sum_i \Delta\lambda_i\{P_{ref}\} \quad (6)$$

$$\text{Displacement array } \{u\}_i = \sum_i \{\Delta u\}_i \quad (7)$$

$$\text{arc length } l_i = \sum_i \Delta l_i \quad (8)$$

The arc length increment Δl contains the information of load increment and displacement increment at the same time. As long as the appropriate iteration step is selected, the arc length method can track the load displacement equilibrium path of the structure in the process of loading and unloading, and effectively solve the 'step' problem in the limit equilibrium path.

2.2 Quasi Static Method

In essence, the quasi–static analysis method is a dynamic solution process of structure. The basic idea of the quasi–static method is to simulate the static problems with the dynamic analysis of slow loading, which is based on the explicit solution of the structural nonlinear motion equation (central difference method).

The central difference method is used to perform explicit time integration for the structural nonlinear motion Eq. (9). The dynamic conditions of the next incremental step are calculated from the dynamic conditions of the previous incremental step until the end of the solution time.

$$[M]\{\ddot{u}\} = \{P\} - \{I\} \quad (9)$$

where $\{P\} - \{I\}$ is the mass matrix, $\{\ddot{u}\}$ is the acceleration array, $\{P\}$ is the load array, and $\{I\}$ is the internal force array.

The key to the solution of quasi–static method is to set a reasonable loading rate. Too fast loading rate leads to severe local deformation of the solution results, which makes the calculation results deviate from the requirements of "quasi–static", and the load displacement curve oscillates. However, too slow loading rate increases the calculation time greatly, which means a long loading time. Therefore, we usually take several different loading rates for comparison and analysis, and then select the appropriate loading rate when analyzing.

By studying the various energies in the model, we can evaluate whether the simulation produces the correct quasi–static response.

$$E_{KE} + E_{FD} + E_I + E_V - E_W = E_{total} \quad (10)$$

where E_{KE} is the structural kinetic energy, E_{FD} is the energy dissipated and absorbed by friction, E_I is the internal energy, including plastic and elastic strain energy, E_V is the energy dissipated and absorbed by viscosity, E_W is the work done by external forces, and E_{total} is the total energy in the system, which is a constant.

If the finite element simulation is quasi−static, the internal energy of the system is almost equal to the work done by the external force. Generally, the viscous dissipation energy is very small, unless material damping, viscoelastic materials or discrete shock absorbers are used. Therefore, we can determine that the inertia force is ignored according to the small flow velocity of the material in the model in the quasi−static process. It can be inferred from these two conditions that the kinetic energy is also very small. In most processes, the ratio of kinetic energy to strain energy of the structural model is an important criterion to judge whether the loading rate is appropriate. As a general rule, the general quasi−static requirement is that the ratio of kinetic energy to strain energy is less than 5%.

The applied loads are required to be as smooth as possible for accurate and efficient quasi−static problems. Stress waves arise from sudden, rapid movements that will cause oscillations or inaccurate results. The loading method shall be as smooth as possible, and the acceleration between two incremental steps shall only change by a small amount. If the acceleration is smooth, the varying velocity and displacement are smooth.

Since the central difference method is a conditionally stable algorithm, the time step Δt must be less than the stability limit Δt_{stable} to ensure the stability of the solution:

$$\Delta t_{stable} = 2/\omega_{\max} \geq \Delta t^e \tag{11}$$

where $\omega_{\max}$ is the highest natural frequency of the structure, Δt^e is the stable time step of the element with the smallest size in the structural model, Δt^e is related to the characteristic scale L^e, elastic modulus E and material density of the element ρ Is the upper bound of the stability limit Δt_{stable}:

$$\Delta t^e = L^e/\sqrt{E/P} \tag{12}$$

In the calculation, since the highest order frequency $\omega_{\max}$ of the structure is not easy to obtain, the time step Δt is taken as Δt^e.

Compared with the arc length method, the quasi−static method has no convergence problem because the central difference method is used for explicit integration, and can solve more complex structural collapse problems such as material failure and structural self−contact. Because the time step Δt is often small, the quasi−static loading is slow and the solution time is very long when solving the limit state problem, which can be adjusted by means of mass amplification.

3 Calculation of Ultimate Bearing Capacity Under Different Working Conditions

According to the classification of marine environmental conditions and the guide for buckling strength assessment of offshore engineering structures, the load distribution characteristics of the hull under different marine conditions are calculated respectively. The finite element model of the floating nuclear power plant is established based on ANSYS software, and then the load borne by the hull is applied to the finite element analysis model of the floating nuclear power plant to calculate and analyze the stress and displacement response of the overall structure of the floating nuclear power plant and the steel containment, and carry out the calculation of the ultimate bearing capacity of the containment of the floating nuclear power plant under different working conditions.

The finite element model of the hull structure is constructed with shell181 and beam188 elements. The mesh size of the bottom and supporting parts of the containment is 0.1 M, the mesh size of the upper part of the containment is 0.2 m, and the mesh size of the rest is 0.8 m. The total number of elements on the whole ship is about 1.56 million (Figs. 2 and 3).

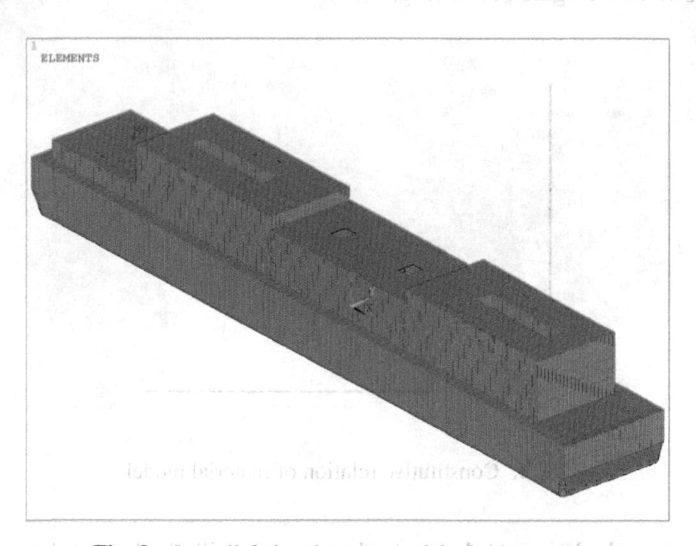

Fig. 2. Overall finite element model of ship structure

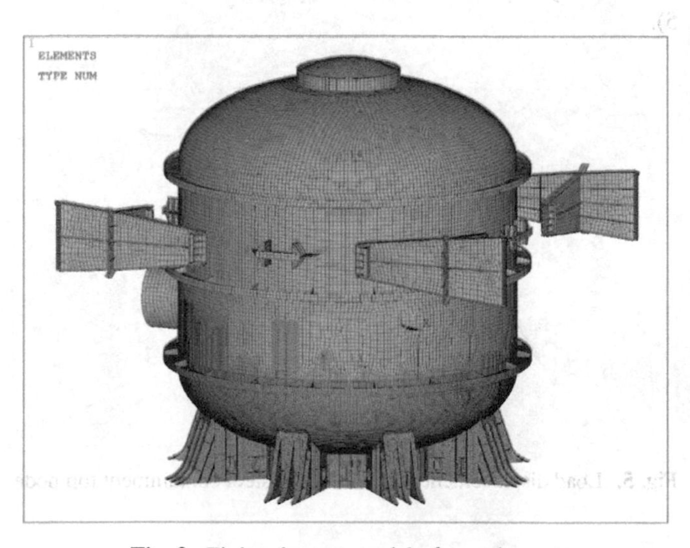

Fig. 3. Finite element model of containment

3.1 Calculation and Analysis of Ultimate Bearing Capacity Under Internal Pressure Condition of Containment

The internal pressure load is loaded inside the containment by uniformly distributing the load in the element, and the fixed support constraint is added at the bottom node of the compartment where the containment is located. Considering the influence of material hardening on the ultimate strength of the structure, the material constitutive relationship is set as shown in the figure below (Fig. 4).

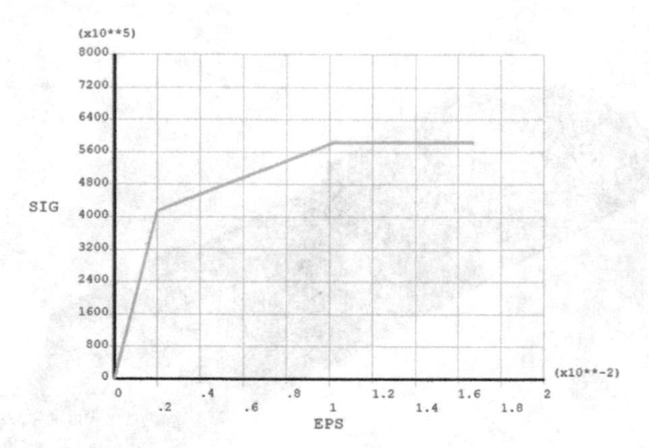

Fig. 4. Constitutive relation of material model

Sort out the calculation results, and select the top with the largest deformation and change rate as the analysis point. The load displacement curve is shown in the figure below (Fig. 5).

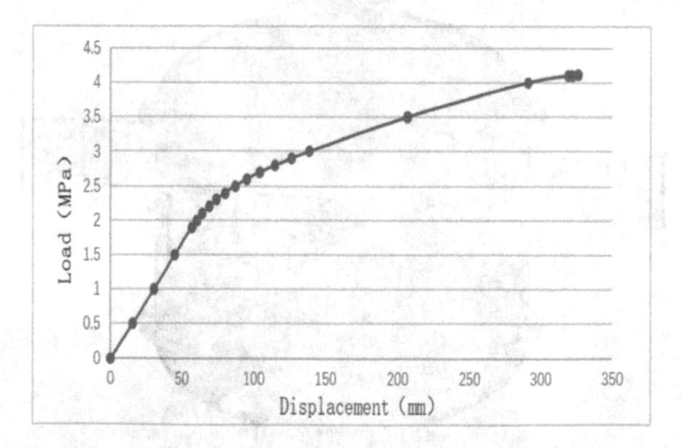

Fig. 5. Load displacement relationship of steel containment top node

The analysis shows that when the internal pressure load is below 2 MPa, the slope of the curve does not change, showing a linear growth trend, indicating that the structure is still in the elastic deformation order, and the equivalent plastic strain is 0; When the internal pressure load increases from 2 MPa, the curvature in the load displacement curve begins to change, indicating that the structure begins to enter the plastic deformation stage.

(1) Double limit method

The double elastic slope method is used to determine the limit load in ASME VIII−2. In this method, the origin of load displacement curve is taken as the starting point to draw the straight line of elastic slope in the elastic stage, and then another straight line is drawn to meet the elastic slope whose slope is equal to twice. The load value projected on the vertical coordinate by the intersection of the straight line and the load displacement curve is the limit load value [10] (Fig. 6).

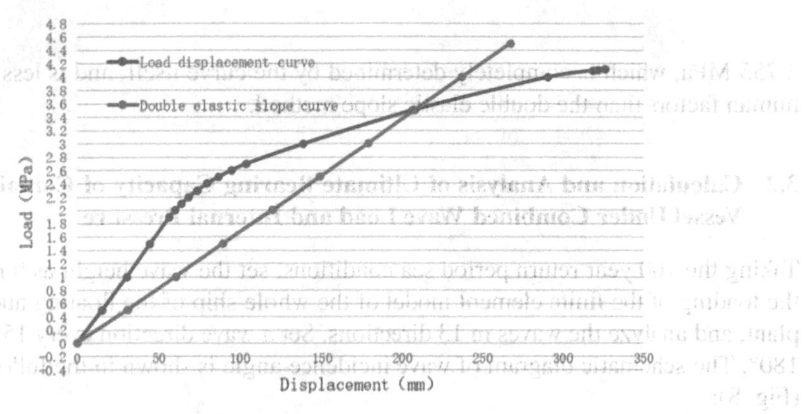

Fig. 6. Load displacement relation of double limit method

It can be seen from the figure that the limit load value determined by this model based on the more accurate load displacement curve after reducing the load increment step and using the double elastic slope method is about 3.442 MPa. This method is an artificial regulation and is greatly affected by human factors. The determined limit load value may have great dispersion and error.

(2) Double tangent intersection method

The double tangent intersection method is the method for determining the limit load in the European standard EN 13445. Based on the load displacement curve, the tangent lines of elastic stage and plastic stage are drawn respectively. The tangent of the plastic section adopts the first−order fitting function obtained from the plastic section data processed by the least square method, and the load value projected on the longitudinal coordinate by the intersection of the two tangents is the limit load value [11] (Fig. 7).

It can be seen from the figure that the limit load value determined by the model based on the load displacement curve and the double tangent intersection method is about

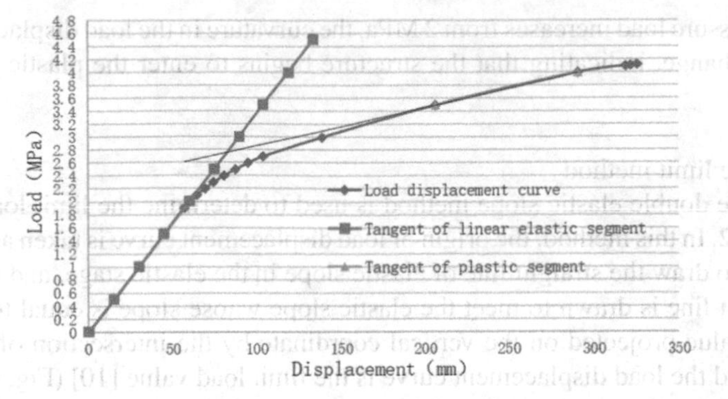

Fig. 7. Load displacement relationship of double tangent intersection method

2.755 MPa, which is completely determined by the curve itself, and is less affected by human factors than the double elastic slope method.

3.2 Calculation and Analysis of Ultimate Bearing Capacity of Containment Vessel Under Combined Wave Load and Internal Pressure

Taking the 100 year return period sea conditions, set the wave height as 9 m, calculate the loading of the finite element model of the whole ship of the floating nuclear power plant, and analyze the waves in 13 directions. Set a wave direction every 15° from 0° to 180°. The schematic diagram of wave incidence angle is shown in the following figure (Fig. 8):

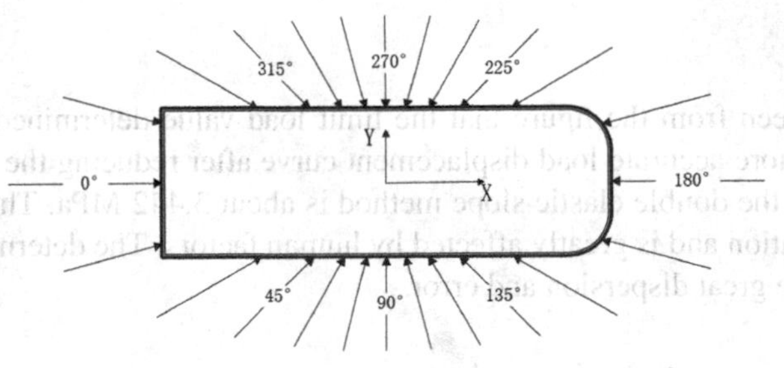

Fig. 8. Schematic diagram of wave incidence angle

In order to improve the calculation efficiency, the displacement structure response in the overall finite element model under wave load is calculated, and then the node displacement of the containment support in the overall model is extracted and loaded at the corresponding node of the containment finite element model as a boundary condition. Take the nodes with the stress value greater than 50 MPa in the containment, and compare the stress value of the containment nodes in the finite element model of the whole ship with the corresponding node stress value on the separate containment model. Some results are shown in the table below (Table 1).

Table 1. Comparison of node stress (zero degree wave load as an example)

NODE	Full ship model	Containment model		
	S1(Pa)	deformation and acceleration	deformation	acceleration
12200162	5.50E+07	3.24E+07	3.35E+07	4.22E+04
12200163	5.52E+07	2.76E+07	2.86E+07	5.25E+04
12200783	5.01E+07	2.69E+07	2.76E+07	−3.67E+04
12200822	5.45E+07	2.64E+07	2.72E+07	5.30E+04
12200823	5.52E+07	3.18E+07	3.28E+07	4.30E+04
12201023	6.81E+07	3.87E+07	4.03E+07	−5.00E+04
12201214	5.52E+07	3.20E+07	3.30E+07	−4.04E+04
12201215	5.56E+07	2.67E+07	2.75E+07	−4.80E+04
12202276	6.24E+07	3.65E+07	3.76E+07	4.21E+04
12202277	6.19E+07	3.05E+07	3.14E+07	5.21E+04
12202354	5.37E+07	2.75E+07	2.83E+07	5.24E+04
12202355	5.75E+07	3.29E+07	3.37E+07	4.51E+04
12202382	6.10E+07	3.50E+07	3.68E+07	−4.15E+04
12202768	5.67E+07	3.29E+07	3.46E+07	56786
12202798	5.31E+07	2.72E+07	2.79E+07	−52338
12202799	5.65E+07	3.19E+07	3.27E+07	−38352
12203178	5.43E+07	2.92E+07	3.01E+07	−39267
12203179	5.43E+07	3.03E+07	3.12E+07	−46813
12204241	6.28E+07	3.63E+07	3.74E+07	41805
12204242	6.28E+07	3.10E+07	3.19E+07	51568
12204764	5.35E+07	2.69E+07	2.75E+07	−49527
12204765	5.53E+07	3.06E+07	3.14E+07	−35320
12204954	6.82E+07	3.89E+07	4.05E+07	−49907
12205145	5.51E+07	3.20E+07	3.30E+07	−40471
12205146	5.54E+07	2.67E+07	2.75E+07	−47568
12206147	5.70E+07	3.09E+07	3.18E+07	45327
12206174	5.57E+07	3.29E+07	3.46E+07	−39438
12206543	6.30E+07	3.52E+07	3.67E+07	55582
12206573	5.78E+07	2.90E+07	2.96E+07	−50465
12206574	6.15E+07	3.39E+07	3.46E+07	−38538

(*continued*)

Table 1. (*continued*)

NODE	Stress ratio (containment node stress value/whole ship node stress value)		
	deformation and acceleration	deformation	acceleration
12200162	58.98%	60.99%	0.08%
12200163	50.02%	51.80%	0.10%
12200783	53.54%	55.00%	0.07%
12200822	48.50%	49.89%	0.10%
12200823	57.69%	59.37%	0.08%
12201023	56.93%	59.25%	0.07%
12201214	57.97%	59.77%	0.07%
12201215	48.10%	49.47%	0.09%
12202276	58.49%	60.21%	0.07%
12202277	49.23%	50.74%	0.08%
12202354	51.26%	52.67%	0.10%
12202355	57.09%	58.63%	0.08%
12202382	57.40%	60.28%	0.07%
12202768	58.00%	60.94%	0.10%
12202798	51.23%	52.56%	0.10%
12202799	56.37%	57.84%	0.07%
12203178	53.72%	55.38%	0.07%
12203179	55.83%	57.41%	0.09%
12204241	57.88%	59.53%	0.07%
12204242	49.37%	50.84%	0.08%
12204764	50.21%	51.41%	0.09%
12204765	55.37%	56.76%	0.06%
12204954	57.07%	59.38%	0.07%
12205145	58.11%	59.91%	0.07%
12205146	48.26%	49.65%	0.09%
12206147	54.21%	55.83%	0.08%
12206174	58.99%	62.17%	0.07%
12206543	55.80%	58.22%	0.09%
12206573	50.10%	51.16%	0.09%
12206574	55.13%	56.33%	0.06%

It is known from the data that the acceleration has little effect on the node stress, and the node stress on the containment mainly comes from the deformation of the structure. In order to simulate the structural response under the wave load in the separate containment model, compare the stress values of the corresponding nodes in the whole ship model and the separate containment model, and load twice the displacement structural response of the containment support node to the corresponding node of the separate containment, As the initial displacement boundary condition, the ultimate bearing capacity of wave load under the combined action of internal pressure is analyzed. The internal pressure loading method is the same as above.

The double limit method is used to analyze the ultimate bearing capacity of the structure as follows (Fig. 9).

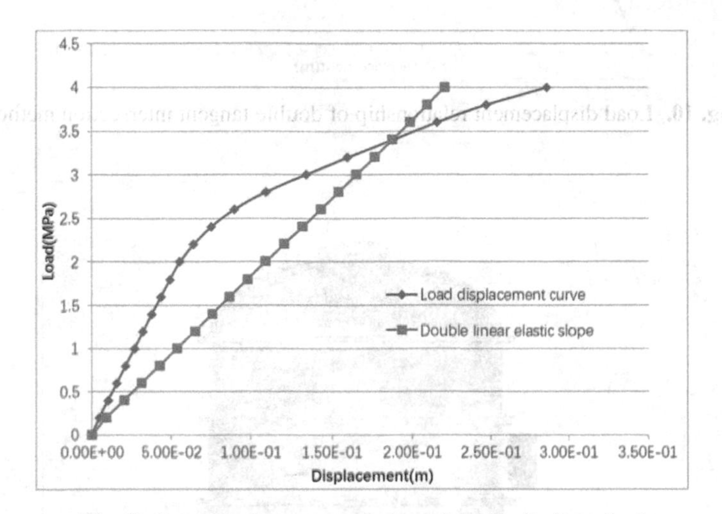

Fig. 9. Load displacement relation of double limit method

It can be seen from the figure that the limit load value determined by the model based on the more accurate load displacement curve after reducing the load increment step and using the double elastic slope method is about 3.45 MPa.

The ultimate bearing capacity of the structure is analyzed by the double tangent intersection method as follows (Fig. 10).

It can be seen from the figure that the limit load value determined by the model based on the load displacement curve and the double tangent intersection method is about 2.56 MPa.

According to the constitutive characteristics of the material, when the stress of the material exceeds the tensile strength of 585 mpa, it is considered that the strength failure of the material occurs there. According to the finite element analysis, when the internal pressure of the structure is 3.1 Mpa, the maximum value of the maximum stress point is 584 mpa. Therefore, it is considered that the structure has strength failure at 3.1 Mpa (the stress nephogram is shown in the figure below) (Fig. 11).

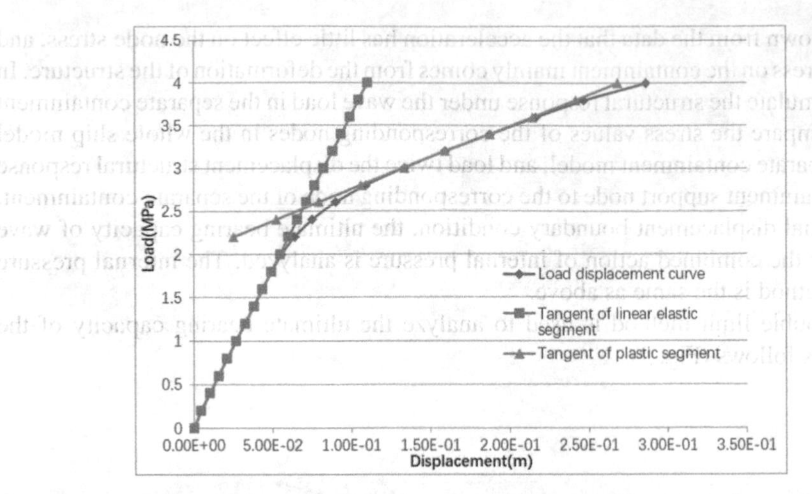

Fig. 10. Load displacement relationship of double tangent intersection method

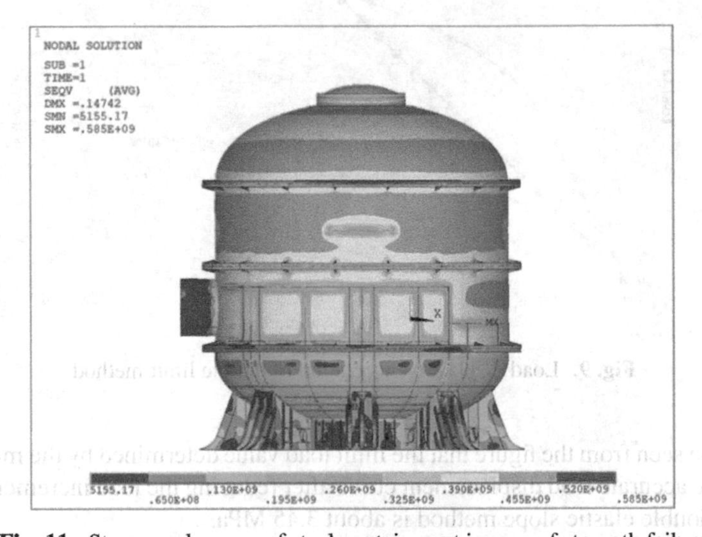

Fig. 11. Stress nephogram of steel containment in case of strength failure

4 Conclusions

Based on an engineering example model, this paper studies the analysis method of the ultimate bearing capacity of the containment of a floating nuclear power plant, and draws the following conclusions.

1. Based on the finite element analysis software, the stress–strain nephogram and load displacement curve are accurately calculated, and the limit load can be judged and determined by the double elastic slope method and the double tangent intersection method. In terms of the accuracy of determining the limit load value, the double

tangent intersection method is within an acceptable range due to the difficult determination of the tangent in the plastic stage, and the double elastic slope method is susceptible to human factors, so the accuracy is conservative.
2. Considering the wave load, compared with the finite element nonlinear analysis of the whole ship model, the method of extracting the load at the containment support in the overall model for linear calculation and loading it on the corresponding node of the local model of the containment is adopted to ensure the accurate load loading, shorten the calculation time and improve the calculation efficiency.
3. In the method of indirect loading, the impact of wave load on the containment of floating nuclear power plant is divided into two parts: the impact of structural deformation and the impact of structural acceleration. Compared with the impact of structural deformation on the ultimate bearing capacity of the containment, the impact of structural acceleration is relatively small and can be ignored.

References

1. Veritas, D.N.: Design of Offshore Steel Structures - General (lrfd Method) (2004)
2. Caldwell, J.B.: Ultimate longitudinal strengt. trans rina (1965)
3. Smith, C.S.: Influence of local compressive failure on ultimate longitudinal strength of a ship's hul. In: Proceedings of International Symposium on Practical Design in Shipbuilding, Tokyo, Japan (1977)
4. Ueda, Y., Rashed, S.M.H., Paik, J.K.: Plate and stiffened plate units of the idealized structural unit method (1st Report) under in plane loading. J. Soc. Naval Arch. Jpn. (1984)
5. Dow, R.S.: Testing and analysis of 1/3-scale welded steel frigate model. In : Proceedings of the International Conference on Advances in Marine Structures, pp. 749–773 (1991)
6. Nishihara, S., Engineering, O.: Ultimate longitudinal strength of mid-ship cross section. Naval Arch. **22**, 200–214 (1984)
7. Chen, Y.K., Kutt, L.M., Piaszezyk, C.M., et al.: Ultimate strength of ship structures. Trans. SNAME, pp. 149–168 (1983)
8. Kutt, L.M., Piaszczyk, C.M., Yung-Kuang, C., et al.: Evaluation of the longitudinal ultimate strength of various ship hull configurations. Trans.-Soc. Naval Arch. Marine Eng. **93**, 33–53 (1985)
9. Peng, D., Zhang, S.: Study on three finite element methods for structural ultimate strength analysis. China Offshore Platform **2**, 5 (2010)
10. Li, J.: Mechanical basis of pressure vessel design and its standard application (2004)
11. Spraragen, W.: Unfired pressure vessels. Indengchem **23**(2), 220–226 (1931)

Decoupling and Coupling Simulation Analysis in Small Nuclear Power Plant

Peiyu Tian[1](✉), Yi Li[1], Tianyang Xing[2], Tiebo Liang[1], and Changshuo Wang[1]

[1] Science and Technology on Reactor System Design Technology Laboratory, Nuclear Power Institute of China, Chengdu, Sichuan, China
805656155@qq.com

[2] School of Energy and Environment, Southeast University, Nanjing, Jiangsu, China

Abstract. As the development of computer science technology and the requirement of thorough research, more researchers are setting their eyes on coupling method because of the tight coupling in NPP (nuclear power plant) system. However there were few researches studying the difference between decoupling and coupling methods and the importance of coupling method. This research respectively establishes the primary and secondary loop decoupling models and the two loops coupling model based on the small NPP by APROS. Then the differences between the decoupling and coupling models is studied under the steady state and dynamic state which contains the ramp load variation and load shedding. The results show that there are small differences between these models in the main parameter values under the steady state. But the differences between decoupling models and coupling model are large. Therefore the NPP system needs be modeled by coupling method as to study its dynamic characteristic.

Keywords: Small NPP · Decoupling simulation · Coupling simulation · APROS · Ramp load variation · Load shedding

Nomenclatures

A	heat transfer area, m^2
$C(t)$	delayed neutron precursor concentration for each period
c_p	specific heat capacity, kJ/(kg °C)
G	mass flow, kg/s
h	specific enthalpy, kJ/kg
K	heat transfer coefficient, W/(m2 °C)
M	mass, kg
$n(t)$	neutron number for each period
T	temperature, °C
t	time, s
V	volume, m^3
β	delayed neutron fraction
λ	disintegration constant, s
$\rho(t)$	reactivity for each period, Δk/k

C. Liu (Ed.): PBNC 2022, SPPHY 283, pp. 827–844, 2023.
https://doi.org/10.1007/978-981-99-1023-6_71

ρ density, kg/m^3
Λ neutron generation time, s
Subscript
a average
c coolant
fw feed water
i ith delayed neutron group
m metal material
s saturation
st steam
w water

1 Introduction

Simulation technique has been the main method used to study the reactor engineering field since the late 1960s because of the objective factors limitation such as measurement and manufacture level, the danger of test involving accident conditions, and the high price of the test. Then when there are more NPP systems with higher capacity and parameter value designing and application, researchers and engineers pay more attention to the simulation technique aftthe er accidents at Three-Mile Island and Chernobyl. But most of the research was decoupling simulation because the NPP system contains many subsystems which make the modeling difficult and there were few simulation softwares that amid at both the primary loop and secondary loop. For instance, Guo Liang [1], etc. used the C++ Builder to study the dynamic characteristic of the Daya Bay NPP's secondary loop system. Huo Binbin [2] etc. used Matlab/Simulink software to model the reactor coolant system of Daya Bay NPP under the positive signal disturbance.

And with the rapid development of computer techniques and the requirement of thorough research in the system characteristic, more and more researchers want to study the NPP system by coupling simulation model because of the tight coupling in the NPP system and think it's necessary to study the system by coupling model which is more complex and time-consuming than decoupling model. For example, LIN Meng [3] etc. expanded RELAP5 by C++ language to achieve the real-time simulation of a nuclear power plant. Xiao Kai [4] etc. coupled RELAP5, 3KEYMASTER and Matlab/Simulink to establish multi-reactors and multi-turbines nuclear power systems and simulated the dynamic condition. Nianci Lu [5] etc. studied the nuclear power system dynamic response under reactor trip by the Gsuite simulation platform. Pack J [6] etc. proposed a coupling model between RELAP5 and LABVIEW to model and simulate the whole nuclear power system.

However although Pack J compared the coupling model with the RELAP5 decoupling model under the LOCA event which showed that the coupling model agreed well with the RELAP5 decoupling model, the paper didn't talk about the necessity and importance of the coupling method. Besides in that paper, the second model in LABVIEW was simplified where the steam generator was calculated by the thermal equilibrium equation and there was no control system, so the comparison between decoupling and coupling models had limitations.

Therefore this research respectively established decoupling system models and coupling system models based on a small NPP. Then on basis of it, steady state and transient state are simulated to compare the coupling method and decoupling method, which direct the difference between the two methods and under what circumstances the decoupling method could be used.

2 Model

This small NPP system is configured with one reactor, double circuits, and double turbine generators. And the primary loop system contains a reactor coolant system and pressure safety relief system. The secondary loop system contains the main steams system, exhausted steams system, two turbine generator units, condensate extraction system and feed water system, and so on. According to the system configuration and flow, this paper establishes respectively the primary and secondary loop decoupling models and the whole system coupling model based on the APROS simulation software.

2.1 Primary Loop Decoupling Model

It supposes that the secondary side of the steam generators is a boundary condition in this decoupling model. So the user should give the steam mass flow value when the dynamic process is simulated. And the feed water temperature is constant during the simulation.

In this case, the steam mass flow is controlled by the control valve which is on the main steam pipe as Fig. 1 showed. And the control schematic diagram is shown in Fig. 2. Then the feed water mass flow is mainly controlled by the flow deviance between the

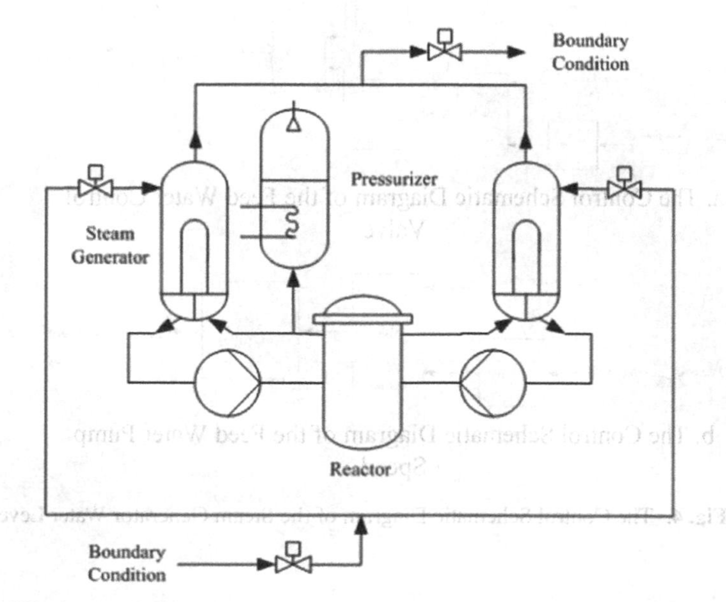

Fig. 1. The Schematic Diagram of the Primary Loop Decoupling Model

steam and feed water, whose control schematic diagram is shown in Fig. 3. It is also controlled by the control system of the steam generator water level, which consists of the feed water control valve control module and the feed water pump control module. And Fig. 4 is the water level control system's schematic diagram. Besides, there is the reactor power control system as Fig. 5 shows.

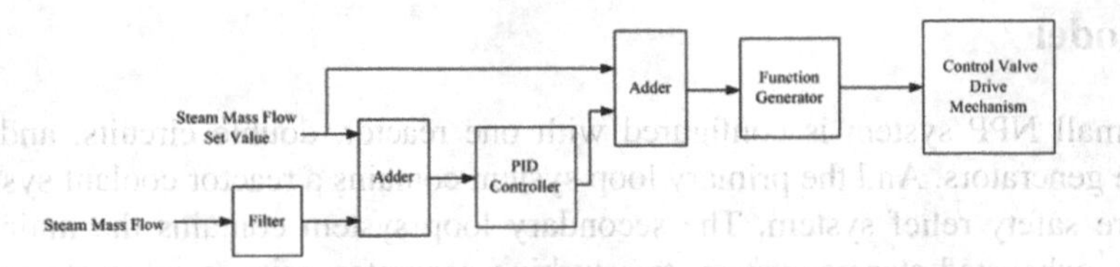

Fig. 2. The Control Schematic Diagram of the Steam Mass Flow

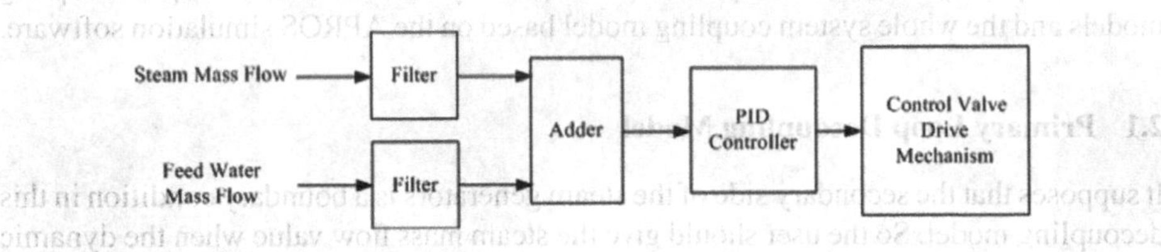

Fig. 3. The Control Schematic Diagram of the Feed Water Mass Flow

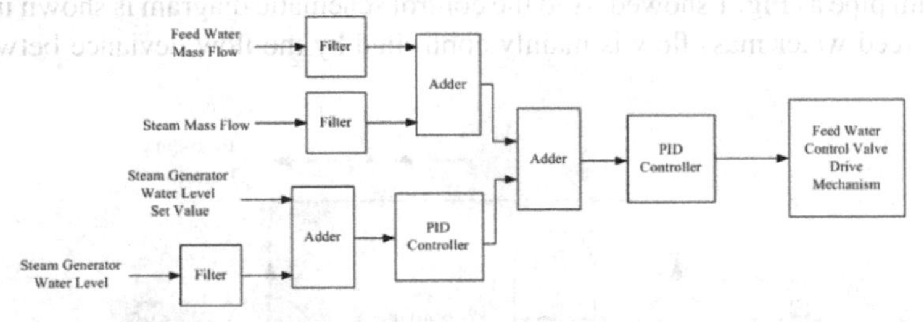

a. The Control Schematic Diagram of the Feed Water Control Valve

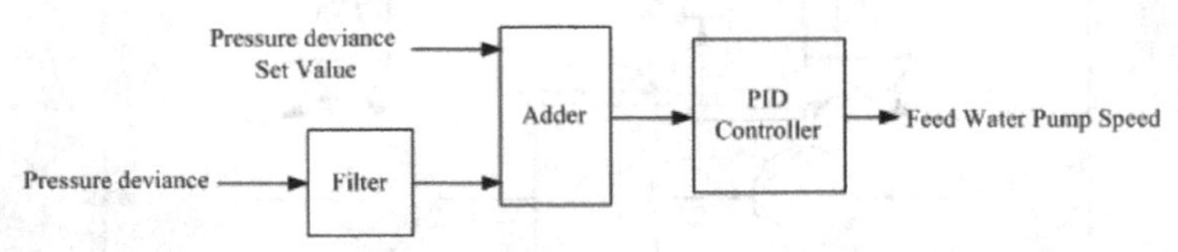

b. The Control Schematic Diagram of the Feed Water Pump Speed

Fig. 4. The Control Schematic Diagram of the Steam Generator Water Level

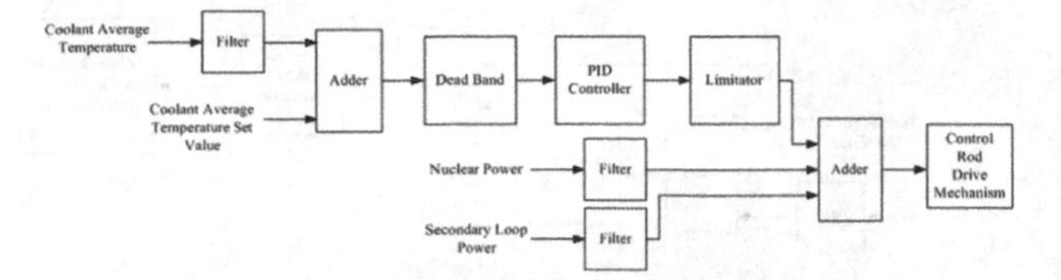

Fig. 5. The Control Schematic Diagram of the Reactor Power

2.2 Secondary Loop Decoupling Model

Although the secondary loop decoupling model usually doesn't consist of the steam generator, in this paper the steam generator is contained in the secondary loop decoupling model because the steam generator is a vital link connecting the primary and secondary loop and contrast to the primary loop decoupling model, the primary side of the steam generator is a boundary condition. Figure 6 is this decoupling system schematic diagram. If the dynamic process is simulated, the coolant enthalpy of the primary side inlet needs to be given. And the primary inlet and outlet pressure are constant during the simulation.

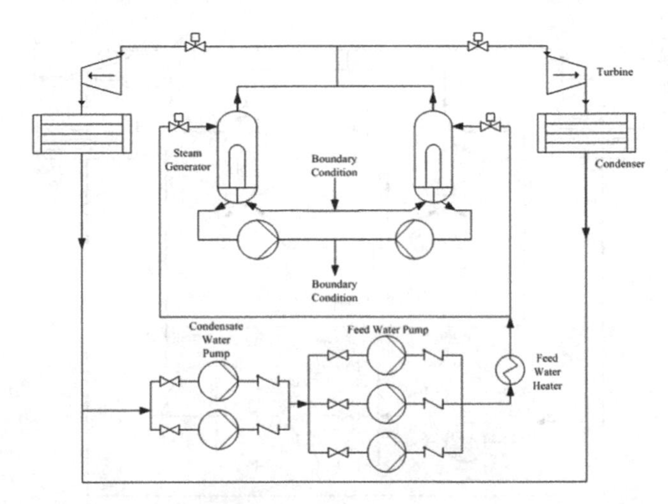

Fig. 6. The Schematic Diagram of the Secondary Loop Decoupling Model

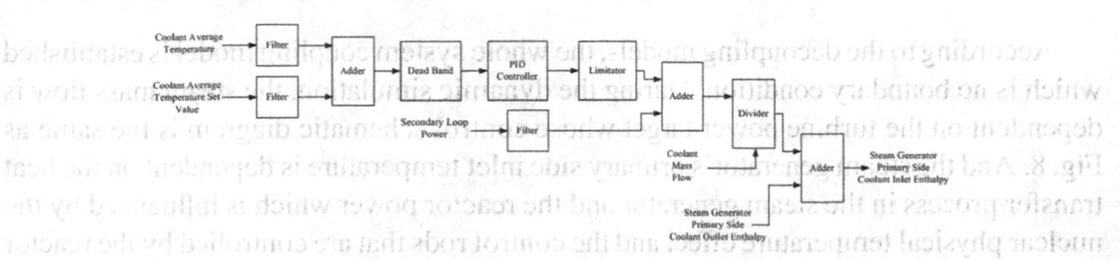

Fig. 7. The Control Schematic Diagram of the Reactor Power

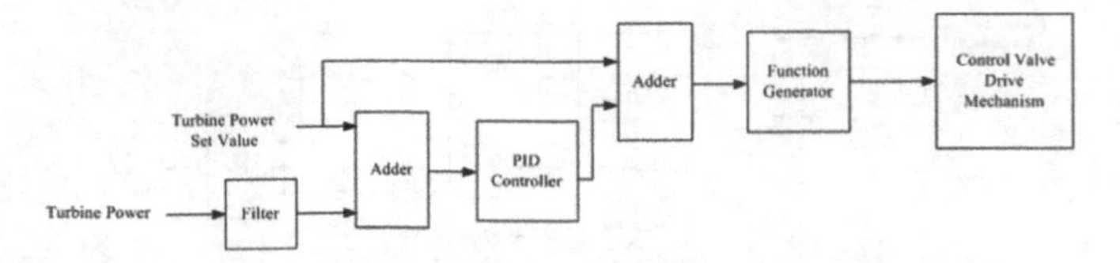

Fig. 8. The Control Schematic Diagram of the Turbine Power

The steam generators' primary side coolant inlet enthalpy is calculated and controlled by the reactor power control module which is shown in Fig. 7. And the feed water mass flow is mainly controlled by the control system of the steam generator water level in this secondary loop decoupling model, whose schematic diagram is the same in Fig. 4 Besides, the turbine generator power control system is established as Fig. 8 shows. So the steam mass flow is dependent on the turbine valve opening.

2.3 Primary and Secondary Loop Coupling Model

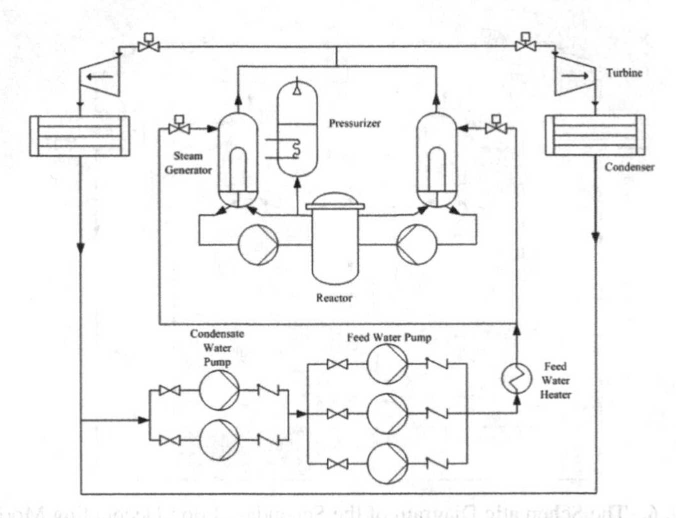

Fig. 9. The Schematic Diagram of the Two Loops Coupling Model

According to the decoupling models, the whole system coupling model is established which is no boundary condition. During the dynamic simulation, the steam mass flow is dependent on the turbine power target whose control schematic diagram is the same as Fig. 8. And the steam generator's primary side inlet temperature is dependent on the heat transfer process in the steam generator and the reactor power which is influenced by the nuclear physical temperature effect and the control rods that are controlled by the reactor power control system under the reactor following the generator mode. The reactor power control system is shown in Fig. 5. And the feed water mass flow is controlled by the steam generator water level control system as Fig. 4 shows (Fig. 9).

3 Simulation Analyses

3.1 Model Verification

The system's nominal design values are used to verify the coupling simulation model's accuracy. And the comparison results are in Table 1. And all the values are normalized by the nominal design values.

Table 1. Comparison of Nominal Steady Simulation Results

Parameter	Design Value	Simulation Value	Relative Error
Reactor Power	1.00000	0.99827	−0.173%
Coolant Mass Flow	1.00000	1.00001	0.001%
Coolant Average Temperature	1.00000	1.00002	0.002%
Pressurizer Pressure	1.00000	0.99493	−0.507%
Feed Water Temperature	1.00000	1.00135	−0.014%
Feed Water Mass Flow	1.00000	0.99986	−0.934%
Steam Mass Flow	1.00000	0.99066	0.065%
Steam Pressure	1.00000	1.00065	−0.013%
Steam Generator Water Lever	1.00000	0.99987	−0.121%
Turbine Power	1.00000	0.99879	0.038%
Condenser Pressure	1.00000	0.99968	−0.042%

The steady state simulation accuracy requirement is selected as 5%. And it shows that the simulation results' absolute relative errors are less than 5%, which meets the precision requirement. Therefore this model is available.

3.2 Steady Simulation Comparison

The full power condition and 20% of full power condition are simulated in order toe difference in steady condition analysis between these models. The main parameters' steady simulation results of these models and their relative errors based on the nominal design values are listed in Table 2. And all the values are normalized by nominal design values.

According to Table 2a, there are apparent differences between these models in the reactor power, feed water mass flow, steam mass flow, and steam pressure under the 20% FP steady condition. And Table 2b shows that the differences between these models in all parameters are small under the nominal steady condition.

There are some reasons accounting for the difference between these models. Firstly, the secondary loop decoupling model doesn't consider the heat caused by the coolant pumps and the heat capacity of the pipes' wall in the reactor coolant system. Secondly, the primary system decoupling model doesn't consider the heat caused by the condensate pumps and the feed water pumps.

However, the relative errors of these models in all parameter values listed in Table 2 are all less than 5%, which meets the simulation precision requirement. And it shows that the decoupling method can be used to study the steady characteristic of the small NPP, which is more convenient and simple than the coupling method.

3.3 Dynamic Simulation Comparison

To study the difference in dynamic analysis between the decoupling method and the coupling method, the ramp load variable condition and the load shedding condition are simulated in different models. And the main system parameters, for instant reactor power, steam mass flow, steam pressure, and so on, are monitored and recorded. Figure 10 is the system dynamic response in these models under the load ascension and Fig. 11 is the system dynamic response in these models under the load reduction. Besides, Fig. 12 is the system dynamic response in these models under the load shedding to 15%FP. All the values are normalized by the nominal design values.

According to Fig. 10 and Fig. 11, it is obvious that there are differences between the coupling system and decoupling system under ramp load variation. But Fig. 11 shows that the differences in steam mass flow and coolant average temperature dynamic response curves are small. Besides it can be found through these figures that the dynamic response of the primary loop decoupling model isn't similar to the secondary loop decoupling model.

And the difference between these models is more apparent under the load shedding in the main parameters.

The reasons that the simulation results of these models aren't similar under different conditions are following:

(1). According to formulas (3-1) and (3-2) called the point reactor model [7], nuclear power mainly depends on the reactivity and delayed neutron precursor concentration which are related to time. So nuclear power is a non-linear parameter. Besides, the most significant feature of the reactor is the temperature effect. But in the secondary loop decoupling model, there is no reactor, and the reactor power is calculated according to the steam mass flow and the coolant temperature operation scheme, which neglects the temperature effect and the control rods that impact the change rate of nuclear power because of the control dead band, its limited moving speed, and its integrated worth. Therefore the differences between the secondary loop decoupling model and the coupling model are larger than primary loop decoupling in reactor power and coolant average temperature.

$$\frac{dn(t)}{dt} = \frac{\rho(t) - \beta}{\Lambda} n(t) + \sum_{i=1}^{6} \lambda_i C_i(t) \tag{3-1}$$

$$\frac{dC_i(t)}{dt} = \frac{\beta_i}{\Lambda} n(t) + \lambda_i C_i(t) \quad i = 1, 2, \ldots 6 \tag{3-2}$$

Table 2. Steady Simulation Results Comparison

a. 20%FP Condition			
Parameter	Primary Loop Decoupling	Secondary Loop Decoupling	Two Loops Coupling
Reactor Power	0.198	0.208	0.205
Coolant Mass Flow	1.003	1.004	1.004
Coolant Average Temperature	0.954	0.956	0.954
Pressurizer Pressure	0.999	0.997	1.002
Feed Water Temperature	1.000	0.997	0.998
Feed Water Mass Flow	0.198	0.206	0.207
Steam Mass Flow	0.200	0.208	0.205
Steam Pressure	1.290	1.302	1.284
Steam Generator Water Lever	1.000	0.998	1.001
Turbine Power	/	0.096	0.096
Condenser Pressure	/	0.293	0.293
b. 100%FP Condition			
Parameter	Primary Loop Decoupling	Secondary Loop Decoupling	Two Loops Coupling
Reactor Power	0.997	1.000	0.998
Coolant Mass Flow	1.000	1.000	1.000
Coolant Average Temperature	1.000	1.000	1.000
Pressurizer Pressure	0.997	0.997	0.998
Feed Water Temperature	1.000	0.999	0.999
Feed Water Mass Flow	0.990	0.991	1.000
Steam Mass Flow	1.000	1.001	0.998
Steam Pressure	1.000	1.000	0.999
Steam Generator Water Lever	1.000	1.001	1.003
Turbine Power	/	1.001	1.000
Condenser Pressure	/	1.002	1.002

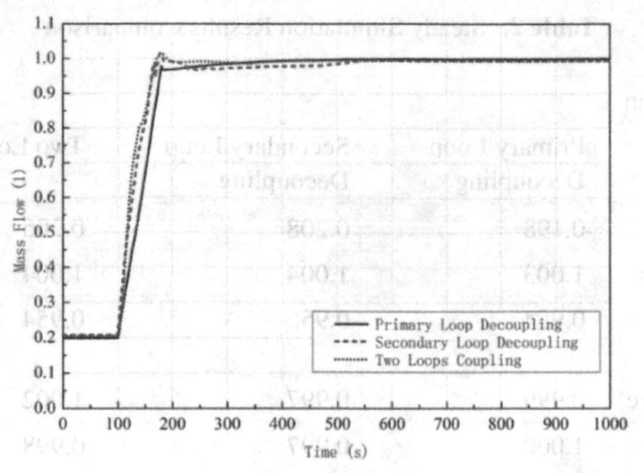

a. Steam Mass Flow Dynamic Response Curve

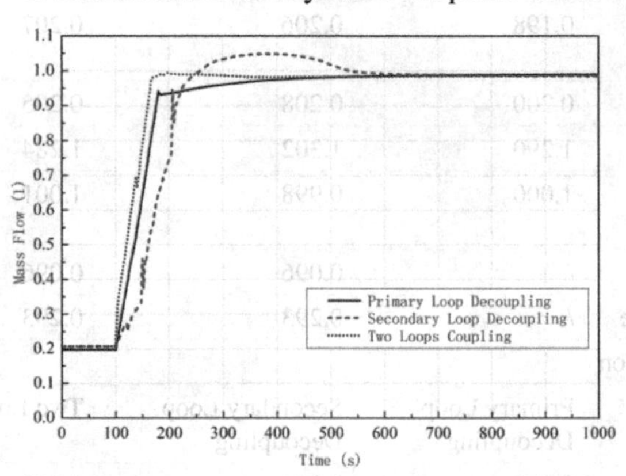

b. Feed Water Mass Flow Dynamic Response Curve

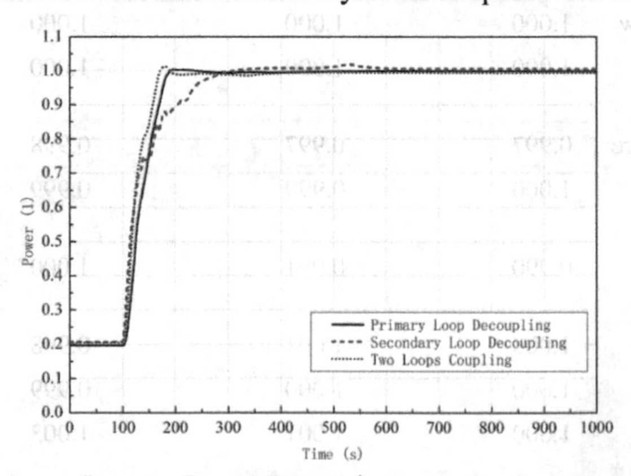

c. Reactor Power Dynamic Response Curve

Fig. 10. System Dynamic Response under the Load Ascension

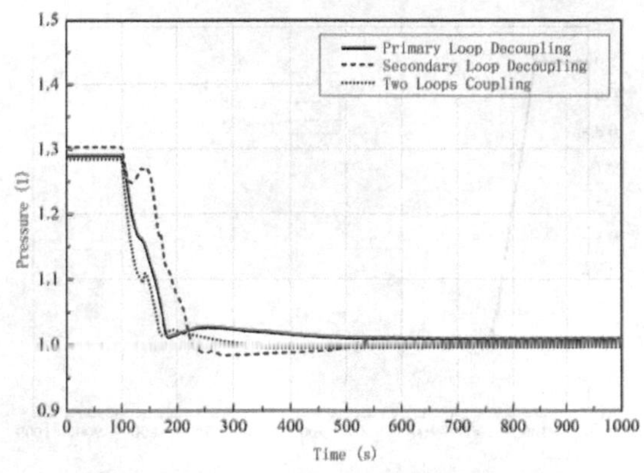

d. Steam Pressure Dynamic Response Curve

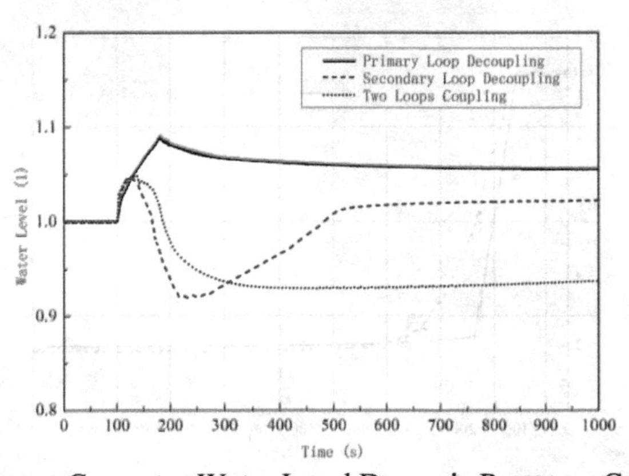

e. Steam Generator Water Level Dynamic Response Curve

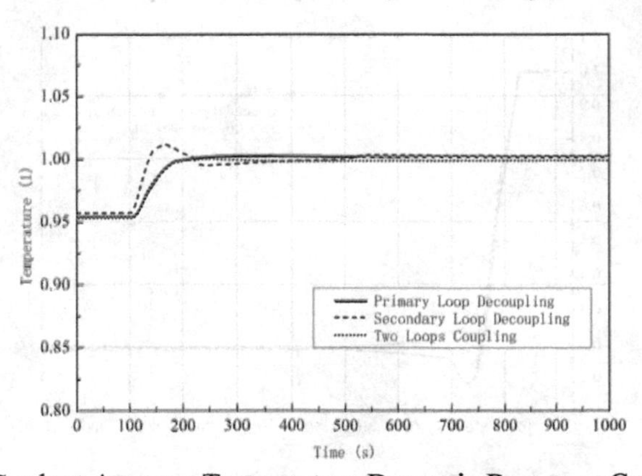

f. Coolant Average Temperature Dynamic Response Curve

Fig. 10. (*continued*)

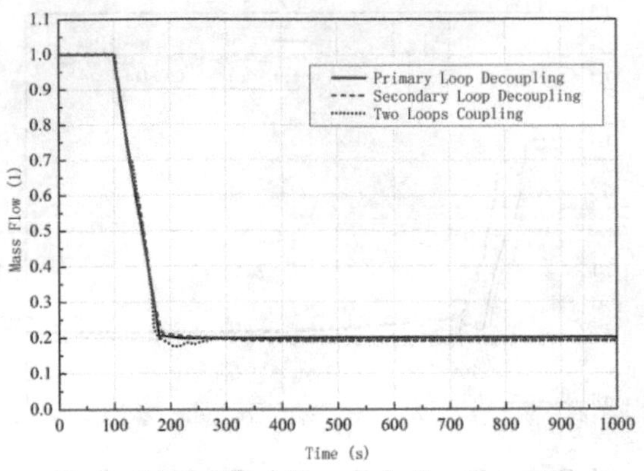

a. Steam Mass Flow Dynamic Response Curve

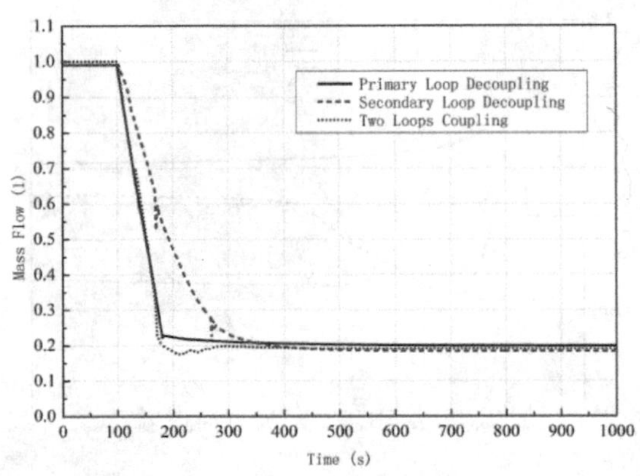

b. Feed Water Mass Flow Dynamic Response Curve

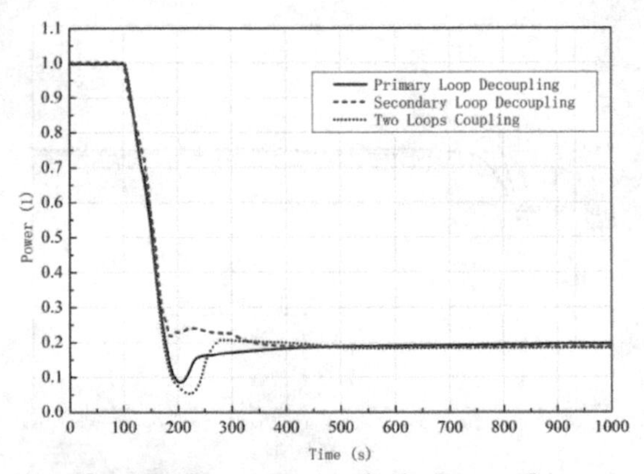

c. Reactor Power Dynamic Response Curve

Fig. 11. System Dynamic Response under the Load Reduction

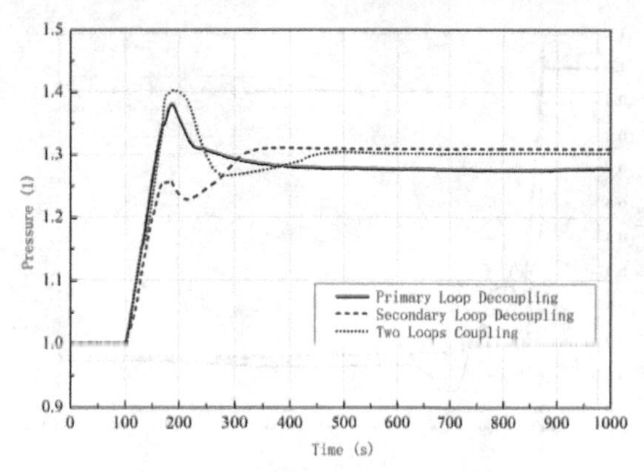

d. Steam Pressure Dynamic Response Curve

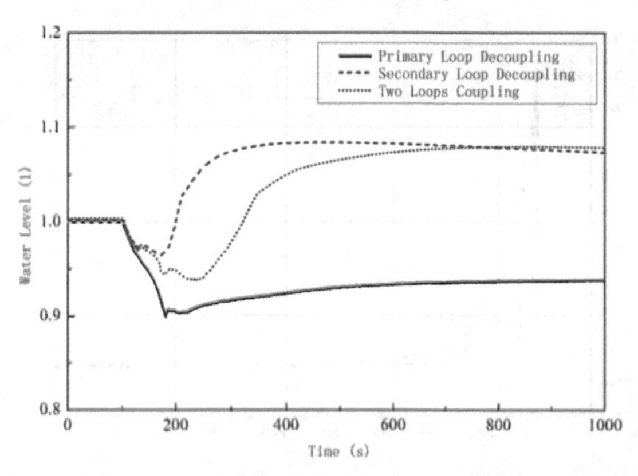

e. Steam Generator Water Level Dynamic Response Curve

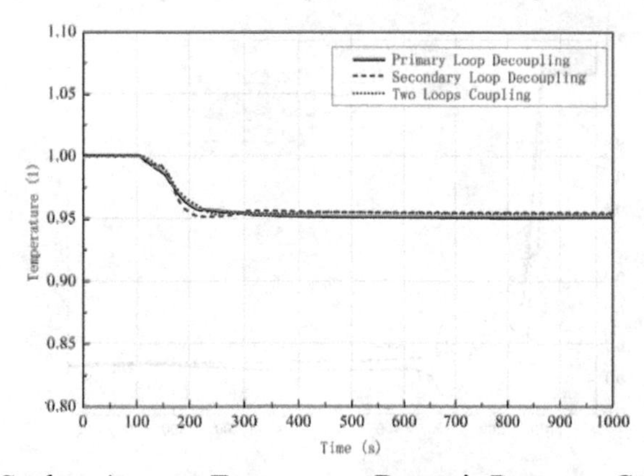

f. Coolant Average Temperature Dynamic Response Curve

Fig. 11. (*continued*)

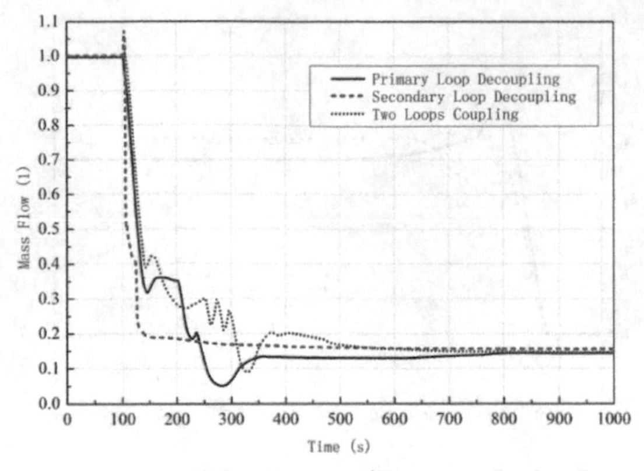

a. Steam Mass Flow Dynamic Response Curve

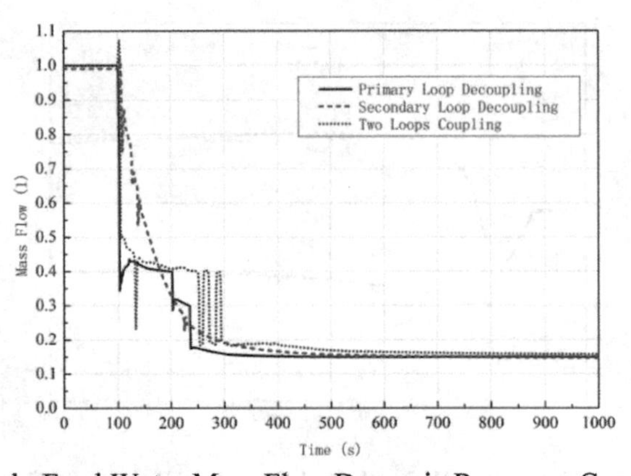

b. Feed Water Mass Flow Dynamic Response Curve

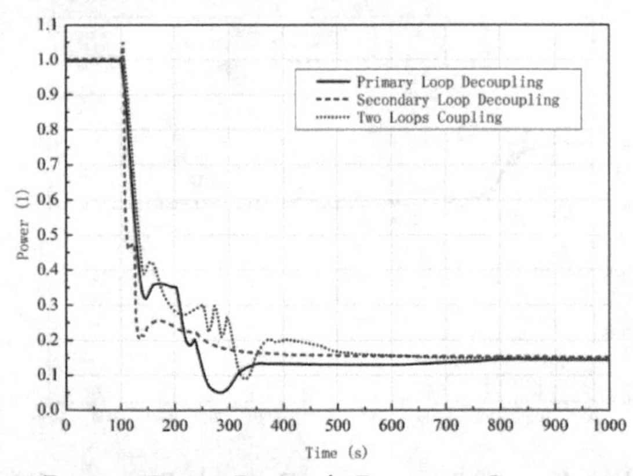

c. Reactor Power Dynamic Response Curve

Fig. 12. System Dynamic Response under the Load Shedding

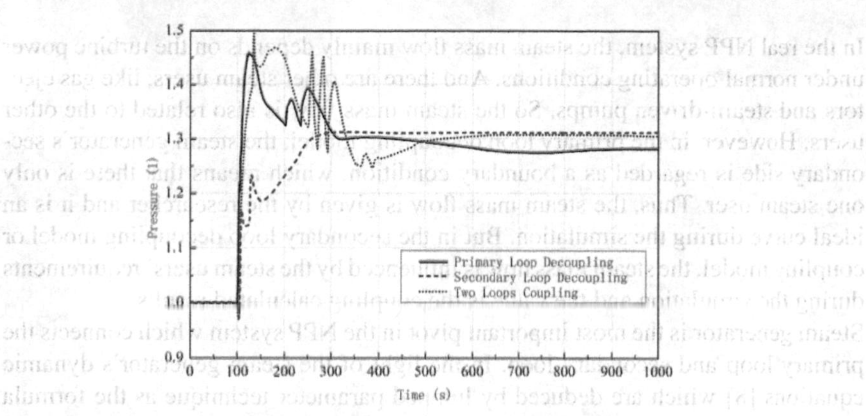

d. Steam Pressure Dynamic Response Curve

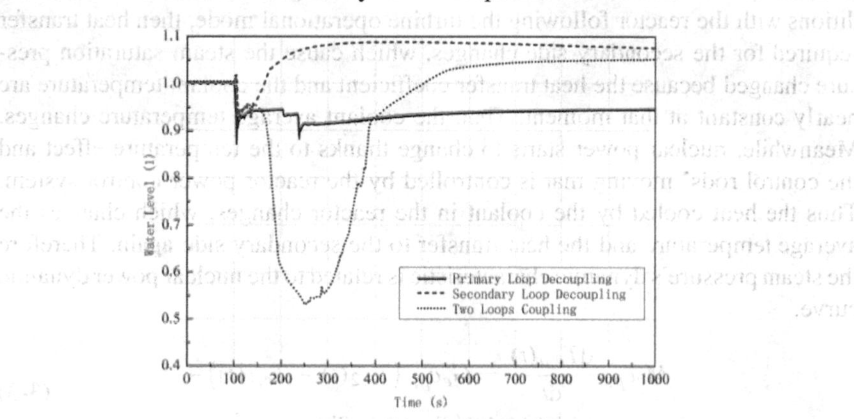

e. Steam Generator Water Level Dynamic Response Curve

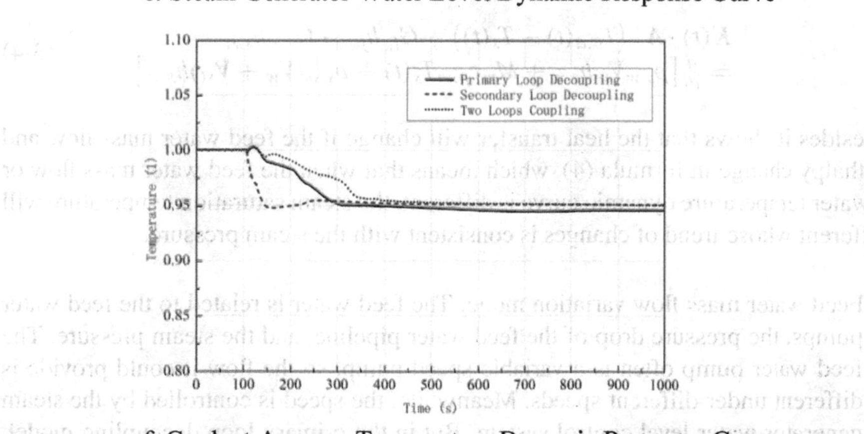

f. Coolant Average Temperature Dynamic Response Curve

Fig. 12. (*continued*)

(2). In the real NPP system, the steam mass flow mainly depends on the turbine power under normal operating conditions. And there are other steam users, like gas ejectors and steam-driven pumps. So the steam mass flow is also related to the other users. However, in the primary loop decoupling model, the steam generator's secondary side is regarded as a boundary condition, which means that there is only one steam user. Thus, the steam mass flow is given by the researcher and it is an ideal curve during the simulation. But in the secondary loop decoupling model or coupling model, the steam mass flow is influenced by the steam users' requirements during the simulation and the value is the coupling calculated results.

(3). Steam generator is the most important pivot in the NPP system which connects the primary loop and secondary loop. In the light of the steam generator's dynamic equations [8] which are deduced by lumped parameter technique as the formula (3-3) and (3-4) show, at first the steam mass flow changes under load variation conditions with the reactor following the turbine operational mode, then heat transfer required for the secondary side changes, which cause the steam saturation pressure changed because the heat transfer coefficient and the coolant temperature are nearly constant at that moment. Then the coolant average temperature changes. Meanwhile, nuclear power starts to change thanks to the temperature effect and the control rods' moving that is controlled by the reactor power control system. Thus the heat cooled by the coolant in the reactor changes, which changes the average temperature and the heat transfer to the secondary side again. Therefore the steam pressure's dynamic characteristic is related to the nuclear power dynamic curve.

$$M_c c_{p,c} \frac{\mathrm{d}T_{c,a}(t)}{\mathrm{d}t} = G_c c_{p,c}\big(T_{c,2}(t) - T_{c,1}(t)\big) - K(t) \cdot A \cdot \big(T_{c,a}(t) - T_s(t)\big) \tag{3-3}$$

$$K(t) \cdot A \cdot \big(T_{c,a}(t) - T_s(t)\big) + G_{fw} h_{fw} - G_{st}, h_{s,st} = \frac{d}{dt}\big[\rho_{s,w} V_w h_{s,w} + M_m c_{p,m} T_s(t) + \rho_{s,st}(V_w + V_{st}) h_{s,st}\big] \tag{3-4}$$

Besides it shows that the heat transfer will change if the feed water mass flow and its enthalpy change in formula (4), which means that when the feed water mass flow or feed water temperature dynamic curve is different, the steam saturation temperature will be different whose trend of changes is consistent with the steam pressure.

(4). Feed water mass flow variation mode. The feed water is related to the feed water pumps, the pressure drop of the feed water pipeline, and the steam pressure. The feed water pump often is a variable speed pump, so the flow it could provide is different under different speeds. Meanwhile, the speed is controlled by the steam generator water level control system. But in the primary loop decoupling model, the feed water pumps models don't establish and the feed water mass flow is mainly controlled through the mass flow deviance. Besides, during operation, the numbers of operating pumps relate to the secondary loop power and current feed water mass flow.

To sum up, because many assumptions neglect the system coupling and the internal disturbances in the decoupling model, its results are more ideal. Besides the nuclear power which is influenced by the moderator and fuel temperature is one of the most important parameters in the nuclear system, so compared with the primary loop decoupling model, the secondary decoupling model is less accurate.

4 Conclusions

The ramp load variation and load shedding simulation in two-power loop a small NPP system is analyzed in this paper based on APROS software. And the main parameter's dynamic response curves are compared in different models, such as steam mass flow, nuclear power, steam pressure, and so on. According to the simulation results, the following results are obtained.

1) Difference between the coupling method and the decoupling method is small in steady-state simulation, so the system's steady characteristic feature could be studied by the decoupling method.
2) Differently observed between the coupling method and decoupling method in dynamic simulation. Therefore the coupling method should be used to study the dynamic characteristic feature of the system.
3) Because the assumptions aren't similar, the simulation results of the primary loop decoupling model and secondary loop decoupling model are different. But compared with the secondary loop decoupling model, the results of the primary loop decoupling model are more close to the coupling model.

These conclusions are crucial and could significantly refer to the system simulation analysis.

References

1. Guo, L., Sun, B., Song, Z., et al.: Modeling and real-time simulation of secondary circuit thermal system for nuclear power station based on C++ builder. Electric Power Constr. **34**(6), 1–6 (2013)
2. Huo, B., Xu, F., Liu, S., et al.: Nuclear power plant reactor coolant system principle simulation based on matlab. J. Nanjing Inst. Technol. (Nat. Sci. Ed.) **12**(2), 55–58 (2014)
3. Lin, M., Yang, Y.-H., Zhang, R.-H., et al.: Development of nuclear power plant real-time engineering simulator. Nucl. Sci. Tech. **16**(3), 177–180 (2005)
4. Kai, X., Liao, L., Zhou, K., et al.: The modeling and simulation research on control scheme of multi-reactor and multi-turbine nuclear power plant. J. Univ. South China (Sci. Technol.) **31**(3), 13–20 (2017)
5. Lu, N., Li, Y., Pan, L., et al.: Study on dynamics of steam dump system in scram condition of nuclear power plant. IFAC-PapersOnLine **52**(4), 360–365 (2019)
6. Pack, J., Fu, Z., Aydogan, F.: Modeling primary and secondary coolant of a nuclear power plant system with a unique framework (MCUF). Prog. Nucl. Energy **3**, 197–211 (2015)
7. Xie, Z.: Nuclear Reactor Physical Analysis. Xi'an Jiaotong University Press, Xi'an (2004)
8. Pang, F., Peng, M.: Marine Nuclear Power Device. Harbin Engineering University Press, Harbin (2000)

Development of Heat Pipe Modeling Capabilities in a Fully-Implicit Solution Framework

Guojun Hu(✉)

University of Science and Technology of China, Heifei Anhui, China
huguojun@ustc.edu.cn

Abstract. One key aspect in analysis of heat pipe microreactors is the efficient modeling of heat pipes and its coupling with the solid reactor core. Various options exist for modeling of heat pipes. Most models require an explicit coupling between the vapor core, which brings in an additional layer of coupling when the heat pipe model is integrated into a system-level safety analysis model. This additional layer of coupling causes both convergence concern and computational burden in practice. This article aims at developing a new heat pipe modeling algorithm, where the heat pipe wall, wick, and vapor core are discretized and coupled in a monolithic fully-implicit manner. The vapor core will be modeled as a one-dimensional compressible flow with the capability of predicting sonic limit inherently; a two-dimensional axisymmetric heat conduction model will be used to model the heat pipe wall and wick region. The heat pipe wick and vapor core are coupled through a conjugate heat transfer interface. Eventually, the coupled system will be solved using the Jacobian-Free Newton-Krylov (JFNK) method. It is demonstrated that the new coupled system works well. Consideration of the vapor compressibility in the two-equation model allows more detailed representation of the vapor core dynamics while remains light-weight in terms of computational complexity. The new model is verified by an approximate analytical solution to the heat pipe vapor core and is validated by a sodium heat pipe experiment.

Keywords: Microreactor · JFNK · Heat Pipe

1 Introduction

The heat pipe cooled MicroRx reduces most moving parts in the reactor design and provides an autonomous operation capability with passive safety systems. Its applications in remote area power stations, space nuclear power supplier, and other non-commercial fields attract much attention from governments, military, and nuclear industry in recent years. The heat pipe cooled MicroRx had been designed and demonstrated in several space exploration projects, KRUSTY [1] for example. Recently, several MWe-scale heat pipe cooled MicroRx were designed for terrestrial applications as well, including the eVinci reactor of Westinghouse and Aurora reactor of Oklo [2]. Before the demonstration and eventual construction of this new type of reactor technology, substantial effort is required in the safety analysis of this type of reactor and the associated modeling and

C. Liu (Ed.): PBNC 2022, SPPHY 283, pp. 845–860, 2023.
https://doi.org/10.1007/978-981-99-1023-6_72

simulations work. One key aspect in this analysis work is the efficient modeling of heat pipes and its coupling with the solid reactor core.

It is necessary to know the liquid and vapor flow phenomena occurring within a heat pipe to determine the heat removal capacity and heat pipe performance. Many analytical and numerical models were proposed for predicting heat pipe operation limits, startup phenomena, steady-state operation, and transient behaviors. Faghri [3] proposed a two-dimensional incompressible vapor flow model for prediction of an annular heat pipe steady-state operation. Later, Faghri and Pavani [4] considered the compressibility of vapor and derived analytical axial pressure drop within the heat pipe. Chen and Faghri [5] studied the importance of conjugate heat transfer within the heat pipe wall and wick by the coupling of a two-dimensional compressible vapor flow model and a two-dimensional heat conduction model. This study showed that the compressible vapor flow model gave more reasonable results than incompressible vapor flow model. Cao and Faghri [6] proposed a two-dimensional transient analysis model which considered the vapor compressibility and coupled the vapor flow with wall and wick. They confirmed that the heat conduction model was sufficient for modeling the heat transfer in the wick when the thermal conductivity of working fluid is high, especially for high-temperature heat pipes where alkali metals are used as the working fluid. Cao and Faghri [7] also proposed a transient model for modeling of nonconventional heat pipes where a one-dimensional compressible vapor flow model is coupled to the multi-dimensional heat conduction model for the wall and wick. A comprehensive review of heat pipe analysis model is referred to Faghri's paper [8].

In this work, the target application of heat pipe analysis model is the transient safety analysis of heat pipe cooled MicroRx. Targeting the nuclear applications, several heat pipe analysis models were proposed, ranging from the simpler superconductor model to the more complex two-phase flow model. The ANL's SAM code implemented a so-called superconductor model where the heat transfer through vapor core was modeled with heat conduction using an extremely high effective thermal conductivity [9]. This heat pipe model was eventually used in a full-core multi-physics simulations of a 5 MWt heat pipe cooled MicroRx with great success. Wang et. al [10] developed a heat pipe heat transfer model based on the quasi-steady compressible one-dimensional laminar flow model and applied it to performance analysis of heat pipe radiator unit for space nuclear power reactor. A more complex heat pipe analysis model was proposed in the Sockeye code based on a two-phase flow model for the vapor core and porous medium flow model for the wick [11].

There are two critical issues when integrating a heat pipe analysis model into a system-level safety analysis model. The first issue is the balance between accuracy and computational complexity of heat pipe analysis model. The simplest thermal resistance model requires user inputs of vapor core thermal resistance and brings in significant uncertainties if without support from experiments. The more detailed multi-dimensional flow model of the vapor core is not applicable to system-level safety analysis due to its computational complexity. From the author's point of view, the balance lies in the one-dimensional vapor flow model. The second issue is the tight coupling between the heat pipe wall and the vapor core models. The fully-implicit technique is best fit for resolving this tight-coupled problem.

This work aims at developing the heat pipe modeling capability in a fully-implicit code framework, named RETA. RETA is a modern object-oriented system thermal-hydraulics analysis software developed in C++ by the author. It aims at providing the capabilities for modeling and simulation (M&S) of advanced reactors. Its underline solution scheme is the Newton-Krylov algorithm developed for the solution of system of nonlinear equations. The heat pipe analysis model is one key module of RETA code for the next-step development of transient safety analysis model of heat pipe cooled MicroRx.

2 Models

This work considers at first the conventional cylindrical heat pipes consisting of heat pipe wall, wick, and vapor core regions.

2.1 Vapor Flow Model

Various models and governing equations were formulated for modeling of the flow field in the vapor core region, including the simple thermal resistance network model, one-dimensional flow model, and three-dimensional flow model. This work follows the transient compressible one-dimensional vapor flow model, as was originally proposed by [7]. The one-dimensional vapor flow model will be coupled to a cylindrical two-dimensional heat conduction model for the heat pipe wall and wick. The major difference is that explicit coupling between the wick and vapor core is avoided in this newly developed model.

The conservation equations for vapor flow with negligible body forces were formulated using mass and momentum equation, i.e., a two-equation model. The conservation of mass is formulated as,

$$\frac{\partial \rho}{\partial t}+\frac{\partial \rho u}{\partial z}-\Gamma=0 \tag{1}$$

In which Γ is the mass generation rate per unit volume. The conservation of momentum equation is,

$$\frac{\partial \rho u}{\partial t}+\frac{\partial \rho u^2}{\partial z}+\frac{\partial p}{\partial z}+\frac{\lambda}{2D_h}\rho u|u|=0 \tag{2}$$

In which D_h is the hydraulic diameter of the vapor core region, and λ is the dimensionless friction coefficient. This work considers the compressible vapor flow model with potential Mach number as high as 1. The ideal gas law is used to describe the vapor equation of state, i.e.

$$p=\rho RT \tag{3}$$

In which T is vapor temperature and R is the specific gas constant. In the current formulation, the vapor is assumed to be in saturated condition, and the Clausius-Clapeyron equation is used to determine the vapor saturation temperature from the vapor pressure,

$$\frac{dp}{p}=\frac{h_{fg}}{R}\frac{dT}{T^2} \tag{4}$$

In which h_{fg} is the specific enthalpy of vaporization. In practice, given the reference temperature T_c and reference pressure p_c, the saturation temperature can be obtained by

$$\frac{1}{T} = \frac{1}{T_c} - \frac{R}{h_{fg}} \ln \frac{p}{p_c} \tag{5}$$

Equation (1), (2), (3), and (5) gives the closed set of governing equation for the one-dimensional vapor flow.

Equation (3) and (5) serve as an algorithmic correlation, and there are 2 partial differential equations (PDEs) for solution of 2 nonlinear variables, yet to be selected. This work aims at developing the heat pipe modeling capability in a fully-implicit solution framework, the first important task is the selection of nonlinear variables. It is natural to use vapor velocity u as the first nonlinear variable to be solved by the momentum Eq. (2). We have the freedom to solve the mass equation using vapor temperature T or vapor pressure p as the nonlinear variable. These two choices should be equivalent mathematically. Using pressure p is more straightforward because the pressure gradient is an important term in the momentum equation. We implemented and tested this choice, however it was observed that the resulting nonlinear system of equations was quite unstable, as minor density perturbation causes significant change in pressure. In contrast, a temperature formulation was found to be much more stable and was selected in this work.

Using the Clausius-Clapeyron equation, the pressure gradient term is transformed into,

$$\frac{\partial p}{\partial z} = \frac{dp}{dT} \frac{\partial T}{\partial z} = \frac{\rho h_{fg}}{T} \frac{\partial T}{\partial z} \tag{6}$$

With vapor temperature as the dependent variable, the vapor density is formulated as

$$\rho = \frac{p_c}{RT} \exp\left[\frac{h_{fg}}{R}\left(\frac{1}{T_c} - \frac{1}{T}\right)\right] \tag{7}$$

Finally, Eq. (1), (2), (6), and (7) are the closed set of governing equations for modeling the one-dimensional vapor flow.

2.2 Heat Conduction Model

This work considers primarily high temperature heat pipe with liquid metal (e.g., Sodium) as the working fluid. Because the thermal conductivity of the liquid metal working fluid is quite high, and wick thickness is thin, the effect of liquid flow in the wick can be neglected without causing significant modeling error in terms of macroscopic average temperature in the wick structure [6]. We will ignore the liquid flow in the wick structure and model the wick structure as a solid heat conduction region, with an effective thermal conductivity evaluated based on wick structure porosity, fluid thermal conductivity, and wick material conductivity. Currently, this effective thermal conductivity is specified as a user-input, which could be a temperature and porosity dependent function.

The heat pipe wall and wick regions are modeled as a two-dimensional axisymmetric solid heat structure. The governing equation for solid temperature is

$$\rho_s c_{p,s}\frac{\partial T_s}{\partial t} - \frac{1}{r}\frac{\partial}{\partial r}\left(rk_s\frac{\partial T_s}{\partial r}\right) - \frac{\partial}{\partial z}\left(k_s\frac{\partial T_s}{\partial z}\right) - q_s''' = 0 \tag{8}$$

In which subscript s represents solid. Note that solid density ρ_s, specific heat capacity $c_{p,s}$, and thermal conductivity k_s are all temperature dependent. The nonlinearity due to this dependency is resolved by the fully-implicit solution scheme.

2.3 Heat and Mass Transfer Model

The heat pipe vapor core is coupled with the heat pipe wick inner surface through a convection-like formulation. A user-specified effective heat transfer coefficient h_v is used to couple the vapor core temperature T and solid temperature T_s by

$$q_s'' = -k_s\frac{\partial T_s}{\partial r} = h_v(T_s - T) \tag{9}$$

In which q_s'' is the heat flux at wick-core interface. For the heat pipe, the heat transfer at the wick-core interface is in fact through the evaporation/condensation of working fluid. The mass generation rate per unit volume Γ is thus modeled as

$$\Gamma = \frac{a_w q_s''}{h_{fg}} \tag{10}$$

In which a_w is the heat transfer surface area density per unit volume.

2.4 Friction Model

A closed form friction coefficient correlation is required to correctly predict the flow field in the vapor core region. The friction coefficient λ in Eq. (2) depends on the Reynolds number. In this study, the friction coefficient is modeled as

$$\lambda = \begin{cases} \frac{64}{\mathrm{Re}} & \mathrm{Re} \le 2200 \\ \frac{0.316}{\mathrm{Re}^{0.25}} & \mathrm{Re} > 3000 \end{cases} \tag{11}$$

This corresponds to the classical friction correlation for pipe flow in laminar and turbulent situation. An interpolation is performed for Reynolds number between 2200 and 3000. The effect of friction pressure drop will be seen clearly in the following verification and validation cases.

3 Numerical Scheme

3.1 Discretization Scheme

Finite Volume Method (FVM) is used to solve the one-dimensional vapor flow equations. A one-dimensional staggered grid is used, as is shown in Fig. 1. This work considers the implicit Backward Euler discretization of temporal terms and upwind discretization of the convection terms. Extension to higher-order temporal and spatial discretization schemes are possible, which will not be discussed in this article. Wall boundary conditions are applied at both ends of the vapor core region. Velocity magnitude and temperature gradient are set to be zero at both ends.

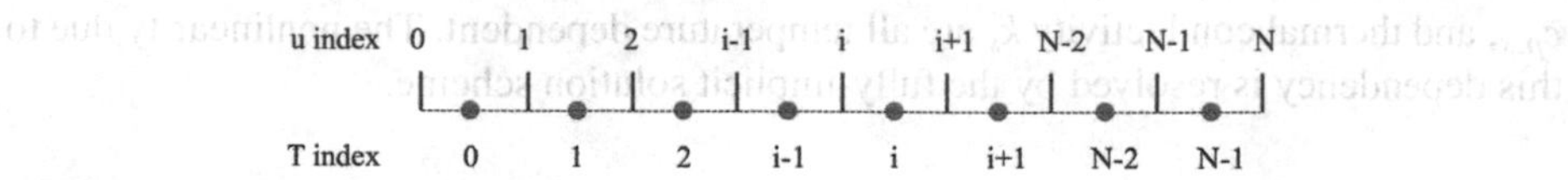

Fig. 1. One-dimensional staggered grid for vapor flow simulation

The final discretized vapor mass and momentum equations are summarized as:

$$R_{1,i} = \Delta x_i \frac{\rho_i - \rho_i^n}{\Delta t} + \left(u_{i+1}\rho_i^+ - u_i\rho_i^-\right) - \Delta x_i \Gamma_i \tag{12}$$

$$R_{2,i} = \Delta x_i \frac{\rho_i u_i - \rho_i^n u_i}{\Delta t} + \left(\rho_i \frac{u_i + u_{i+1}}{2} u_i^+ - \rho_{i-1} \frac{u_{i-1} + u_i}{2} u_i^-\right) + \frac{\rho(\overline{T}) \cdot h_{fg}(\overline{T})}{\overline{T}} (T_i - T_{i-1}) + \Delta x_i \frac{\lambda_i}{2D_h} \rho(\overline{T}) \cdot u_i |u_i| \tag{13}$$

Let $\mathcal{U}(u, a, b)$ be the operator for selecting the upwind value based on the velocity, which is defined as

$$\mathcal{U}(u, a, b) = \begin{cases} a \; u \geq 0 \\ b \; u < 0 \end{cases} \tag{14}$$

In Eq. (12) and (13), ρ_i^+ and ρ_i^- are evaluated based on the upwind concept as

$$\rho_i^+ = \mathcal{U}(u_{i+1}, \rho_i, \rho_{i+1}), \rho_i^- = \mathcal{U}(u_i, \rho_{i-1}, \rho_i) \tag{15}$$

u_i^+ and u_i^- are evaluated based on the upwind concept as

$$u_i^+ = \mathcal{U}\left(\frac{u_i + u_{i+1}}{2}, u_i, u_{i+1}\right) \tag{16}$$

$$u_i^- = \mathcal{U}\left(\frac{u_{i-1} + u_i}{2}, u_{i-1}, u_i\right) \tag{17}$$

$\overline{T}$ is vapor temperature evaluated at velocity cell center

$$\overline{T} = \frac{T_{i-1} + T_i}{2} \tag{18}$$

The heat conduction equation of wall and wick region is also solved with FVM in an orthogonal two-dimensional grid. This discretization is rather straightforward and is thus ignored in this article, except for two notes. The first note is that non-uniform mesh size is allowed in both radial and axial directions. This flexibility is quite useful in distinguishing the wall and wick regions in radial direction; or in distinguishing the evaporator, adiabatic, and condenser regions in axial direction. The second note is that solid density, specific heat capacity, and thermal conductivity are all temperature dependent. The nonlinearity is resolved by the fully-implicit solution scheme.

The discretized vapor flow equations and heat conduction equations are combined to form a monolithic system of nonlinear algebraic equations. The vapor temperature and solid temperature are coupled using the conjugate heat transfer Eq. (9). This coupling is achieved internally with respect to nonlinear degree of freedom (DOF), which avoids the explicit boundary data exchange, as in the case where the vapor flow equations and solid heat conductions equations are solved separately.

3.2 Jacobian-Free Newton-Krylov Method

The system of nonlinear algebraic equations will be solved with the JFNK method, with an approximate Jacobian matrix serving as the preconditioner. Let **U** the vector of nonlinear DOFs and **R** be the system residual vector generated from the discretized vapor flow equations and solid heat conduction equations. The system of nonlinear algebraic equations is expressed as

$$\mathbf{R}(\mathbf{U}) = 0 \tag{19}$$

The JFNK method is based on the Newton's method, which solves the nonlinear algebraic equation iteratively by

$$\mathbb{J}\left(\mathbf{U}^k\right) \cdot \delta\mathbf{U} = -\mathbf{R}\left(\mathbf{U}^k\right) \tag{20}$$

$$\mathbf{U}^{k+1} = \mathbf{U}^k + \alpha \cdot \delta\mathbf{U} \tag{21}$$

In which $\mathbb{J}$ represents the system Jacobian matrix and α is a relaxation parameter in [0, 1] to improve the stability of Newton's iteration. In this work, α is calculated with a line-search algorithm. In the JFNK algorithm, the linear Eq. (20) is solved with the Krylov subspace method. The Krylov subspace method tries to find the iterative solution of the linear equation using the truncated Krylov subspace defined as

$$\mathcal{K}_m(\mathbb{J}, \mathbf{v}) := \left\{\mathbf{v}, \mathbb{J}\mathbf{v}, \mathbb{J}^2\mathbf{v}, \cdots, \mathbb{J}^{m-1}\mathbf{v}\right\} \tag{22}$$

The fundamental operation in constructing the Krylov subspace is the matrix-vector product. The essential idea of the JFNK method is that the matrix-vector product in Eq. (22) is approximated using a finite difference scheme by

$$\mathbb{J}\mathbf{v} \approx \frac{\mathbf{R}(\mathbf{U} + \varepsilon\mathbf{v}) - \mathbf{R}(\mathbf{U})}{\varepsilon} \tag{23}$$

In which ε is a small scalar parameter. In this work, the JFNK scheme is implemented with the open-source scientific toolkit PETSc [12] developed by Argonne National Laboratory. The generalized minimum residual (GMRES) Krylov subspace linear solver is used.

It is well-known that the Krylov subspace linear solver in general needs a good preconditioner for better and faster convergence behavior. In this work, an approximate system Jacobian matrix is provided to serve as the preconditioner. The complexity and computational burden of constructing an approximate system Jacobian matrix could be significantly reduced compared with constructing the exact system Jacobian matrix as required in the conventional Newton's method. In practice, one approximate system Jacobian matrix is constructed for each time step. If deemed necessary and efficient, this JFNK method can be converted to the Newton's method in a rather simple way, specified by an end user.

4 V & V Tests

4.1 Heat Conduction Model Verification

A simple cylinder with uniform internal heating power density and constant solid properties is used to verify the two-dimensional axisymmetric heat conduction model. Dimensionless heat conduction equation is solved. The radius of the cylinder is set to be 1. The radial outer surface is fixed at a boundary value of 0 and the axial boundaries are adiabatic. The density, specific heat capacity, and thermal conductivity is set to 1. A uniform heating power density of 50.0 is applied in the cylinder. The verification is shown in Fig. 2. Additional verifications of the heat conduction model were conducted but ignored here for brevity.

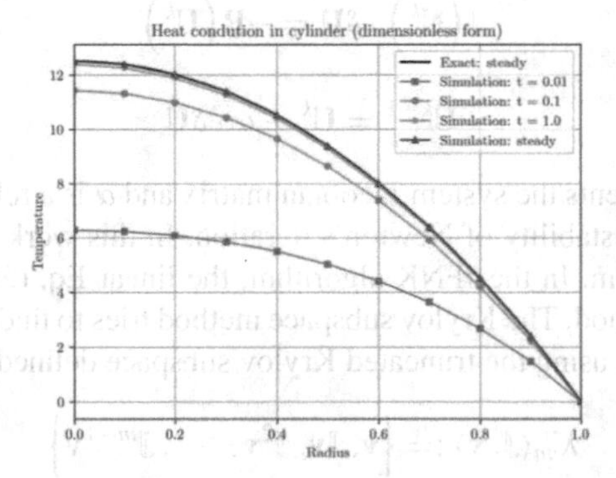

Fig. 2. Verification of two-dimensional axisymmetric heat conduction model in dimensionless form

4.2 Vapor Flow Model Verification

The verification of vapor flow model is performed using a heat pipe with uniform heating at the evaporator wall surface and a fixed temperature at the condenser wall surface, as is shown in Fig. 3.

We at first derive an approximate analytical solution of vapor for verification purposes. The following assumptions are made:

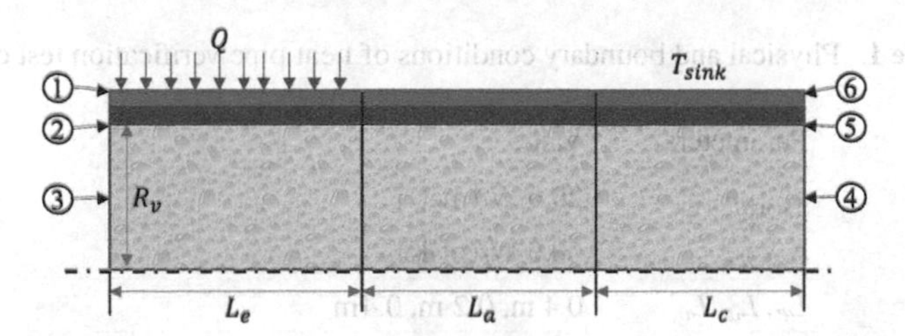

Fig. 3. Heat pipe verification model

- Heat flux at the interface between wick inner surface and vapor core is uniform in both evaporator and condenser section.
- The specific enthalpy of evaporation is a constant value.
- Heat removal rate in the vapor core is low, the pressure and temperature variation in the vapor core are thus small, vapor density is approximated to be a constant value.
- Friction coefficient is a constant value. This is in general not a valid approximation in practical simulations but should not be a problem in a verification test.

Let $M = \rho u$ be the mass flux of vapor flow. At steady-state, with the assumption that Γ is a piecewise uniform function of axial location z, the mass flux and pressure drop are found to be:

$$M = \int_0^z \Gamma dz \tag{24}$$

$$\Delta p = -\frac{1}{\rho_c}\left(M^2 + \frac{\lambda}{2D_h}\int_0^z M^2 dz\right) \tag{25}$$

The vapor temperature can be calculated with Eq. (25) and the Clausius-Clapeyron Eq. (4). Physical and boundary conditions for this verification test are listed in Table 1.

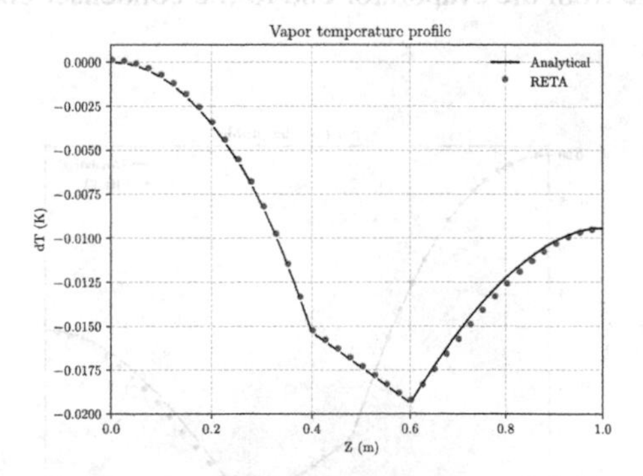

Fig. 4. Verification of vapor temperature

The total heat removal rate in this verification test is 10 W, which is relatively low such that the vapor speed is low. The maximum Mach number in this test is about

Table 1. Physical and boundary conditions of heat pipe verification test cases

Parameters	Value
$k_{s,wall}$	20.0 W/(m-K)
$k_{s,wick}$	60.0 W/(m-K)
L_e, L_a, L_c	0.4 m, 0.2 m, 0.4m
$\delta_{wall}, \delta_{wick}$	0.001 m, 0.001 m
R, h_{fg}	361.66 J/(kg-K), 4.0E + 06 J/kg
R_v, D_h	0.005 m, 0.01 m
λ, h_v	0.032, 1.0E + 06 W/(m^2-K)
Q, T_{sink}	10 W, 800 K
p_c, T_c, ρ_c	1157 Pa, 800 K, 0.004 kg/m^3

0.015, thus the incompressible assumption is valid. The verification results for vapor temperature, vapor pressure, and vapor mass flux are shown in Fig. 4, Fig. 5, and Fig. 6. The simulation results agree very well with the analytical solutions in the evaporator and adiabatic sections. Minor difference is seen in the condenser section, which is caused by the non-uniform heat flux distribution (and thus vapor condensation rate) in this section. The heat conduction model in the heat pipe wall and wick considers two-dimensional effect. While uniform heat flux is added at the evaporator outer wall, the heat flux at the interface of wick and vapor core still contains non-uniform profile. In the evaporator section, vapor pressure drops due to both evaporation of working fluid and friction; in the adiabatic section, vapor mass flux is a constant value, vapor pressure drops linearly due to friction; in the condenser section, the condensation of vapor brings in pressure recovery, but because of the frictional losses, only partial pressure recovery is achieved. Similar trend is seen in vapor temperature. Overall, there is a positive pressure and temperature drop in the vapor core from the evaporator end to the condenser end, as required by the thermodynamics law.

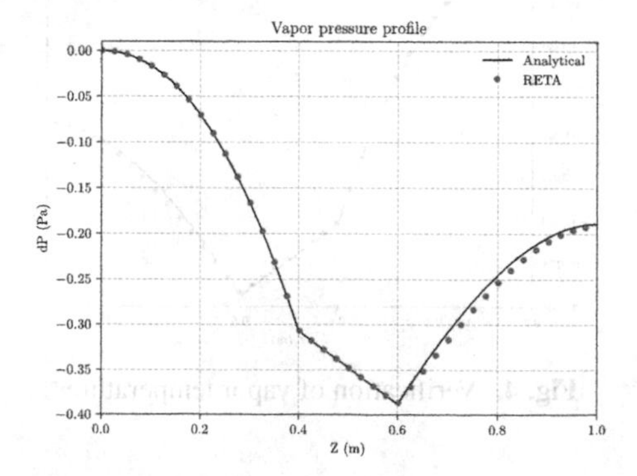

Fig. 5. Verification of vapor pressure

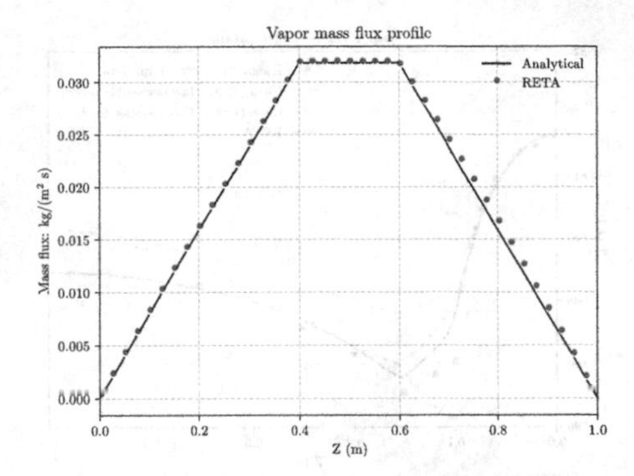

Fig. 6. Verification of vapor mass flux

A temperature drop analysis is also conducted for this verification test. The results are shown in Table 2. The subscript number is corresponding to location as labeled in Fig. 3. The simulation results match analytical results very well. The newly implemented fully-implicit model is thoroughly verified.

Table 2. Temperature drop analysis results

Variable	Analytical: K	Simulation: K	Error: K
$T_1 - T_2$	4.28E-2	4.16E-2	-1.2E-3
$T_2 - T_3$	7.95E-4	7.96E-4	1.0E-6
$T_3 - T_4$	9.44E-3	9.59E-3	1.5E-5
$T_4 - T_5$	7.95E-4	8.58E-4	6.3E-5
$T_5 - T_6$	4.28E-2	4.30E-2	2.0E-4

4.3 Validations

In this subsection, we will conduct two steady-state validation studies using experimental data and reference results from other codes available in literatures. The validation cases are based on cylindrical sodium heat pipe experiments conducted by Ivanovskii et al. [13], where vapor temperatures were measured and reported. Besides the experimental data, simulation results from Chen and Faghri [5] are also used as reference results for a code-to-code comparison, which includes results from both incompressible and compressible models.

Two cylindrical sodium heat pipes are modeled, the details of physical and boundary conditions are listed in Table 3. In the evaporator wall surface, a uniform heat flux is applied as the boundary condition; in the condenser wall surface, a convective boundary condition is applied with a sink temperature (T_{sink}) of 300 K. For this model, the thermal resistance due to convective heat transfer in the condenser wall surface is much larger

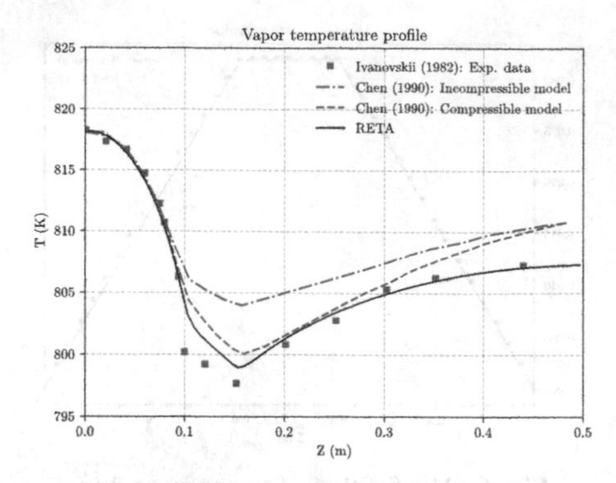

Fig. 7. The axial vapor temperature profile for sodium heat pipe with Q = 560 W

than that of the heat pipe. The heat pipe working temperature is very sensitive to the heat transfer coefficient (h_{sink}) at this surface. For this study, h_{sink} is iteratively calibrated such that the vapor pressure and temperature at the evaporator end match the reference values. The heating power in these two test cases are 560 W and 1000 W, respectively. An axial mesh size of 6.25E-03 m is used in the simulations. Steady-state results are obtained and compared with experimental/reference values.

Table 3. Physical and boundary conditions of heat pipe validation test cases

Parameters	Case 1	Case 2
Fluid	Sodium	Sodium
p_C: Pa	1300	2476
T_C: K	818	856
$k_{s,wall}$: W/(m-K)	19.0	19.0
$k_{s,wick}$: W/(m-K)	66.2	66.2
L_e, L_a, L_c: m	0.1, 0.05, 0.35	0.1, 0.05, 0.55
δ_{wall}: m	0.001	0.001
δ_{wick}: m	0.0005	0.0005
h_{fg}: J/kg	4.182E + 06	4.182E + 06
μ_v: Pa s	1.80E-05	1.80E-05
R_v: m	0.007	0.007
h_v: W/(m^2-K)	1.0E + 06	1.0E + 06
Q: W	560	1000
h_{sink}: W/(m^2-K)	59.6	62.6
T_{sink}: K	300	300
Axial elements	80	112

Figure 7 shows the comparison between numerical results from this study and reference results, including experiment data and simulation results, for the vapor temperature for Case 1. The results from RETA match the experimental data well, with a maximum deviation of about 4 K near the evaporator-adiabatic interface. Similar order of deviation was observed in the reference simulation results as well. This deviation is acceptable considering that uncertainty in vapor temperature measurement can be quite big.

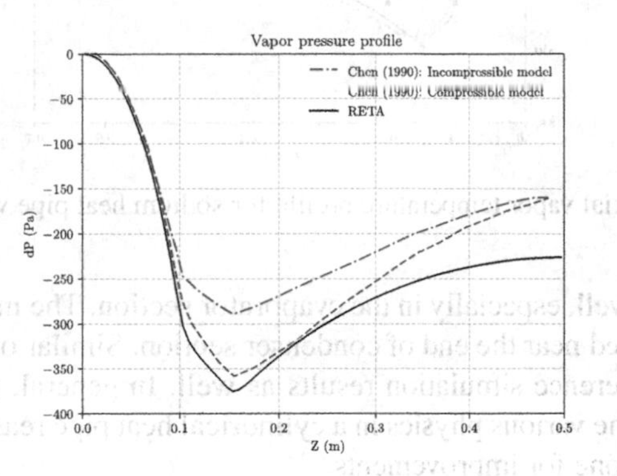

Fig. 8. The axial vapor pressure profile for sodium heat pipe with Q = 560 W

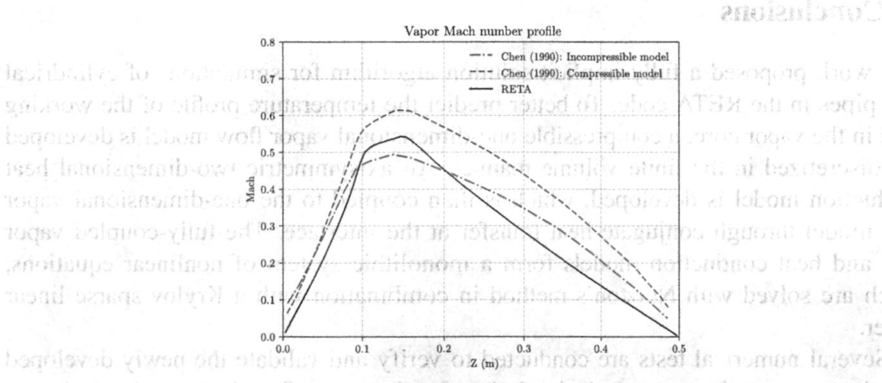

Fig. 9. The axial vapor Mach number profile for sodium heat pipe with Q = 560 W

Figure 8 shows the vapor pressure profile for Case 1. It is seen that results from RETA match the reference results and is closer to the reference compressible model results. This is expected because the current model considers the compressibility as well. Figure 9 shows the Mach number profile for Case 1. The maximum Mach number, happening at the adiabatic section, is around 0.55 from RETA prediction. In general, results from RETA match the reference results well in the evaporator section but shows deviation in the condenser section. This is determined by the difference in the vapor flow model (e.g., one-dimensional vs two-dimensional) and the difference in the friction model, etc.

Figure 10 shows the comparison between numerical results from RETA and reference results for the vapor temperature for Case 2. The results from RETA match the

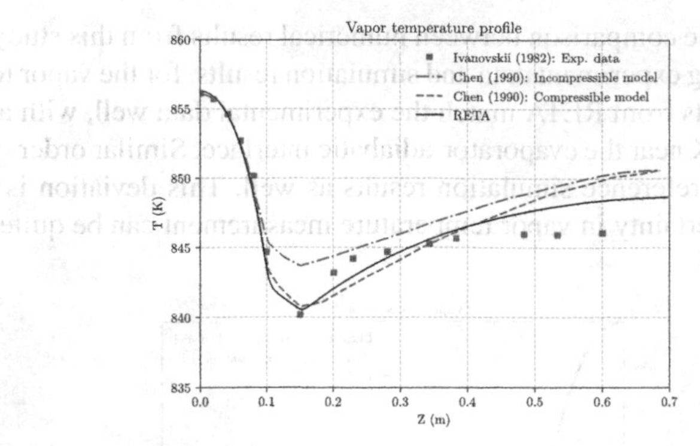

Fig. 10. The axial vapor temperature profile for sodium heat pipe with Q = 1000 W

experimental data well, especially in the evaporator section. The maximum deviation of about 3 K is observed near the end of condenser section. Similar order of deviation was observed in the reference simulation results as well. In general, the newly developed model can predict the various physics in a cylindrical heat pipe reasonably well but with future work to be done for improvements.

5 Conclusions

This work proposed a fully-implicit solution algorithm for simulations of cylindrical heat pipes in the RETA code. To better predict the temperature profile of the working fluid in the vapor core, a compressible one-dimensional vapor flow model is developed and discretized in the finite volume manner. An axisymmetric two-dimensional heat conduction model is developed, which is then coupled to the one-dimensional vapor flow model through conjugate heat transfer at the interface. The fully-coupled vapor flow and heat conduction models form a monolithic system of nonlinear equations, which are solved with Newton's method in combination with a Krylov sparse linear solver.

Several numerical tests are conducted to verify and validate the newly developed model. An approximate analytical solution for the vapor flow is derived based on a few assumptions. The excellent agreement of RETA simulation results with analytical solution confirms the model setting, derivation, and code implementation of the newly developed model. Two steady-state validation tests are conducted. Comparison of vapor temperature with experimental data and reference simulation results shows that the newly developed model has a good prediction accuracy. It is observed that the prediction accuracy is worse in the condensation section, likely caused by the rather simple friction factor correlations. Future work is needed to improve this.

To conclude, this new model considers the vapor flow with a compressible one-dimensional flow model in the vapor core. Compared with other simpler models, for example thermal resistance model, this new model can predict the vapor pressure, temperature, and velocity with good accuracy, which should help reduce the uncertainty in predicting the performance of a heat pipe. It is expected that this new heat pipe model will play an important role in the design and safety analysis of heat pipe cooled microreactors.

There are several future works for improvement of the newly developed model, including but not limited to:

- Transient simulation capability is available in the current code; however, it requires more verification and validation tests.
- Development and implementation of various heat transfer limits of heat pipe operation.
- Improvement of heat pipe solution algorithms for simulation of vapor flow at higher Mach number. At a test study, the solution algorithm is found to be struggle at supersonic conditions where discontinuities exist in vapor region.
- Development of heat pipe microreactor modeling capabilities by coupling of heat pipe models with heat conduction model of the reactor core, system loop model of the heat pipe heat exchangers, etc.

References

1. McClure, P.R., et al.: Kilopower project: the KRUSTY fission power experiment and potential missions. Nucl. Technol. **206**(sup1), S1–S12 (2020)
2. Mueller, C., Tsvetkov, P.: A review of heat-pipe modeling and simulation approaches in nuclear systems design and analysis. Ann. Nucl. Energy **160**, 16 (2021)
3. Faghri, A.: Vapor flow analysis in a double-walled concentric heat pipe. Numerical Heat Transf. **10**(6), 583–595 (1986)
4. Faghri, A., Parvani, S.: Numerical analysis of laminar flow in a double-walled annular heat pipe. J. Thermophys. Heat Transf. **2**(2), 165–171 (1988)
5. Chen, M.-M., Faghri, A.: An analysis of the vapor flow and the heat conduction through the liquid-wick and pipe wall in a heat pipe with single or multiple heat sources. Int. J. Heat Mass Transf. **33**(9), 1945–1955 (1990)
6. Cao, Y., Faghri, A.: Transient two-dimensional compressible analysis for high-temperature heat pipes with pulsed heat input. Numerical Heat Transf. **18**(4), 483–502 (1991)
7. Cao, Y., Faghri, A.: Transient multidimensional analysis of nonconventional heat pipes with uniform and nonuniform heat distributions (1991)
8. Faghri, A.: Review and Advances in Heat Pipe Science and Technology. J. Heat Transfer-Trans. Asme, **134**(12) (2012)
9. Hu, G., et al.: Multi-physics simulations of heat pipe micro reactor (2019)
10. Wang, C.L., et al.: Performance analysis of heat pipe radiator unit for space nuclear power reactor. Ann. Nucl. Energy **103**, 74–84 (2017)
11. Hansel, J.E., et al.: Sockeye theory manual. Idaho National Lab.(INL), Idaho Falls, ID (United States) (2020)
12. Balay, S., et al.: PETSc users manual (2019)
13. Ivanovski, M., Sorokin, V., Yagodkin, I.: Physical principles of heat pipes (1982)

Analysis and Qualification Control of Welding Defects of Coated 15-15Ti Cladding Tube

Junling Han[1,2](✉), Guannan Ren[1,2], Limei Peng[1,2], Hongyu Tian[1,2], and Pengbo Ji[1,3]

[1] CNNC Key Laboratory on New Materials Research and Application Development, Baotou, Inner Mongolia, China
hanjunling0805@163.com

[2] China North Nuclear Fuel Co., Ltd., Baotou, Inner Mongolia, China

[3] CNNC Key Laboratory on Fabrication Technology of Reactor Irradiation Special Fuel Assembly, Baotou, Inner Mongolia, China

Abstract. Physical property test fuel rod is used for the engineering test and thermal comprehensive experimental verification of lead-cooled reactors. Preliminary electron beam welding (EBW) trials showed that the welding quality of coated 15-15Ti tube and 316L end plug were significantly affected by welding defects. By studying the welding defects with optical microscopy (OM) and scanning electron microscopy (SEM), it is showed that the inclusions in the coating of the cladding tube enter the welding line during EBW, increasing the tendency to form cracks and leading to welding cracks; the excessively long mating surface between the cladding tube and end plug results in welding gas expansion. Through the design of orthogonal tests with influential parameters including the length of the mating surface of the end plug, the removal amount of the inner wall of the cladding tube and the interference amount on the quality characteristics, the comprehensive effects of these parameters were studied and the best matching structure was determined, which breaks through the difficulty in the welding between 316L end plug and coated 15-15Ti tube, and the welding qualification has been verified to be improved.

Keywords: fuel rod · electron beam welding (EBW) · welding with coating material · welding defect · orthogonal experiment qualification control · 15-15Ti cladding tube

1 Introduction

The lead-based reactor is one of six fourth-generation reactors recommended by the International Forum on Energy Systems (GIF), it has significant advantage in safety, miniaturization and feasibility. Therefore, the research on Physical property test fuel rod promotes the progress of ADS transmutation system and related technology of lead based reactor, and lays a solid foundation for the engineering of lead based reactor in China. Physical characteristics fuel rod welding is completed by ring welding of upper end plug, lower end plug and clad tube. The upper and lower end plugs are machined from 316L stainless steel and the coated tube is 15-15Ti material.

C. Liu (Ed.): PBNC 2022, SPPHY 283, pp. 861–871, 2023.
https://doi.org/10.1007/978-981-99-1023-6_73

The main defects of fuel rod welding are incomplete penetration, porosity, gas expansion and crack, etc. Porosity is one of the commonly occurring welding defects in welding, it exists inside or on the surface of the weld metal, and its defect form is round porosity, columnar porosity and round dense porosity. The harmfulness of porosity is mainly manifested as a reduction in the load-bearing capacity of the welded joints, resulting in increased probability of local corrosion perforation. Incomplete penetration is the incompletely penetration of root or interlayer of the joint during welding. Stress concentration is easily generated at the incomplete ends and gaps, which may lead to cracking under external forces. Cracks are gaps that exist inside or on the surface of the weld or heat-affected zone, and there is a significant stress concentration at the sharp root. When the stress level exceeds the strength limit of the sharp root, the crack will expand, which will aggravate fatigue failure and stress corrosion failure.

2 Statistics and Analysis of Welding Defects

2.1 Statistics of Qualification Rate

According to the statistics of welding test results in the early stage of the test, the average welding qualification rate was 81.9%, and the batch welding qualification rate was shown in Fig. 1. The welding defects leading to unqualified fuel rods in the above experimental batches include gas expansion, crack, incomplete penetration, porosity, non-fusion and unqualified appearance, as shown in Fig. 2.

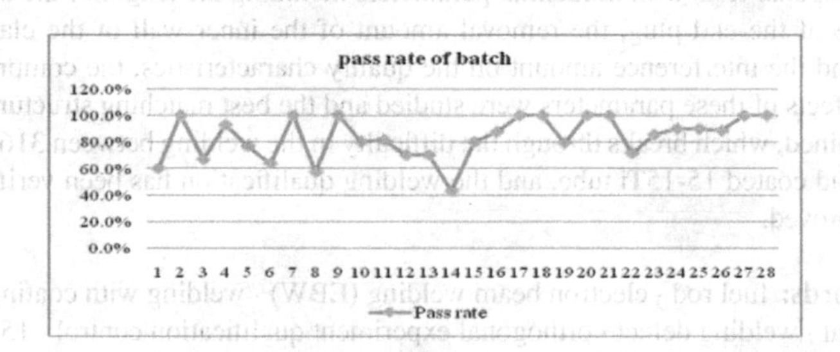

Fig. 1. Fuel rod welding batch pass rate statistics during the experimental phase Statistics

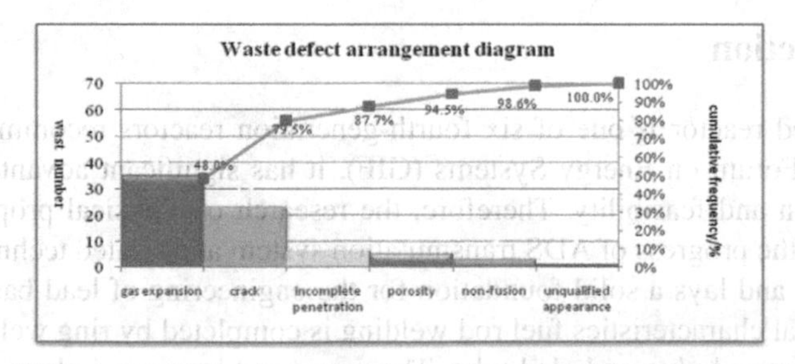

Fig. 2. Experimental phase fuel rod welding failure statistics

Analysis of the arrangement in Fig. 2 shown that the cause of the passing rate plummeting is the generation of gass expansion and cracks; By the statistical date, it was found that gas expansion and cracking accounted for 79.5% of the total number of defects, through the theoretical calculations, if we reduce the number of gas expansion and cracking defects of 85.1%; Though the analysis of two kinds of defects, take the corresponding control measures, and improve the welding pass rate.

2.2 Defect 1-Crack Analysis

In the welding experiments, the fuel rods that are not qualified in NDT were examined by means of penetration test, metallurgical inspection, scanning electron microscopy and energy spectrum analysis, and the welds were further analyzed.

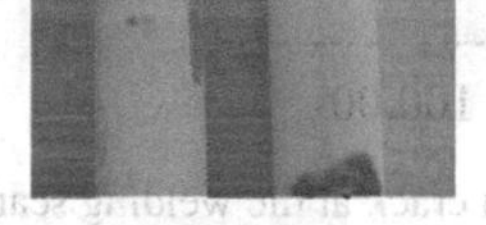

Fig. 3. Penetration testing-crack

Fig. 4. Metallographic testing-crack

For the analysis of the test welded fuel rods for the experimental process generated by the X-ray inspection failed fuel rods, and targeted further destructive testing of the welding seam is carried out to analyze the root cause of its generation. Figure 3 shows the penetration test, and the test results shown that there was a crack about 5 mm in the lower plug ring welding area on the left side of the fuel rod, and there was a crack about 2 mm in the upper plug ring welding area on the right side of the fuel rod. The metallographic test specimens were prepared for the penetration detection cracks; Fig. 4 shown the metallographic test at the corresponding cracks, and the figure shown that the cracks were found in the metallographic test corresponding to the welding cracks of the fuel rod on the left, and no cracks were found at the welding cracks of the fuel rod on the, right, but one crack was found in the cladding matrix material.

Figure 5 shown the scanning electron microscopy analysis of the crack at the welding, combined with the energy spectrum analysis can be obtained, the crack is mainly composed of C, O and Si elements, and the C is the highest content of elements. Figure 6 shown the scanning electron microscopy analysis of the crack at the substrate, combined with the energy spectrum analysis, it can be seen that in addition to the high content of C and O elements, Ti, Mn and Cr elements with high composition were been detected (Fig. 7).

In view of the metallographic results founded cracks from the cladding matrix, though the cladding analysis, it is founded that the fuel rod cladding is a kind of coated 15-15Ti cladding tube, combined with the analysis about scanning electron microscopy and energy spectrum of cladding tube coating in Fig. 9, it was obtained the inner wall coating C, O, Ti, Mn, Cr elements on the high side. It is inferred that the inner wall coating of the tube is mainly composed of oxides and carbides of Mn, Ti and Cr etc.

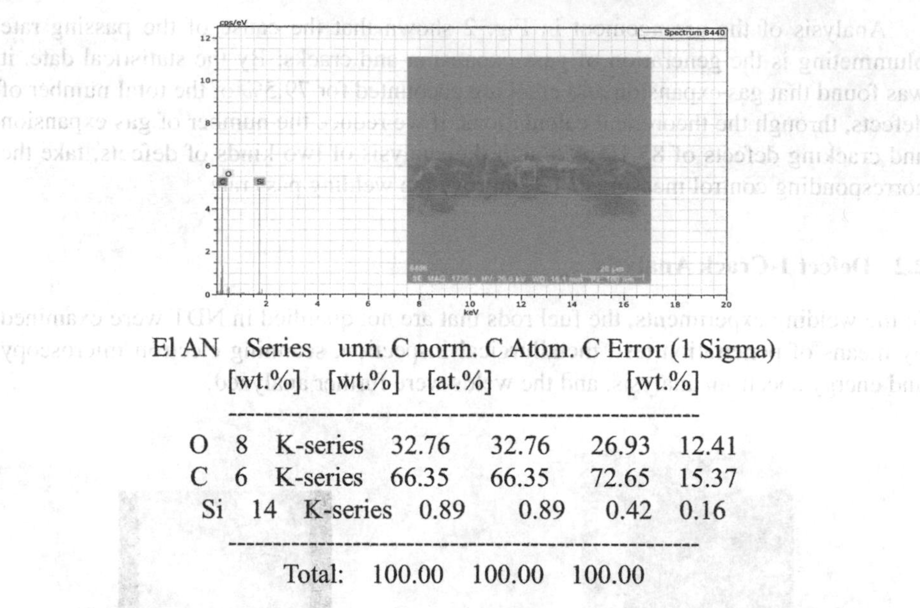

El	AN	Series	unn. C [wt.%]	norm. C [wt.%]	Atom. C [at.%]	Error (1 Sigma) [wt.%]
O	8	K-series	32.76	32.76	26.93	12.41
C	6	K-series	66.35	66.35	72.65	15.37
Si	14	K-series	0.89	0.89	0.42	0.16
		Total:	100.00	100.00	100.00	

Fig. 5. The results of electron microscopy detection crack at the welding scanning

Referred to the relevant literatures, it was known that the cracking mechanism can be divided into hot cracks and cold cracks; Hot cracking is generated during the transition of the weld metal from liquid to solid state, such as sulfur and phosphorus in the weld is prone to thermal cracking. Cold cracking is generated during the cooling process of the weld, such as higher carbon content or more alloy elements is prone to cold cracking. When the carbon equivalent of the material is less than 0.4%, the basic will not produce cracks, while the carbon equivalent of the material is greater than 0.6%, it is difficult to completely avoid the cracks.

Combined with the analysis of the above test results could be obtained, the inclusions of inner coating lead to the cracks; the oxides and carbides of inner coating enter the weld increases the cracking tendency during the welding, so that the cracks were occured to some welds, at the same time, the high content of C elements in the coating can lead to the formation of saturation at the grain boundaries, further formation of carbon precipitation, increased the cracking tendency of the weld.

2.3 Defect 2-Gas Expansion Analysis

Selected X-ray detection of the failed fuel rod, the further analysis of metallurgical testing about the weld, the results were shown in Fig. 8, the cavity between the weld and the substrate is the gas expansion.

Reference to the relevant literatures [4] can be seen, gas expansion mainly occurs in the heat-affected zone close to the fusion zone, a slight gas expansion lead to the mating surface of the end plug and the inner wall to form a small gap along the circumferential direction of the clad tube, a serious gas expansion can make the wall thickness of the clad tube thinning.

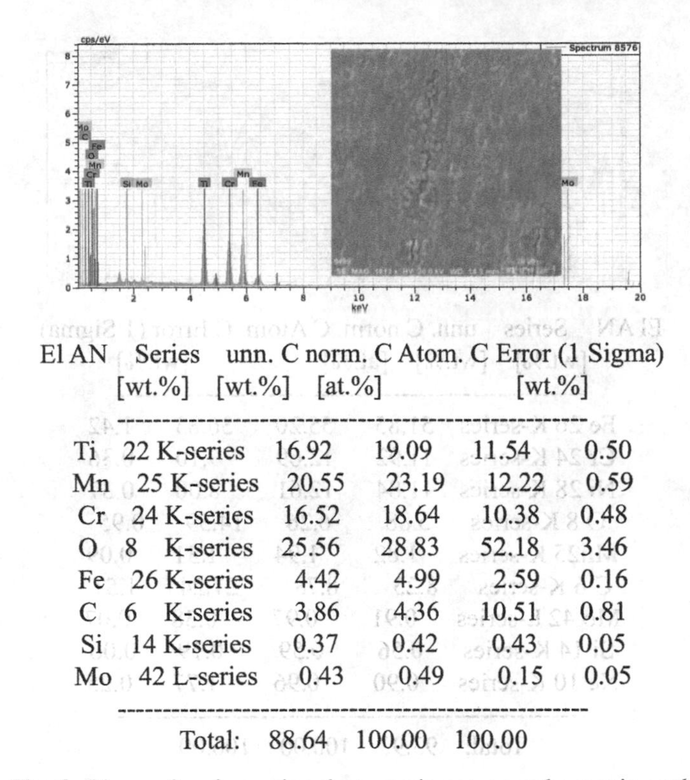

El	AN	Series	unn. C [wt.%]	norm. C [wt.%]	Atom. C [at.%]	Error (1 Sigma) [wt.%]
Ti	22	K-series	16.92	19.09	11.54	0.50
Mn	25	K-series	20.55	23.19	12.22	0.59
Cr	24	K-series	16.52	18.64	10.38	0.48
O	8	K-series	25.56	28.83	52.18	3.46
Fe	26	K-series	4.42	4.99	2.59	0.16
C	6	K-series	3.86	4.36	10.51	0.81
Si	14	K-series	0.37	0.42	0.43	0.05
Mo	42	L-series	0.43	0.49	0.15	0.05
		Total:	88.64	100.00	100.00	

Fig. 6. The results of scanning electron microscopy at the matrix crack

During the vacuum electron beam welding process, Due to the rapid crystallization of metal, electron beam welding seam is usually very narrow, rapid cooling and solidification, litter chance of gas overflow from the weld, thus increased the possibility of porosity and gas expansion. The analysis of gas expansion in electron beam welding was complicated, the main reasons include: 1) the cleanliness of workpiece is not enough; 2) The coating tube and the end plug are not properly matched; 3) The gas generated during the welding process without drainage channels; 4) Excessive heat input during welding.

Through the observation of a lot of X-ray photos found that gas expansion has the following characteristics: 1) the gas expansion and porosity produced at the same time, different batches of ring welds, the probability and size of gas expansion and porosity fluctuates greatly; 2) the location of gas expansion is fixed, it is that the fusion zone near the end plug side, the end plug and the clad tube with the gap area, as shown as Fig. 9.

In view of relevant research and tests has been carried out in the experimental stage, the best welding parameters had been developed, and the welding heat input could be well controlled; In order to eliminate the porosity and gas expansion, the measures were be taken as followed: Strictly cleaning into and drying for the clad tube and end plug before welding, maintaining the cleanliness of the welding area; Choose a qualified end plug interference structure to reduce the residual gas in the mating gap.

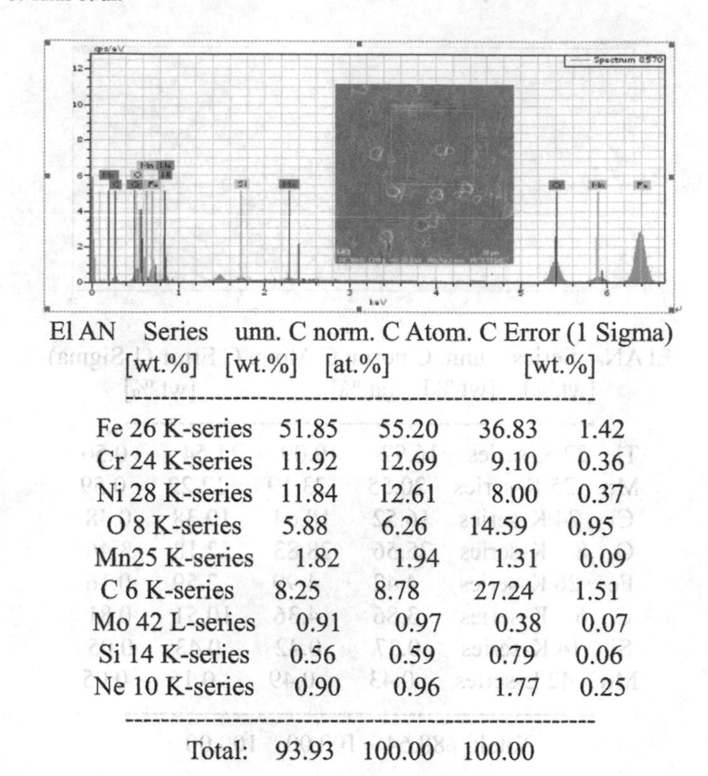

El	AN	Series	unn. C [wt.%]	norm. C [wt.%]	Atom. C [at.%]	Error (1 Sigma) [wt.%]
Fe	26	K-series	51.85	55.20	36.83	1.42
Cr	24	K-series	11.92	12.69	9.10	0.36
Ni	28	K-series	11.84	12.61	8.00	0.37
O	8	K-series	5.88	6.26	14.59	0.95
Mn	25	K-series	1.82	1.94	1.31	0.09
C	6	K-series	8.25	8.78	27.24	1.51
Mo	42	L-series	0.91	0.97	0.38	0.07
Si	14	K-series	0.56	0.59	0.79	0.06
Ne	10	K-series	0.90	0.96	1.77	0.25
		Total:	93.93	100.00	100.00	

Fig. 7. The scanning electron microscopy detection results of cladding coating

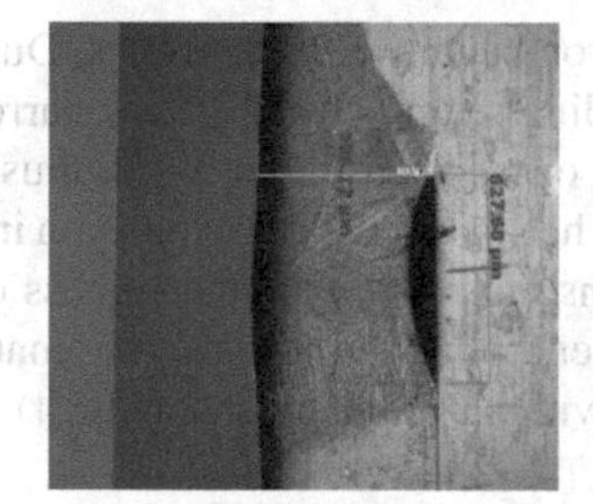

Fig. 8. The defect of gas expansion

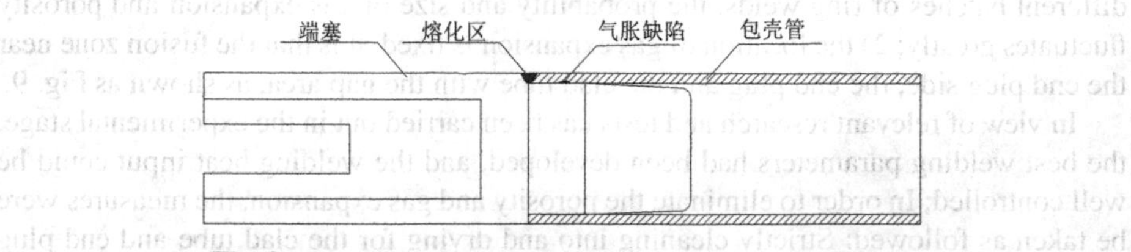

Fig. 9. Schematic diagram of fuel rod gas expansion defect

3 Test Program

Summarize above statistics can be shown that the main reason for the low welding pass rate is cracks and gas expansion, the main reason affecting the generation of cracks in the weld is the inner wall coating of the casing tube; The gas expansion during the welding process is caused by the longer surface between the end plug and the casing tube contact.

Therefore, the impact of welding defects is the inner wall removal amount of cladding tube, end plug and clad mating surface length., The amount of interference fit between them will be changed, when the clad and end plug structure changes at the same time.

Select orthogonal table of $L_9(3^4)$, makes the factor level Table 1.

Table 1. Factor level table of welding test

levels	Factors		
	Removal amount of inner wall of cladding(mm)	Interference between end plug and the tube(mm)	Length of mating surface between end plug and cladding(mm)
1	0	0.02	3.5
2	0.02	0.03	2.0
3	0.04	0.04	0.8

According to the above three influencing factors, orthogonal experiments were designed, and 10 fuel rod welding tests were carried out according to different test groups. As the X-ray inspection, penetration detection and weld appearance detection results for judge reference, it was shown in Table 2.

Table 2. Test criteria items

Record	Item		
	X-ray result	penetration detection	weld appearance detection
1	pass	flawless	pass
0	defects of porosity and gas expansion	crack	defects of rough weld, obvious fish-lock lines, appearance holes

Analysis of test data can be obtained, the crack defects could been effectively controlled by removed the internal coating, the generation of gas expansion defects could been effectively suppressed for the length about 0.8 mm of mating surface. The best test scheme is to remove the internal coating of the cladding tube about 0.04 mm, the mating surface length between the cladding tube and end plug is 0.8 mm, and the interference fitting amount of the cladding tube and end plug is 0.03 mm.

4 Verification of Process Parameters

4.1 Experimental Data Verification

In order to further verify the feasibility of the orthogonal test results, experimental verification was carried out for the above parameters. The cladding tube with 0.04 mm removal of the inner wall coating, re-processed the end plug contact surface with 0.8 mm, and the interference amount of 0.03 mm were used to carry out welding tests, The welding of 10 fuel rods were completed, and the helium leak detection of welding seam, X-ray inspection, metallographic testing and penetration testing were carried out.

The helium leak detection results were all passed, the test results of X-ray inspection of the weld shown that 9 roots qualified, while one root had small pores, Which may be caused by the uncleaned pores before welding; Combined with the analysis of the results of metallographic testing and penetration testing, it can be found that there is no welding cracks. Figure 10 shown the metallographic testing results of a few randomly selected welds, metallographic testing results shown that no cracks and no gas expansion; Fig. 11 shown the results of penetration testing of the weld, It was could be seen no cracks from the figure.

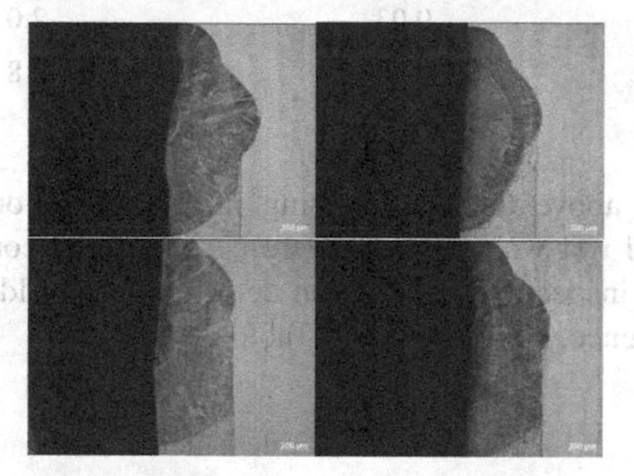

Fig. 10. Weld metallographic result

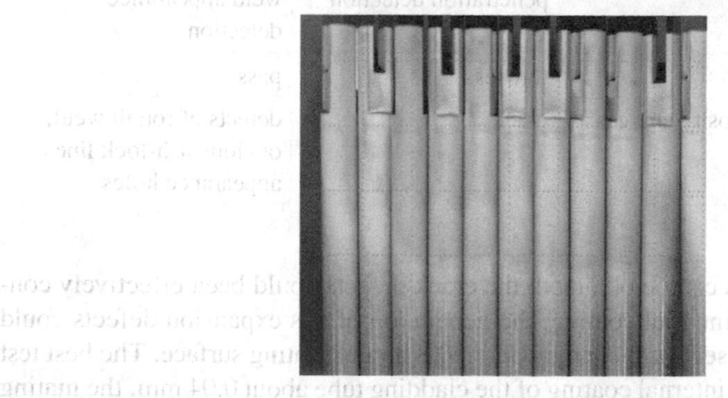

Fig. 11. Weld penetration result

In order to further confirm whether there were cracks in the above-mentioned weld seam with qualified test results, 1 fuel rod with qualified test results was randomly selected for metallographic layer grinding test, and the grinding was carried out and photographed according to the grinding volume of 0.15 mm each time, and the test results were shown in Fig. 11. Through the layer grinding metallographic inspection photos can be seen, there was no cracks in the welding area (Fig. 12).

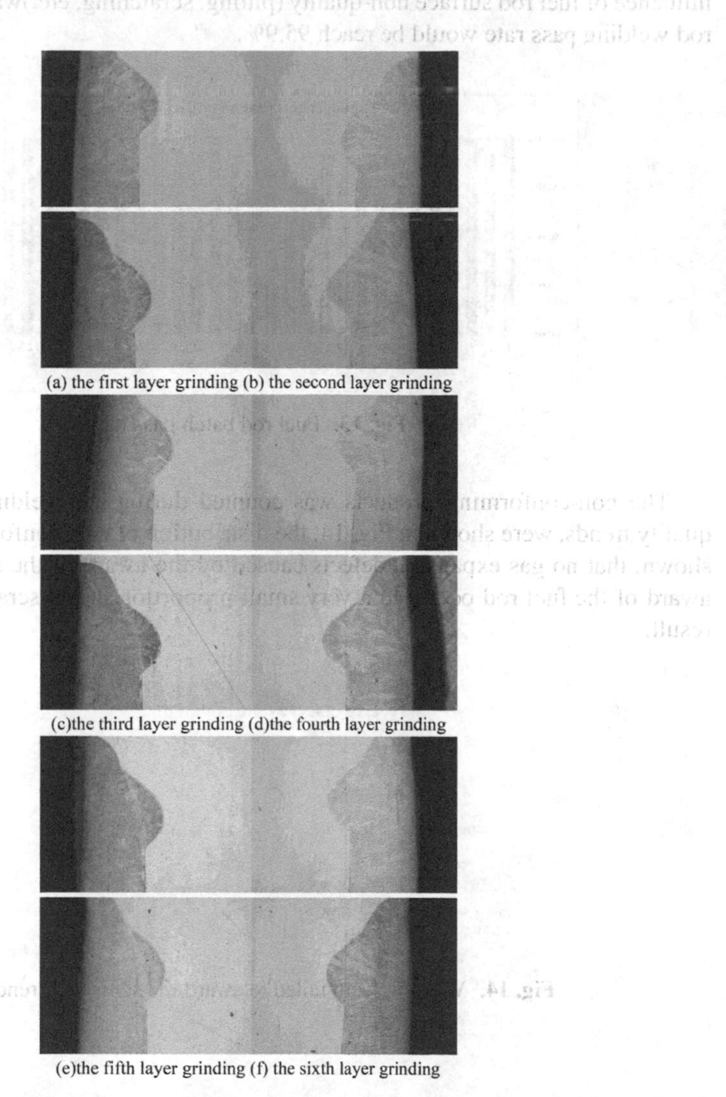

(a) the first layer grinding (b) the second layer grinding

(c)the third layer grinding (d)the fourth layer grinding

(e)the fifth layer grinding (f) the sixth layer grinding

Fig. 12. Layer grinding test metallographic test results

4.2 Production Data Verification

The process program was carried out to subsequent welding work, the welding about 876 fuel rods had no helium leak detection and penetration testing failed fuel rods. The quality data of the production phase was analyzed, the qualification rate of fuel rods reached 94.1%, and the qualification rate statistics were shown in Fig. 13. IF the influence of fuel rod surface non-quality (pitting, scratching, etc.)was excluded, the fuel rod welding pass rate would be reach 95.9%.

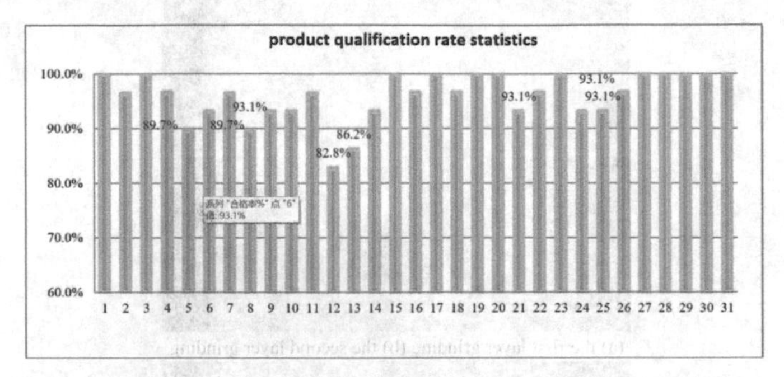

Fig. 13. Fuel rod batch pass rate

The non-conforming products was counted during the welding process, and the quality trends, were shown in Fig. 14, the distribution of non-conforming product types shown, that no gas expansion defects caused by the award of the scrap, and the crack award of the fuel rod occupied a very small proportion of the scrap, achieved a better result.

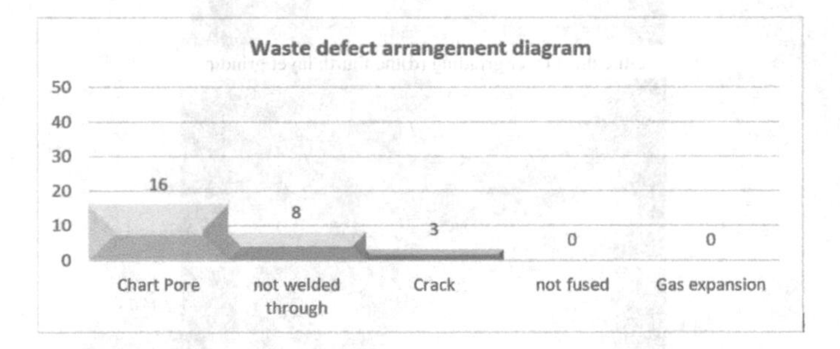

Fig. 14. Welding seam failed to award the statistical trend chart

5 Conclusions

(1) Cracking and gas expansion is the root cause of the fuel rod welding quality degradation on the welding between 316L end plug and coated 15-15Ti tube, the inner wall coating of the clad tube is the root cause of cracking, end plug mating surface length is the root cause of gas expansion.

(2) The best process parameters for the inner wall coating removal amount of 0.04 mm, end plug mating surface length with 0.8 mm and the amount of overfill is 0.03 mm. It could effectively reduce the generation of cracks and gas expansion defects, the pass rate of fuel rod welding increase to 95.9%.

References

1. Yang, J.: Exploring the defects and prevention measures in the welding press of metallic materials. Management and others 125–126 (2021)
2. Jia, Z., Jiang, Z., Li, Y., Wang, F.: Analysis of welding defects and prevention of austenitic stainless steel tubes. Weld. Technol. **47**(9), 26–29 (2018)
3. ASTM Technical Committee: ASTM B338. Standard Specification for Seamless and Welded Titanium and Titanium Alloy Tubes for Condensers and Heat Exchangers. ASTM International, United States (2010)
4. Ma, C.: Analysis of welding defects and prevention measures of austenitic stainless steel. Heat Process. Technol. **44**(17), 243–246 (2015)
5. Liu, Y.: Using orthogonal tests to optimize the welding process parameters of Gr2 titanium tube (T0.5 mm). 05, 8 (2014)
6. Zhang, J.: Selection of welding parameters for the bottom frame floor based on the orthogonal test method. Des. Res. **43**(04), 12–16 (2016)

Experimental Study on the Intrusion/Erosion Behavior of GMZ Bentonite Considering Fracture Aperture Effects

Libo Xu, Weimin Ye(✉), Qiong Wang, and Hewen Luo

Tongji University, Shanghai, China
{xulb,ye_tju}@tongji.edu.cn

Abstract. Intrusion/erosion will cause mass loss of bentonite, resulting in endangering the operation safety of the repository for disposal of high-level radioactive waste. In this work, intrusion/erosion tests were conducted on GMZ bentonite specimens with a fracture aperture of 0.1, 0.5, 0.75 and 1.0 mm, as well as a duration of 60 days. Images were regularly captured with a digital camera. The intrusion distance and width of accessory-mineral ring were obtained. The erosion rate was determined by a turbidimeter. Meanwhile, parallel tests with a fracture aperture of 0.5 and 1.0 mm were repeated with a duration of 5, 15 and 30 days. After tests stopped, dry weight of specimens was measured. Results show that the intrusion distance increased with increasing fracture apertures, resulting in the increase of accessory-mineral-ring width, which will in turn restrict intrusion and erosion. Actually, the erosion rate in the early period decreased obviously with increasing accessory-mineral-ring widths, due to its filtration effects. However, when erosion reached stabilization, the erosion rate increased with increasing fracture apertures and intrusion distances but not independent of the accessory-mineral ring. The majority part of specimen mass loss ratio in the early period was from intrusion. But when intrusion reached stabilization, the mass loss ratio of specimen will be mainly contributed by erosion and would keep on increasing. Moreover, the mass loss ratios of specimens after 60 days by intrusion and erosion both increased with increasing fracture apertures. Meanwhile, the difference between the mass loss ratio by intrusion and erosion increased with increasing fracture apertures.

Keywords: Bentonite · Intrusion/erosion · Accessory-mineral ring · Fracture aperture · Mass loss

1 Introduction

Deep geological disposal is considered as an effective way to solve the problem of safe disposal of high-level radioactive waste (HLW). According to the concept of this method, high-level radioactive waste is designed to store in a natural geological body (500–1000 m below the ground surface), in order to isolate it from the ecological environment for a long time or permanently. In this regard, compacted bentonite has been considered as a candidate buffer material for sealing the gaps and fractures in the repository

C. Liu (Ed.): PBNC 2022, SPPHY 283, pp. 872–886, 2023.
https://doi.org/10.1007/978-981-99-1023-6_74

[1–4]. Researches reveal that numerous fractures with various sizes would be inevitably encountered in the surrounding rock including those naturally originated or generated during construction activities, etc. [1, 13, 16]. Theses fractures could serve as channels for groundwater flow. Once compacted bentonite in the core area contacted with groundwater in the fractures, it will expand and squeeze into the fractures. Meanwhile, when the cation concentration in the groundwater is below the critical coagulation concentration (CCC) [8, 15], bentonite colloidal particles could be generated at the clay/water interface and carried away, resulting in bentonite erosion by flowing groundwater. The phenomenon that bentonite buffer material hydrated into fractures in the surrounding rock and then eroded away by groundwater was defined as bentonite intrusion/erosion. This intrusion/erosion behavior may cause significant mass loss of bentonite during the long-term operation of the repository, especially for the fractures with larger sizes. Therefore, it is of great importance to figure out the intrusion/erosion behavior of bentonite with consideration fracture aperture effects.

During past decades, numerous work has been conducted on the intrusion/erosion process of montmorillonite in artificial fractures. A two-dimension fracture formed by two transparent plates is designed and the intrusion/erosion tests were performed to investigate effects of the types of montmorillonite, water velocity and slope angle of the fracture on the intrusion distance and erosion rate. Vilks et al. (2010) found that the intrusion distance and erosion rate of Na-rich montmorillonite was significantly higher than Ca-rich montmorillonite [14]. Schatz et al. (2013) found that there was not enough test data to prove the relationship between the intrusion distance and flow velocity [12]. While the erosion rate of Na-rich montmorillonite increased with the increase of the flow velocity under the most dilute conditions (ionic strength < 1 mM). Subsequently, with the same apparatus, the influence of fracture angles on the intrusion distance and erosion rate was investigated by Schatz et al. (2016) [11]. It was observed that the intrusion distance was larger in the bottom half than that in the upper one, with the fracture angle of 45° and 90°. Moreover, an increase in erosion rate of Na-rich montmorillonite was observed with an increasing fracture slope angle from 0° to 90°.

In comparison with montmorillonite materials, the intrusion/erosion process of bentonite also has attracted much attention. The intrusion/erosion behavior of MX-80 bentonite with a naturally varying fracture aperture was investigated by Reid et al. (2015) [10]. It was found that an accessory-mineral ring, mainly composed of accessory minerals (such as quartz, felspar, etc.) was generated at the intrusion edge during the process of erosion. A two-stage cyclic intrusion/erosion mechanism was derived, that confirms the mitigating effects of accessory-mineral ring on the intrusion and erosion.

Previous studies have focused on the intrusion/erosion behavior with one single fracture aperture. However, the variability of the fracture aperture may play an important role in the development of the accessory-mineral ring, which may in turn inhibit intrusion and erosion [10]. Moreover, research of influence of intrusion/erosion on mass loss of bentonite has rarely been reported, which is necessary for assessing the deterioration process of buffer materials with fractures.

In this study, tests on intrusion/erosion of GMZ bentonite into artificial fractures were carried out with a fracture aperture of 0.1, 0.5, 0.75 and 1.0 mm, as well as a duration of 60 days, in order to evaluate the influence of fracture apertures on bentonite intrusion and erosion. Meanwhile, parallel tests with a fracture aperture of 0.5 and 1.0 mm were repeated with a duration of 5, 15 and 30 days. After tests stopped, the mass loss of specimens for each test could be determined. Based on these results, influences of the intrusion/erosion behavior and on the mass loss of bentonite were analyzed.

2 Methodlogy and Test Procedures

2.1 Apparatus

In this study, an experimental apparatus was developed (Fig. 1) for conducting intrusion/erosion tests. By inserting stainless-steel gasket into two transparent acrylic plates, a flow-through cell was generated between the two plates. Then, fractures with different sizes could be formed by adjusting the thickness of the gasket. Meanwhile, a hole for placing the specimen was set in the center of the apparatus. A cross-type counterforce frame was mounted on the upside of the acrylic plate (Fig. 1), to restrict the vertical deformation of the acrylic plate due to the specimen swelling. Finally, with a high-precision camera fixed above the test apparatus, the bentonite intrusion process could be monitored by taking images automatically at given time intervals.

(a) Image of the test apparatus

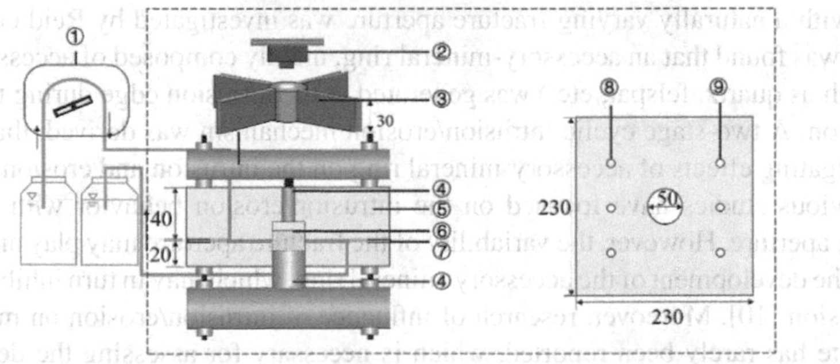

(b) Schematic diagram of the test apparatus

① Peristaltic pump; ② Camera; ③ Counterforce frame;
④ Piston; ⑤ Acrylic plate; ⑥ Specimen;
⑦ Crack; ⑧ Inlet; ⑨ Outlet

Fig. 1. Setup for conducting the intrusion/erosion test

2.2 Materials and Specimen Preparation

1) Materials

Gaomiaozi (GMZ) bentonite was tested in this study, which was extracted from Inner Mongolia autonomous region of northern China [5, 6]. Its basic properties are listed in Table 1.

2) Specimen preparation

For specimen preparation, 77.5 g of GMZ bentonite powder with an initial water content of 10.0% and filtrate conductivity below 10 μs/cm was statically compacted into a cylindrical specimen with a target height of 20 ± 0.5 mm, a diameter of 50.4 mm and a dry density of 1.70 ± 0.05 Mg/m^3.

Table 1. Basic properties of GMZ bentonite [5]

Properties	Description
Specific gravity of soil grain	2.66
Mesh	200
Liquid limit (%)	276
Plastic limit (%)	37
Total specific surface area (m^2/g)	570
Cation exchange capacity (mmol/g)	0.773
Main minerals	Montmorillonite (75.4%), Quartz (11.7%), Cristobalite (7.3%), Feldspar (4.3%)

Table 2. Specifications of the tests conducted

Tests	Fracture aperture (mm)	Test termination time (day)
T1	0.1	60
T2-5	0.5	5, 15, 30 and 60
T6	0.75	60
T7-10	1.0	5, 15, 30 and 60

2.3 Test Procedures

In order to investigate the influence of fracture apertures on intrusion/erosion behavior, as well as the intrusion/erosion process on the mass loss of buffer materials, a total of 10 tests were conducted (Table 2).

First of all, after installation of the compacted bentonite specimen into the central compartment, a peristaltic pump was used to provide deionized water injected into the three inlets. A constant velocity of 8.3 × 10^{-5} m/s could be produced. Then, the camera was activated and images were regularly obtained. According to the relationship between the circular area and the radius, the intrusion distance (l) or width of the accessory-mineral ring (l_1) could be derived,

$$l = \left(\sqrt{\frac{2(S_1 + S_2)}{\theta} + r^2} \right) - r \quad (1)$$

$$l_1 = l - \left[\left(\sqrt{\frac{2S_1}{\theta} + r^2} \right) - r \right] \quad (2)$$

where, S_1 and S_2 are the circular area of non-accessory and accessory-mineral ring, θ is the angle of fan-shape area and r is the radius of specimen. (Fig. 2).

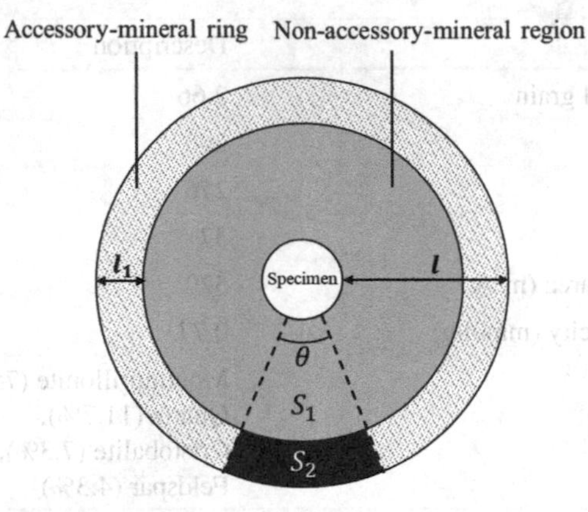

Fig. 2. A bird view of the test apparatus with bentonite intruded

Meanwhile, with a portable turbidimeter (TN-100 from Eutech Instruments), the turbidity of outlet effluent was measured every 24 h. Then, according to the calibration curve shown in Fig. 3, erosion rate could be determined [2]. After the test, the specimen was pushed out and its dry mass was weighed.

New specimens were installed and procedures above were repeated. Until all the 10 specimens with respect to the various fracture apertures and test termination time were tested corresponding to their own specifications (Table 2), all the tests were completed.

3 Results

3.1 Intrusion Behavior

The intrusion stage with a fracture aperture of 1.0 mm during 60 days is presented in Fig. 4. As shown in Fig. 4, a ring of layered material (accessory-mineral ring) was

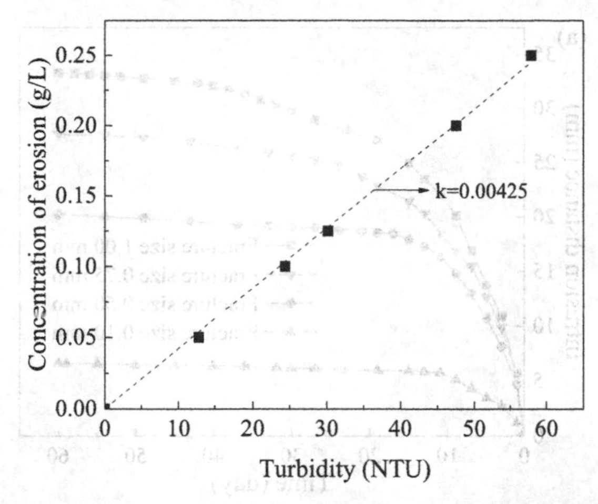

Fig. 3. Calibration curve for the turbidity and concentration of erosion

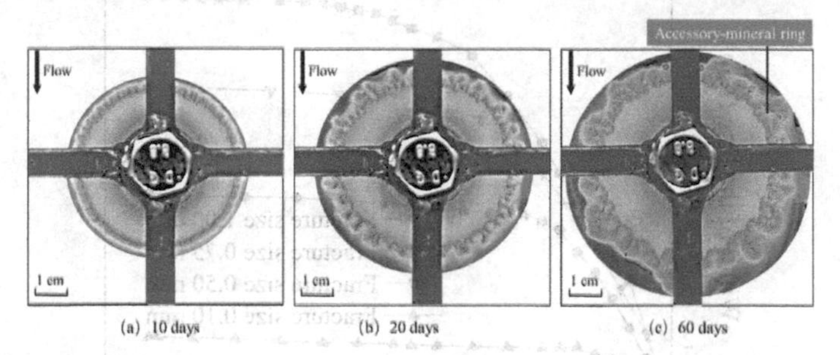

Fig. 4. Scenarios after different intrusion times for the test with a fracture aperture of 1.0 mm

formed and developed at the clay-water interface. This observation is consistent with the research by Reid et al. (2015) [10].

According to the ImageJ software, evolutions of intrusion parameters (intrusion distance and width of accessory-mineral ring) are obtained in Figs. 5 (a) and (b). It clearly shows that the intrusion distance and width of accessory-mineral ring both increased significantly in the early stage. Then, the two intrusion parameters increased more gradually and reached stability after 60 days. Moreover, for one given day, the two intrusion parameters increased with increasing fracture apertures.

3.2 Erosion Behavior

The erosion curve in terms of erosion rate versus time is shown in Figs. 6 (a). Results in Fig. 6 (a) show that the erosion rates with a fracture aperture of 1.0, 0.75, 0.5 and 0.1 mm decreased rapidly before 5 days. Then the erosion rates increased and reached a second peak value after 10, 9, 8 and 9 days, with a fracture aperture of 1.0, 0.75, 0.5 and 0.1 mm, respectively. Finally, their corresponding erosion rates decreased and kept stable around after 60 days.

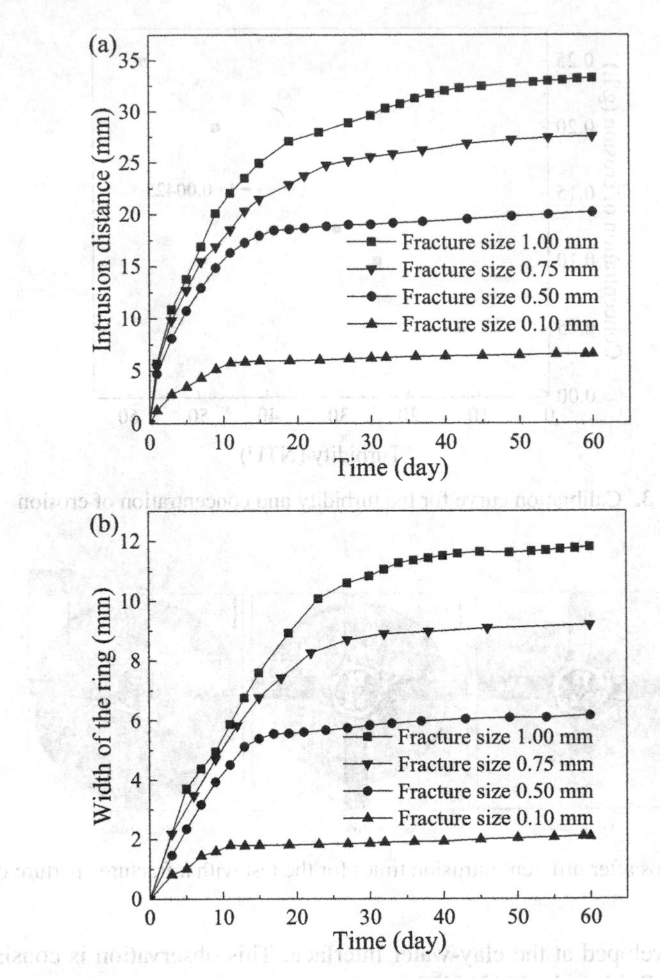

Fig. 5. Evolutions of intrusion distance (a) and width of accessory-mineral ring (b) with time

Meanwhile, the evolutions of the accumulated erosion mass with time are depicted in Fig. 6 (b). It can be observed that the accumulated erosion mass increased significantly in the early period. Then, an obvious turning point of the erosion curve was observed. Moreover, for one given day, the accumulated erosion mass increased obviously with increasing fracture apertures.

3.3 Mass Loss Ratio of Specimens

With tests stopped, the specimens were pushed and weighed to obtain the dry mass. Then, for tests with a fracture aperture of 0.5 and 1.0 mm, the evolutions of mass loss ratio of specimens with time are obtained in Fig. 7. It appears in Fig. 7 that once the intrusion/erosion occurred, the mass loss ratios of specimens both increased rapidly and then followed by an obvious reduction in the increase rate after about 15 days.

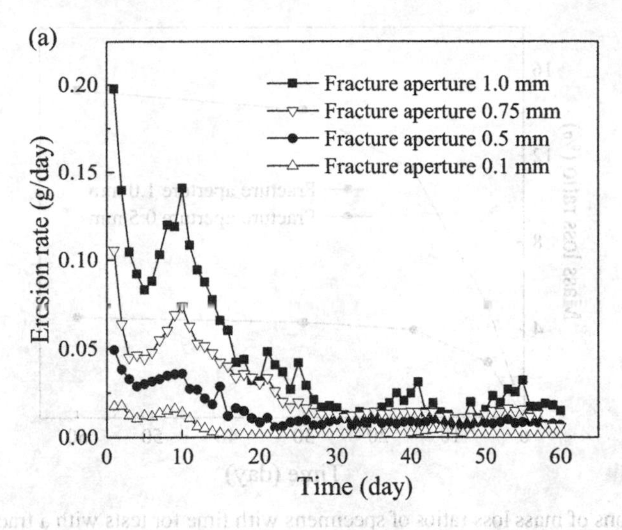

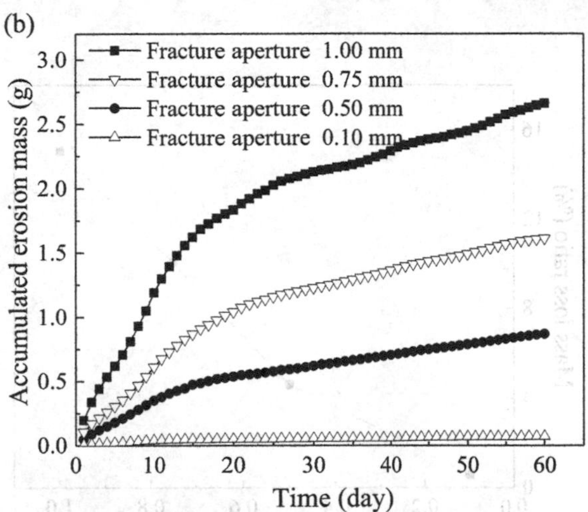

Fig. 6. Evolutions of erosion rate (a) and accumulated erosion mass (b) with time

Subsequently, the mass loss ratio of specimens after 60 days is plotted versus the fracture aperture in Fig. 8. It can be observed in Fig. 8 that the mass loss ratio of specimens increased significantly with increasing fracture apertures.

4 Discussion

4.1 Influences of Fracture Apertures on Bentonite Intrusion

According to the intrusion model by Svobody (2013) [13], the sidewall friction factor decreases with increasing fracture apertures. Therefore, the intrusion distance in the steady state is proportional to the fracture aperture. Meanwhile, based on this theory and

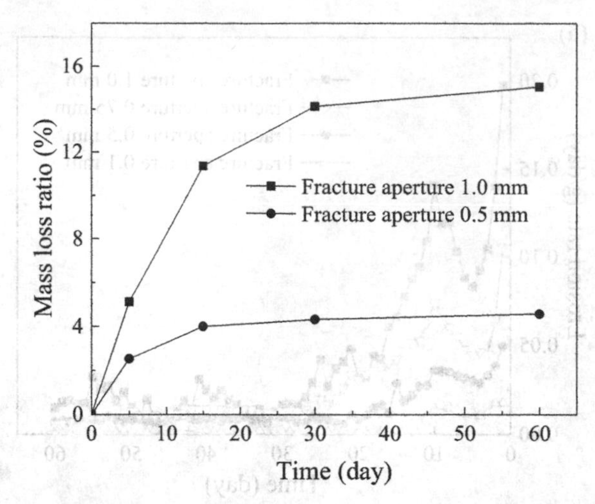

Fig. 7. Evolutions of mass loss ratios of specimens with time for tests with a fracture aperture of 0.5 and 1.0 mm

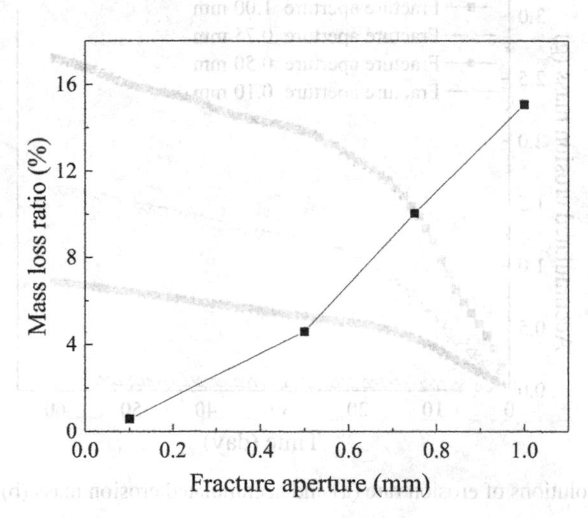

Fig. 8. Relationship between mass loss ratios of specimens and fracture apertures after 60 days

the test result with fracture aperture of 0.1 mm, the predicted intrusion distance after 60 days versus fracture apertures could be obtained in Fig. 9. However, it appears in Fig. 9 that the predicted value was obviously larger than the measured one, especially for tests with a larger fracture aperture. This observation could be due to the mitigating influence of the accessory-mineral ring on intrusion (Reid et al., 2015) [10] and its explanation could be given as follows. According to the research by Moreno et al. (2011) [7], the erosion rate would increase with increasing intrusion distances, leading to more accessory minerals left at the clay/water interface. Accordingly, the width of accessory-mineral ring would increase correspondingly with increasing intrusion distances and more significantly with a larger fracture aperture. Indeed, according to Figs. 5 (a) and (b), the relationship between the width of accessory-mineral ring and intrusion distance could be obtained in Fig. 10. The width of accessory-mineral ring increased with increasing intrusion distances and was larger with increasing fracture apertures, which confirmed the conclusion above. Therefore, as the fracture aperture and intrusion distance increase, the larger width of accessory-mineral ring may lead to the more restrictive effects on intrusion. In this regard, the intrusion model proposed by Svobody (2013) [13] could not be used for simulating the intrusion behavior in this study, especially with a larger fracture aperture.

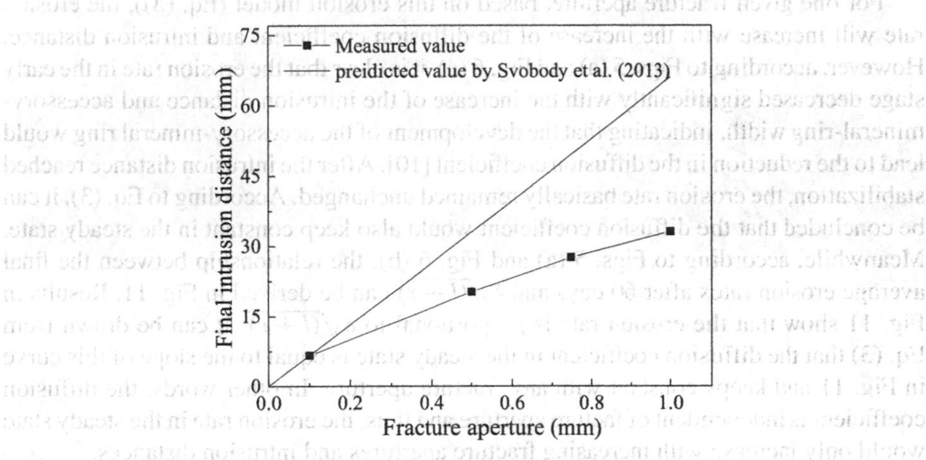

Fig. 9. Relationship between the final intrusion distance and fracture aperture after 60 days

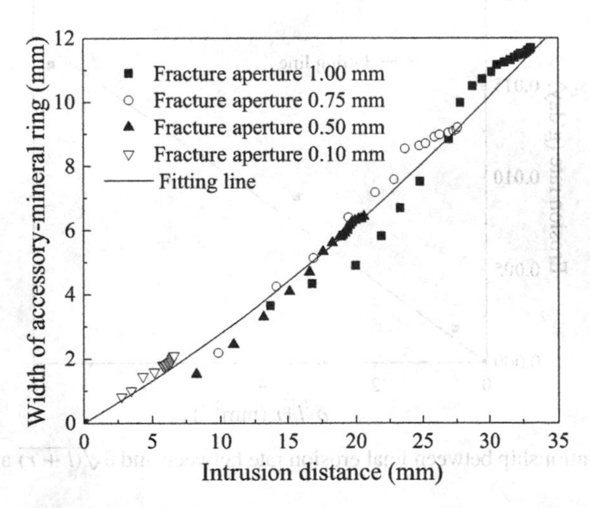

Fig. 10. Relationship between width of the accessory-mineral ring and intrusion distance

4.2 Influences of Fracture Apertures on Bentonite Erosion

According to research by Neretnieks et al. (2017) [9], the erosion rate ($N_{erosion}$) could be calculated as,

$$N_{erosion} = 4\rho_s \delta \phi_c \sqrt{Du(l + r)} \quad (3)$$

where, ρ_s is the density of montmorillonite, δ is the fracture aperture, ϕ_c is the volume fraction of montmorillonite at the intrusion front, D is the diffusion coefficient, u is the flow velocity, l is the intrusion distance and r is the radius of specimens.

For one given fracture aperture, based on this erosion model (Eq. (3)), the erosion rate will increase with the increase of the diffusion coefficient and intrusion distance. However, according to Figs. 5 (a) and Fig. 6 (a), it is clear that the erosion rate in the early stage decreased significantly with the increase of the intrusion distance and accessory-mineral-ring width, indicating that the development of the accessory-mineral ring would lead to the reduction in the diffusion coefficient [10]. After the intrusion distance reached stabilization, the erosion rate basically remained unchanged. According to Eq. (3), it can be concluded that the diffusion coefficient would also keep constant in the steady state. Meanwhile, according to Figs. 5 (a) and Fig. 6 (b), the relationship between the final average erosion rates after 60 days and $\delta\sqrt{(l + r)}$ can be derived in Fig. 11. Results in Fig. 11 show that the erosion rate is proportional to $\delta\sqrt{(l + r)}$. It can be drawn from Eq. (3) that the diffusion coefficient in the steady state is equal to the slope of this curve in Fig. 11 and keeps constant with any fracture aperture. In other words, the diffusion coefficient is independent of facture aperture and thus, the erosion rate in the steady state would only increase with increasing fracture apertures and intrusion distances.

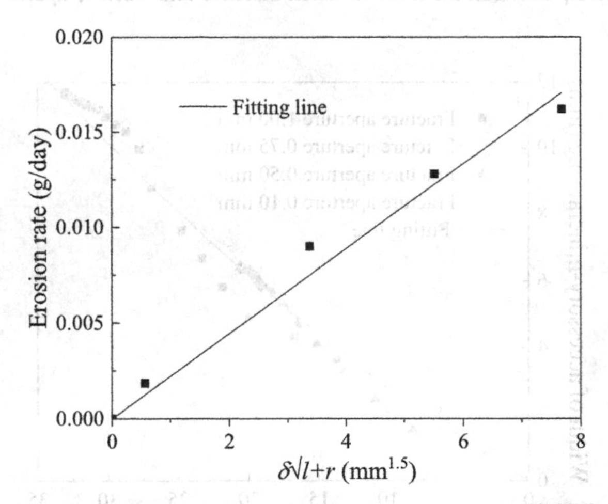

Fig. 11. Relationship between final erosion rate between and $\delta\sqrt{(l + r)}$ after 60 days

4.3 Influences of Intrusion/erosion on the Mass Loss of Specimens

According to Fig. 6 (b), the mass loss ratio of specimens by erosion could be calculated and was plotted versus time in Fig. 12. It appears in Fig. 12 that the mass loss ratio by erosion with a fracture aperture of 0.5 and 1.0 mm increased quickly but then followed by an obvious reduction in the increase rate after 15 days, which has a good agreement with that in the erosion curve (Fig. 6 (b)). Meanwhile, the evolution of mass loss ratio of specimens by intrusion with time could also be determined by the difference between total mass loss ratio of specimens (Fig. 7) and that by erosion (Fig. 12). Results in Fig. 12 show that the mass loss ratio by intrusion with a fracture aperture of 0.5 and 1.0 mm both started with a significant increase and then reached a stable value after 15 and 30 days, respectively. The turning points are consistent with that in the intrusion curve (Fig. 5 (a)). Moreover, it could also be observed in Fig. 12 that the mass loss ratios of specimens with a fracture aperture of 0.5 and 1.0 mm were mainly contributed by intrusion before 15 and 30 days, respectively. Subsequently, the specimen mass loss ratios kept on increasing and would be only dependent on the erosion rate. Indeed, as long as time for erosion is enough, the mass loss ratios of specimens by erosion may exceed that by intrusion.

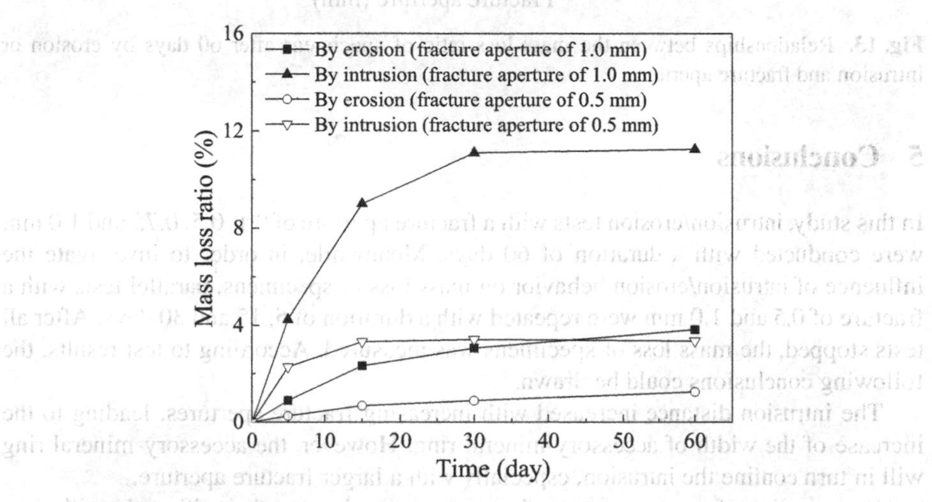

Fig. 12. Evolutions of mass loss ratio of specimens by erosion or intrusion with time foe tests with a fracture aperture of 0.5 and 1.0 mm

According to Fig. 6 (b), the mass loss ratio of specimens by erosion after 60 days could be calculated and was plotted versus the fracture aperture in Fig. 13. Results in Fig. 13 shows that the mass loss ratio by erosion increased with increasing fracture apertures. Meanwhile, according to Figs. 8 and 13, mass loss ratio of specimens by intrusion could also be determined. Then, the relationship between mass loss ratio of specimens by intrusion and fracture aperture after 60 days was depicted in Fig. 13. It could be observed in Fig. 13 that the mass loss ratio after 60 days by intrusion increased with increasing fracture apertures. Moreover, the difference between the mass loss ratio by intrusion and erosion increased with increasing fracture apertures.

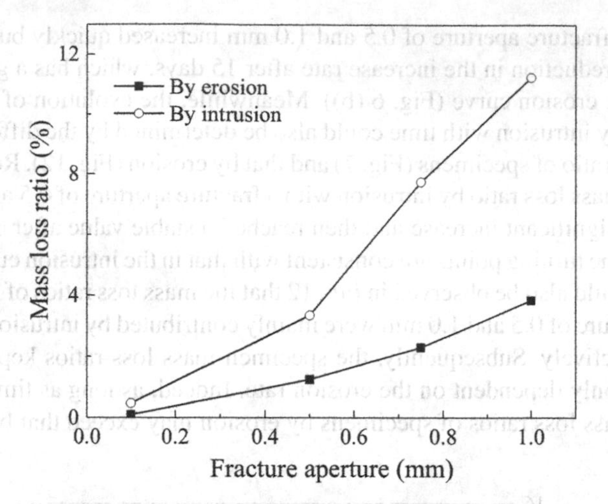

Fig. 13. Relationships between the mass loss ratio of specimens after 60 days by erosion or intrusion and fracture aperture

5 Conclusions

In this study, intrusion/erosion tests with a fracture aperture of 0.1, 0.5, 0.75 and 1.0 mm were conducted with a duration of 60 days. Meanwhile, in order to investigate the influence of intrusion/erosion behavior on mass loss of specimens, parallel tests with a fracture of 0.5 and 1.0 mm were repeated with a duration of 5, 15 and 30 days. After all tests stopped, the mass loss of specimens was measured. According to test results, the following conclusions could be drawn.

The intrusion distance increased with increasing fracture apertures, leading to the increase of the width of accessory-mineral ring. However, the accessory-mineral ring will in turn confine the intrusion, especially with a larger fracture aperture.

For one given fracture aperture, the erosion rate decreased significantly with time in the early period due to reduction in the diffusion coefficient by the filtration of the accessory-mineral ring. But when the erosion reached stabilization, the erosion rate increased with increasing fracture apertures and intrusion distances, irrespective of the accessory-mineral ring.

The majority of mass loss ratio of specimens was from intrusion in the early period. But when intrusion reached stabilization, the mass loss ratio of specimen will be mainly contributed by erosion. Moreover, the mass loss ratios of specimens after 60 days by intrusion and erosion both increased with increasing fracture apertures. The difference between the mass loss ratio by intrusion and erosion increased with increasing fracture apertures.

Acknowledgements. The financial supports of the National Nature Science Foundation of China (42030714 and 41807237) are greatly acknowledged.

References

1. Bian, X., Cui, Y.J., Zeng, L.L., Li, X.Z.: State of compacted bentonite inside a fractured granite cylinder after infiltration. Appl. Clay Sci. **186**, 105438 (2020)
2. Birgersson, M., Boergesson, L., Hedstroem, M., Karnland, O., Nilsson, U.: Bentonite erosion. Final report (No. SKB-TR-09-34). Swedish Nuclear Fuel and Waste Management Co (2009)
3. Cui, L.Y., Ye, W.M., Wang, Q., Chen, Y.G., Chen, B., Cui, Y.J.: Investigation on gas migration in saturated bentonite using the residual capillary pressure technique with consideration of temperature. Process Saf. Environ. Protect **125**, 269–278 (2019)
4. Nguyen, T.S., et al.: A case study on the influence of THM coupling on the near field safety of a spent fuel repository in sparsely fractured granite. Environ. Geol. **57**(6), 1239–1254 (2008)
5. Xu, L.B., Ye, W.M., Liu, Z.R., Wang, Q., Chen, Y.G.: Extrusion behavior of bentonite-based materials considering pore size and sand content effects. Constr. Build Mater. **347**, 128580 (2022)
6. Luo, H.W., Ye, W.M., Wang, Q., Chen, Y.G., Chen, B.: Pore fluid chemistry effects on the swelling behavior of compacted GMZ bentonite with an artificial annular gap. Bull. Eng. Geol. Env. **80**(7), 5633–5644 (2021)
7. Moreno, L., Liu, L., Neretnieks, I.: Erosion of sodium bentonite by flow and colloid diffusion. Phys. Chem. Earth, Parts A/B/C **36**(17–18), 1600–1606 (2011)
8. Missana, T., Alonso, Ú., Turrero, M.J.: Generation and stability of bentonite colloids at the bentonite/granite interface of a deep geological radioactive waste repository. J. Contam. Hydrol. **61**(1–4), 17–31 (2003)
9. Neretnieks, I., Moreno, L., Liu, L.: Clay Erosion: Impact of Flocculation and Gravitation. Svensk kärnbränslehantering AB. Swedish Nuclear Fuel and Waste Management Company (2017)
10. Reid, C., Lunn, R., El Mountassir, G., Tarantino, A.: A mechanism for bentonite buffer erosion in a fracture with a naturally varying aperture. Mineral. Mag. **79**(6), 1485–1494 (2015)
11. Schatz, T., Akhanoba, N.: Bentonite buffer erosion in sloped fracture environments (No. POSIVA-2016-13). Posiva Oy (2017)
12. Schatz, T., et al.: Buffer erosion in dilute groundwater (No. POSIVA-12-44). Posiva Oy (2013)
13. Svoboda, J.: The experimental study of bentonite swelling into fissures. Clay Miner. **48**(2), 383–389 (2013)
14. Vilks, P., Miller, N.H.: Laboratory bentonite erosion experiments in a synthetic and a natural fracture. NWMO Technical Report TR-2010-16 (2010)
15. Xu, L.B., Ye, W.M., Wang, Q., Chen, Y.G., Chen, B.: Investigation on intrusion of bentonite–sand mixtures in fractures with consideration of sand content andseepage effects. Bull. Eng. Geol. Environ. **81**(1), 1–16 (2022)
16. Ye, W.M., Xu, L.B., Wang, Q., Chen, Y.G., Chen, B.: Bentonite-sand mixture intrusion process and its model in rock fissures with consideration of lateral wall friction. Chin. J. Geotech. Eng. **44**(4), 613–621 (2022). (in Chinese)

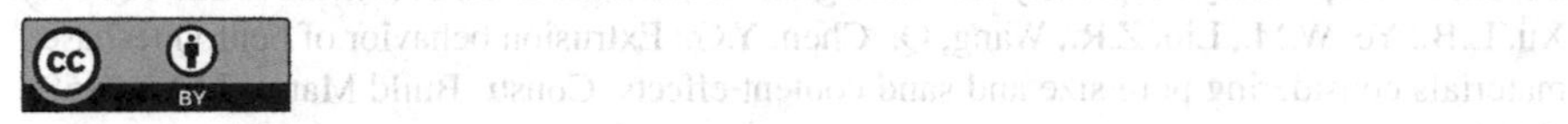

Research on Ultra High Flux Research Reactor

Xue Zhang, Hongxing Yu(✉), Bangyang Xia, Wenjie Li, and Xilin Zhang

Science and Technology on Reactor System Design Technology Laboratory, Nuclear Power Institute of China, Chengdu 610213, China
Yuhong_xing@126.com

Abstract. The lack of fast neutron irradiation test equipment is a key practical factor that restricts the development of the new generation nuclear energy technology with fast reactor as the core in China. Scarce isotopes such as californium 242, which are urgently needed in industry and medicine, are also highly dependent on imports. In order to provide a fast reactor material irradiation environment and isotope production environment, significantly accelerate the development of fast reactor fuel and structural material technology, and improve the efficiency of rare isotope production, this paper proposes a multifunctional inherently safe ultra-high flux reactor design with fast/thermal neutron flux levels up to 10^{16} n/(cm^2.s), namely UFR-10^{16}. The neutron energy spectrum covers fast spectrum, superheat spectrum and thermal spectrum; several material irradiation test orifices, isotope production orifices and test loop orifices are arranged in the core, and the test loop supports many new reactor coolants such as H_2O, Na, He, He-Xr, molten salt, etc., and supports the simulation of typical transient processes and accident conditions. This paper discusses the overall construction scheme of the reactor, and the current construction scheme shows that each performance index can meet the target requirements and achieve the expected functions while ensuring the safety performance of the reactor. This study can provide a basis for the development of ultra-high flux reactors and provide strong support for the cause of nuclear energy, nuclear technology, and nuclear medicine in China.

Keywords: Ultra-high flux · Research reactor · Fast reactor · UFR

1 Introduction

The research and development of cleaner, more efficient, and safer new nuclear energy systems is of great significance to the sustainable development of nuclear energy, and the performance of nuclear fuel and structural materials, especially their irradiation performance, has always been an important basis for the development of new nuclear energy systems. The irradiation experiment of materials is mainly carried out in the research reactor. The irradiation test ability of the research reactor is mainly determined by the neutron flux level. The higher the neutron flux level, the faster the irradiation test of materials and shorten the research and development cycle of materials.

The construction of high-flux reactors is of great benefit to the entire nuclear energy field. In the 1960s, thermal spectrum and fast spectrum research reactors were built and

C. Liu (Ed.): PBNC 2022, SPPHY 283, pp. 887–897, 2023.
https://doi.org/10.1007/978-981-99-1023-6_75

put into use in various countries around the world. In terms of thermal spectrum research reactors, Russia and Belgium have built high-flux research reactors SM-2 [1] and BR-2 [2] respectively, and the United States has built HFIR [3] and ATR [4]. Among them, the HFIR thermal neutron flux is about 2.5×10^{15} $cm^{-2}s^{-1}$, which is one of the research reactors with the highest steady-state thermal neutron flux. Most of the 252 Cf in the world comes from here; the ATR thermal neutron flux is about 1×10^{15} $cm^{-2}s^{-1}$, with a power level of 250 MW, is one of the research reactors with the highest power level and has a strong material irradiation capability. In terms of fast-spectrum research reactors, Russia's BOR-60 [4] has not yet been retired, and the United States' EBR-II [6] and FFTF [7] were retired in the 1990s for political and economic reasons, but also has left quite a wealth of experience and relatively mature technology to American fast-spectrum reactors (especially sodium-cooled fast reactors). The development of high flux research reactors in China is relatively late compared to other countries. The thermal spectrum research reactor HFETR [8] was critical in 1979, and the fast spectrum research reactor CEFR [9] was critical in 2010.

In 2002, the International Forum on Generation IV Nuclear Energy Systems proposed six priority development of Generation IV nuclear energy systems, most of which are fast reactors. The development of fast reactors is inseparable from the corresponding material irradiation experiments. However, the existing thermal spectroscopy research reactors have limited irradiation capabilities, and it is difficult to match the development and research speed of fast reactors. France started construction of the water-cooled research reactor JHR [10] in 2007, Russia started the construction of the sodium-cooled research reactor MBIR [11] in 2015, and the U.S. sodium-cooled fast reactor VTR [12] has also been put on the agenda. It can be seen that the overall development trend of foreign research reactors is to develop towards high-flux fast reactors. At the same time, more advanced irradiation methods are required, such as independent coolant circuits, on-line monitoring equipment in the reactor, and particle beam pipelines, etc.

Among the fourth-generation advanced nuclear power systems, fast reactor is one of the most promising reactor types. At present, there is an extreme lack of data on fast neutron irradiation of reactor materials in China, and there is also a lack of fast neutron irradiation research reactor. In addition, scarce isotopes such as californium 242, which are urgently needed in industry and medicine, are also highly dependent on imports. In order to provide fast reactor material irradiation environment and isotope production environment, greatly accelerate the research and development of fast reactor fuel and structural materials technology, and improve the production efficiency of scarce isotopes, this paper proposes a multifunctional intrinsically safe ultra-high flux reactor design with fast/thermal neutron flux level up to 10^{16} n/(cm^2.s), namely UFR-10^{16}, and discusses the overall construction scheme.

2 The Design Goals of the Ultra-high Flux Reactor

According to the current demand for research reactors, the overall design goal of UFR is to cover the fast spectrum, super thermal spectrum and thermal spectrum with fast/thermal neutron flux up to 10^{16} n/(cm2.s); A number of material irradiation test channels, isotope production channels and test circuit channels are arranged in the core.

Table 1. Comparison of key parameters

	VTR	MBIR	JHR	PIK	UFR
Power/MW	150–300	150	100	100	200
Type	SFR	SFR	LWR	LWR	LFR
Neutron spectrum	Fast	Fast	Thermal	Thermal	Fast
Neutron Flux $n/cm^2.s$	4.0×10^{15}	5.3×10^{15}	1.0×10^{15}	4.5×10^{15}	1.0×10^{16}
Coolant	Na、He、Pb-Bi、Pb、molten salt	Na、He、Pb-Bi、Pb、molten salt	light water, gas	light water, gas	Na、Pb-Bi、Pb、molten salt、light water、CO_2、He
Loop Number	3	3	3	/	3

The test circuit supports many new reactor coolants such as H_2O, Na, he, He-Xr, molten salt, etc., supports the simulation of typical transient processes and accident conditions, and meets the strong needs of radiation test and performance test of nuclear fuel and materials, transient and typical accident simulation. The ultra-high-flux multi-function reactor is facing the frontier of world science and technology, and aims to surpass the comprehensive research facilities such as the VTR under construction in the United States and the MBIR under construction in Russia. The key parameters are shown in Table 1.

3 Ultra-high Flux Reactor Construction Scheme

The overall construction plan of the research reactor includes: a key system, namely, ultra-high flux inherently safe nuclear reactor system; Three large-scale test loop systems, namely, advanced nuclear fuel and material steady-state test loop (coolant could be Na, Pb, Pb Bi, CO_2, He, H_2O, etc.), reactor transient behavior simulation test loop (through radial movement, realize rapid and controllable adjustment of fuel power, and carry out research on transient characteristics of fuel elements under the conditions of power jump, load follow, reactivity introduction, etc.) and accident simulation test circuit (meet the functional requirements of operating condition simulation, triggering of the accident condition, accident mitigation, fuel failure and fission product monitoring); A neutron science experiment platform, which can provide horizontal channels of neutron beams with different energies; Carry out experimental research on irradiation and neutron activation of small-size materials. Figure 1 shows the general layout of ultra-high flux reactor. The construction schemes of these systems will be discussed below.

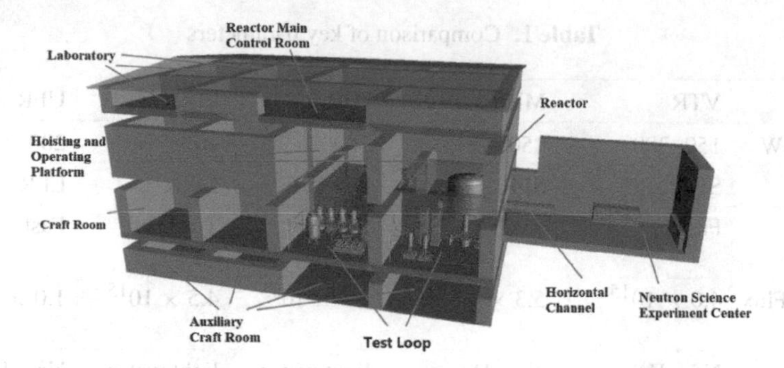

Fig. 1. Schematic diagram of Ultra-high flux reactor

3.1 Inherent Safety Nuclear Reactor System Scheme of Ultra-high Flux Reactor

Table 2 shows the overall design parameters and design choices of the intrinsically safe nuclear reactor system of ultra-high flux reactor. The rated thermal power of the reactor is set at 200 MW, and the refueling cycle is 90 days. Lead-bismuth eutectic (LBE) is selected as the core coolant, which has good neutron physical properties, thermophysical properties and chemical properties. The reactor core is arranged as an atmospheric pool with a height of 450 mm, a temperature of 165/350 °C at the inlet and outlet of the core, and a maximum coolant flow rate of 4.0 m/s. As for the selection of the fuel, considering the characteristics of the core, such as high fast neutron flux, hard neutron energy spectrum and high power density, the U-Zr metal fuel with high U density, few moderating elements and good heat conduction is selected, zirconium alloy with excellent neutron property, mechanical property and irradiation property is a good choice as cladding material.

Table 2. Overall design parameters

Parameters	Value
Rated Power	200 MW
Core arrangement	atmospheric pool
Coolant	LBE
Fuel	U-Zr metal fuel
Cladding	Zirconium alloys
Height of active zone	450 mm
Core Inlet Temperature	165 °C
Coolant max velocity	4.0 m/s
Reload cycle	90 EFPD

The natural circulation of coolant in the circuit is sufficient to cool the reactor without the risk of residual heat export from the reactor core, and the reactor protection vessel is arranged in the passive heat conduction system tank filled with water. The primary circuit system mainly includes: reactor core, steam generator module, main pump, and internal radiation shielding, which is located in the reactor vessel. The secondary circuit system mainly includes: steam generator module, feedwater and steam pipelines, steam-water separator, and independent cooler. The LBE coolant is heated by the core, enters the core outlet chamber and flows out laterally, enters the steam generator from the bottom of the primary circuit, and is cooled by the secondary loop at the same time. When the coolant reaches the top of the primary side of the steam generator, it turns over and enters the main pump, and then the coolant is transported by the main pump from top to bottom to the reactor inlet chamber. The protective gas system mainly includes: gas system condenser, membrane safety device, pressure relief device and pipeline. The coolant process system mainly includes: LBE filling and discharging system, purification system and real-time online monitoring system, which are used to maintain the quality of LBE in the system during operation. The safety system mainly includes reactive accident protection system, steam generator leakage suppression system, independent cooling system and passive residual heat removal system. In addition, the refueling system needs to be set up.

Among them, the reactor core is the key part. The layout of the reactor core is shown in Fig. 2. The high-power density standard fuel assembly is located in the central area, and the periphery is the shielding assembly. In order to improve the reactivity control ability and reliability, two kinds of control rod systems with different principles are set up. The holes for the irradiation test of advanced nuclear fuel and materials are located in the central zone of the core with ultra-high neutron flux. The holes for the nuclear reactor transient test and the typical accident simulation test are located in the reflector area of the core. And several material irradiation test holes are set in the high-power density fuel area and emitter area of the core. Several horizontal and vertical experimental channels are set in the radial direction of the reactor (mainly used for neutron scattering experiments, neutron photography and other basic research of Neutron Science).

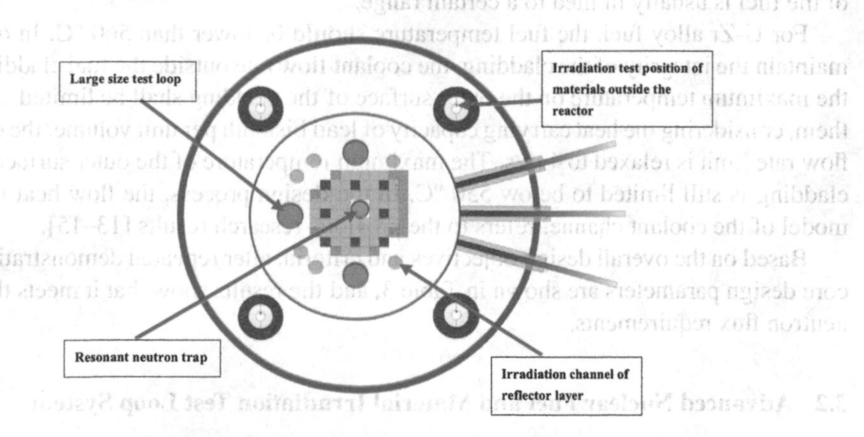

Fig. 2. Schematic diagram of the core layout

In order to improve the power density of the reactor core and reduce the problem of fuel core, considering the swelling of fuel elements caused by axial and radial temperature differences, irradiation and other factors, and referring to the design ideas of fuel elements of European JHR, China's HFETR and other test reactors, the standard fuel assemblies in the high-power density area of the reactor core initially adopt the narrow rectangular fuel elements shown in Fig. 3. The core adopts high-performance U-10Zr metal fuel with an U235, enrichment of 64.4%., and the cladding material adopts T91.

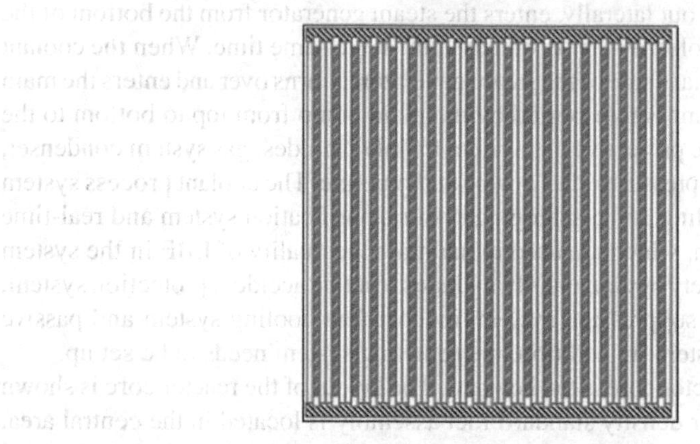

Fig. 3. Narrow rectangular fuel assembly

In terms of thermal design criteria, due to the high boiling point of LBE, it is very difficult to boil, and the design criteria related to boiling threshold do not need to be considered. It is only necessary to consider operating temperature limits of the core and integrity requirements design limits of the cladding.

For narrow rectangular fuel elements, there is usually no gap to contain fission gas. In order to avoid the fission gas causing the fuel rod to swell and burst, the temperature of the fuel is usually limited to a certain range.

For U-Zr alloy fuel, the fuel temperature should be lower than 560 °C. In order to maintain the integrity of the cladding, the coolant flow rate outside the fuel cladding and the maximum temperature on the outer surface of the cladding shall be limited. Among them, considering the heat carrying capacity of lead bismuth per unit volume, the coolant flow rate limit is relaxed to 4 m/s. The maximum temperature of the outer surface of the cladding is still limited to below 550 °C. In the design process, the flow heat transfer model of the coolant channel refers to the previous research results [13–15].

Based on the overall design objectives and criteria, after repeated demonstration, the core design parameters are shown in Table 3, and the results show that it meets the core neutron flux requirements.

3.2 Advanced Nuclear Fuel and Material Irradiation Test Loop System

In order to speed up the research on the radiation mechanism of advanced nuclear fuels and materials, three forms of nuclear fuel and material irradiation tests can be carried out on ultra-high flux reactors, including static container irradiation test, instrumented irradiation test and loop irradiation test. Among them, the loop irradiation test can accommodate nuclear materials or fuels with large heat release and scaled fuel components

Table 3. Core design parameters

	Value	Units
Power	200	MW
Fuel Assembly number	56	
Diameter	954.33	mm
Active height	450.00	mm
Pitch	83.70	mm
Power density	1132.87	MW/m^3
Uranium quality	400.16	kg
U-235 quality	279.83	kg
Upper reflector	400.00	mm
Lower reflector	400.00	mm
Total length	1450.00	mm
Flow direction	upward	
Coolant velocity	4.00	m/s
Coolant flow rate	10609.91	kg/s
Inlet temperature	165.00	°C
Outlet temperature of active zone	293.84	°C
Maximum fast neutron flux	$\geq 1.0 \times 10^{16}$	$n/cm^2.s$

and cladding materials requiring irradiation under actual operating conditions. The most significant advantage is that it can simulate the actual thermal hydraulic environment and hydro chemical environment, and accurately monitor and control irradiation parameters. This is one of the most complex and important test devices. The loop device is located in the center of the nuclear reactor and has ultra-high fast neutron flux. According to the different test tasks, the coolant can choose Na, Pb, Pb Bi, CO2, he, etc., and the inlet and outlet temperature can be adjusted according to the coolant type to meet the requirements of rapid screening and performance testing of advanced nuclear energy materials and nuclear fuel samples.

The loop device has an independent coolant circuit, which can independently control the coolant parameters in the device, such as pressure, temperature, flow rate, chemical composition, etc., and take away the heat generated by the test piece. By connecting with the computer control system, the emergency control and alarm functions are realized, and a variety of irradiation parameters are monitored online, including flow, temperature, pressure, differential pressure, fission products and water chemistry. The loop irradiation device needs to occupy multiple core cells, and the empty cell space outside the irradiation device is filled with special-shaped components (Fig. 4).

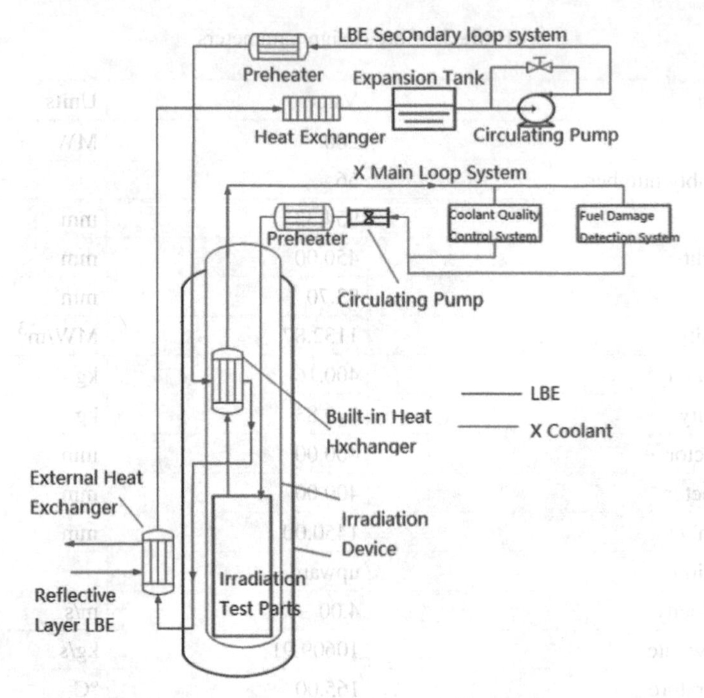

Fig. 4. Advanced nuclear fuel and material irradiation test device

3.3 Nuclear Reactor Transient Behavior Simulation Test Loop System

In order to study the structural integrity and material irradiation behavior of fuel elements in advanced nuclear power system under transient conditions, the transient behavior simulation test of nuclear fuel with fast power up and down can be carried out on ultra-high flux reactor. The loop device is located in the reactor core reflector area and is mainly used for the power jump experiment in the irradiation environment to realize various hypothetical irradiation operating conditions for new fuel elements or irradiated fuel elements, such as power increase experiment, fuel rod internal pressure overpressure experiment, fuel rod free gas elimination experiment, fuel core melting experiment, etc. Rapid changes in neutron flux are achieved using device translation and/or He-3 loops, which in turn rapidly regulate the heat release rate of the test fuel element and its axial distribution. The loop device also has a separate coolant loop so that the coolant parameters in the device, such as pressure, temperature and flow rate, chemical composition, etc., can be controlled individually and take away the heat generated by the test piece. The connection to the computer control system enables emergency control and alarm functions, and online monitoring of a variety of irradiation parameters, including flow rate, temperature, pressure, differential pressure, fission products and water chemistry. The coolant circuit also features fission gas collection and is capable of on-line analysis of gas composition, taking into account the leakage of radioactive material after the breakage of the fuel cladding in transient tests. A fully automatic quick-disassembly joint is located above the unit to enable remote disassembly and assembly of fuel elements and to avoid high dose irradiation to the operator.

3.4 Accident Simulation Test Loop System

This loop device is located in the core reflector region and is mainly used for LOCA experimental simulations of LWRs to test the thermal-mechanical behavior of fuel elements and their cladding, and the radioactive consequences in case of breakage in the event of a LOCA accident at LWRs nuclear power plants.

The loop device is divided into two parts: the internal system and the external system (compartment). The internal reactor system is partially installed in the channel of the reflective layer (210 mm in diameter), and the internal reactor system is moved by the moving device (which can be moved radially by about 0.5 m). Its nuclear power is controlled by the distance between the fuel rod of the reactor system and the ultra-high flux multifunctional reactor vessel (close to the reactor system, the power is high, far away from the reactor vessel, the power is low) and by increasing or reducing the blocks between them. At the same time, the minimum distance between the mobile device and the reactor vessel and the blocks are limited to limit the maximum nuclear power. In addition, the anti-fly-out system is installed on the top of the reactor system and fixed on the mobile system to prevent the axial movement of the reactor system.

The reactor system is located in a double-layered bottle body made of stainless steel. The inner layer bottle is a pressure bottle body, which is used to bear the internal pressure. An air gap is formed between the inner and outer bottle bodies, which is filled with helium gas for thermal insulation. The flow separation tube is located in the inner bottle, forming two concentric channels, the hot channel surrounding the fuel sample, and the cold channel between the separation tube and the inner bottle. Thus, thermosiphon can be established to ensure fuel rod cooling before LOCA transient starts. The flow separation tube is integrated with the surrounding heater to form an adiabatic condition for the fuel in the adiabatic stage. The gap between the support structure of the in-reactor system and the outer bottle body forms a cooling channel, and the pump is used to provide forced circulation. The neutron shield (hafnium) is fixed on the inner side of the support structure of the in-reactor system to flatten the axial neutron flux. The test instruments are used to measure temperature and pressure.

The external reactor system (compartment) is connected to the internal system by cables. The main function of the external system is to ensure the initial structure of the internal system loop to control the emptying and re-flooding phases. The in-reactor system is connected to the fission product laboratory, where the contaminated fluid will be contained, analyzed and partially sampled during the final phase of the test. The sampling line at the bottom will be for water, while the top is for gas. The external reactor system (compartment) are also used to store contaminated gases and liquids (from the fission product laboratory), and the fluids will be stored in different containers depending on the level of contamination.

3.5 Neutron Science Experiment Platform

The device is located in the external area of the reactor, which is mainly used for neutron photography of highly reflective nuclear fuel elements, materials and large-scale equipment, analyzing its internal organization structure, and providing basic data for material performance evaluation and modification. The reactor horizontal channel provides neutron sources to the device.

The cooling pool outside the reactor provides a measurement environment that is closer to the operating parameters for the study of the behavior of nuclear materials in the reactor. Under this condition, the camera platform for highly radioactive materials has its unique advantages. Due to the difference in the interaction characteristics of neutrons and photons with matter, neutrons are sensitive to light materials, fissile materials and strong absorbing materials, while photons are sensitive to heavy materials; the attenuation of neutrons in matter is much smaller than that of photons, which is beneficial to Photography of heavy samples. Therefore, the neutron photography platform and the photon photography platform complement each other and have good complementarity in material detection.

The platform is located in the reactor pool and is used for neutron photography of new fuel and irradiated nuclear fuel with strong radioactivity. In order to achieve the neutron detection of large samples, indirect neutron imaging based on wedge-shaped neutron beam is adopted, and water provides effective neutron shielding for the photography platform. The underwater neutron and photon photography platform can meet the load-bearing requirements of most samples, which can not only realize rapid imaging of short-lived nuclides in irradiated samples, but also grasp the distribution of nuclides in the samples (such as distribution of special nuclides in fuel rods). Under the three-degree-of-freedom spatial motion adjustment of the sample holding mechanism, γ-ray passive tomography imaging of samples of any shape can be realized. In order to achieve different scanning effects, various shapes and types of gamma-ray front-end collimators are preset. Since the collimator penetrates the reactor pool wall, it is necessary to set up a shielding structure at the corresponding position outside the pool wall to achieve effective protection for devices and staff.

4 Conclusion

This paper proposes an ultra-high flux research reactor and discusses the reactor construction scheme, including the nuclear reactor system scheme and the design of each test loop. After demonstration and design, on the premise of ensuring the safety performance of the reactor, each performance index can meet the target requirements and achieve the expected functions:

1) It can build a high-intensity, fast neutron spectrum and high-energy neutron field irradiation testing environment for the research and development of nuclear fuels and materials for the next generation and future advanced nuclear power systems, fusion reactors and other new reactors;
2) It can significantly shorten the irradiation test time of nuclear fuel and nuclear materials, and improve the research and development efficiency of new nuclear fuel and nuclear materials; It can realize the deep depletion of nuclear fuel, and effectively meet the basic theoretical data required by researches such as closed fuel cycle and nuclear fuel transmutation theory;
3) It can significantly improve the industrial production capacity and efficiency of 238 Pu, 252 Cf and other high value-added or scarce isotopes;
4) It can provide a more powerful, stable, continuous horizontal neutron beam that can be used for neutron scientific research;

5) It can provide the super-strong neutron field required by the fuel element to realize the step change of power for simulation experiments such as transient behavior and typical accidents of nuclear power plants.

References

1. Tsykanov, V.A., Korotkov, R.I., Kormushk, Y.P.: Some Physical Features of SM-2 Reactor, and Comparison of SM-2 with Other High-Flux Reactors. Meeting Abstract (1971)
2. Ponsard, B.: The BR2 high-flux reactor. ATW **57**(10):, 612–613 and 583 (2012)
3. Cheverton, R.D., Sims, T.M.: Hfir Core Nuclear Design (1971)
4. O'Kelly, D.S.: The Advanced Test Reactor. Reference Module in Earth Systems and Environmental Sciences (2020)
5. Izhutov, A.L., Krasheninnikov, Y.M., Zhemkov, I.Y., et al.: Prolongation of the BOR-60 reactor operation. Nucl. Eng. Technol. **47**(3), 253–259 (2015)
6. Koch, L.J.: Experimental Breeder Reactor-II (EBR-II) (1967)
7. Tobin, J.C.: FFTF and the ASME Code (1978)
8. Hong, Y., Bu, Y., Lin, J. : 10 Years Safety Operation of the High Flux Engineering Test Reactor(HFETR). Nuclear Power Engineering (1990)
9. Gu, C.X., Liu, et al.: Progress in China Experimental Fast Reactor. Annual Report of China Institute of Atomic Energy **00**, 46–49 (2014)
10. Gilles, B., Christian, C., Jocelyn, P., et al. : The Jules Horowitz Reactor Research Project: A New High Performance Material Testing Reactor Working as an International User Facility – First Developments to Address R&D on Material. EPJ Web of Conferences **115**, 01003-(2016)
11. Tuzov, A.: MBIR International Research Center: Current Progress and Prospects (2015)
12. Pasamehmetoglu, K.: Versatile Test Reactor Overview. Advanced Reactors Summit VI (2019)
13. Jaeger, W.H.W.L. : Liquid metal thermal hydraulics in rectangular ducts: Review, proposal and validation of empirical models: International Conference on Nuclear Engineering, Japan, JSME (2015)
14. Ushakov, P.A., Zhukov, A.V., Matyukhin, N.M. : Heat transfer to liquid metals in regular arrays of fuel elements. High Temp. (USSR)(Engl. Transl.); (United States) **15**(5) (1978)
15. Mikityuk, K. : Heat transfer to liquid metal: Review of data and correlations for tube bundles. **239**(4), 680–687 (2009)

Numerical Simulation of Convective Heat Transfer of CO_2 in a Tube Under Supercritical Pressure at Low Reynolds Numbers

Zhihui Li(✉)

State Power Investment Corporation Research Institute, Beijing, China

Abstract. The supercritical carbon dioxide (S-CO_2) Brayton cycle has the advantages of compact layout, simple structure, high thermal efficiency, clean working quality, its application in lead-cooled fast reactor power conversion system helps the miniaturization and modularization of the whole system. The development of an efficient and compact supercritical CO_2 heat exchanger has important reference significance for improving the thermal efficiency of the system of lead-cooled fast reactor. Supercritical CO_2 can operate in high Reynolds number turbulence and low Reynolds number turbulence in heat exchanger. The convective heat transfer of supercritical pressure CO_2 flowing upward and downward in a vertical circular tube (d = 2 mm) at low inlet Reynolds number ($Re_{in} = 1970$) was numerically simulated by different turbulence models to study the effects of variable properties, buoyancy and thermal acceleration on wall temperature and turbulent kinetic energy. The results showed that the heat transfer deterioration and enhancement occurred at the entrance of the heating section during the upward flow, which was mainly attributed to the influence of buoyancy and heat acceleration on the turbulent kinetic energy distribution. The LB turbulence model was used to simulate the heat transfer phenomenon, which was occurred in the downward flow.

Keywords: Supercritical Pressure · Buoyancy · Thermal Acceleration · Heat Transfer Deterioration · The Numerical Simulation

1 Introduction

In recent years, researchers at home and abroad have proposed to use supercritical CO_2 Brayton cycle in the power cycle system of the fourth generation reactor (lead-cooled fast reactor) and the energy production of solar energy system [1, 2]. The advantage of this cycle is that the whole cycle will not undergo phase change. Compared with supercritical water, supercritical helium and other circulating refrigerants, supercritical CO_2 has the advantages of high density and large specific heat, which helps to simplify the cycle process and reduce the size of the whole cycle. The development of a high efficiency and compact supercritical CO_2 heat exchanger has important reference significance for improving thermal efficiency of the lead-cooled fast reactor power cycle system.

C. Liu (Ed.): PBNC 2022, SPPHY 283, pp. 898–908, 2023.
https://doi.org/10.1007/978-981-99-1023-6_76

Supercritical CO_2 can operate in high Reynolds number turbulence and low Reynolds number turbulence in heat exchanger. Domestic and foreign scholars have carried out a lot of experimental research and theoretical analysis on the turbulent flow of supercritical pressure fluid at high Reynolds number [3–7], but there are few studies on the turbulent mixed convection heat transfer in this small structure heat exchanger at low Reynolds number. Therefore, in this paper, the turbulent mixed convective heat transfer of supercritical CO_2 under the condition of low Reynolds number in a tube with an inner diameter of 2 mm is numerically simulated and compared with the experimental results. The effects of variable physical properties, buoyancy and thermal acceleration on flow and heat transfer are analyzed to provide theoretical guidance for the design of supercritical CO_2 heat exchanger.

2 Physical Model and Governing Equation

The physical model and coordinate system are shown in Fig. 1. The origin of the coordinates is the center point of the supercritical pressure CO_2 at inlet section. The axial coordinate x axis is the same as the flow direction, and the radial coordinate r axis is perpendicular to the flow direction and points to the pipe wall. The physical model used for numerical calculation is the same as the material and size of the experimental section in the experiment [8]. The experimental section has an inner diameter of 2.0 mm, an outer diameter of 3.14 mm and a length of 500 mm. In the middle is a heating section that is directly energized, with a length of 290 mm. As the resistivity changes with temperature during the experiment, the local heat flux density is not completely uniform, but its unevenness does not exceed 1%. Therefore, in the process of numerical calculation, it can be regarded as a uniform internal heat source q_v. There are 105 mm long inlet and outlet sections at both ends of the heating section. The material of the experimental section is stainless steel, and the thermal conductivity constant is 16.38 W/m.k. This is a coupled problem of heat conduction and convection. Since the convection heat transfer of supercritical pressure CO_2 in a vertical circular tube is symmetrical, in the process of numerical calculation, half of the circular tube is taken for two-dimensional calculation for convenience. The whole flow is axisymmetric two-dimensional steady flow.

Governing equations in cylindrical coordinates are as follows:

Heat conduction equation:

$$\frac{1}{r}\frac{\partial}{\partial r}(\lambda r\frac{\partial T}{\partial r}) + \frac{1}{x}(\lambda\frac{\partial T}{\partial x}) + \dot{\phi} = 0 \tag{1}$$

Continuity equation:

$$|\frac{1}{r}\left\{\frac{\partial}{\partial x}(prU) + \frac{\partial}{\partial r}(rpV)\right\} = 0 \tag{2}$$

Momentum equation in *U* direction:

$$\frac{1}{r}\left\{\frac{\partial}{\partial x}(\rho r U^2)+\frac{\partial}{\partial r}(\rho r V U)\right\}=-\frac{\partial p}{\partial x}+\rho g+\frac{1}{r}\left\{\begin{array}{l}2\frac{\partial}{\partial x}\left[r\mu_e\left(\frac{\partial U}{\partial x}\right)\right]+\\ \frac{\partial}{\partial r}\left[r\mu_e\left(\frac{\partial U}{\partial r}+\frac{\partial V}{\partial x}\right)\right]\end{array}\right\} \tag{3}$$

Momentum equation in *V* direction

$$\begin{aligned}\frac{1}{r}\left\{\frac{\partial}{\partial x}(\rho r U V)+\frac{\partial}{\partial r}(\rho r V^2)\right\}&=-\frac{\partial p}{\partial r}+\\ \frac{1}{r}\left\{\frac{\partial}{\partial x}\left[r\mu_e\left(\frac{\partial U}{\partial r}+\frac{\partial V}{\partial x}\right)\right]+2\frac{\partial}{\partial r}\left[r\mu_e\left(\frac{\partial V}{\partial r}\right)\right]\right\}\\ -2\frac{\mu_e V}{r^2}\end{aligned} \tag{4}$$

Energy equation:

$$\frac{1}{r}\left\{\begin{array}{l}\frac{\partial}{\partial x}(\rho C_p r U T)+\\ \frac{\partial}{\partial r}(\rho C_p r V T)\end{array}\right\}=\frac{1}{r}\left\{\begin{array}{l}\frac{\partial}{\partial x}\left[rC_p\left(\frac{\mu}{\mathrm{Pr}}+\frac{\mu_T}{\sigma_T}\right)\frac{\partial T}{\partial x}\right]+\\ \frac{\partial}{\partial r}\left[rC_p\left(\frac{\mu}{\mathrm{Pr}}+\frac{\mu_T}{\sigma_T}\right)\frac{\partial T}{\partial r}\right]\end{array}\right\} \tag{5}$$

The control equations are discretized by the control volume integral method using the ANSYS FLUENT software for numerical calculation. In discretizing the equations, the fluid region and the solid wall region adopt the uniform grid in the axial direction, the non-uniform grid in the radial direction, and the pressure velocity coupling is carried out by the SIMPLEC algorithm. The momentum equation and the energy equation first adopt the first-order upwind scheme, and then change to the second-order upwind scheme after reaching convergence, and then iterate until convergence. For the numerical turbulence models, LB, LS and RNG, standard and realizable turbulence models with enhanced wall function method are used in this paper. Due to the drastic changes in physical properties, relaxation factors in the range of 0.1–0.3 are used for all independent variables. In order to ensure the grid independence, a relatively precise grid is divided near the wall during the calculation process. Through adaptive grid adjustment, the condition $y+ < 0.6$ is satisfied to obtain an approximate grid independent solution. When all variables meet the following criteria, the numerical solution is considered to be convergent.

$$\left|\left(\phi^{i+1}-\varphi^{i}\right)/\phi^{i}\right|\leq 10^{-6}\quad \phi, U, V, T, \varepsilon$$

Figure 2 shows the physical properties change of supercritical CO_2 at the pressure of 8.8 MPa. Because the pressure difference between the inlet and outlet of the vertical circular pipe is very small, the change of CO_2 physical properties with pressure has little impact on the results. Therefore, the physical properties of CO_2 in the circular pipe are selected as the corresponding physical properties under different inlet pressures, and the change of physical properties with temperature is processed by piecewise linear

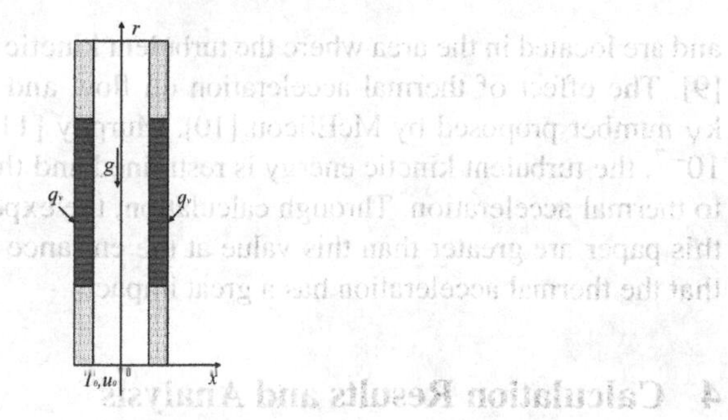

Fig. 1. Physical model and coordinate system

interpolation. After verification, the maximum deviation between the processing method and the physical property value calculated by NIST is not more than 1%.

The boundary conditions of numerical calculation are selected according to the boundary conditions of experimental conditions. The boundary conditions at the inlet are the velocity inlet boundary conditions of uniform incoming flow. At the same time, given the inlet temperature, the boundary conditions at the outlet are the pressure outlet boundary conditions. The heating section of the experimental section is a solid wall with internal heat source, and the boundaries of the rest are adiabatic boundary conditions. Coupled solution of heat conduction in solid wall and convection heat transfer in fluid region.

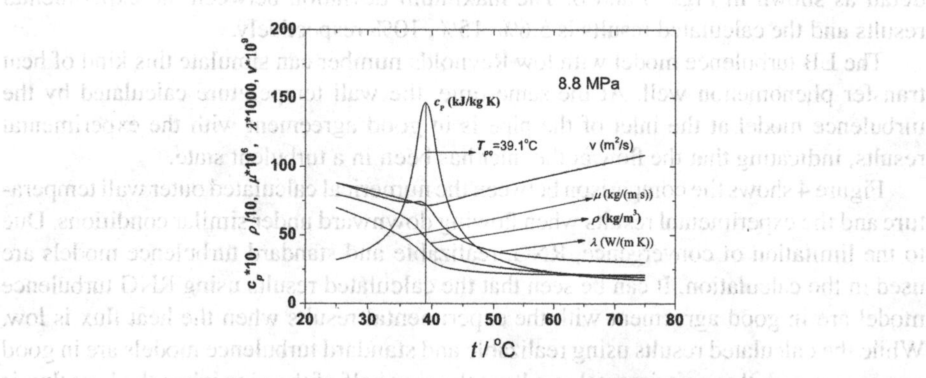

Fig. 2. CO_2 Physical properties at the pressure of 8.8 MPa

3 Effects of Buoyancy and Thermal Acceleration on Flow and Heat Transfer

In this paper, Bo* number proposed by Jackson hall [5] is used to evaluate the effect of buoyancy on flow and heat transfer. The experimental conditions taken in this paper are calculated as $Bo^* > 8 \times 10^{-6}$, indicating that the buoyancy force has a great influence

and are located in the area where the turbulent kinetic energy is enhanced. See literature [9]. The effect of thermal acceleration on flow and heat transfer is evaluated by the k_V number proposed by McElicott [10]. Murphy [11] believes that when $Kv > 9.5 \times 10^{-7}$, the turbulent kinetic energy is restrained and the heat transfer is deteriorated due to thermal acceleration. Through calculation, the experimental conditions calculated in this paper are greater than this value at the entrance of the heating section, indicating that the thermal acceleration has a great impact.

4 Calculation Results and Analysis

For the convenience of verification, the conditions of numerical calculation are exactly the same as those of experiment. The inlet pressure is 8.8 MPa, the inlet temperature is 25 °C, the mass flow is 0.77 kg/h, and the inlet Re is about 1970.

Figure 3 shows the comparison between the numerical calculated outside wall temperature and the experimental results for upward flow under three heat flux conditions. Different turbulence models are used in the calculation. It can be seen from the figure that local wall temperature peaks and valleys occur at the inlet of the pipe when flowing upward, that is, local heat transfer deterioration and heat transfer enhancement occur, and this phenomenon becomes more obvious with the increase of heat flux. The heat transfer enhancement and heat transfer deterioration at the inlet of the tube are mainly attributed to the changes of thermal physical properties and turbulent kinetic energy caused by buoyancy and thermal acceleration. The later part of this paper will be explained in detail as shown in Figs. 5 and 6. The maximum deviation between the experimental results and the calculated results is 5.6%, 15%, 10% respectively.

The LB turbulence model with low Reynolds number can simulate this kind of heat transfer phenomenon well. At the same time, the wall temperature calculated by the turbulence model at the inlet of the pipe is in good agreement with the experimental results, indicating that the flow at the inlet has been in a turbulent state.

Figure 4 shows the comparison between the numerical calculated outer wall temperature and the experimental results when flowing downward under similar conditions. Due to the limitation of convergence, RNG, realizable and standard turbulence models are used in the calculation. It can be seen that the calculated results using RNG turbulence model are in good agreement with the experimental results when the heat flux is low. While the calculated results using realizable and standard turbulence models are in good agreement with the experimental results at the rear half of the pipe when the heat flux is high. It may be that the choice of turbulence model is related to heat flux which needs to be further studied. The wall temperature rises continuously along the flow direction, and there is no abnormal distribution phenomenon in the upward flow. The maximum deviation between the experimental results and the calculated results is less then 2%.

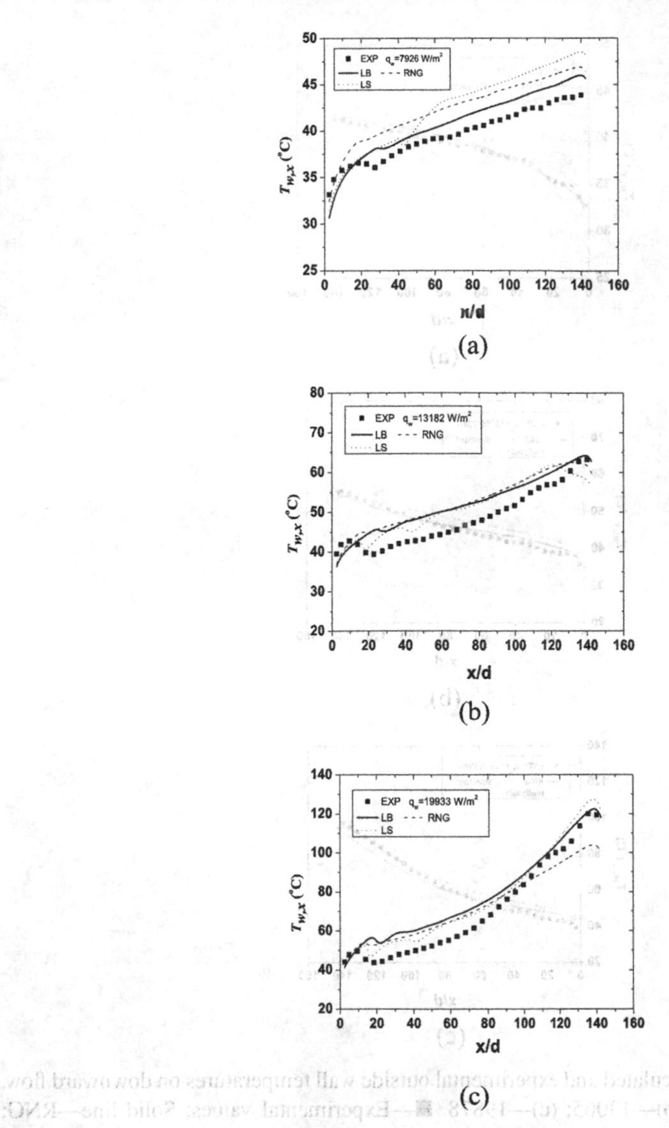

Fig. 3. Comparison of calculated and experimental outside wall temperatures on upward flow. q_w, W/m^2: (a)—7 926; (b)—13182; (c)—19933 ■—Experimental values; Solid line—LB; Dashed line—RNG; Dot line—LS

Figure 5 shows the comparison of turbulent kinetic energy at r/R = 0.9 calculated by LB model with and without buoyancy under the three heat flux conditions on upward flow and downward flow. It can be seen from the figure that: (1) the turbulent kinetic energy of upward flow and downward flow with buoyancy considered is higher than that without buoyancy considered, indicating that buoyancy enhances the turbulent kinetic energy of upward flow and downward flow, which is consistent with the experimental results; (2) The turbulent kinetic energy of downward flow is significantly higher than

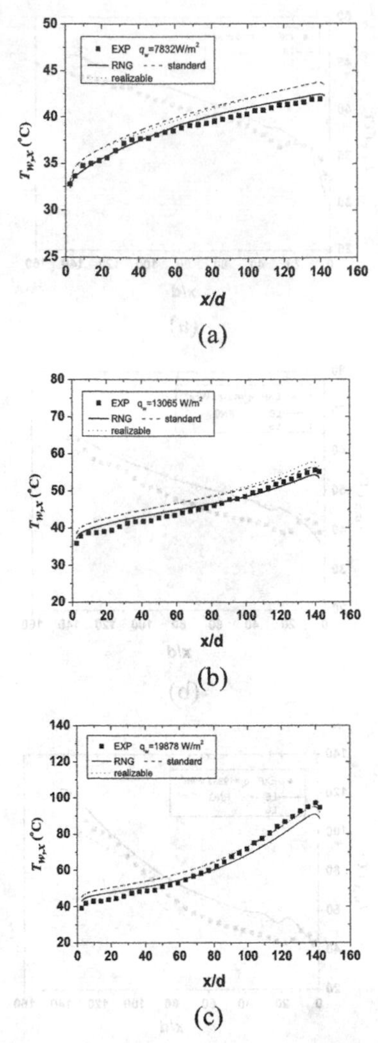

Fig. 4. Comparison of calculated and experimental outside wall temperatures on downward flow. q_w, W/m^2: (a)—7 832; (b)—13065; (c)—19878. ■—Experimental values; Solid line—RNG; Dashed line—Standard; Dot line—Realizable

that of upward flow (except for local positions), indicating that the buoyancy force has a stronger effect on heat transfer enhancement of downward flow than upward flow; (3) When flowing upward, the turbulent kinetic energy begins to decrease to 0 at the inlet of the tube for a short distance, and the heat transfer deteriorates. This may be because the physical property change and thermal acceleration lead to the weakening effect of the turbulent kinetic energy being greater than the enhancement of the turbulent kinetic energy caused by the buoyancy, which will be discussed further below. Later, due to the physical property change and the weakening of the thermal acceleration effect, the buoyancy force changed the flow from laminar flow to turbulent flow again. At a certain position of the pipe, the turbulent kinetic energy increased sharply, and the corresponding wall temperature appeared a valley as shown in Fig. 5, that is, the heat transfer appeared local enhancement; (5) The turbulent kinetic energy in downward flow rises continuously along the path, and there is no abnormal change in upward flow. This may be because the effect of buoyancy on the enhancement of turbulent kinetic energy is stronger than that caused by variable physical properties and thermal acceleration.

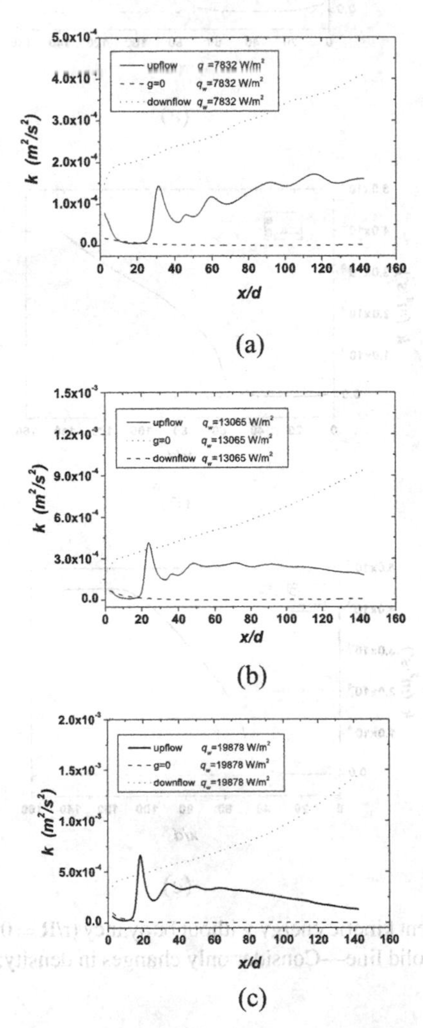

Fig. 5. The change of turbulent kinetic energy under the action of upward, downward and no gravity (R/R = 0.9). q_w, W/m^2: (a)—7 832; (b)—13065; (c)—19878 Solid line—LB(upward); Dashed line—LB(g = 0); Dot line—LB(downward)

Figure 6 shows the axial distribution of turbulent kinetic energy at the radial position r/R = 0.9 obtained by considering the density change and all physical parameters without considering the buoyancy force. LB turbulence model is used in the calculation. It can be

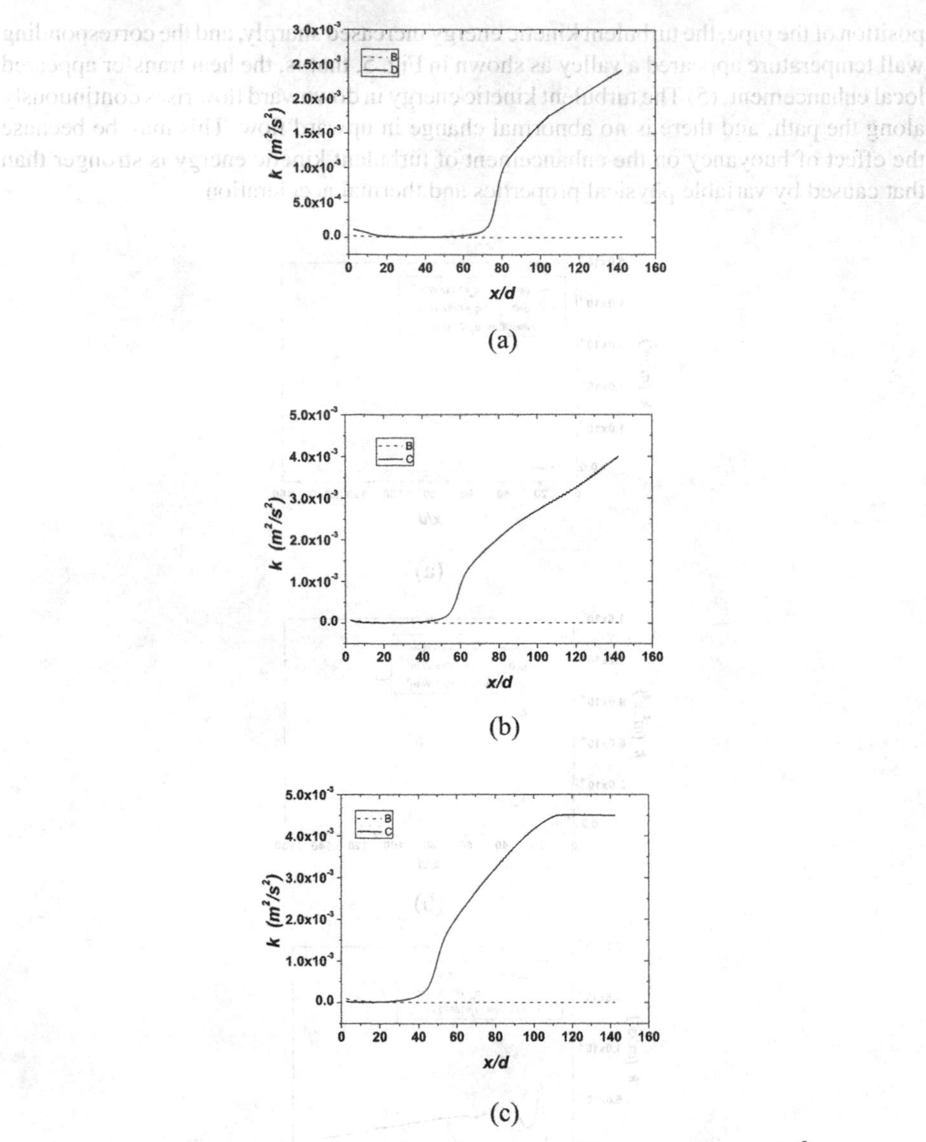

Fig. 6. Variation of turbulent kinetic energy without buoyancy (r/R = 0.9) q_w, W/m^2: (a)—7926; (b)—13182; (c)—19933 Solid line—Consider only changes in density; Dashed line—Change of all physical parameters

seen that at the inlet of the pipe, the change trend of turbulent kinetic energy obtained by considering only the change of density and all physical parameters basically coincides, and both decrease to 0, That is, laminar fluidization occurs in turbulence. We know that the thermal acceleration is mainly caused by the axial fluid density difference, which can further explain that the heat transfer deterioration at the pipe inlet in the upward flow is mainly caused by the thermal acceleration, which is also in good agreement with the experimental results.

5 Conclusions

(1) LB turbulence model can better simulate the local wall temperature peaks and valleys of upward flow at low Reynolds number, while RNG turbulence model can better simulate the wall temperature of downward flow.
(2) The enhancement effect of buoyancy on downward flow heat transfer is greater than that of upward flow.
(3) The local heat transfer deterioration and enhancement in the upward flow and the heat transfer enhancement in the downward flow are mainly due to the influence of buoyancy and thermal acceleration on the turbulent kinetic energy.
(4) For the design of supercritical CO_2 heat exchanger of lead-cooled fast reactor, the subsequent research on convective heat transfer under higher pressure and temperature will be carried out in the future.

References

1. Huang, Y., Wang, J.: Applications of supercritical carbon dioxide in nuclear reactor system. Nuclear Power Engineering **33**(3), 21–27 (2012)
2. Li, M.J., Zhu, H.H., Guo, J.Q.: The development technology and applications of supercritical CO_2 power cycle in nuclear energy, solar energy and other energy industries. Applied Thermal Engineering **126**, 255–275 (2017)
3. Shitsman, M.E.: Impairment of the heat transmission at supercritical pressures. High Temp. **1**, 237–244 (1963)
4. Krasnoshchekov, E.A., Protopopov, V.S.: Experimental study of heat exchange in carbon dioxide in the supercritical range at high temperature drops(in Russian). Teplofizika Vysokikh Temperature **4**(3), 389–398 (1963)
5. Jackson, J.D., Cotton, M.A., Axcell, B.P.: Studies of mixed convection in vertical tubes. Int. J. Heat Fluid Flow **10**(1), 2–15 (1989)
6. Kurganov, V.A., Kaptilnyi, A.G.: Flow structure and turbulent transport of a supercritical pressure fluid in a vertical heated tube under the conditions of mixed convection. Int. J. Heat Mass Transfer **36**(13), 3383–3392 (1993)
7. Jiang, P.X., Zhang, Y., Xu, Y.J., Shi, R.F.: Experimental and numerical investigation of convection heat transfer of CO_2 at supercritical pressures in a vertical tube at low Reynolds numbers. Int. J. Thermal Sciences **47**, 998–1011 (2008)
8. Zhi-hui, L., Pei-xue, J., Chen-ru, Z., et al.: Experimental Investigation of Convection of Heat Transfer of CO_2 at supercritical pressures in vertical circular tube. J. Engineering Thermophysics **29**(3), 461–464 (2008)
9. McEligot, D.M., Jackson, J.D.: "Deterioration" criteria for convective heat transfer in gas flow through non-circular ducts. Nucl. Eng. Des. **232**, 327–333 (2004)
10. McEligot, D.M., Coon, C.W., Perkins, H.C.: Relaminarization in tubes. Int. J. Heat Mass Transfer **13**, 431–433 (1970)
11. Murphy, H.D., Chambers, F.W., McEligot, D.M.: Laterally converging flow. I. Mean flow. J. Fluid Mech **127**, 379–401 (1983)

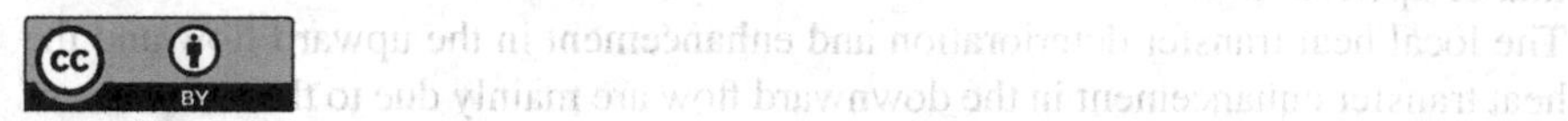

Verification of Solver for Coupled Simulation of Fluid and Fuel Pin in LFR Based on Openfoam

Wenlan Ou, Zhengyu Gong, Qiwen Pan, Ling Zhang, Jianing Dai, and Zhixing Gu(✉)

College of Nuclear Technology and Automation Engineering, Chengdu University of Technology, Chengdu, Sichuan, China
guzhixing17@cdut.edu.cn

Abstract. The Computational Fluid Dynamics (CFD)-based thermal-hydraulics and safety analyses of Lead-based Fast Reactor (LFR) have attracted great attentions in recent years. Commercial CFD tools have been widely used in the 3D simulations of pool-type reactors owing to their powerful abilities in geometric modeling and meshing. Compared with the commercial CFD tools, OpenFOAM is a free open-source CFD code, which is more flexible to perform multi-physics coupling activities. In this paper, in order to develop a solver for simulating the coupled flow and heat transfer behaviors of fluid (coolant) and fuel pin in LFR, the fuel pin Heat Conduction (HC) model was coupled to the modified icoFoam solver of OpenFOAM. Verifications were conducted by the steady-state coupled simulation of fluid and fuel pin heat transfer behaviors, comparing with the MPC-LBE code which has been verified by the benchmarks for LFR fuel pin channel. The results simulated by the coupled solver proposed in this paper agreed well with the ones provided by the MPC-LBE code. This study lays a foundation for the further development of transient safety analysis code for LFR in our future work.

Keyword: OpenFOAM · IcoFoam · LFR · Multi-physics coupling · Benchmark

1 Introduction

As one of the Generation-IV advanced nuclear energy systems, Lead-based Fast Reactor (LFR), has appealed to many international research institutions owing to its excellent inherent safety and nuclear sustainability [1]. The Lead Bismuth Eutectic (LBE) coolant employed in LFR with special characteristics (such as better heat conduction) different from the water (conventional coolant), as well as its integrative pool-type configuration may give rise to complicated three-dimensional thermal hydraulic phenomena in the large space plenum, such as thermal stratification and coolant mixing [2], making the thermal-hydraulics and safety problems of LFR always the research highlights. Computational fluid dynamics (CFD)-based commercial programs for simulating, owing to its mature technologies (plentiful models and algorithms) as well as the powerful preprocessing (modeling and mesh generation of complex geometric structures) capability,

C. Liu (Ed.): PBNC 2022, SPPHY 283, pp. 909–918, 2023.
https://doi.org/10.1007/978-981-99-1023-6_77

is considered as an effective method to overcome the multi-dimensional complicated thermal hydraulic problems involved in the liquid metal pool-type reactors [3].

At present, some commercial CFD tools, such as ANSYS Fluent, Star-CCM + , ABAQUS, have been widely used for simulations of liquid metal pool-type reactors, particularly the multi-scale coupling simulations with other reactor core physical models. For example, Gu Z. developed an advanced two-dimensional fuel pin heat transfer model, then integrated it with the self-developed PK model into ANSYS Fluent to conduct multi-physics coupling [4]. Deng J. developed a three-dimensional transient nuclear thermal coupling solution program based on OpenFOAM platform [5]. Narayanan developed the numerical models that deal with lead thermal hydraulics and solidification with ANSYS Fluent and Star-CCM + software [2].

As we summarized above, CFD tools are popularly used for multi-physics coupling simulation on liquid-metal-cooled reactors, but such commercial CFD programs are incapable to conduct the advanced coupling algorithms due to their closed source codes. OpenFOAM, as a free open-source CFD code, has a unique advantage of direct contact with the source code of existing solvers, making it relatively convenient for us to develop a new solver by modifying its source code from existing solver to vitally design its core computing functions.

In order to conduct the multi-physics coupling simulation with a high degree of customization based on complex coupling algorithms, a solver for simulating the coupled flow and heat transfer behaviors of fluid (coolant) and fuel pin in LFR was developed, and the fuel pin HC model was coupled to the modified icoFoam solver of OpenFOAM in this study. Verifications were conducted by the steady-state coupled simulation of coolant and fuel pin heat transfer behaviors, comparing with the MPC-LBE code which has been verified by the benchmarks for LFR fuel pin channel [6]. The simulation results agree well with the results provided by the MPC-LBE code, which shows the feasibility and accuracy of the solver.

2 Coupling of HC Code with Modified IcoFoam Solver

For LFR, the fuel pin channel consists of a slender cylindrical fuel pin and LBE coolant. Especially, the fuel pin constitutes of a centra hole, pin fuel pellet made of mixed oxides of uranium and its proliferator plutonium, gas plenum filled with Helium (under high pressure) and cladding mainly made of Zr-4 alloy. All the features above can be found in the benchmark reports [7].

2.1 CFD-Based Modifications of IcoFoam Solver

The new solver developed in this paper is based on the icoFoam solver of OpenFOAM, which can simulate the laminar flow by solving the incompressible Newtonian fluid N-S equations shown in Eq. (1) with the finite volume method.

$$\frac{\partial \rho \boldsymbol{U}}{\partial t} + \nabla \cdot (\rho \boldsymbol{U}\boldsymbol{U}) = \rho \boldsymbol{g} - \nabla p + \nabla \cdot (\mu \nabla \boldsymbol{U}) \quad (1)$$

In icoFoam, the fluid flow model is simplified by the neglect of gravity as well as the assumption of constant fluid density and kinematic viscosity. As shown in Fig. 1, the coupling solution of velocity and pressure, based on the simplified N-S equations and continuity equations, is obtained by using the transient PISO algorithm.

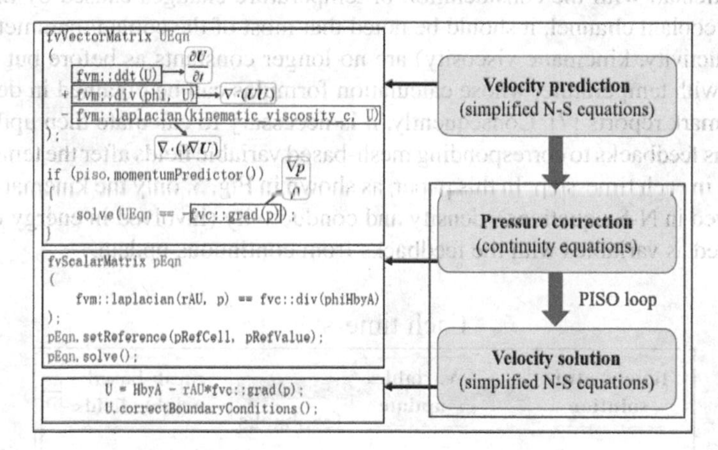

Fig. 1. Coupling solution of velocity and pressure

However, icoFoam is incapable of simulating the heat transfer phenomena in fluid (coolant) region due to the neglecting of energy conservative equation for fluid. To conduct the simulation coupled flow and heat transfer behaviors in coolant region, corresponding energy equation containing the heat source term, shown as Eq. (2), is added to icoFoam source code.

$$\frac{\partial \rho_{LBE} c_{p,LBE} T_{LBE}}{\partial t} + \nabla \cdot \left(\rho_{LBE} c_{p,LBE} T_{LBE} U\right) = \nabla \cdot (\lambda_{LBE} \nabla T_{LBE}) + q_V \tag{2}$$

The velocity-based solution of energy conservative equation using the existing solver in OpenFOAM, is carried out after the coupling solution of velocity and pressure, as shown in Fig. 2.

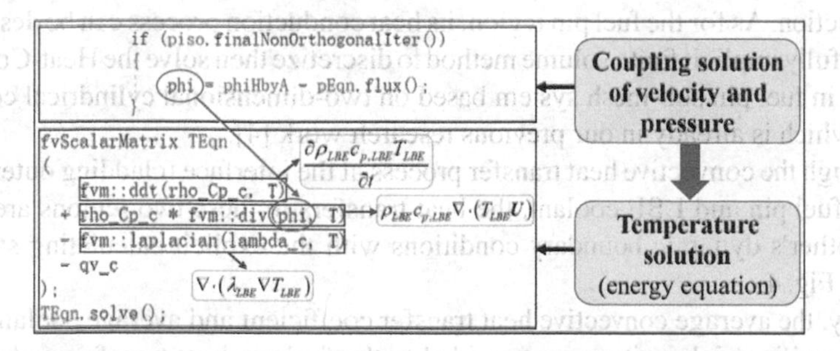

Fig. 2. Velocity-based solution of temperature

What's more, the new variables involved in this new equation, such as temperature, specific heat, conductivity etc., indispensably need to be added to the solver as mesh-based variable fields just, which is similar to the pressure and velocity. Meanwhile, these newly added variable fields should be defined and initialized in include files, source code and running environment, including internal meshes and boundary conditions.

In particular, with the consideration of temperature changes caused by the energy change in coolant channel, it should be noted that most of the coolant parameters (density, conductivity, kinematic viscosity) are no longer constants as before but variables changing with temperature, whose calculation formulas can be obtained in detail from the benchmark reports [7]. Consequently, it is necessary to calculate then update these variables as feedbacks to corresponding mesh-based variable fields after the temperatures are solved in each time-step. In this paper, as shown in Fig. 3, only the kinematic viscosity (involved in N-S equations), density and conductivity (involved in energy equation) are regarded as variables with the feedbacks from continuous update.

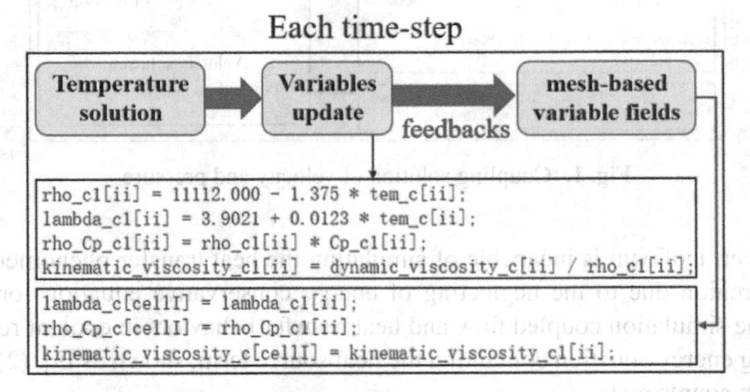

Fig. 3. Mesh-based variable fields update

2.2 Coupling Strategies with HC Code and Modified IcoFoam Solver

The modified icoFoam can simulate both the flow and heat transfer phenomena in coolant region, a three-dimensional mesh system based on OpenFOAM, after the works we did in the last section. As for the fuel pin region, its heat conduction process can be described by using the fully implicit finite volume method to discretize then solve the Heat-Conduction Equation in fuel pin sub-mesh system based on two-dimensional cylindrical coordinate system, which is already in our previous research work [4].

Through the convective heat transfer process at the interface (cladding outer surface) between fuel pin and LBE coolant, the heat transfers in these two regions are coupled as each other's dynamic boundary conditions with the explicit calculating strategy as shown in Fig. 4.

Firstly, the average convective heat transfer coefficient and average coolant temperature at specific axial position are provided to the fuel pin heat transfer module within the same axial position after the temperatures are solved in each time-step. Especially, the calculations of these two average variables are conducted by modifying icoFoam source code.

After that, the fuel pin region gives a heat source to coolant region in return. Generally, this heat source is transferred in the form of heat flux as a boundary condition,

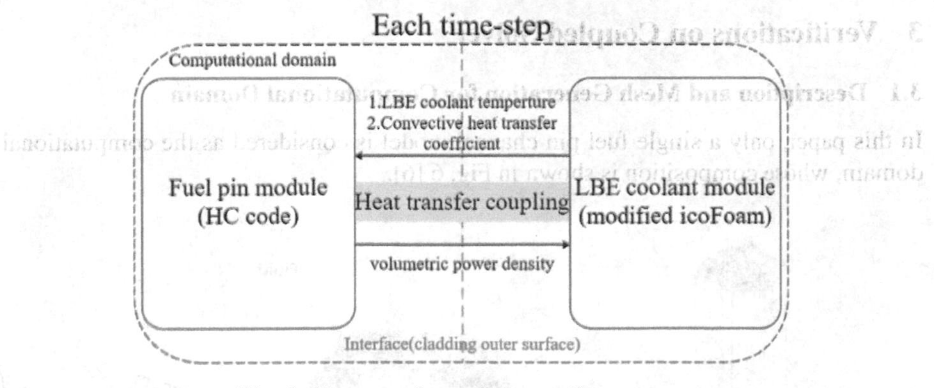

Fig. 4. Heat transfer explicit coupling strategy

nevertheless, in this study we tried a new method to conduct this transfer. By the heat flux calculated in fuel pin region, the power transferred from cladding outer surface to LBE coolant can be also obtained, which is applied to the meshes at the inner wall of coolant channel as its volume power density (external heat source). Many mesh related parameters need to be obtained then used for calculations above.

Importantly, the fuel pin heat transfer module (HC code) is added after the coolant module to icoFoam source code based on the integral coupling framework shown in Fig. 5, so as to realize the coupling calculations in each time-step. These two modules use the same time-step size (0.02s) to coupled calculate, which has been already proved that this size of time-step can guarantee numerical stability. In each time-step, the temperatures of fuel pin module and coolant module are calculated iteratively, until the calculated time-steps meet the steady-state establishment.

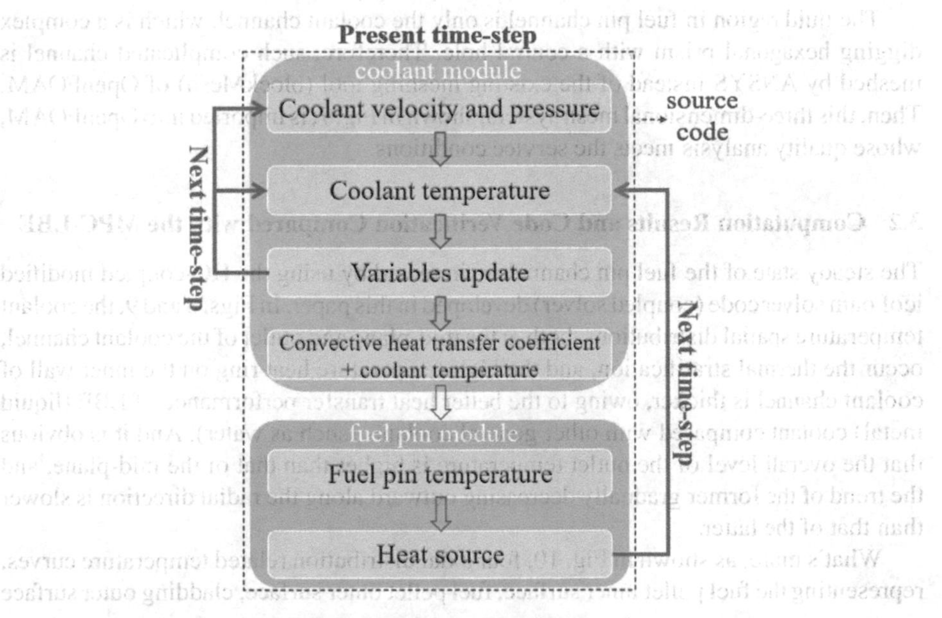

Fig. 5. Integral coupling framework

3 Verifications on Coupled Solver

3.1 Description and Mesh Generation for Computational Domain

In this paper, only a single fuel pin channel model is considered as the computational domain, whose composition is shown in Fig. 6 [6].

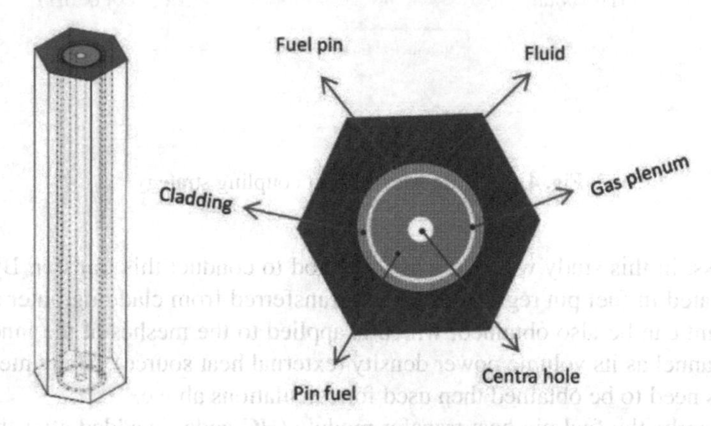

Fig. 6. Model of fuel pin channel [6]

The computational domain is treated as a hexagon prism by its symmetrical characteristic. For the convenience of calculations of steady-state process, the inlet coolant temperature and velocity are constantly equal to the initial value, and the heat decay power is neglected. The geometry data and material properties of the fuel pin channel can be found with details in the benchmark report [6].

The fluid region in fuel pin channel is only the coolant channel, which is a complex digging hexagonal prism with a central hole. Therefore, such complicated channel is meshed by ANSYS instead of the existing meshing tool (blockMesh) of OpenFOAM. Then, this three-dimensional mesh system, shown in Fig. 7, is imported into OpenFOAM, whose quality analysis meets the service conditions.

3.2 Computation Results and Code Verification Compared with the MPC-LBE

The steady state of the fuel pin channel is simulated by using the HC-coupled modified icoFoam solver code (coupled solver) developed in this paper. In Figs. 8 and 9, the coolant temperature spatial distributions, both at the mid-plane and outlet of the coolant channel, occur the thermal stratification, and the high-temperature heat ring on the inner wall of coolant channel is thicker, owing to the better heat transfer performance of LBE (liquid metal) coolant compared with other general coolants (such as water). And it is obvious that the overall level of the outlet temperature is higher than that of the mid-plane, and the trend of the former gradually decreasing outward along the radial direction is slower than that of the latter.

What's more, as shown in Fig. 10, four axial distribution related temperature curves, representing the fuel pellet inner surface, fuel pellet outer surface, cladding outer surface

Fig. 7. Mesh generation of coolant channel

and coolant temperatures (arranged from high to low), are drawn and compared with the ones by the MPC-LBE code whose feasibility and accuracy have already been verified [4]. Since the power generated from fuel pellet will heat the coolant channel around with coolant continuously flowing through the channel, the coolant temperature will gradually rise during the process from inlet to outlet. Obviously, such good agreements are achieved in terms of both the trends and values among the results provided by the HC-coupled modified icoFoam solver code and by the MPC-LBE code [4] in Fig. 10.

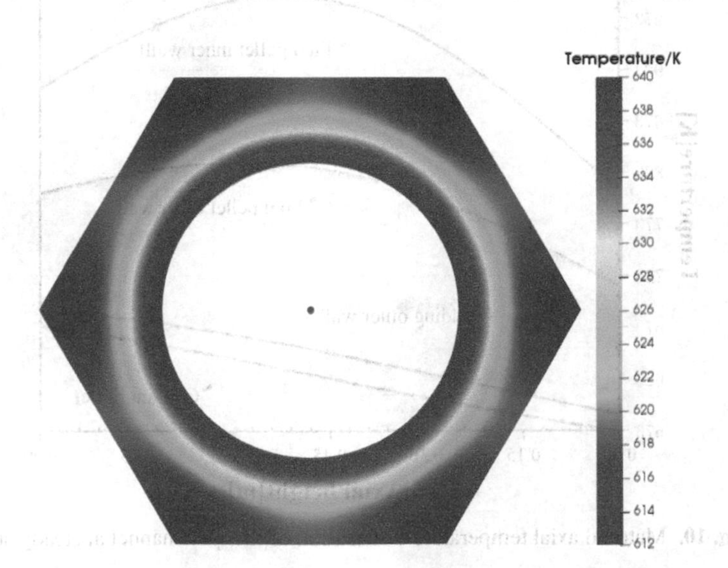

Fig. 8. Temperature spatial distribution at mid-plane of coolant channel

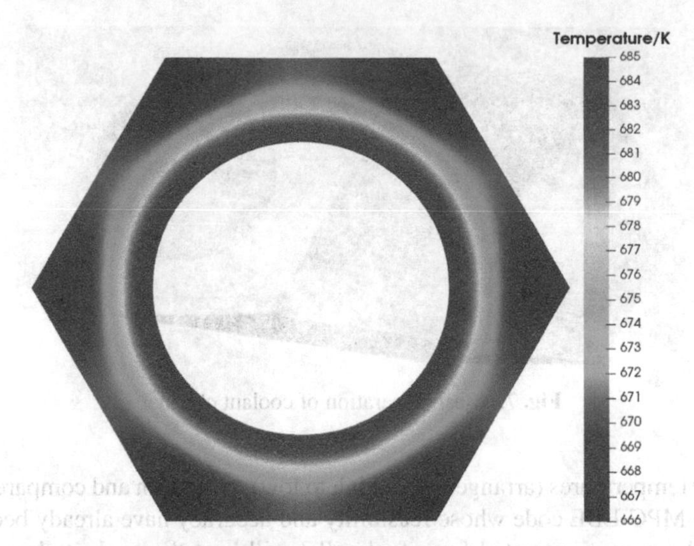

Fig. 9. Temperature spatial distribution at outlet of coolant channel

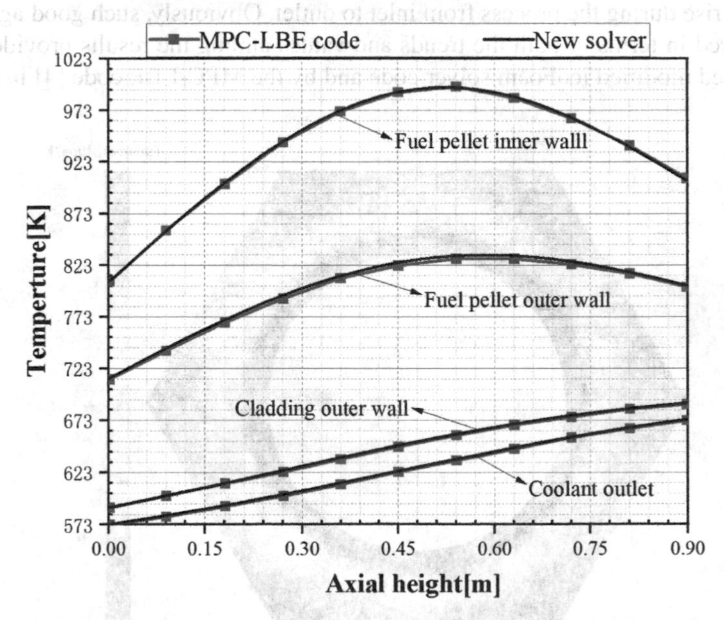

Fig. 10. Material axial temperature distribution of fuel pin channel at steady state

4 Conclusions

In this paper, a single fuel pin channel model is established for the reactor core of LFR, so as to analyse the coupled flow and heat transfer behaviors in coolant and fuel pin. Multi-physics simulation system is built by coupling the fuel pin heat transfer model (two-dimensional HC module) to the three-dimensional open-source CFD code OpenFOAM by its icoFoam. This coupled simulation system for reactor thermal-hydraulics and safety analyses is abbreviated as the HC-coupled modified icoFoam solver code (coupled solver), which copes with the key data transmission problems of multi-physics coupling strategy within the framework of OpenFOAM platform. Mathematics models and methods as well as the coupling strategy of two modules are elaborated.

Verifications were conducted by the steady-state coupled simulation of flow and heat transfer behaviors between coolant and fuel pin regions in fuel pin channel of LFR, including the comparison with the benchmarks-verified MPC-LBE code [4]. The results simulated by the HC-coupled modified icoFoam solver code proposed in this paper agreed well with the ones provided by benchmarks-verified MPC-LBE code. It is testified that the HC-coupled modified icoFoam solver code is capable to perform the multi-physics simulations involved in thermal-hydraulics and safety analyses of LFR with enough accuracy.

However, only a simple single fuel pin model is considered to preliminarily verify the basic coupling strategies and schemes of reactor core heat transfer behaviors. Therefore, the coupling simulation of the whole fuel assembly channel of reactor core should be further studied in the future work, which can describe the reactor core more carefully and comprehensively. Meanwhile, for the heat transfer calculation coupled with coolant and fuel pin regions, the implicit iteration can be considered instead of the explicit calculating strategy used in this study. All in all, this study already verified the initial steady state, laying a foundation for the further development of transient safety analysis code to conduct transient simulation for LFR in our future work.

References

1. 罗晓, 张喜林, 陈红丽, 王帅, 郭超 & 王驰 (2021). 铅冷快堆多物理耦合分析方法及三维瞬态特性研究. 核动力工程(S1), 11–16.
2. Achuthan, N., et al.: Computational fluid dynamics modelling of lead natural convection and solidification in a pool type geometry. Nuclear Engineering and Design, p. 376 (2021).
3. Tenchine, D.: Some thermal hydraulic challenges in sodium cooled fast reactors. Nucl. Eng. Des. **240**(5), 1195–1217 (2010)
4. Gu, Z., Li, F., Ge, L., et al.: Verification of a HC-PK-CFD coupled program based a benchmark on beam trip transients for XADS reactor. Ann. Nucl. Energy **133**, 491–500 (2019)
5. Deng, J., Deng, J., Li, Z., et al.: Study on transient characteristics of nuclear ramjet reactor based on three-dimensional nuclear thermal coupling method. Nucl. Sci. Eng. **42**(01), 18–27 (2022)
6. Gu, Z.-X., et al.: Verification of a self-developed CFD-based multi-physics coupled code MPC-LBE for LBE-cooled reactor. Nucl. Sci. Tech. **32**(5), 1–17 (2021). https://doi.org/10.1007/s41365-021-00887-x
7. D'Angelo, A., Arien, B., Sobolev, V., et al.: Benchmark on Beam Interruptions in an Accelerator-Driven System Final Report on Phase I Calculations, Technical Report NEA/NSC/DOC(2003)17, NEA (2003)
8. Behar, C.: Technology roadmap update for generation IV nuclear energy systems. In: OECD Nuclear Energy Agency for the Generation IV International Forum, accessed Jan. 2014, 17, 2014–03

9. Alemberti, A., Smirnov, V., Smith, C.F., et al.: Overview of leadcooled fast reactor activities. Prog. Nucl. Energy **77**, 300–307 (2014)
10. Chen, H., Chen, Z., Chen, C., et al.: Conceptual design of a small modular natural circulation lead cooled fast reactor SNCLFR100. Inter. J. Hydrog. Energy **41**(17), 7158–7168 (2016)
11. Chen, H., Zhang, X., Zhao, Y., et al.: Preliminary design of a medium-power modular lead-cooled fast reactor with the application of optimization methods. Inter. J. Hydrog. Energy **42**(11), 3643–3657 (2018)
12. Zou Zeren et al. (2020). 3D thermal hydraulic characteristics analysis of pool-type upper plenum for lead-cooled fast reactor with multi-scale coupling program. Nuclear Engineering and Design, pp. 370 110892-
13. 魏诗颖, 王成龙, 田文喜, 秋穗正 & 苏光辉 (2019). 铅基快堆关键热工水力问题研究综述. 原子能科学技术(02), 326-336.
14. 雷洲阳 (2021). 基于 CFD 物理热工耦合的池式快堆 UTOP 事故不确定性分析研究(硕士学位论文,南华大学)
15. 李尚卿, 王伟民 & 李玉同 (2022). 基于 OpenFOAM 的磁流体求解器的开发和应用. 物理学报(11), 424–433.
16. T.H. Fanning, A.J. Brunett, T. Sumner, The SAS4A/SASSYS-1 Safety Analysis Code System, Version 5 (Argonne national lab. (ANL), argonne, IL United States, 2017)
17. Mesina, G.L.: A history of RELAP computer codes. Nucl. Sci. Eng. **182**(1), v–ix (2016)
18. Lerchl, G., Austregesilo, H., Glaeser, H., et al.: ATHLET Mod 3.0— Cycle A. Validation. Gesellschaft fu¨r Anlagen-und Reaktorsicherheit (GRS) gGmbH, Garching bei Mu¨nchen, Germany, Report No. GRS-P-1, 2012, 3
19. Fanning, T.H., Thomas, J.W.: Advances in Coupled Safety Modeling Using Systems Analysis and High-Fidelity Methods (Argonne national lab. (ANL) argonne, IL United States (2010)
20. Pialla, D., Tenchine, D., Li, S., et al.: Overview of the system alone and system/CFD coupled calculations of the PHENIX natural circulation test within the THINS project. Nucl. Eng. Des. **290**, 78–86 (2015)
21. Bandini, G., Polidori, M., Gerschenfeld, A., et al.: Assessment of systems codes and their coupling with CFD codes in thermal– hydraulic applications to innovative reactors. Nucl. Eng. Des. **281**, 22–38 (2015)
22. Chen, Z., Chen, X.N., Rineiski, A., et al.: Coupling a CFD code with neutron kinetics and pin thermal models for nuclear reactor safety analyses. Ann. Nucl. Energy **83**, 41–49 (2015)

Research of Helium Thermal Power System Based on Lead-Cooled Fast Reactor

Zhihui Li(✉)

State Power Investment Corporation Research Institute, Beijing, China

Abstract. The Helium Brayton cycle with re-compression has the advantages of compact layout, simple structure, thermal high efficiency, good heat transfer characteristics and small friction characteristics, its application in the power conversion system of the lead-cooled fast reactor and the high temperature gas-cooled reactor helps the miniaturization of the whole system. In this paper, the mathematical model was established for Helium Brayton cycle with re-compression and the 100 MWt lead-cooled fast reactor power system was calculated. The effects of several key factors such as the turbine inlet temperature, the turbine outlet pressure, the high pressure compressor outlet pressure, the low pressure compressor outlet pressure and the recuperator outlet temperature were analyzed. The results show that the turbine outlet pressure, the turbine inlet temperature and the high/low pressure compressor outlet pressure have remarkable effects on thermal efficiency of the system. Thermal efficiency of the system increases first and then decreases with the turbine outlet pressure increasing as well as increases with turbine inlet temperature. The research results of this paper could provide important theoretical reference both for thermal cycle parameters for 100 MWt lead-cooled fast reactor and system design of power cycle based on the lead-cooled fast reactor.

Keywords: LEAD-COOLED FAST REACTOR · HELIUM · RE-COMPRESSION · BRAYTON CYCLE

1 Introduction

In the traditional nuclear reactor, the Rankine steam cycle is used in the energy conversion system. Because the reactor can only provide saturated steam at about 320 °C, it can not provide superheated steam at a higher temperature to meet the need of Rankine cycle to improve efficiency, so the efficiency is relatively low. The lead-cooled fast reactor gas turbine cycle combines the gas turbine with the modular lead-cooled fast reactor, and uses the high-temperature gas generated by the lead- cooled fast reactor to directly drive the gas turbine to do work for high-efficiency power generation. It can break through the temperature limit of the steam cycle, and use intermediate cooling and regenerative technologies to improve efficiency, thus becoming an important direction for the study of high-efficiency power generation of the lead-cooled fast reactor.

The earliest gas-cooled reactor power cycle device combined with helium turbine began to develop in the late 1960's. From 1968 to 1981, Germany cooperated with

C. Liu (Ed.): PBNC 2022, SPPHY 283, pp. 919–929, 2023.
https://doi.org/10.1007/978-981-99-1023-6_78

the United States and Switzerland to complete the HHT (HTR with helium turbine) experimental program for the helium turbine power conversion system [1]. The purpose of the plan was to study the technology of high temperature gas-cooled reactor power generation with helium turbine, including turbine, compressor, hot gas duct, materials, heat exchanger and other component technologies.

In the early 1990's, with the support of the U.S. Department of energy, MIT carried out the research on modular high temperature gas-cooled reactor (MGR) helium turbine cycle power plant [2], and proposed two technical schemes, namely direct helium turbine cycle scheme (MGR-GT) and indirect helium turbine cycle scheme (MGR-GTI). Necsa of South Africa has carried out research on PBWR of pebble bed modular high temperature gas-cooled reactor [3]. PBWR adopts standard Brayton cycle with closed water-cooled precooler and intercooler. The United States and Russia cooperated in the research of GT-MHR [4] and adopted the design of closed Brayton cycle power conversion system.

In recent years, foreign scholars have carried out a lot of theoretical research on the scheme and system characteristics of helium turbine cycle system [5–12]. Professor Wang Jie from the Institute of nuclear energy technology design and research, Tsinghua University, China [13–15] respectively carried out thermodynamic analysis and optimization calculation for the direct helium cycle, open air cycle and indirect nitrogen cycle of the high temperature gas-cooled reactor, and carried out aerodynamic design for the turbine compressor. The results show that helium direct circulation is an ideal choice, but it is difficult based on the existing technical level. The closed indirect circulation of helium or nitrogen is a relatively realistic scheme at present, which can realize the idea of gas turbine circulation and accumulate technology for the direct circulation in the future. State power corporation research institute are carrying out the research on 100 MW lead-cooled fast reactor (BLESS-D) to meet the public's demand for a safer, more economical and more environmentally friendly nuclear power system [16–19]. It is a pool-style reactor with LBE (lead and bismuth eutectic alloy) as the coolant. It is mainly used to study the key problems and technologies of lead cooled fast reactor, verify and demonstrate relevant solutions. Among, The thermal power cycle system adopts different working fluid scheme, such as water, supercritical carbon dioxide, Helium.

In this paper, the thermodynamic calculation of helium re-compression Brayton cycle is carried out for 100 MW_t lead-cooled fast reactor. The effects of core outlet temperature, turbine outlet pressure, high-pressure compressor outlet pressure, low pressure compressor outlet pressure and recuperator outlet temperature on thermal efficiency of the system are analyzed. The design parameters of helium re-compression Brayton cycle for 100 MWt lead-cooled fast reactor are optimized, which can provide an important theoretical reference for the design of power cycle system of lead-cooled fast reactor.

2 Helium Brayton Cycle Introduction

2.1 Thermophysical Properties of Helium

Table 1 shows the comparison of thermophysical properties between helium and other working fluids under standard conditions. It can be seen from Table 1 that, compared with other working fluids, helium has the following physical properties: strong thermal conductivity, large specific heat, small gas density and large is entricity index.

The effects of these special thermophysical properties of helium on the helium Brayton cycle are as follows:

(1) The gas density is small, so a high circulating pressure is required to increase its density;
(2) The isentropic index is large, so under the same temperature difference, the pressure ratio is small, so it is difficult to compress;
(3) The specific heat is large, so under the same temperature difference, the cycle specific work is large, and the expansion work of the turbine and the compression work of the compressor are larger than those of other gases.

Helium is very close to ideal gas in the range of 0–1000 °C, 0.1–10 MPa, and its specific heat and adiabatic index are almost constant. Compared with air, helium has a higher specific heat (about 5 times that of air), so the compression ratio of helium is small under the same temperature difference, and the mass flow of helium is small under the same output power. Helium also has good heat transfer characteristics and small friction characteristics, which is conducive to improve the efficiency of the heat exchanger and reduce the volume of the heat exchanger.

Table 1. THE COMPARISON OF THERMO PHYSICAL PROPERTIES BETWEEN HELIUM AND OTHER WORKING FLUIDS (1ATM, 0 °C)

Fluids	Thermal conductivity (W/(m.K))	Density (kg/m^3)	Specific heat (kJ/(kg.K))	Isentri-ity index
He	0.14	0.17	5.19	1.67
N_2	0.02	1.25	1.04	1.4
Air	0.02	1.29	1.06	1.4
CO_2	0.01	1.64	0.82	1.3
Ar	0.01	1.78	0.52	1.67
H_2	0.16	0.08	14.19	1.41
O_2	0.02	1.42	0.92	1.34

2.2 Helium Closed Re-compression Brayton Cycle Process

There are many specific circulation modes in the helium turbine circulation scheme. From the point of view of improving power generation efficiency, helium turbine direct circulation is an ideal choice. In other words, the high-temperature and high-pressure gas produced by the reactor core is used to directly drive the turbine to generate electricity.

Figure 1 shows the flow chart of helium closed re-compression Brayton cycle [14] which is the scheme to realize the direct cycle of helium turbine. This cycle scheme adopts recuperator and intermediate cooling of gas flow in the process of compression to improve thermal efficiency of the system. The system is composed of turbine, recuperator, high-pressure compressor, low-pressure compressor, intercooler, precooler, generator and corresponding pipelines.

The working process is as follows: 1-2a refers to the isentropic compression process of low-temperature and low-pressure helium in the low-pressure compressor; 2a-2b refers to the constant pressure heat release process of helium in the intercooler (ICL); 2b-2 refers to the isentropic compression process of cooled helium in the high-pressure compressor; 2–3 shows the heat absorption process of high pressure helium at the high pressure side of the recuperator (RPT); 3–4 refers to the constant pressure heat absorption process of high temperature and high pressure helium in the reactor; 4–5 refers to the isentropic expansion process of high-temperature helium in the turbine. The expansion work is used to drive the high-pressure compressor, low-pressure compressor and generator. 5–6 shows the constant pressure heat release process of helium at the low pressure side of the recuperator (RPT). 6–1 shows the constant pressure heat release process of cooled helium in the precooler (PCL). Finally, the low-temperature and low-pressure helium enters the low-pressure compressor for isentropic compression to complete the whole cycle.

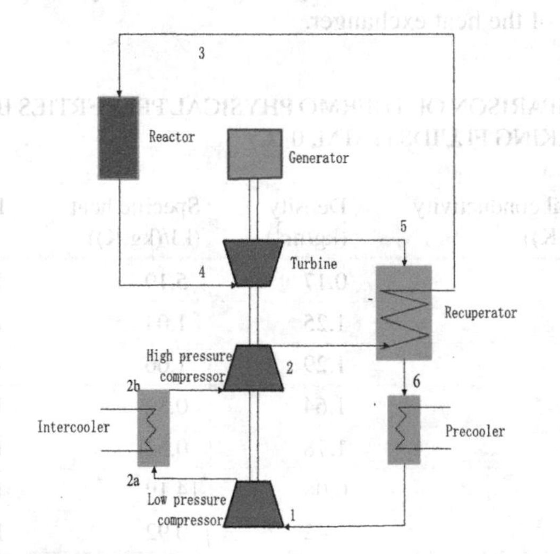

Fig. 1. HELIUM BRAYTON CYCLE WITH RE-COMPRESSION

3 Influence of Key Parameters on Thermal Efficiency of the System

The mathematical model of helium closed recompression Brayton cycle of 100 Wt lead cooled fast reactor is established by using MATLAB software. Thermal efficiency of the system is expressed by the following formula:

$$\eta = f(\tau, \gamma, \overline{\theta}, \overline{\eta}_c$$

where, τ refers to the circulating temperature ratio which is the ratio of the absolute temperature of the high-temperature heat source to the low-temperature heat source of the system; γ refers to the circulating pressure ratio which is ratio of the inlet pressure to the outlet pressure of high-pressure compressor and low-pressure compressor; The physical property vector $\overline{\theta}$ is composed of the ratio of the specific heat at constant pressure for the compression process and the endothermic process, the ratio of the specific heat at constant pressure for the expansion process and the endothermic process, and the isentropic index. For helium, the specific heat at constant pressure and specific heat at constant volume are constant, the ratio of specific heat at constant pressure is 1, and the isentropic index is 1.67. Component efficiency vector is composed of recuperator efficiency, turbine efficiency, compressor efficiency and pressure loss. Limited by today's manufacturing level, after taking a group of reasonable values according to literature [13], as shown in Table 2, the component efficiency vector becomes a constant vector. The whole cycle is steady-state calculation. Therefore, thermal efficiency of the system is mainly determined by the circulating temperature ratio and circulating pressure ratio.

Table 2. COMPONENT EFFICIENCY AND RESSURE LOSS

Component parameters	Values
Turbine efficiency	90%
High-pressure Compressor efficiency	89%
Low-pressure Compressor efficiency	89%
Generator efficiency	98%
Recuperator efficiency	95%
Pressure loss	5%

3.1 Influence of the Core Outlet Temperature

Figure 2 shows the influence of different core outlet temperature on the thermal efficiency of the system under the conditions of high-pressure compressor outlet pressure of 5.6 MPa, low-pressure compressor outlet pressure of 4.9 MPa and recuperator outlet temperature of 80 °C. With the increase of core outlet temperature, thermal efficiency of the system increases. When the core outlet temperature increases from 400 °C to 500 °C, thermal efficiency of the system increases by 86%. Therefore, under the conditions of reactor thermal conditions and core materials, increasing the core outlet temperature can increase the output power of the system.

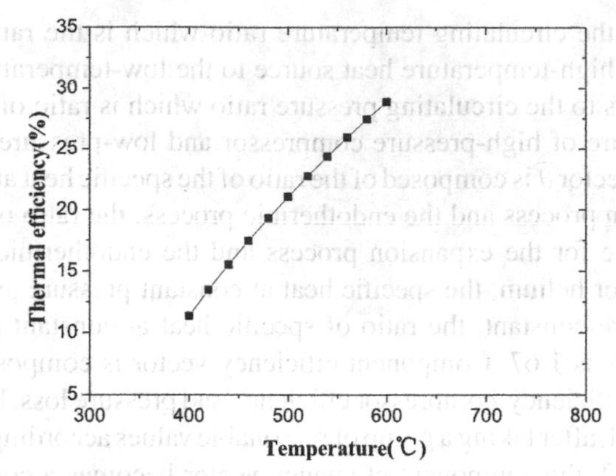

Fig. 2. VARIATION OF THERMAL EFFICIENCY OF THE SYSTEM WITH CORE OUTLET TEMPERATURE

3.2 Influence of the Turbine Outlet Pressure

Figure 3 shows the influence of turbine outlet pressure on thermal efficiency of the system under the conditions of high-pressure compressor outlet pressure of 5 MPa, reactor core outlet temperature of 400 °C and low-pressure compressor outlet pressure of 3.5 MPa. With the increase of the turbine outlet pressure, thermal efficiency of the system first increases and then decreases. The reason is that with the increases of turbine pressure, the power consumption of the low-pressure compressor decreasing and the net power of the system increasing. However, with the further increase of the pressure, the output power of the turbine decreases rapidly, resulting in the decrease of the net output power and the decrease of thermal efficiency of the system. That is, there is an optimal turbine outlet pressure, which well matches the turbine output power and compressor power consumption, so as to maximize thermal efficiency of the system.

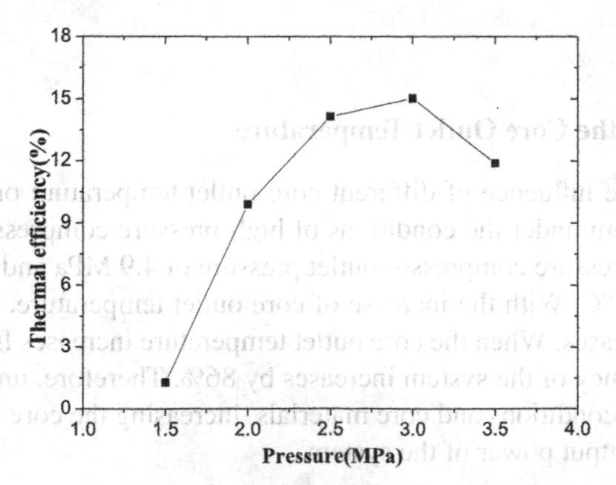

Fig. 3. VARIATION OF THERMAL EFFICIENCY OF THE SYSTEM WITH TURBINE OUTLET PRESSURE

3.3 Influence of the High-Pressure Compressor Outlet Pressure

Figure 4 shows the influence of high-pressure compressor outlet pressure on thermal efficiency of the system under the conditions of low-pressure compressor outlet pressure of 4.18 MPa, reactor core outlet temperature of 500 °C and recuperator outlet temperature of 142.5 °C. With the increase of the high-pressure compressor outlet pressure, thermal efficiency of the system increases. Therefore, increasing the high-pressure compressor outlet pressure can improve thermal efficiency of the system, but subject to the current technical conditions, the high-pressure compressor outlet pressure is usually lower than 7 MPa.

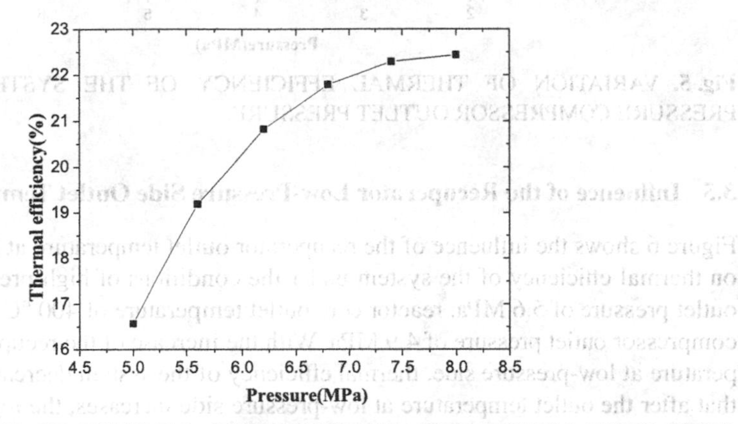

Fig. 4. VARIATION OF THERMAL EFFICIENCY OF THE SYSTEM WITH HIGH-PRESSURE COMPRESSOR OUTLET PRESSURE

3.4 Influence of the Low-Pressure Compressor Outlet Pressure

Figure 5 shows the influence of low-pressure compressor outlet pressure on thermal efficiency of the system under the conditions of high-pressure compressor outlet pressure of 5 MPa, reactor core outlet temperature of 400 °C and recuperator outlet temperature of 167.5 °C. With the increase of low-pressure compressor outlet pressure, thermal efficiency of the system decreases. The reason is that the specific capacity at the inlet of the low-pressure compressor is large. Under the same pressure ratio, the compression work is much greater than that of the high-pressure compressor. At the beginning, with the increase of the low-pressure compressor outlet pressure, the power consumption of the low-pressure compressor increases relatively slowly, which is less than the power consumption reduced by the high-pressure compressor, resulting in the increase of thermal efficiency of the system. When the low-pressure compressor outlet pressure is continuously increased, although the power consumption of the high-pressure compressor decreases, the power consumption of the low-pressure compressor increases faster, resulting in the decline of thermal efficiency of the system.

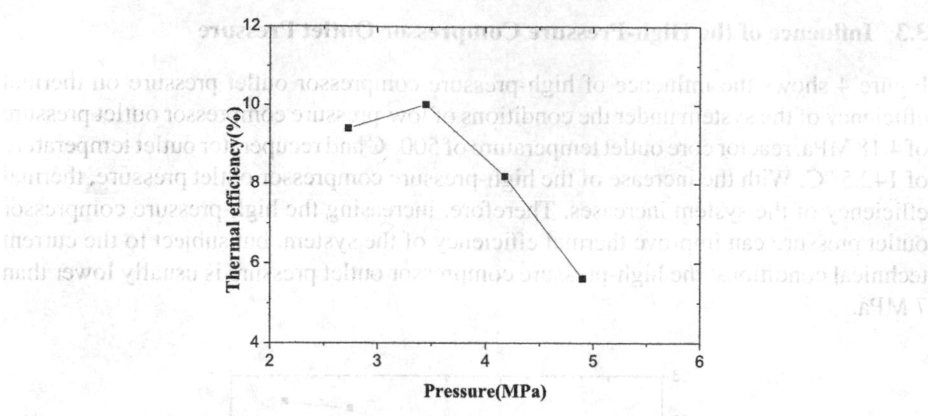

Fig. 5. VARIATION OF THERMAL EFFICIENCY OF THE SYSTEM WITH LOW-PRESSURE COMPRESSOR OUTLET PRESSURE.

3.5 Influence of the Recuperator Low-Pressure Side Outlet Temperature

Figure 6 shows the influence of the recuperator outlet temperature at low-pressure side on thermal efficiency of the system under the conditions of high-pressure compressor outlet pressure of 5.6 MPa, reactor core outlet temperature of 400 °C and low-pressure compressor outlet pressure of 4.9 MPa. With the increase of the recuperator outlet temperature at low-pressure side, thermal efficiency of the system decreases. The reason is that after the outlet temperature at low-pressure side increases, the regenerative degree of the recuperator decreases, the cooling capacity of the precooler increases, and the high-temperature heat generated by the compressor is not fully utilized, resulting in the increase of heat absorption of the system and the decrease of thermal efficiency of the system.

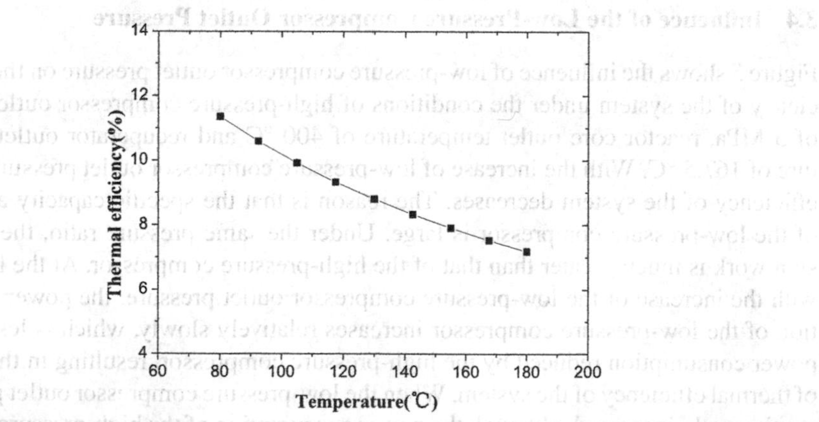

Fig. 6. VARIATION OF THERMAL EFFICIENCY OF THE SYSTEM WITH RECUPERATOR OUTLET TEMPERATURE AT LOW-PRESSURE SIDE

To sum up, in order to improve the thermal efficiency of the helium Brayton cycle, the reactor core outlet temperature should be increased as much as possible, the design scheme of the recuperator should be reasonably optimized, the system regenerative efficiency should be increased, the compression ratio should be reasonably distributed in combination with the characteristics of the high and low pressure compressor, and the of the high pressure compressor outlet pressure should be increased as much as possible.

4 Optimization Results of He Closed Re-compression Brayton Cycle

Through the analysis of the influence of different key parameters on helium closed re-compression Brayton cycle, it can be concluded that the turbine inlet temperature can be taken as 400–600 °C, the low-pressure compressor inlet temperature is taken as 35 °C for lead-cooled fast reactor. Considering that the maximum pressure limit of the system is 7.0 MPa, the high-pressure compressor outlet pressure is set as 5.0 MPa. Table 3 shows the optimization results of he closed re-compression Brayton cycle of lead-cooled fast reactor with thermal power of 100 MW under three different working conditions. Thermal efficiency of the system. When the core outlet temperature reaches 600 °C,Thermal efficiency of the system reaches 34.8%.

Table 3. OPTIMIZATION RESULTS OF HE CLOSED RE-COMPRESSION BRAYTON CYCLE

Cycle category	He (Condition 1)	He (Condition 2)	He (Condition 3)
Initial parameters	$t_{min} = 35$ °C $t_{max} = 400$ °C	$t_{min} = 35$ °C $t_{max} = 500$ °C	$t_{min} = 35$ °C $t_{max} = 600$ °C
Restrictions	$P_{max} = 7.0$MPa	$P_{max} = 7.0$MPa	$P_{max} = 7.0$MPa
High-pressure compressor outlet ressure (MPa)	5.0	5.0	5.0
Optimal pressure ratio of high-pressure compressor	1.44	1.44	1.44
	1.73	1.73	1.73
Mass flow (kg/s)	153.4	133.6	118.4
Thermal efficiency of the system (%)	16.03	26.6	34.8
Heat absorption per unit mass (kJ/kg)	651.8	748.4	844.9
Turbine power (MW)	98.55	98.48	98.42
Turbine inlet volume flow (m^3/s)	43.2	43.2	43.2
Turbine outlet volume flow (m^3/s)	64.8	64.8	64.9
Low-pressure compressor power (MW)	36.42	31.72	28.1
High-pressure compressor power (MW)	46.01	40.16	35.5

5 Conclusions

(1) Thermal efficiency of helium re-compression Brayton cycle system increases with the increase of high-pressure compressor outlet pressure, increases with the increase of core outlet temperature, decreases with the increase of recuperator outlet temperature and increases first and then decreases with the increase of turbine outlet pressure.
(2) The optimized parameters of helium re-compression Brayton cycle system of lead-cooled fast reactor with thermal power of 100 MW under three different core outlet temperatures are obtained, which provides an important theoretical reference for the design of power cycle system of lead-cooled fast reactor. When the core outlet temperature reaches 600 °C, thermal efficiency of the system reaches 34.8%.
(3) At present, the technical level of helium re-compression Brayton cycle still has some limitations. The research on the materials of helium turbine device, the deposition of radioactive products on turbine blades and the integrated layout are still need to be carried out in the future.

References

1. Weibrodt, I.A.: Summary Report on Technical Experiences from High-temperature Helium Turbomachinery Testing in Germany. In: Proceedings of a Technical Committee Meeting, IAEA-TECDOC-899. Beijing (1995)
2. Brey, H.L.: Historical background and Future Development of the High Temperature Gas-cooled Reactor[A]. In: Proceedings of the Seminar on HTGR Application and Development. Beijing (2001)
3. Nicholls.: Pebble Bed Modular Reactor. In: Proceedings of Seminar on HTGR Application and Development Beijing, China (2001)
4. Ball, S.: Status of the Gas Turbine Modular Helium Reactor for Plutonium Disposition. In: Proceedings of Seminar on HTGR Application and Development, Beijing, China (2001)
5. Forsberg, C.W., Peterson, P.F., Pickard, P.S.: Study on characteristic of helium turbine with the high temperature gas-cooled react0r. Nucl. Technol. **144**(3), 289–302 (2003)
6. Fujikawa, S., Hayashi, H., Nakazawa, T., et al.: Achievement of reactor-outlet coolant temperature of 950° C in HTTR. J. Nucl. Sci. Technol. **41**(12), 1245–1254 (2004)
7. Yari, M., Mahmoudi, S.M.S.: Utilization of waste heat from GT-MHR for power generation in organic Rankine cycles. Appl. Therm. Eng. **30**(4), 366–375 (2010)
8. Kunitomi, K., Yan, X., Nishihara, T., et al.: JAEA's VHTR for hydrogen and electricity cogeneration: GTHTR300C. Nucl. Eng. Technol. **39**(1), 9–20 (2007)
9. Meyer, M.K., Fielding, R., Gan, J.: Fuel development for gas-cooled fast reactors. J. Nucl. Mater. **371**(1), 281–287 (2007)
10. Kissane, M.P.: A review of radionuclide behaviour in the primary system of a very-high-temperature reactor. Nuclear Eng. Design **239**(12), 3076–3091 (2009).
11. El-Genk, M.S., Tournier, J.M.: Noble gas binary mixtures for gas-cooled reactor power plants. Nucl. Eng. Des. **238**(6), 1353–1372 (2008)
12. Talamo, A., Gudowski, W., Venneri, F.: The burnup capabilities of the deep burn modular helium reactor analyzed by the Monte Carlo continuous energy code MCB. Ann. Nucl. Energy **31**(2), 173–196 (2004)

13. Verfondern, K., Nabielek, H., Kendall, J.M.: Coated particle fuel for high temperature gas cooled reactors. Nucl. Eng. Technol. **39**(5), 603–616 (2007)
14. Jie, W.: Preliminary study on thermal features for high temperature gas-cooled reactor gas turbine cycle. Chinese High Technology Letters **12**(9), 91–95 (2002)
15. Wang, J., Gu, Y.: Study on fundamental features of helium turbo machine for high temperature gas-cooled reactor. Chinese J. Nuclear Sci. Eng. **24**(3), 218–223 (2004)
16. Wang, J., Yihua, G.: Parametric studies on different gas turbine cycles for a high temperature gas-cooled reactor. Nucl. Eng. Des. **235**, 1761–1772 (2005)
17. Mian, X., Linsen, L., Gang, Z., et al. Preliminary Transient Analysis for LBE-cooled Fast Reactor BLESS D. In: Proceedings of the ICONE28 international Conference on Nuclear Engineering, August4–6, 2021, Virtual Conference (2021)
18. Wang, Z.G., Zhang, L.Y., Yeoh, E.Y., et al.: Pre-conceptual core design of a LBE-cooled fast reactor (BLESS). In: Proceeding of International Conference on Mathematics & Computational Methods Applied to Nuclear Science & Engineering (M&C 2017), Jeju, Korea, April 16–20, (2017)
19. Wang, Z., Li, L., Yeoh, E.Y., et al. Research on accumulation of high level radioactive waste for a LBE-cooled fast reactor. Atomic Energy Sci. Technol. **51**(12), 2294–2299 (2017)
20. Yeoh, E.Y., Li, L., Chen, X., et al.: Calculation of DPA in the main components of a LBE-cooled fast reactor (BLESS-D). Nuclear Techniques **43**(6), 37–42 (2020)

Study on the 3-D Natural Circulation Characteristics of LFR Under Steady State by Using Ansys Fluent

Jianing Dai, Yulin Yan, Erhao Li, Zhengyu Gong, Ling Zhang, and Zhixing Gu(✉)

College of Nuclear Technology and Automation Engineering, Chengdu University of Technology, Chengdu, Sichuan, China

guzhixing17@cdut.edu.cn

Abstract. As one of the Generation IV reactors, Lead-based Fast Reactor (LFR) has been considered to be great promising owing to its advantages in nuclear safety, sustainable development of nuclear energy and nuclear waste disposal. Owing to the excellent thermal expansion characteristics of Lead-based coolant materials, the primary cooling system of LFR can operate in natural circulation driven mode. The CFD (Computational Fluid Dynamics)-based thermal-hydraulics and safety analyses of nuclear reactors, especially liquid metal pool-type reactors have attracted great attentions in recent years. In this paper, the entire 3-D geometric model of a 10 MWth natural circulation driven LFR primary cooling system was established and simulated by ANSYS Fluent, in which the mesh was partitioned by utilizing structured meshing technology, and the porous medium model was utilized to fine the reactor core simulation. The results showed that the above LFR can operate safely in natural circulation mode, and has excellent natural circulation characteristics for the primary cooling system.

Keywords: Natural Circulation · LFR · CFD · ANSYS Fluent · Structured meshing technology

1 Introduction

As one type of the much anticipated fourth generation reactors, Lead or Lead-bismuth cooled Fast Reactor (LFR) was developed rapidly, its excellent capabilities in miniaturization, modularization and waste transmutation made it has great prospects in district heating, electricity and accelerator driven sub-critical system (ADS). Numerous of research institutes have great enthusiasm for LFRs, and a lot of experimental and teaching LFRs were developed, such as ELSY [1], SSTAR [2], SNCLFR [3], CLEAR-I [4] and MYRRHA [5]. Owing to the large thermal expansion coefficient and stable physical properties of Lead-based materials, LFRs have well natural circulation capability. At the same time, natural circulation capacity also guaranteed forced circulation passive security. Therefore, it is important to study the natural circulation thermal-hydraulic characteristics. With the development of computer hardware and software, computational fluid dynamics (CFD) methods was recognized as an accurate and efficient way,

C. Liu (Ed.): PBNC 2022, SPPHY 283, pp. 930–940, 2023.
https://doi.org/10.1007/978-981-99-1023-6_79

and it had been utilized widely. Nowadays, numerous of thermal-hydraulic programs or codes were developed. On the one hand, 1-D system analysis codes, such as RELAP5 and ATHLET, had been applied extensively. These codes can simulate the entire reactor system rapidly and precisely. However, system analysis codes were not good at reflecting thermal-hydraulic phenomena elaborately in LFRs, such as thermal stratification and flow details. On the other hand, numerous of CFD programs were adopted, the popular ones are ANSYS FLUENT, STAR-CCM . Through the simulation by these programs costs higher than system analysis codes, they can show heat transfer and flow characteristic details. In recent years, CFD programs became more and more extensive in investigating the thermal-hydraulic characteristics of LFRs.

In the past decades, a great deal of 2-D and 3-D simulations were carried out. In 2012, Jin [6] built a quarter 3-D model of CLEAR-I, carried out 3-D simulations by using ANSYS FLUENT to investigate the natural circulation capacity and thermal-hydraulic characteristics of CLEAR-I under steady state and loss of heat sink (LOHS) accident condition. In2013, in order to study the thermal stratification phenomena in CLEAR-I, Zhao [7] established a 2-D axisymmetric model and carried out 2-D simulations by utilizing ANSYS FLUENT, in which different power density was given based on the neutronics analysis. From the result, apparent thermal stratification appeared in the regions of the hot pool below the inlet window of heat exchanger under reactor scram conditions, but it never stopped the natural circulation of the primary circuit. In 2015, a self-developed CFD code namely NTC-2D was utilized by Gu [8] to investigate CLEAR-I under steady state, unprotected loss of heat sink (ULOHS) and unprotected transient overpower (UTOP) conditions. NTC-2D is a 2-D CFD code coupled with neutron transport kinetics model. The results demonstrated that the nice natural circulation capabilities contributed to the accident mitigation process. In 2015, to investigate the natural circulation characteristics of a small modular natural circulation LBE (Lead-Bismuth Eutectic) cooled fast reactor, 3-D simulations was conducted by using ANSYS FLUENT [9], in which a 3-D quarter reactor model was established, and the core power distribution was realized by UDF tools according to the reactor core layout. In 2015, to study the thermal-hydraulic characteristics of SNCLFR under UTOP accident, Chen [10] performed a 2-D simulation by using FLUENT coupled with neutron kinetics and pin thermal transfer models. In 2016, 3-D FLUENT simulations which aimed to evaluate two types of cooling systems, RVACS (reactor vessel air cooling system) and PHXs (primary heat exchangers), was conducted by Wang [11]. The results showed that both two cooling systems have excellent heat exchange capability, while PHXs is stronger than the RVACS. In 2017, Martelli [12] developed a RELAP5-ANSYS FLUENT coupling code to investigate the thermal-hydraulic characteristics of NACIE experimental loop. Natural circulation condition, isothermal gas enhanced circulation and unprotected loss of flow (ULOF) accident were simulated, in which the fuel pin was simulated by FLUENT while other parts were simulated by RELAP5. In 2020, aimed to study the hydraulic phenomena in M2LFR-1000 reactor, Zou [13] performed the simulations of steady state and ULOF accident conditions by using coupling ATHLET with OpenFOAM, in which a 3-D 1/8 model of hot pool was adopted. In 2021, Achuthan [14] studied the natural circulation characteristics of SESAME facility in steady state and several transient conditions.

At present, on the one hand, most researches on thermal-hydraulic characteristics of LFRs in the whole reactor scale did not consider the core power distribution. One the other hand, simulations with considerations of core power distribution were restricted to 2-D or symmetrical 3-D scales. Comparing with the above scales, the whole 3-D simulations can reveal the thermal-hydraulic characteristics more elaborately and precisely. In this paper, a 3-D CFD model of a pool type LFR primary cooling system was established, and the mesh was structured by using block-structed strategy. The steady state simulation which considered power distribution in the core was performed, and the evaluation and discussion of velocity and temperature distribution were also conducted.

2 Calculation Model

2.1 Geometry Model

The geometry model was established based on a typical 10 MW_{th} pool type experimental reactor. To guarantee the simulation efficiency and enhance the visuality of the simulation, some subordinate parts and unnecessary details were removed or simplified. After abundant rational consideration, the concrete components of interior structure in heat exchangers and subassemblies in reactor core was simplified. The CFD simulation geometry includes hot pool, cold pool, core, above core structure (simplified based on control rod driven system) and HXs. The hot pool and the cold pool were separated by a heat barrier. Four inlets and four outlets of each HX were located in the side of them. The geometry model was showed in Fig. 1.

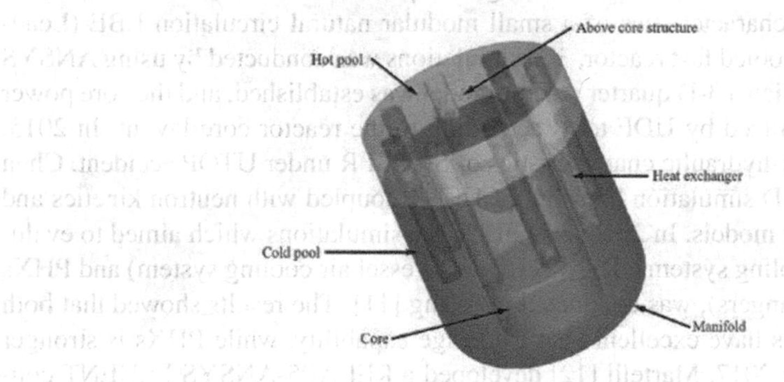

Fig. 1. The calculated geometry model of the primary cooling system

2.2 Mesh Construction and Sensitivity Analysis

To guarantee the accuracy of results and reduce simulation cost, the mesh was established by utilizing block-structed strategy. Specifically, mesh at the hot pool and core were more dense than other parts. To make the simulation results independent with the mesh quantity, the mesh sensitivity analysis was also carried out. Three mesh sizes, including 1.5 million, 5 million and 8 million were selected. Three cases distinguished by different mesh quantity were simulated, and the corresponding temperatures of core outlet were interpolated linearly and plotted in Fig. 2. Obviously, it can be found from this figure that the 1.5 million mesh agreed bad with the two others, while the result of 5 million mesh had tiny difference with the 8 million mesh ones. Therefore, 5 million mesh was selected for the subsequent simulations.

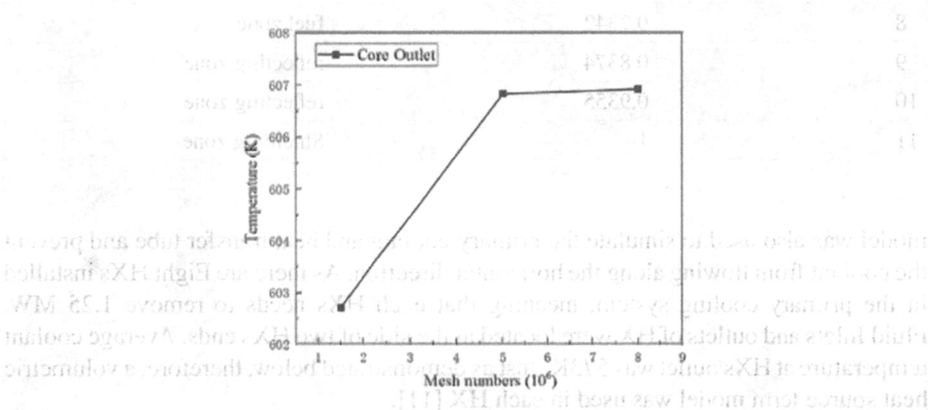

Fig. 2. The temperatures of core outlet for three cases

2.3 Construction of Reactor Core Model

Limited to the simulation cost and complexity, the reactor core was divided into eleven annular sections instead of simulating all the sub-assemblies, which was illustrated in Table 1. Each section contained certain number of sub-assemblies which had similar characteristics, such as power density. Considering the existence of sub-assembly walls, the internal boundary conditions were used to prevent above sections from heat and mass transfer. For each section, the porous medium model was employed in ANSYS FLUENT to simulate the reactor core configuration, in which the volume fraction for structures was set to be 0.7, and the volume fraction of fluid was set to be 0.3. As the detail configuration for sub-assembly was not considered, the viscous resistance in the x and y direction was set to be extremely large to ignore the cross flow of fluid. Moreover, the power density distributions in each annular section was considered by using the UDF techniques of Fluent [15].

2.4 The HXs Model

The HX is one of the significant components of LFR to establish the natural circulation, it determined the temperature level of primary cooling system. In HXs, the porous medium

Table 1. Radial distribution of the reactor core

Layer Number	Radius of loops (m)	Zone
1	0.0761	Neutron source zone
2	0.2014	Neutron source zone
3	0.3318	fuel zone
4	0.4239	fuel zone
5	0.4439	Control rod zone
6	0.5646	fuel zone
7	0.5897	Control rod zone
8	0.7342	fuel zone
9	0.8374	reflecting zone
10	0.9355	reflecting zone
11	1	Shielding zone

model was also used to simulate the primary coolant and heat transfer tube and prevent the coolant from flowing along the horizontal direction. As there are Eight HXs installed in the primary cooling system, meaning that each HXs needs to remove 1.25 MW. Fluid Inlets and outlets of HX were located at the side of two HXs ends. Average coolant temperature at HXs outlet was 573K. Just as demonstrated below, therefore, a volumetric heat source term model was used in each HX [11].

$$Q = \frac{1.25\,\mathrm{MW}}{0.73} \times \frac{T_{LBE} - 573}{119} \tag{1}$$

2.5 Physical Properties

Density difference provided the driving force of natural circulation. The temperature dependent lead properties equations referred to the lead or lead-alloy properties handbook edited by OECD/NEA (2007) was used here. The equation of LBE density was given as Eq. (2). And other 2 key physical properties, conductivity and viscosity were also given. UDF tools were utilized to realize these items.

$$\rho_{LBE} = 11065 - 1.293 * T \tag{2}$$

$$C_{pLBE} = 164.8 - 0.0394 * T + 0.0000125 * T^2 - 4.56 \tag{3}$$

$$\mu_{LBE} = 0.000494 * e^{\frac{754.1}{T}} \tag{4}$$

3 Results and Discussion

Aiming to obtain the velocity field and temperature distribution under full power conditions, the steady state analysis was implemented. For evaluating and discussing the simulation results, a plane y = 0 m, which crossed hot pool, cold pool, two HXs, above core structure and core was established, as shown in Fig. 3.

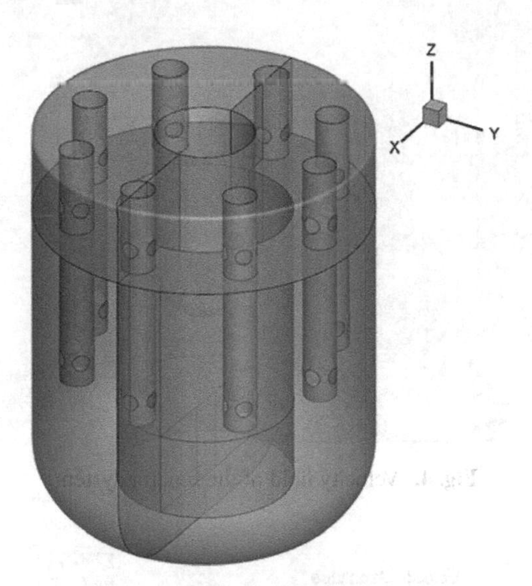

Fig. 3. The schematic diagram of the plane

3.1 Discussion of Velocity Field

The velocity field was given in Fig. 4, it can be seen that a stable natural circulation was established in the primary cooling system. In hot pool, hot coolant raised perpendicularly until arrived at the bottom of above core structure, then it dispersed around, flow upward and entered HXs inlets at last. The larger velocity along wall of above core structure and swirl at the upper part of hot pool promoted coolant mixing. The maximum velocity in the core was 0.127 m/s presented at ring 4 (fuel zone). Figure 5 showed the velocity field approaching a HX. Obvious swirl can be seen near the right HX inlet, and velocities near inlets and outlets of HXs were significantly larger, the maximum velocity in primary cooling system was 0.44060713 m/s, located in a HX. However, velocities at two ends of HXs were slightly small, that may lead poor heat transfer capability in these areas. In cold pool, velocity in upper part of cold pool was relatively small, and occurred swirl that promoted coolant mixing. While at the lower part, owing to the reduction of flowing area at lower part of cold pool, the velocity increased obviously. In addition, it can be seen that velocity vectors above and below the core were extremely intensive, this was due to the slender shape of mesh blocks and big nodes number.

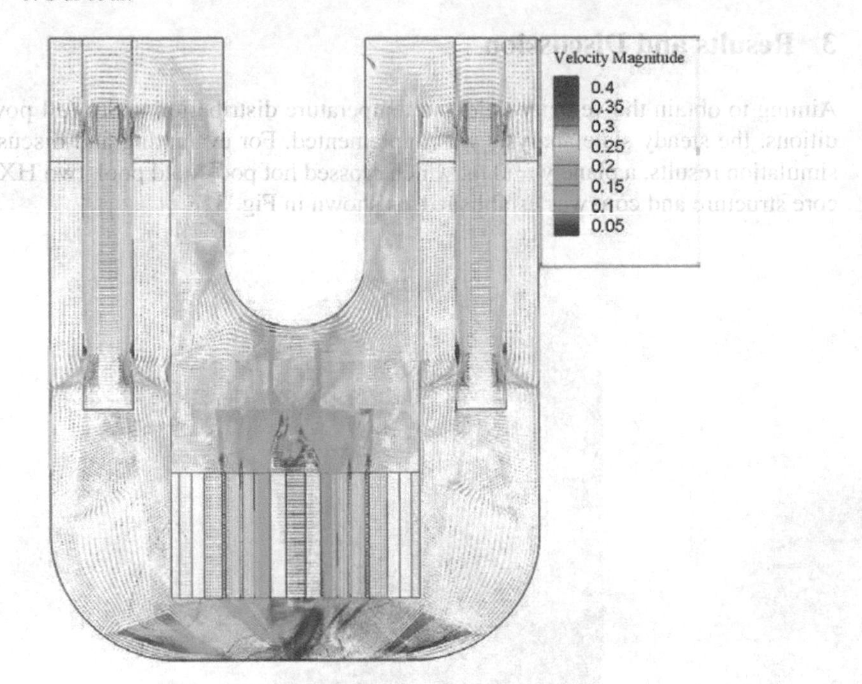

Fig. 4. Velocity field of the reactor system

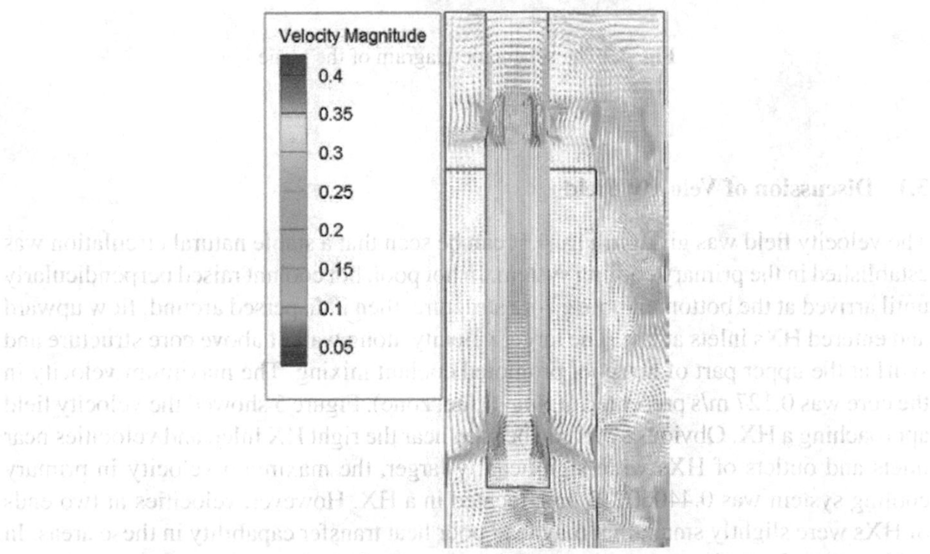

Fig. 5. Velocity field at HXs

3.2 Discussion of Temperature Distribution

Temperature distributions under steady state at y = 0 m was illustrated in Fig. 6. In hot pool, the temperature of coolant reached maximum value of 710.51K at ring 8 (fuel zone). Owing to the heat exchange in reactor core, it can be seen apparent thermal stratification in vertical direction. In radial direction, the figure indicated that temperature at fuel rings was significantly higher than other parts. However, the temperature at lower parts of fuel zones was relatively small, the reason can be concluded combing with Fig. 7. The figure indicated that velocity at fuel zones was much higher than other parts, means that the temperature and density changed most drastically, and caused the mass flow rate at these zones was larger than other parts. In the upper part of hot pool, owing to the reasonable arrangement of subassemblies and hot pool structure, no obvious thermal stratifications can be observed. In cold pool, the temperature distribution was more homogeneous than in hot pool ones, and the maximum temperature was 578.323K occurred at an outlet of HX. However, the symmetry of temperature distribution was imperfect. In the preliminary analysis, this was due to the mesh was not symmetrical entirely.

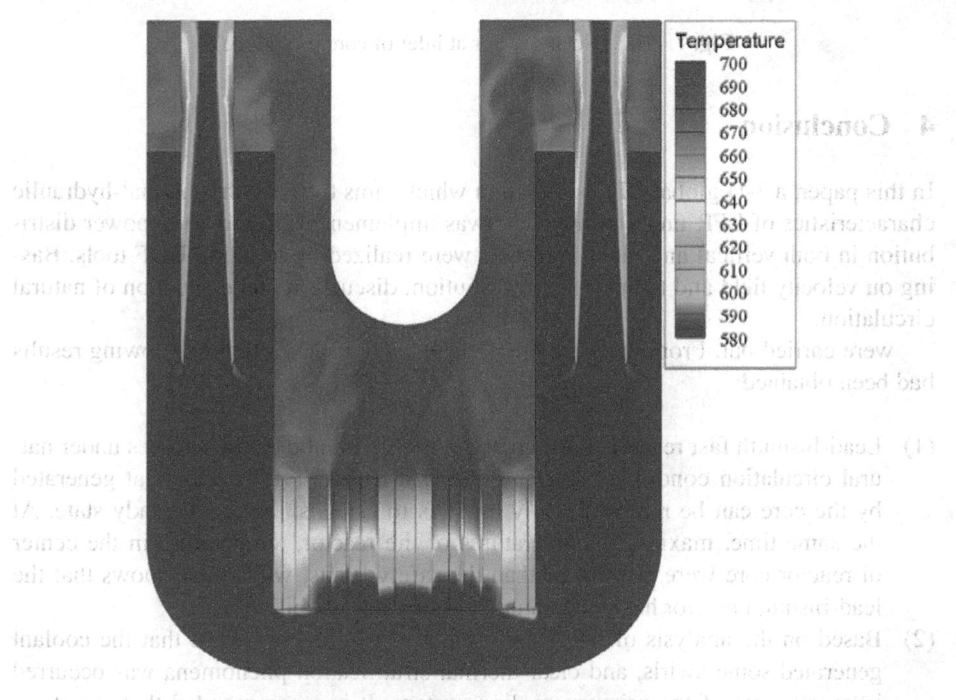

Fig. 6. The temperature distribution of the reactor system

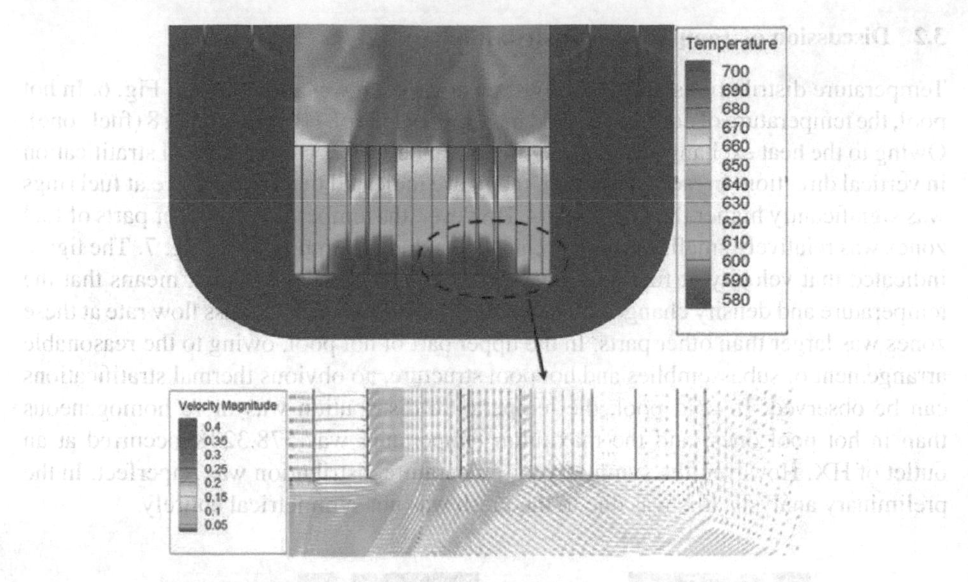

Fig. 7. The velocity vector at inlet of core (right side)

4 Conclusion

In this paper, a 3-D global CFD simulation which aims to study the thermal-hydraulic characteristics of LFR under steady state was implemented. In the core, power distribution in both vertical and radial direction were realized by utilizing UDF tools. Basing on velocity field and temperature distribution, discussion and evaluation of natural circulation.

were carried out. From the simulation results and discussion, the following results had been obtained:

(1) Lead-bismuth fast reactors have good thermal-hydraulic characteristics under natural circulation conditions. Under steady state conditions, decay heat generated by the core can be removed welly by HXs to establish an ideal steady state. At the same time, maximum temperatures of the reactor, temperature in the center of reactor core were also lower than the safety limits, which also shows that the lead-bismuth reactor has good natural circulation capability.
(2) Based on the analysis of the velocity field, it can be concluded that the coolant generated some swirls, and clear thermal stratification phenomena was occurred in many parts of the primary cooling system. It is recommended that structural design optimization or material reinforcement be carried out, such as decrease cavity volume between end face and inlets or outlets of HXs.
(3) Referred to other relevant literature, the simulation results in this paper were accurate generally. However, under the condition that the mesh quality was acceptable, the mesh symmetry and mesh shape may still lead obvious difference of simulation results.

In our future work, a more detailed model will be established, a more regular and symmetrical mesh partition strategy will be utilized. Furthermore, point kinetic model in core will be considered.

References

1. Alemberti, A., et al.: European lead fast reactor—ELSY. Nucl. Eng. Des. **241**(9), 3470–3480 (2011)
2. Smith, C.F., Halsey, W.G., Brown, N.W., Sienicki, J.J., Moisseytsev, A., Wade, D.C.: SSTAR: the US lead-cooled fast reactor (LFR). J. Nucl. Mater. **376**(3), 255–259 (2008)
3. Chen, H., et al.: Conceptual design of a small modular natural circulation lead cooled fast reactor SNCLFR-100. Int. J. Hydrogen Energy **41**(17), 7158–7168 (2016)
4. Wu, Y., Bai, Y., Song, Y., Huang, Q., Zhao, Z., Hu, L.: Development strategy and conceptual design of China Lead-based Research Reactor. Ann. Nucl. Energy **87**, 511–516 (2016)
5. Hamid, A.A., Baeten, P., De Bruyn, D., Fernandez, R.: MYRRHA – a multi-purpose fast spectrum research reactor. Energy Conv. Manag. **63**, 4–10 (2012)
6. Jin, M., Chen, Z., Chen, H., Zhou, T., Bai, Y.: Natural circulation characteristics of China lead alloy cooled research. In: The 9th International Topical Meeting on Nuclear Thermal-Hydraulics, Operation and Safety (NUTHOS-9), Kaohsiung, Taiwan (2012)
7. Zhao, P., Chen, Z., Zhou, T., Chen, H.: CFD analysis of thermal stratification of China lead alloy cooled research reactor (CLEAR-I). In: Proceedings of the 2013 21st International Conference on Nuclear Engineering (2013)
8. Gu, Z., Wang, G., Wang, Z., Jin, M., Wu, Y.: Transient analyses on loss of heat sink and overpower transient of natural circulation LBE-cooled fast reactor. Prog. Nucl. Energy **81**, 60–66 (2015)
9. Pengcheng, Z., Shuzhou, L., Zhao, C., Jie, Z., Hongli, C.: Natural circulation characteristics analysis of a small modular natural circulation lead–bismuth eutectic cooled fast reactor. Prog. Nucl. Energy **83**, 220–228 (2015)
10. Chen, Z., Chen, X.-N., Rineiski, A., Zhao, P., Chen, H.: Coupling a CFD code with neutron kinetics and pin thermal models for nuclear reactor safety analyses. Ann. Nucl. Energy **83**, 41–49 (2015)
11. Wang, X., Jin, M., Wu, G., Song, Y., Li, Y., Bai, Y.: Natural circulation characteristics of lead-based reactor under long-term decay heat removal. Prog. Nucl. Energy **90**, 11–18 (2016)
12. Zhao, P., Shi, K., Li, S., Feng, J., Chen, H.: CFD analysis of the primary cooling system for the small modular natural circulation lead cooled fast reactor SNRLFR-100. Sci. Technol. Nucl. Install. **2016**, 1–12 (2016)
13. Zou, Z., Shen, C., Zhang, X., Wang, S., Chen, H.: 3D thermal hydraulic characteristics analysis of pool-type upper plenum for lead-cooled fast reactor with multi-scale coupling program. Nucl. Eng. Design **370**, 110892 (2020)
14. Achuthan, N., Melichar, T., Profir, M., Moreau, V.: Computational fluid dynamics modelling of lead natural convection and solidification in a pool type geometry. Nucl. Eng. Des. **376**, 111104 (2021)
15. Incorporated, A.: ANSYS Fluent UDF Manual15 (2013)

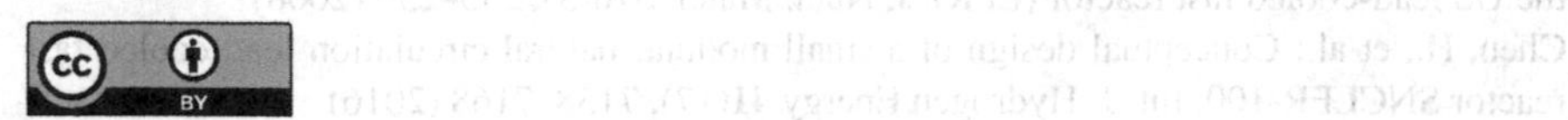

Multi-physics Coupling Analyses of Nuclear Thermal Propulsion Reactor

Wenbin Han[1,2], Zechuan Guan[1,3], Shanfang Huang[1(✉)], and Jian Deng[2]

[1] Department of Engineering Physics, Tsinghua University, Beijing, China
sfhuang@mail.tsinghua.edu.cn

[2] Science and Technology on Reactor System Design Technology Laboratory, Nuclear Power Institute of China, Chengdu, China

[3] China Institute of Atomic Energy, Beijing, China

Abstract. Nuclear thermal propulsion (NTP) reactors have high-temperature solid-state characteristics and significant thermal expansion, which therefore require multi-physics coupling analyses. In this paper, the framework of Neutronics, Thermal-Hydraulics and Mechanics coupling (N/T-H/M) of nuclear thermal propulsion reactor is developed, and the typical reactor XE-2 is analyzed with this method. The results show that the N/T-H/M coupling will bring -1049 pcm negative reactivity, of which the thermal expansion effect accounts for 22%, indicating that the nuclear thermal propulsion reactor has a certain capacity for self-regulation. However, thermal expansion will lead to 0.88 mm peak deformation and 233 MPa peak stress, which will severely threaten the mechanical tolerance of the materials. Therefore, there is a trade-off between the advantages and disadvantages of the high-temperature solid-state core while designing NTP reactors.

Keywords: Nuclear Thermal Propulsion Reactor · Neutronics and Thermal-Hydraulics Coupling · Neutronics · Thermal-Hydraulics and Mechanics Coupling

1 Introduction

Deep space exploration and interstellar travel are the persistent pursuits of humankind. The propulsion system is the key to further investigation of the universe. Currently, the chemical propulsion system have been widely used in rockets and spacecraft, but its low energy density and small specific impulse makes it difficult to be applied in deep space exploration and interstellar navigation. Nuclear Thermal Propulsion (NTP) uses nuclear fission energy to heat the working medium flowing through the core to high temperature, and then the hot propellant flows through the nozzle, expands and accelerates to provide thrust for the rockets. NTP has advantages of high specific impulse, high thrust, and long service life, therefore is the preferred propulsion choice for deep space exploration.

The concept of nuclear thermal propulsion could date back to the "space race" between the United States and the Soviet Union in the 1950s [1]. From the 1950s to 1970s, the US successively carried out ROVER and NERVA programs. The US built

C. Liu (Ed.): PBNC 2022, SPPHY 283, pp. 941–954, 2023.
https://doi.org/10.1007/978-981-99-1023-6_80

and tested more than 20 NTP reactors, including KIWI, Phoebus, Pewee, NRX, and XE series, generating more than 100,000 technical reports and memos [2]. The achievements in ROVER/NERVA program laid the foundation for later NTP research and designs.

Although there are also liquid and gaseous designs [3], the solid-core nuclear thermal propulsion reactor is the most general design, of which the typical structure is shown in Fig. 1. The specific impulse I_{sp} is a key performance of a rocket engine, which is determined by the operation conditions, propellant properties and exhaust nozzle geometry according to Eq. (1).

$$I_{sp} = \frac{1}{g}\sqrt{\frac{2\gamma}{\gamma - 1}\frac{RT}{M}\left(1 - \left(\frac{P_e}{P_i}\right)^{\frac{\gamma-1}{\gamma}}\right)} \tag{1}$$

where i and e represent the inlet and exit conditions, γ is the specific heat ratio, R is the universal gas constant, T is the inlet temperature of the nozzle, P is the pressure of the propellant, g is the gravitational acceleration, and M is the molecular weight of the propellant.

To increase the specific impulse as much as possible, a NTP reactor needs high operation temperatures (usually ~ 3000 K in solid-state core) and small propellant molecular weight (usually hydrogen). Therefore, a typical NTP reactor is actually a high-temperature hydrogen-cooled reactor.

Compared with traditional light water reactors, NTP reactors have mainly the following features:

(1) Solid-state core design. Except for the hydrogen coolant, the body of the reactor is in a solid state, including fuel and moderator elements.
(2) High operation temperature. The operation temperature can reach 3000 K.
(3) Hard neutron energy spectrum. To reduce the size of the reactor, the loading of the moderator is limited, thus the reactor is generally designed with a hard neutron energy spectrum.

The high-temperature solid core can experience significant thermal expansion, which will bring in negative reactivity feedback and thermal stress. The hard spectrum design and small core size lead to the tight coupling of different physical fields.

In light water reactors, the effect of thermal expansion is usually negligible, only the neutronics and thermal-hydraulics (N/T-H) coupling is considered. However, for high-temperature solid-state reactors, such as heat pipe cooled reactors and NTP reactors, thermal expansion is non-negligible. Previous studies show that in the heat pipe cooled reactor KRUSTY, the thermal expansion accounts for ~ 90% of the reactivity feedback [5]. Therefore, the neutronics, thermal-hydraulics and mechanics (N/T-H/M) coupling analyses are needed for core design and safety analysis in high-temperature solid-state reactors.

For the heat pipe cooled reactor, the coolant is replaced by heat pipes, so it should consider neutronics, thermal-mechanics and heat pipe (N/T-M/HP) coupling analysis [6, 7]. However, for NTP reactors, the operation temperature is much higher, and there is a

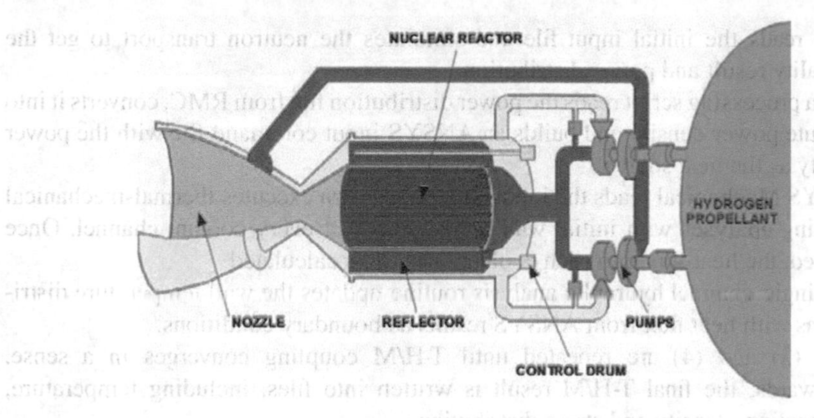

Fig. 1. Schematics of a typical NTP system [4]

hydraulic effect of hydrogen propellant. Therefore, a new coupling code is needed for NTP reactor analysis.

In this work, a neutronics, thermal-hydraulics and mechanics coupling method is developed to analyze the NTP reactor. The Reactor Monte Carlo code (RMC) [8] and the commercial finite element analysis software ANSYS Mechanical are used for the neutronic and thermal-mechanical analyses, respectively. A single channel code is developed for the hydrogen flow and heat transfer analysis.

2 Methodology

2.1 Computational Methods

RMC (Reactor Monte Carlo code) is a Monte Carlo neutron and photon transport code developed by the Department of Engineering Physics at Tsinghua University [8], which has been validated for criticality calculation, burnup calculation, neutron and photon coupled transport calculation, full-core refueling simulation, randomly dispersed fuel calculation, and neutronic-thermal-mechanical coupling analysis [6, 7, 9–13].

ANSYS Mechanical is a commercial finite element mechanical analysis software that includes structural mechanics analysis, thermal analysis, and coupling analysis. APDL, also known as ANSYS parametric design language, enables users to organize ANSYS commands and write parametric user-defined programs.

A single channel code is developed for the hydrogen hydraulic analysis, which will be described detailly in Sect. 2.5.

2.2 Coupled N/T-H/M Framework

The neutronic, thermal-hydraulic and mechanical coupling framework is shown in Fig. 2. An outer iteration strategy is used to schedule different codes with a main scheduling routine. The Picard iteration method is used to couple the various physical fields. The main pipeline in the iteration is as follows:

(1) RMC reads the initial input file and simulates the neutron transport to get the criticality result and power distribution.
(2) A data processing script reads the power distribution file from RMC, converts it into absolute power density and builds an ANSYS input command file with the power density as the heat source.
(3) ANSYS Mechanical reads the input commands then executes thermal-mechanical coupling analyses with initial wall temperatures of every coolant channel. Once finished, the heat flux into each coolant channel is calculated.
(4) The single-channel hydraulic analysis routine updates the wall temperature distributions with heat flux from ANSYS results as boundary conditions.
(5) Steps (3) and (4) are repeated until T-H/M coupling converges in a sense. Afterwards, the final T-H/M result is written into files, including temperature, deformation, density and stress distributions.
(6) A data processing script reads the T-H/M result and builds new geometry, and sets new densities and new temperatures for the RMC input file.
(7) RMC reads the new input file, updates the cross-sections, geometry and material properties, and starts new neutronic calculations.
(8) Steps (1)–(7) are repeated until the N/T-H/M coupling convergence criteria are satisfied.

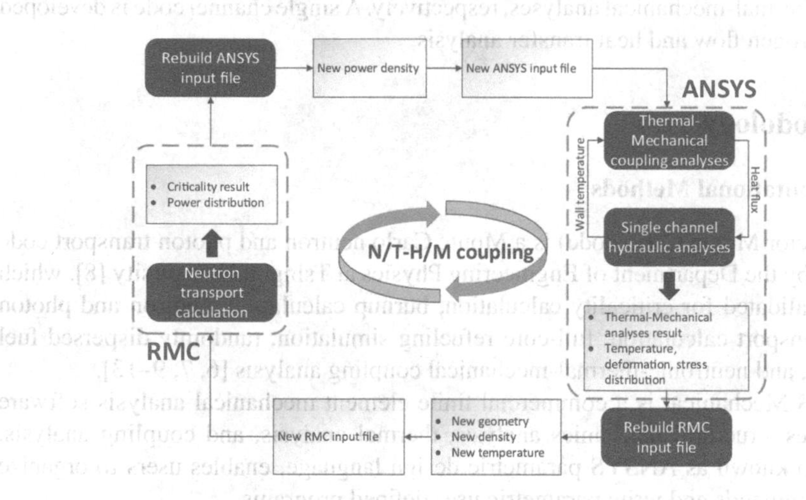

Fig. 2. Coupled N/T-H/M framework

2.3 Neutronic Model

To improve the computing efficiency, a 1/6 model of the XE-2 reactor core is built with RMC constructive solid geometry as shown in Fig. 3. The geometry, materials and nuclide compositions mainly refer to the XE-2 reactor design manual [14]. The fuel element is a 19-channel hexagonal prism dispersed with UC_2 particles. To tally the power distribution and receive T-H/M feedbacks, the active core is axially divided into 26 segments.

In criticality calculations, the simulations use 100,000 particles per cycle with 30 inactive cycles and 270 active cycles, resulting in a standard error of k_{eff} less than 0.0002.

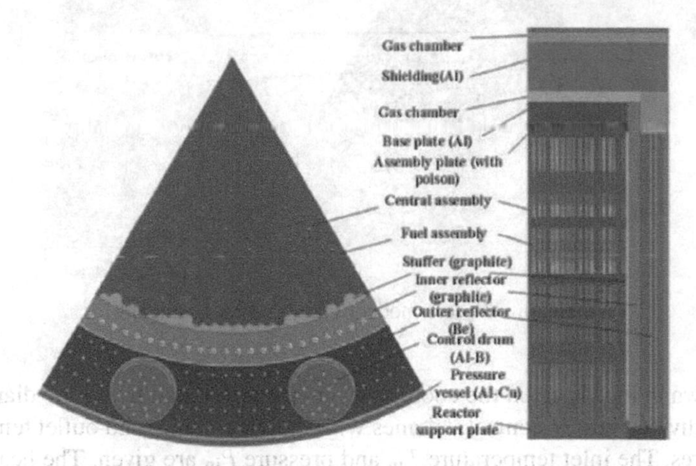

Fig. 3. 1/6 core model of RMC

2.4 Thermal-Mechanical Model

A 1/6 core finite element model is constructed with ANSYS Mechanical APDL as shown in Fig. 4. The thermal-mechanical properties refer to the NERVA material manual [15]. The triangular mesh is used in the radial direction and then sweeping in the axial direction. For thermal boundary conditions, the outer boundary is set to be adiabatic and the walls of each channel are set to be in constant temperatures given by the single-channel analysis code, which will be updated every T-M/H iteration. For mechanical boundary conditions, the center line and the bottom are set to be fixed, while the others can expand freely.

2.5 Hydraulic Model

There are numerous coolant channels in the XE-2 reactor. In the 1/6 model, 500 ~ 600 coolant channels can transfer heat with the fuel. Because the channels are independent without any coolant mixing, they can be treated one by one with single channel analysis. Therefore, a single-channel hydraulic analysis code for hydrogen is developed with APDL. The main correlations used in this algorithm comes from the thermal hydraulic calculation program ELM [16] developed by NASA for nuclear thermal propulsion reactors.

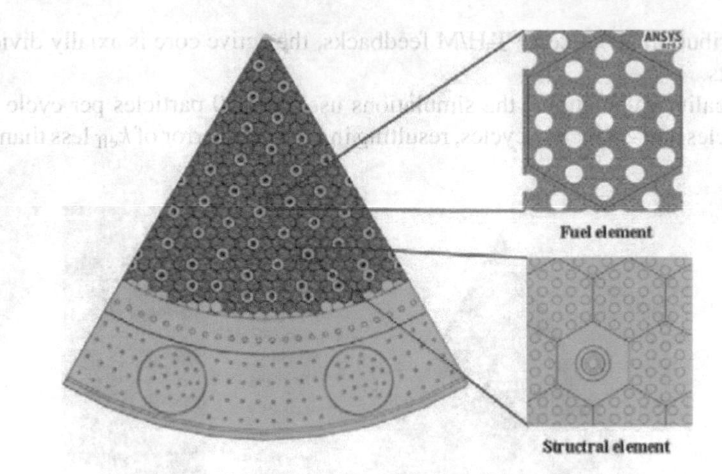

Fig. 4. 1/6 core model of ANSYS Mechanical

As shown in Fig. 5, given the cooling channel length L and hydraulic diameter D, a channel is divided into N control volumes with individual inlet and outlet temperatures and pressures. The inlet temperature T_{in} and pressure P_{in} are given. The heat flux into each control volume Q_1 ~ Q_{N} is known from ANSYS T-M coupling results. The flow rate is known as W_{CH}. The outlet pressure of the nth control volume is assumed as Eq. (2).

$$P_n = \frac{P_N - P_{n-1}}{N - n + 1} + P_{n-1} \tag{2}$$

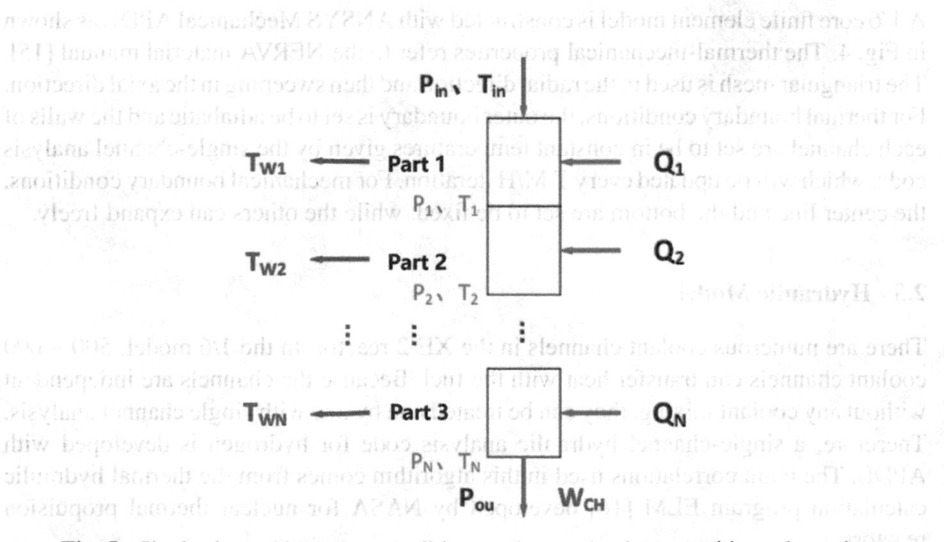

Fig. 5. Single channel boundary conditions and control volume partition schematics

From the inlet control volume on, the calculation proceeds in sequence, and the outlet temperature and pressure obtained from the previous control volume are used as the inlet condition of the next control volume. When the calculation arrives at the last volume, according to the known outlet pressure, the outlet temperature and mass flow are calculated. The iteration continues until convergence. The detailed process is as follows:

(1) The density of hydrogen ρ_{n-1} is looked up in the hydrogen property table according to P_{n-1} and T_{n-1}.
(2) H_n and ρ_n are obtained iteratively by trial and error with Eq. (3) and referring to the hydrogen physical properties look-up table.

$$H_{n-1}+\frac{\left(\frac{4W_{CH}}{\pi D^2\rho_{n-1}}\right)^2}{2}+\frac{Q_n}{W_{ch}}=H_n+\frac{\left(\frac{4W_{CH}}{\pi D^2\rho_n}\right)^2}{2}. \tag{3}$$

(3) Determine the qualitative temperature and pressure:

$$T_b=\frac{T_{n-1}+T_n}{2} \tag{4}$$

$$P_b=\frac{P_{n-1}+P_n}{2}. \tag{5}$$

(4) According to T_b and P_b, look up the hydrogen physical property table and get μ_b, ρ_b, C_{pb}, C_{vb}, λ_b.
(5) Calculate the Reynolds number and Prandtl number:

$$Re_b=\frac{v_b D\rho_b}{\mu_b}=\frac{4W_{CH}}{\pi D\mu_b} \tag{6}$$

$$Pr_b=\frac{C_{pb}\mu_b}{\lambda_b} \tag{7}$$

(6) Find the fluid resistance coefficient:

$$f=\left(\frac{T_b}{T_{wn}}\right)^{0.5}\left(0.0014+\frac{0.125}{Re_b^{0.32}}\right)(1.85\times 10^{-5}Re_b+0.73)^{0.5} \tag{8}$$

(7) If it is the 1st ~ (N-1)th control volume, update the outlet pressure:

$$P_n=P_{n-1}-\left(\frac{4W_{CH}}{\pi D^2}\right)^2\frac{fL}{ND}\left(\frac{1}{\rho_{n-1}}+\frac{1}{\rho_n}\right)-\left(\frac{4W_{CH}}{\pi D^2}\right)^2\left(\frac{1}{\rho_n}-\frac{1}{\rho_{n-1}}\right) \tag{9}$$

If it is the N th control volume, skip.

(8) Solve the Nusselt number Nu_D with Eq. (10)–(14).

$$E_t=(1.82\times\log_{10}Re_b-1.64)^2 \tag{10}$$

$$K_1 = 1 + 3.4E_t \tag{11}$$

$$K_2 = 11.7 + 1.8 \times Pr_b^{-\frac{1}{3}} \tag{12}$$

$$Nu_0 = E_t Re_b Pr_b / 8 / \left(K_1 + K_2 \left(\frac{E_t}{8} \right)^{0.5} \left(Pr_b^{\frac{2}{3}} - 1 \right) \right) \tag{13}$$

$$Nu_D = Nu_0 \left(\frac{T_w}{T_b} \right)^{-0.31 \times \ln(T_w - T_b) - 0.36} \tag{14}$$

(9) Solve the convection transfer coefficient:

$$h = \frac{Nu_D \lambda_b}{D} \tag{15}$$

(10) Update wall temperature T_{wn} according to Eq. (16)–(19):

$$T_{AW} = T_b \times \left(1 + 0.5 \times Pr_b^{\frac{1}{3}} (\gamma_b - 1) M_b^2 \right) \tag{16}$$

$$\gamma_b = \frac{C_{pb}}{C_{vb}} \tag{17}$$

$$M_b = \frac{v_b}{(\gamma_b R T_b)^{0.5}} \tag{18}$$

$$T_{Wn} = \frac{NQ}{h \pi DL} + T_{AW} \tag{19}$$

(11) Iterate and update the channel mass flow:

$$W_{CH} = \frac{\pi D^2}{4} \times \sqrt{\frac{P_{n-1} - P_n}{\frac{fL}{2D\rho_b} + \frac{1}{\rho_n} - \frac{1}{\rho_{n-1}}}} \tag{20}$$

3 Performance Analyses

3.1 Neutronic Performance

To study the significant influence of thermal expansion, the results of N/T-H coupling and N/T-H/M coupling are compared. The variation of k_{eff} with iterations is shown in Fig. 6. There is a sharp decrease of k_{eff} in both N/T-H coupling and N/T-H/M coupling. The k_{eff} without feedback is 1.00792. After several iterations, the k_{eff} converges. The k_{eff} is 1.00035 with N/T-H coupling, and 0.99737 with N/T-H/M coupling. The reactivity feedbacks are summarized in Table 1. Both the doppler effect and thermal expansion effect can lead to a negative reactivity feedback, in which the thermal expansion counts for 22%. This result reveals the fact that in high-temperature solid state reactors, thermal expansion is a significant source of negative reactivity feedback, which enables the reactor certain self-regulating characteristics.

Table 1. Reactivity feedbacks in N/T-H/M coupling

	Doppler effect	Expansion effect	Total effect	Expansion fraction
Reactivity feedback (pcm)	−821	−228	−1049	22%

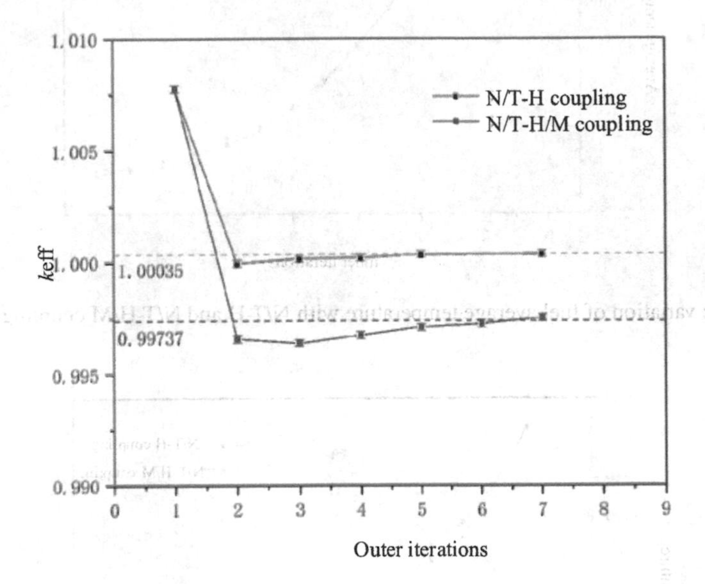

Fig. 6. The variation of k_{eff} with N/T-H and N/T-H/M coupling iterations

3.2 Thermal Performance

The fuel average and peak temperature converge during the coupling iterations as shown in Fig. 7 and Fig. 8. The temperatures gradually decrease from the initial value and tend to be stable. After convergence, the difference in fuel average temperature between N/T-H/M and N/T-H is very small, while the difference in peak temperature is much more. The peak temperature decreases by 164 K with N/T-H coupling, and 175 K with N/T-H/M coupling. The final temperature distribution is shown in Fig. 9.

3.3 Mechanical Performance

The thermal expansion will lead to displacement and thermal stress. The linear expansion variation with iterations is shown in Fig. 10. The final average linear expansion reaches 0.293%. The displacement and stress distributions are shown in Fig. 11 and Fig. 12. The maximum displacement is 0.882 mm. The stress peaks near the wall of coolant channels with the peaking stress as 233 MPa, which will threaten the strength of structural materials.

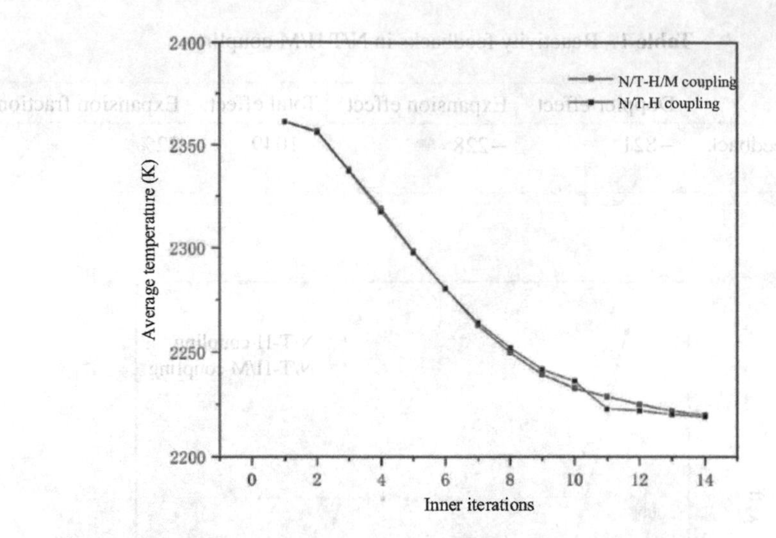

Fig. 7. The variation of fuel average temperature with N/T-H and N/T-H/M coupling iterations

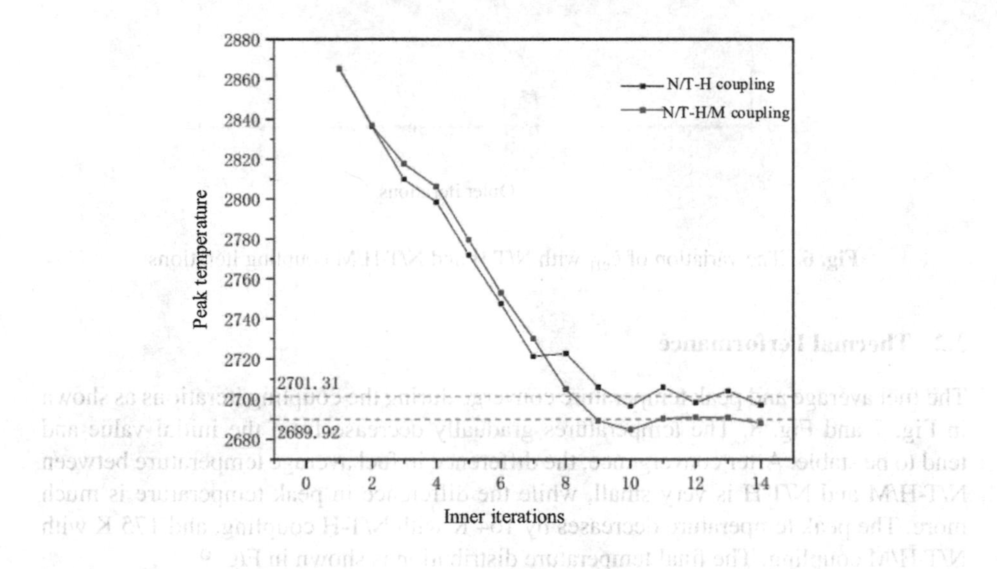

Fig. 8. The variation of fuel peak temperature with N/T-H and N/T-H/M coupling iterations

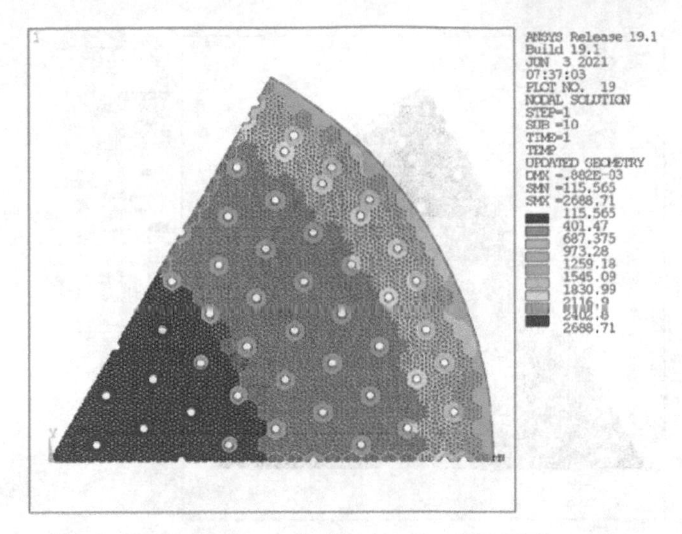

Fig. 9. The temperature distribution after N/T-H/M coupling

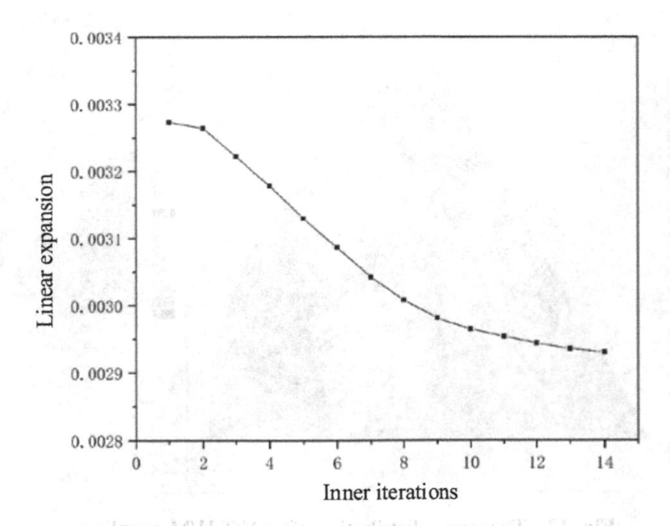

Fig. 10. The variation of expansion with N/T-H/M iterations

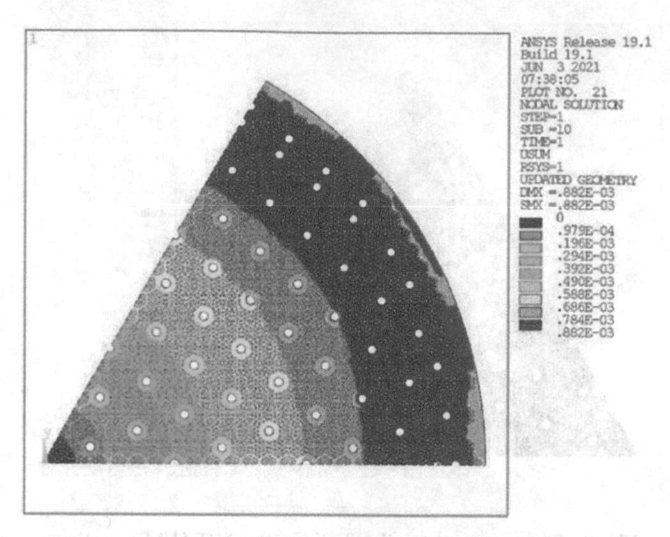

Fig. 11. The displacement distribution after N/T-H/M coupling

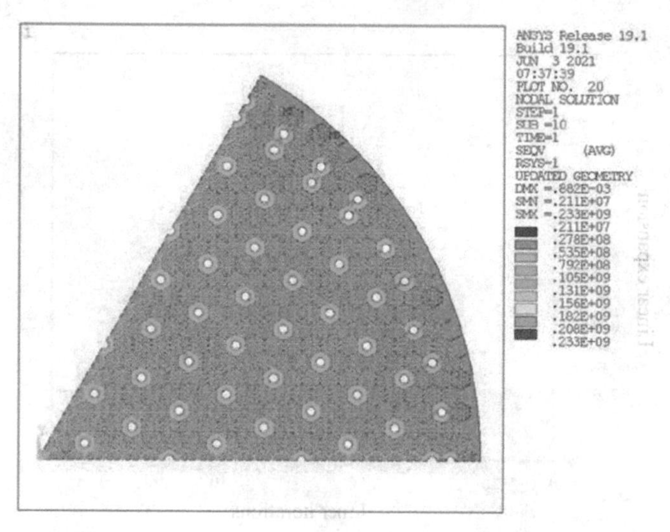

Fig. 12. The stress distribution after N/T-H/M coupling

4 Conclusions

In this work, a neutronic, thermal-hydraulic and mechanical coupling framework is developed for typical thermal propulsion reactors with RMC and ANSYS Mechanical. The XE-2 reactor is analyzed with this method. The effects of multi-physics coupling on analysis results are summarized in Table 2. The results show that the N/T-H/M coupling will bring −1049 pcm negative reactivity, of which the thermal expansion effect accounts for 22%, which indicates that the nuclear thermal propulsion reactor has a certain capacity of self-regulation. However, thermal expansion will lead to 0.88 mm peak deformation and 233 MPa peak stress, which will severely threaten the mechanical tolerance of the materials. Therefore, there is a trade-off between the advantages and disadvantages of the high-temperature solid-state core while designing NTP reactors.

Table 2. Effects of multi-physics coupling on analysis results

	Without coupling	With N/T-H coupling	With N/T-H/M coupling
k_{eff}	1.007918	1.00035	0.99737
Peaking temperature (K)	2865	2701	2690
Peaking stress (MPa)	–	–	233
Peaking displacement (mm)	–	–	0.88

References

1. Zhang, W., et al.: Technical research on nuclear thermal propulsion ground tests. Astron. Syst. Eng. Technology **2**, 10 (2019)
2. Robbins, W.: An historical perspective of the NERVA nuclear rocket engine technology program. In: Conference on Advanced SEI Technologies (1991)
3. Gabrielli, R.A., Herdrich, G.: Review of nuclear thermal propulsion systems. Prog. Aeros. Sci. **79**, 92–113 (2015)
4. Houts, M.: High Performance Nuclear Thermal Propulsion NASA (2020)
5. Poston, D.I., et al.: KRUSTY reactor design. Nucl. Technol. **206**(sup1), S13–S30 (2020)
6. Ma, Y., et al.: Neutronic and thermal-mechanical coupling analyses in a solid-state reactor using Monte Carlo and finite element methods. Ann. Nucl. Energy **151**, 10792 (2021)
7. Ma, Y., et al.: Coupled neutronic, thermal-mechanical and heat pipe analysis of a heat pipe cooled reactor. Nucl. Eng. Des. **384**, 111473 (2021)
8. Wang, K., et al.: RMC–A Monte Carlo code for reactor core analysis. Ann. Nucl. Energy **82**, 121–129 (2015)
9. Wang, K., et al.: Analysis of BEAVRS two-cycle benchmark using RMC based on full core detailed model. Prog. Nucl. Energy **98**, 301–312 (2017)
10. Liu, S., et al.: BEAVRS full core burnup calculation in hot full power condition by RMC code. Ann. Nucl. Energy **101**, 434–446 (2017)
11. Ma, Y., et al.: RMC/CTF multiphysics solutions to VERA core physics benchmark problem 9. Ann. Nucl. Energy **133**, 837–852 (2019)
12. Liu, S., et al.: Random geometry capability in RMC code for explicit analysis of polytype particle/pebble and applications to HTR-10 benchmark. Ann. Nucl. Energy **111**, 41–49 (2018)
13. Wu, Y., et al.: Monte Carlo simulation of dispersed coated particles in accident tolerant fuel for innovative nuclear reactors. Int. J. Energy Res. **45**(8), 12110–12123 (2021)
14. Hill, D.J.: XE-2 nuclear subsystem thermal and nuclear design data book. No. WANL-TME-1805. Westinghouse Electric Corp., Pittsburgh, Pa.(USA). Astronuclear Lab (1975)
15. Hladdun, K.R., et al.: Materials properties data book. Volume II. Ferrous alloys. Environ. Toxicol. Chem. **32**, 2584–2592 (2013)
16. Walton, J.T.: Program ELM: a tool for rapid thermal-hydraulic analysis of solid-core nuclear rocket fuel elements (1992)

Effects of Inlet Conditions on the Two-Phase Flow Water Hammer Transients in Elastic Tube

Zixiang Zhao[1], Zhongdi Duan[1](✉), Hongxiang Xue[1], Yuchao Yuan[1], and Shiwen Liu[2]

[1] State Key Laboratory of Ocean Engineering, Shanghai Jiao Tong University, Shanghai, China
duanzhongdi@sjtu.edu.cn

[2] Science and Technology on Reactor System Design Technology Laboratory, Nuclear Power Institute of China, Chengdu, China

Abstract. Two-phase flow water hammer events occur in the pipelines of the nuclear power systems and lead to transient and violent pressure shock to tube structures. For the sake of operation safety, the occurrence and severity of the two-phase water hammer should be carefully assessed. This paper presents a parameter analysis of the inlet conditions on the two-phase flow water hammer transients, with considering the elastic effect of the tube walls. A numerical model is established for the vapor-liquid two-phase flow based on the two-fluid six-equation modelling approach, with incorporating correlations and criterions for two-phase flow regime, interfacial interactions and heat transfer. The governing equations are transformed to matrix form expressed by characteristic variables, and solved using the splitting operator method and the total variation diminishing scheme. The accuracy of the model is verified against the experimental data in open literature. Then, the model is applied to investigate the effect of inlet velocity and inlet water temperature on the two-phase flow water hammer transients. The simulation results show that the increase of inlet velocity increases the pressure peak values and brings forward the onset of water hammer, and the increase of inlet temperature decreases the pressure shock. A comparison of the water hammer results between the elastic tube and rigid tube is further presented, and the effect of the elastic modulus on the water hammer is analyzed. The results also show that the pressure peak is largely affected by the tube diameter.

Keywords: nuclear power plants · condensation induced water hammer · fluid-structure-interaction · direct contact condensation · shock wave

1 Introduction

The passive heat removal system is one of the important systems to ensure the safety of ocean nuclear power plants. While using seawater as the infinite heat sink to cool the reactor, it also suffers the risk of reverse flow of subcooled seawater into the steam tube, which gives rise to the direct contact between the steam and subcooled water, and triggers condensation-induced water hammer (CIWH), resulting in high peak pressure pulses and consequently pipeline damage. CIWH is a common phenomenon in nuclear

C. Liu (Ed.): PBNC 2022, SPPHY 283, pp. 955–972, 2023.
https://doi.org/10.1007/978-981-99-1023-6_81

power plants, steam power plants, and other energy power plants, and has received widespread attention from domestic and international research institutions and scholars.

Regarding CFD numerical simulation methods for CIWH, there are three main types of representative models: one is the use of surface renewal theory to calculate condensation heat transfer, such as Tiselj [1], Strubelj [2], and Ceuca [3], which established a large interface model for direct contact condensation based on surface renewal theory, which considered the turbulence effect between the vapor-liquid phase and showed that the model can predict the transient temperature change better by comparing with the PMK-2 experimental data. Second, the temperature gradient of the vapor-liquid phase is calculated directly, and the vapor condensation is calculated based on the thermal conductivity heat transfer [4], and this method can also predict the fluid temperature variation of the direct contact condensation process better. Third, a traditional evaporative condensation model is used for the calculation, such as Pham et al. [5], who obtained the CIWH pressure wave at the MPa level through Lee model simulations. In addition, the two-fluid model used by Höhne et al. [6] and the Large Eddy Simulation (LES) method used by Li et al. [7] also allow the simulation of direct contact condensation processes with a good agreement with experiments. These method give visual access to two-phase flow and phase interface information, however, most of the results obtained by commercial CFD software are not very satisfactory in the calculation of CIWH pressure waves, and it is difficult to capture the peak pressure.

For the prediction of condensation-induced water hammer pressure waves, several research institutions and scholars have developed computational models specifically for CIWH, and representative ones include the WAHA3 numerical computational program developed by Barna [8] and Tiselj [9] based on the WAHALoads project, as well as the computational program ATHLET [10]. The above models use a one-dimensional, two-fluid, six-equation model, and a surge capture algorithm for pressure wave prediction of CIWH. Hibiki [11] also computationally captured the transient process of CIWH based on the RELAP5 code, but the validity of this approach is controversial. Milivojevic et al. [12] developed a one-dimensional, three-equation single fluid model based on the HEM (Homogeneous Equilibrium Model) model that captured the steam-filled CIWH phenomenon in a vertical pipe. However, this method has not been applied in CIWH calculations for stratified flow in horizontal pipes because it cannot characterize the heat mass transfer relationship between the gas and liquid phases.

2 Mathematical Model Development

2.1 Basic Control Equations

For the vapor-liquid two-phase flow model, the basic idea for establishing the control equation is that regard the two-phase mixed flow as two single-phase flow, and their characteristics in terms of continuity, momentum and energy are examined separately, and then the inter-phase effects of mass exchange, momentum exchange and energy exchange are taken into account to form the basic control equation of the two-phase flow.

The mass, momentum and energy balance equations for both phases are the following:

$$\frac{\partial A\alpha_l\rho_l}{\partial t}+\frac{\partial A\alpha_l\rho_l v_l}{\partial x}=A\Gamma_l \tag{1}$$

$$\frac{\partial A\alpha_v\rho_v}{\partial t}+\frac{\partial A\alpha_v\rho_v v_v}{\partial x}=A\Gamma_v \tag{2}$$

$$\frac{\partial \alpha_l\rho_l}{\partial t}+\alpha_l\rho_l K\frac{\partial P}{\partial t}+\frac{\partial \alpha_l\rho_l v_l}{\partial x}+\alpha_l\rho_l v_l K\frac{\partial P}{\partial x}=\Gamma_l \tag{3}$$

$$\frac{\partial \alpha_v\rho_v}{\partial t}+\alpha_v\rho_v K\frac{\partial P}{\partial t}+\frac{\partial \alpha_v\rho_v v_v}{\partial x}+\alpha_v\rho_v v_v K\frac{\partial P}{\partial x}=\Gamma_v \tag{4}$$

$$\begin{aligned}\alpha_l\rho_l\frac{\partial v_l}{\partial t}+\alpha_l\frac{\partial P}{\partial x}-P_i\frac{\partial \alpha_v}{\partial x}+\alpha_l\rho_l v_l\frac{\partial v_l}{\partial x}-VM\\ =F_D+\Gamma_l(v_i-v_l)+\alpha_l\rho_l g\sin\theta-F_{l,\ \text{wall}}\end{aligned} \tag{5}$$

$$\begin{aligned}\alpha_v\rho_v\frac{\partial v_v}{\partial t}+\alpha_v\frac{\partial P}{\partial x}+P_i\frac{\partial \alpha_v}{\partial x}+\alpha_v\rho_v v_v\frac{\partial v_v}{\partial x}+VM\\ =-F_D+\Gamma_v(v_i-v_v)+\alpha_v\rho_v g\sin\theta-F_{v,\ \text{wall}}\end{aligned} \tag{6}$$

where index l refers to the liquid phase and index v to the vapor phase. α is volume fraction, Γ is mass transfer rate, P_i is interfacial pressure, F_D is inter-phasic drag force and F_{wall} is wall friction force.

To consider the geometric deformation of the pipe cross-section under pressure pulses, the cross-sectional area term A in Eqs. (1)–(2) is considered as a function of time and space, i.e., the elasticity of the pipe is considered. The cross-sectional area is of the form

$$A(x,\ t)=A(x)+A_{el}[P(x,\ t)] \tag{7}$$

where $A(x)$ is the nominal cross-sectional area of the pipe, and A_{el} represents the amount of change in cross-sectional area produced by a pressure pulse acting on the elastic pipe. The radial movement of the pipe wall is considered free in this paper. The study of Wylie and Streeter [13] gives the relationship between A_{el} and pressure

$$\frac{dA_{el}}{A(x)}=KdP \tag{8}$$

where $K=D/E\chi$, D is the tube cross-sectional diameter (m), E is the modulus of elasticity of the pipe material (N/m^2), χ is the pipe wall thickness (m). Combining Eqs. (1)–(2) and (7)–(8), the mass equation for two-phase flow in an elastic pipe can be written as

$$\frac{\partial \alpha_l\rho_l}{\partial t}+\alpha_l\rho_l K\frac{\partial P}{\partial t}+\frac{\partial \alpha_l\rho_l v_l}{\partial x}+\alpha_l\rho_l v_l K\frac{\partial P}{\partial x}=\Gamma_l \tag{9}$$

$$\frac{\partial \alpha_v\rho_v}{\partial t}+\alpha_v\rho_v K\frac{\partial P}{\partial t}+\frac{\partial \alpha_v\rho_v v_v}{\partial x}+\alpha_v\rho_v v_v K\frac{\partial P}{\partial x}=\Gamma_v \tag{10}$$

The Eqs. (3)–(6),(9)–(10) are still not closed equations and the unknown variables in them need to be determined by supplementary equations.

2.2 Closure Models for Control Equations

2.2.1 Equation of State

There is a differential relationship between the density of the workpiece and the pressure, internal energy and temperature as follows

$$d\rho_k = \left(\frac{\partial \rho_k}{\partial P}\right)_{u_k} dP + \left(\frac{\partial \rho_k}{\partial u}\right)_P du_k \quad (11)$$

where the subscript k represents the vapor phase (v) or liquid phase (l). The thermodynamic coefficients involved in the above equation are constant volume compressibility k_k, constant pressure specific heat capacity C_{Pk}, constant pressure volume thermal expansion coefficient β_k. Differential form written as

$$k_k = -\frac{1}{\forall_k}\left(\frac{\partial \forall_k}{\partial P}\right)_{T_k} \quad (12)$$

$$C_{Pk} = \left(\frac{\partial h_k}{\partial T_k}\right)_P \quad (13)$$

$$\beta_k = \frac{1}{\forall_k}\left(\frac{\partial \forall_k}{\partial T_k}\right)_P \quad (14)$$

where $\forall_k = 1/\rho_k$. The partial differential term of density with respect to pressure and internal energy can be further written as

$$\left(\frac{\partial \rho_k}{\partial u_k}\right)_P = \frac{\beta_k}{(C_{Pk} - \forall_k \beta_k \rho_k)\forall_k} \quad (15)$$

$$\left(\frac{\partial \rho_k}{\partial P}\right)_{u_k} = \frac{C_{Pk} k_k - T_k \beta_k^2 \forall_k}{(C_{Pk} - \forall_k \beta_k P)\forall_k} \quad (16)$$

2.2.2 Two-Phase Flow Regime Criteria

Two-phase flow at different flow rates and vapor-liquid ratios exhibit different inter-phase interactions and interactions between the fluid and the pipe wall. The two characteristic factors, vapor volume fraction α_v and relative velocity $v_r = v_v - v_l$ are selected to classify the two-phase flow regime, as shown in Fig. 1. The flow is divided into five states, including dispersed bubbly flow, dispersed droplet flow, horizontal stratified flow and transition flow, covering the main flow characteristics of horizontal pipeline two-phase flow.

When the relative velocity is low, the vapor-liquid two phases exhibit horizontal stratified flow regardless of the vapor volume fraction. When the relative velocity is higher than the critical velocity, the velocity difference between the two phases is larger and shows the characteristics of dispersed flow; at a lower vapor volume fraction, the liquid phase dominates and the flow process is diffusive bubble flow; at a higher vapor volume fraction, the vapor phase dominates and the flow process is diffusive droplet flow. There is a mutual transformation between these flow types.

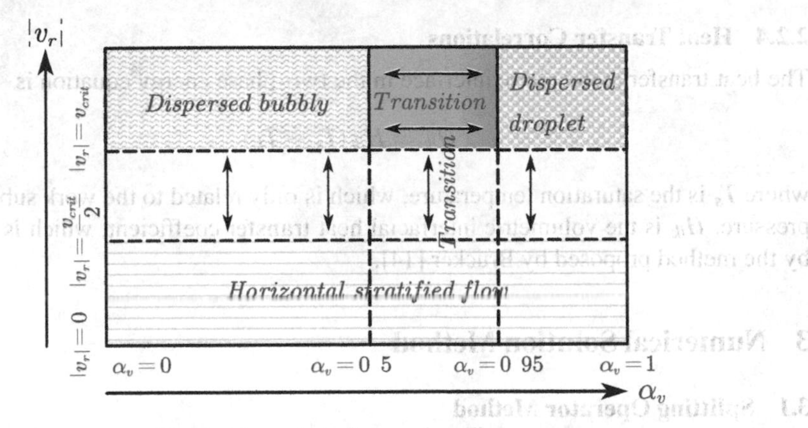

Fig. 1. Two-phase flow regime diagram

The flow regime parameter R is also one of the indicators describing the flow characteristics, reflecting the degree of stratification and dispersion of the two-phase flow, proposed and validated by Tiselj [15], and contains five factors

$$R = R_{K-H} R_{\theta} R_{\rho v} R_{v} R_{\alpha} \tag{17}$$

where R_{K-H} 1 is Kelvin–Helmholtz instability factor, which is assumed to be the primary reason for the transition of flow regime in CIWH phenomenon.

2.2.3 Virtual Mass

During the flow process, forces exist between the phases due to the spatial and temporal variation of the relative velocities of the two phases. For the part of the interphase force caused by the acceleration of the vapor or liquid phase, the virtual mass is used to characterize it. The virtual mass of a solid sphere (bubble or droplet) in an ideal fluid has a theoretical analytical solution, but the form of the virtual mass term in a real two-phase flow is not known. In the WAHA3 method, the virtual mass term in the momentum equation takes the form given by Drew [16]

$$VM = C_{VM}\left(\frac{\partial v_v}{\partial t} + v_l\frac{\partial v_v}{\partial x} - \frac{\partial v_l}{\partial t} - v_v\frac{\partial v_l}{\partial x}\right) \tag{18}$$

where C_{VM} is the virtual mass coefficient, which is determined by the flow regime parameters, mixture density, and vapor volume fraction

$$C_{VM} = \begin{cases} C_1\frac{1+2\alpha_v}{2\alpha_l} & \alpha_v \leq 0.4 \\ C_1[1.5 - 10(\alpha_v - 0.4)(\alpha_v - 0.6)] & 0.4 < \alpha_v < 0.6 \\ C_1\sqrt{\left(\frac{3-2\alpha_v}{2\alpha_v}\right)^2 + \frac{\alpha_l(2\alpha_v - 2)}{\left(\alpha_l + \frac{\alpha_l\rho_v}{\rho_l}\right)^2}} & \alpha_v \geq 0.6 \end{cases} \tag{19}$$

2.2.4 Heat Transfer Correlations

The heat transfer at the phase interface in the two-phase energy equation is

$$Q_{ik} = H_{ik}(T_s - T_k) \tag{20}$$

where T_s is the saturation temperature, which is only related to the work substance and pressure. H_{ik} is the volumetric interfacial heat transfer coefficient, which is calculated by the method proposed by Brucker [14].

3 Numerical Solution Method

3.1 Splitting Operator Method

The basic governing equations of mass, momentum and energy for individual phases can be transformed to the following matrix form

$$\boldsymbol{A}\frac{\partial \Phi}{\partial t} + \boldsymbol{B}\frac{\partial \Phi}{\partial x} = \boldsymbol{S} \tag{21}$$

where ϕ represents a vector of the variables for solving $\phi = (P\alpha_v v_l v_v u_l u_v)^\top$ and $\boldsymbol{A}$, $\boldsymbol{B}$ are 6-times-6 matrices and $\boldsymbol{S}$ is the source vector. Depending on the time scale of the source term, the source is split into a non-relaxation source and a relaxation source

$$\boldsymbol{S} = \boldsymbol{S}_{NR} + \boldsymbol{S}_R \tag{22}$$

Non-relaxation source, which contains friction and volume force terms, is closely related to the convective term of the model. The relaxation source, which establishes the heat and mechanical equilibrium between two phases, reflects the heat, mass and momentum transfer between the phases.

$$\boldsymbol{A} = \begin{pmatrix} G_l\left(\frac{\partial \rho_l}{\partial P}\right)_u + \rho_l K & -\frac{\rho_l}{\alpha_l} & 0 & 0 & 0 & 0 \\ G_v\left(\frac{\partial \rho_v}{\partial P}\right)_u + \rho_k K & \frac{\rho_v}{\alpha_v} & 0 & 0 & 0 & 0 \\ 0 & 0 & \rho_l + \frac{C_{VM}}{\alpha_l} & -\frac{C_{VM}}{\alpha_l} & 0 & 0 \\ 0 & 0 & -\frac{C_{VM}}{\alpha_v} & \rho_v + \frac{C_{VM}}{\alpha_v} & 0 & 0 \\ PK & -\frac{P}{\alpha_l} & 0 & 0 & \rho_l & 0 \\ PK & \frac{P}{\alpha_v} & 0 & 0 & 0 & \rho_v \end{pmatrix} \quad G_k = \frac{1}{1 - \frac{P}{\rho_k^2}\left(\frac{\partial \rho_k}{\partial u_k}\right)_P}$$

$$\boldsymbol{B} = \begin{pmatrix} v_l\left[G_l\left(\frac{\partial \rho_l}{\partial P}\right)_u + \rho_l K\right] & -\frac{\rho_l v_l}{\alpha_l} & \rho_l & 0 & 0 & 0 \\ v_v\left[G_v\left(\frac{\partial \rho_v}{\partial P}\right)_u + \rho_k K\right] & \frac{\rho_v v_v}{\alpha_v} & 0 & \rho_v & 0 & 0 \\ 1 & -\frac{P_i}{\alpha_l} & \rho_l v_l + \frac{C_{VM}}{\alpha_l} v_v & -\frac{C_{VM}}{\alpha_l} v_l & 0 & 0 \\ 1 & \frac{P_i}{\alpha_v} & -\frac{C_{VM}}{\alpha_v} v_v & \rho_v v_v + \frac{C_{VM}}{\alpha_v} v_l & 0 & 0 \\ P v_l K & -\frac{P v_l}{\alpha_l} & P & 0 & \rho_l v_l & 0 \\ P v_v K & \frac{P v_v}{\alpha_v} & 0 & P & 0 & \rho_v v_v \end{pmatrix}$$

$$S_R = \begin{pmatrix} \frac{G_l}{\alpha_l}\left\{\Gamma_l - \left[Q_{il} + \Gamma_l(h_l - u_l)\right]\frac{1}{\rho_l}\left(\frac{\partial \rho_l}{\partial u_l}\right)_P\right\} \\ \frac{G_v}{\alpha_v}\left\{\Gamma_v - \left[Q_{iv} + \Gamma_v(h_v - u_v)\right]\frac{1}{\rho_v}\left(\frac{\partial \rho_v}{\partial u_v}\right)_P\right\} \\ \frac{1}{\alpha_l}F_D + \frac{1}{\alpha_l}\Gamma_l(v_i - v_l) + \rho_l g \sin\theta \\ -\frac{1}{\alpha_v}F_D + \frac{1}{\alpha_v}\Gamma_v(v_i - v_v) + \rho_v g \sin\theta \\ \frac{1}{\alpha_l}\left[Q_{il} + \Gamma_l(h_l - u_l)\right] \\ \frac{1}{\alpha_v}\left[Q_{iv} + \Gamma_v(h_v - u_v)\right] \end{pmatrix} \quad S_{NR} = \begin{pmatrix} -v_l F_{l,\text{ wall}} \frac{G_l}{\rho_l \alpha_l}\left(\frac{\partial \rho_l}{\partial u_l}\right)_P \\ -v_v F_{v,\text{ wall}} \frac{G_v}{\rho_v \alpha_v}\left(\frac{\partial \rho_v}{\partial u_v}\right)_P \\ -\frac{1}{\alpha_l}F_{l,\text{ wall}} \\ -\frac{1}{\alpha_v}F_{v,\text{ wall}} \\ \frac{1}{\alpha_l}v_l F_{l,\text{ wall}} \\ \frac{1}{\alpha_v}v_v F_{v,\text{ wall}} \end{pmatrix}$$

3.2 Two-Step Iterative Algorithm

The final model with all the control and closure equations is

$$\boldsymbol{A}\frac{\partial \boldsymbol{\Phi}}{\partial t} + \boldsymbol{B}\frac{\partial \boldsymbol{\Phi}}{\partial x} = \boldsymbol{S}_{NR} + \boldsymbol{S}_R \tag{23}$$

After applying the operator splitting method, the solution of $\boldsymbol{\Phi}$ is iterated in two steps, as shown in Fig. 2.

$$\boldsymbol{\Phi}_j^{(n)} \underset{S_{NR}\ works\ as\ source}{\overset{\Delta t_{NR}}{\Longrightarrow}} \boldsymbol{\Phi}_j^{*\,(n)} \underset{S_R\ works\ as\ source}{\overset{\Delta t_R}{\Longrightarrow}} \boldsymbol{\Phi}_j^{(n+1)}$$

Fig. 2. Schematic of the two-step iterative method

The superscript n in $\Phi_j^{(n)}$ indicates the time node where Φ is currently located, and the subscript j indicates the spatial node where Φ is currently located. The iteration of a time step is split into two parts, and the variable $\Phi_j^{(n)}$ at a certain spatial position needs to experience an intermediate time *(n) when crossing from time (n) to time (n + 1).

3.2.1 Convective Iterative

Only consider the effect of non-relaxation sources in the convection step, Eq. (23) can be re-written as

$$\frac{\partial \boldsymbol{\Phi}}{\partial t} + \boldsymbol{C}\frac{\partial \boldsymbol{\Phi}}{\partial x} = \boldsymbol{A}^{-1}\boldsymbol{S}_{NR} \tag{24}$$

where the Jacobain matrix $C = A^{-1}B$ can be diagonalized in terms of its eigen vector $\boldsymbol{V}$ and eigen value matrix $\boldsymbol{\Lambda}$ as follows

$$\boldsymbol{C} = \boldsymbol{V}\boldsymbol{\Lambda}\boldsymbol{V}^{-1} \tag{25}$$

Applying the diagonalization result to Eq. (23) and setting s $R_F = -V^{-1}S_{NR}$ yields

$$\boldsymbol{V}^{-1}\left(\frac{\partial \boldsymbol{\Phi}}{\partial t} + \boldsymbol{V}\boldsymbol{\Lambda}\boldsymbol{V}^{-1}\frac{\partial \boldsymbol{\Phi}}{\partial x} + \boldsymbol{R}_F\right) = 0 \tag{26}$$

Introduce the characteristic variable $\boldsymbol{W}_{6\times1}$, which is related to the original variable $\boldsymbol{\Phi}$ as

$$\partial \boldsymbol{W} = \boldsymbol{V}^{-1}\partial\boldsymbol{\Phi} + \boldsymbol{\Lambda}^{-1}\boldsymbol{V}^{-1}\boldsymbol{R}_F\partial x \tag{27}$$

Equation (26) can be expressed by the characteristic variables

$$\frac{\partial \boldsymbol{W}}{\partial t} + \boldsymbol{\Lambda}\frac{\partial \boldsymbol{W}}{\partial x} = 0 \tag{28}$$

The TVD [17] (Total Variation Diminishing) format with restriction factors is used to differential Eq. (28) in the space of eigenvariables

$$\frac{\boldsymbol{W}_j^{*(n)} - \boldsymbol{W}_j^{(n)}}{\Delta t_{NR}} + \boldsymbol{\Lambda}^+\frac{\boldsymbol{W}_j^{(n)} - \boldsymbol{W}_{j-1}^{(n)}}{\Delta x} + \boldsymbol{\Lambda}^-\frac{\boldsymbol{W}_{j+1}^{(n)} - \boldsymbol{W}_j^{(n)}}{\Delta x} = 0 \tag{29}$$

where the eigenvalue diagonal array in Eq. (29) splits into two parts, and the split $\boldsymbol{\Lambda}$ is still a diagonal array, and its relationship in spatial location is shown in Fig. 3.

$$\boldsymbol{\Lambda}_j^{(n)} = \boldsymbol{\Lambda}_{j-1/2}^{(n)} + \boldsymbol{\Lambda}_{j+1/2}^{(n)} = \boldsymbol{\Lambda}^+ + \boldsymbol{\Lambda}^- \tag{30}$$

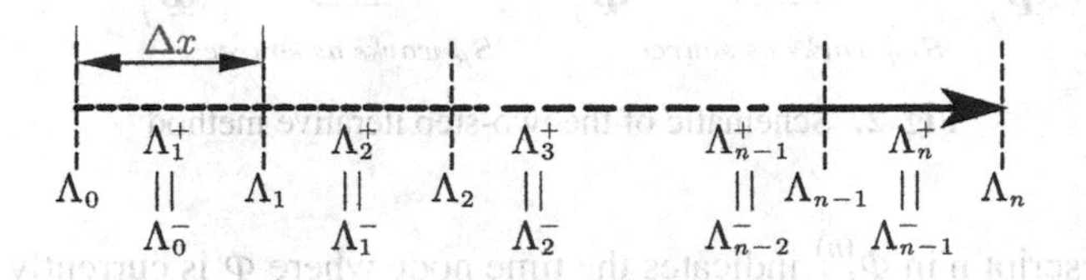

Fig. 3. Spatial discrete schematic

Replacing the characteristic variables with the original variables $\boldsymbol{\Phi}$, $\boldsymbol{\Phi}_j^{*(n)}$ can be deduced from Eq. (28) as

$$\begin{aligned}\boldsymbol{\Phi}_j^{*(n)} = \boldsymbol{\Phi}_j^{(n)} - \Delta t_{NR}\cdot\Bigg(&\boldsymbol{V\Lambda}^+\boldsymbol{\Lambda}^{-1}\boldsymbol{V}^{-1}\boldsymbol{R}_F + \boldsymbol{V\Lambda}^-\boldsymbol{\Lambda}^{-1}\boldsymbol{V}^{-1}\boldsymbol{R}_F \\ &+\boldsymbol{V\Lambda}^+\boldsymbol{V}^{-1}\frac{\boldsymbol{\Phi}_j^{(n)} - \boldsymbol{\Phi}_{j-1}^{(n)}}{\Delta x} + \boldsymbol{V\Lambda}^-\boldsymbol{V}^{-1}\frac{\boldsymbol{\Phi}_{j+1}^{(n)} - \boldsymbol{\Phi}_j^{(n)}}{\Delta x}\Bigg)\end{aligned} \tag{31}$$

The time step size Δt_{NR} is evaluated based on the CFL [18] (Courant-Friedrichs-Levy) criterion as follows

$$\Delta t_{NR} \le \frac{\Delta x}{\max|\lambda_t|} \tag{32}$$

3.2.2 Relaxation Iterative

In the relaxation step only consider the relaxation source. The velocity terms in the variables (corresponding to the momentum equation) are calculated from the assumptions on the relative and mixed velocities. In the calculation of the other variables, momentum transfer is not considered, but only the effects of heat and mass transfer at the phase interface.

$$\boldsymbol{A_T}\frac{\partial \psi}{\partial t} = \boldsymbol{S_{R-HMT}} \tag{33}$$

$$\boldsymbol{A_T} = \begin{pmatrix} \left(\frac{\partial \rho_l}{\partial P}\right)_{T_l} + \rho_l K & -\frac{\rho_l}{\alpha_l} & \left(\frac{\partial \rho_l}{\partial T_l}\right)_P & 0 \\ \left(\frac{\partial \rho_v}{\partial P}\right)_{T_v} + \rho_v K & \frac{\rho_v}{\alpha_v} & 0 & \left(\frac{\partial \rho_v}{\partial T_v}\right)_P \\ PK + \rho_l\left(\frac{\partial u_l}{\partial P}\right)_{T_l} & -\frac{P}{\alpha_l} & \rho_l\left(\frac{\partial u_l}{\partial T_l}\right)_P & 0 \\ PK + \rho_v\left(\frac{\partial u_v}{\partial P}\right)_{T_v} & \frac{P}{\alpha_v} & 0 & \rho_v\left(\frac{\partial u_v}{\partial T_v}\right)_P \end{pmatrix}$$

$$\boldsymbol{S_{R-HMT}} = \begin{pmatrix} \frac{G_l}{\alpha_l}\left\{\Gamma_l - \left[Q_{il} + \Gamma_l(h_l - u_l)\right]\frac{1}{\rho_l}\left(\frac{\partial \rho_l}{\partial u_l}\right)_P\right\} \\ \frac{G_v}{\alpha_v}\left\{\Gamma_v - \left[Q_{iv} + \Gamma_v(h_v - u_v)\right]\frac{1}{\rho_v}\left(\frac{\partial \rho_v}{\partial u_v}\right)_P\right\} \\ \frac{1}{\alpha_l}\left[Q_{il} + \Gamma_l(h_l - u_l)\right] \\ \frac{1}{\alpha_v}\left[Q_{iv} + \Gamma_v(h_v - u_v)\right] \end{pmatrix} \quad \psi = \begin{pmatrix} P \\ \alpha_v \\ T_l \\ T_v \end{pmatrix}$$

The time step of the relaxation iterative is related to the relative rate of change of the variable ψ

$$\frac{\partial \psi}{\partial t} = \boldsymbol{A_T^{-1}}\boldsymbol{S_{R-HMT}} = \left(\dot{P}\ \dot{\alpha}_v\ \dot{T}_l\ \dot{T}_v\right)^{\mathsf{T}} \tag{34}$$

The relaxation time step is evaluated as follows

$$\Delta t_R = 0.01\min(t_1, t_2, t_3, t_4) \tag{35}$$

$$\begin{cases} t_1 = \frac{P}{\max(\in, |\dot{P}|)} \\ t_2 = \frac{\max(0.01, \min(\alpha_v, \alpha_l))}{\max(\in, |\dot{\alpha}_v|)} \\ t_3 = \frac{\max(|T_l - T_s|)}{\max(\in, |\dot{T}_l|)} \\ t_4 = \frac{\max(|T_v - T_s|)}{\max(\in, |\dot{T}_v|)} \end{cases} \tag{36}$$

The relative velocity is iterated in the relaxation step as

$$v_r^{(n+1)} = \frac{v_r^{*(n)}}{1 - \Delta t_R S_{vr}} \tag{37}$$

where

$$S_{vr} = \begin{cases} \frac{-C_i|v_r|\rho_m+\alpha_v\rho_v\Gamma_v}{\alpha_v\alpha_l\rho_l\rho_v+C_{VM}\rho_m} & \Gamma_v < 0 \\ \frac{-C_i|v_r|\rho_m-\alpha_l\rho_l\Gamma_v}{\alpha_v\alpha_l\rho_l\rho_v+C_{VM}\rho_m} & \Gamma_v > 0 \end{cases} \tag{38}$$

4 Results and Discussions

4.1 Verification with Benchmark Problem (Two-Phase Shock Tube Case)

The one-dimensional Riemann two-phase shock wave tube problem is a typical benchmark problem for condensation-induced water hammer, reflecting the ability of the model to capture discontinuous physical states and millisecond pressure transients.

A straight tube of length 1 m is filled with vapor-liquid mixture, the diameter of the tube is 19 mm, the wall thickness is 1.6 mm, and the modulus of elasticity of the tube material is 760 MPa, as shown in Fig. 4. In the initial state, a thin film at the center of the tube divides the tube into two parts, both with an initial vapor volume fraction of 0.3. The densities of the left and right side of the mixture are 998.64 and 998.23 kg/m^3.

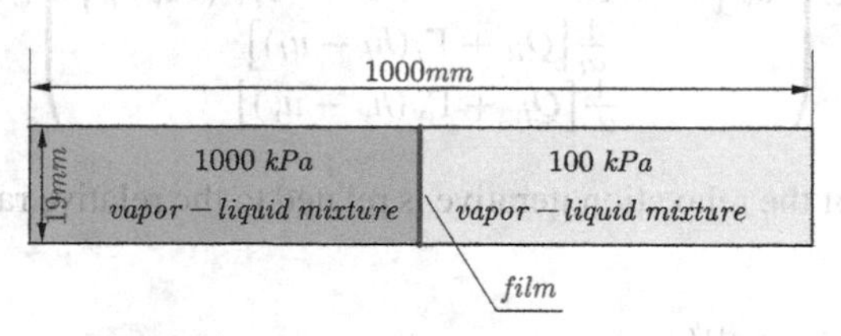

Fig. 4. Schematic of the initial state of the shock wave tube

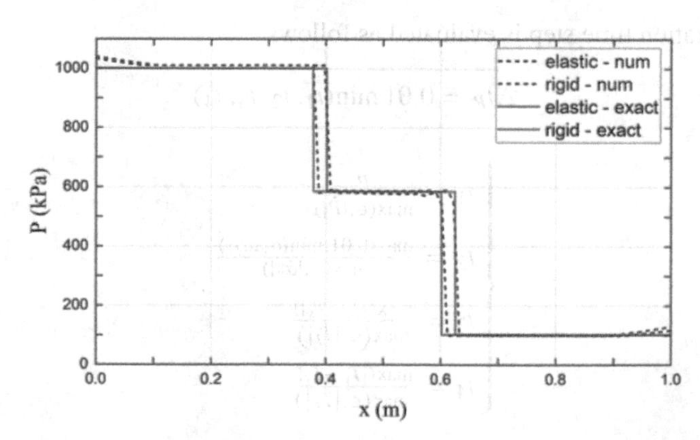

Fig. 5. Numerical solution of the Riemann shock wave tube problem for pressure in elastic and rigid tubes (at 0.42 ms)

The test starts after the film is completely ruptured and the vapor-liquid mixture diffuses from high to low pressure, forming a surge inside the tube. The surge tube is discretized into 100 nodes, and the calculated results are shown in Fig. 5, reflecting the pressure distribution inside the tube at 0.42 ms after the vapor-liquid mixture with different pressure starts to contact. The calculated results show good agreement with the analytical values.

4.2 Verification with Experimental Results of PMK-2

In order to study the CIWH phenomenon that occurs during the entry of cold water into steam-filled pipes in the water circuit of a nuclear power facility, the Hungarian KFKI Institute constructed a model steam pipe and conducted the PMK-2 test.

The test section of the PMK-2 test is a tube with a length-to-diameter ratio of about 39. The pipe is filled with saturated water vapor at a pressure of 1.45 MPa at the initial moment, and the saturation temperature of the water vapor at this pressure is 470 K. The pipe and boundary conditions can be simplified as shown in Fig. 6. The volume of the steam kettle at the end is sufficiently large that the outlet pressure can be considered to be approximately constant.

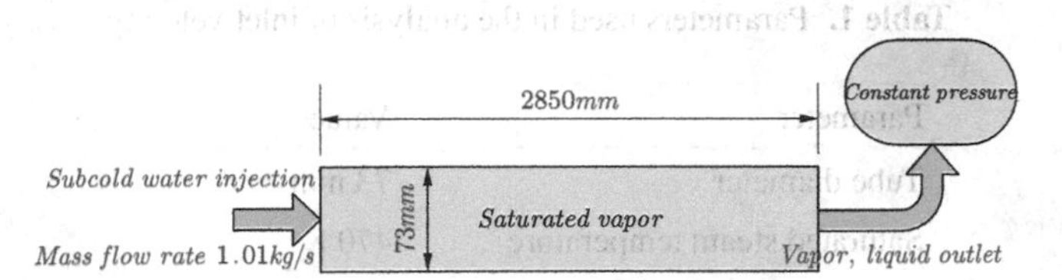

Fig. 6. PMK-2 test section tube schematic

The numerical solution method of this paper is used to calculate the PMK-2 test conditions, and the pipeline is divided into a total of 10 intervals (11 nodes), and the occurrence of pressure pulses is successfully captured. A comparison of the calculation results in this paper with the PMK-2 test results is shown in Fig. 7.

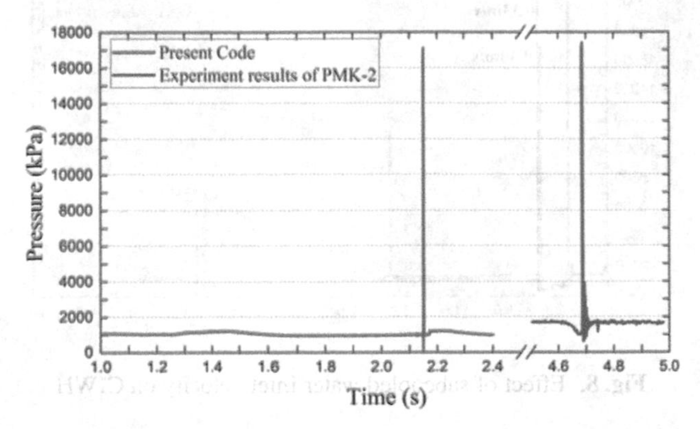

Fig. 7. Comparison of pressure time history between PMK-2 test and the results calculated in this paper

In terms of peak pressure, the pressure pulse amplitude of 17.1 MPa obtained by the calculation method in this paper, compared with the highest value of 17.4 MPa recorded in the test, the relative error is 1.7%, indicating that the calculation method can be more accurate in forecasting the impact pressure when water hammer occurs under the test conditions.

4.3 Analysis of the Effect of Inlet Velocity on CIWH

The temperature and pressure parameters of the PMK-2 test are applied, and the injection velocity of subcooled water is varied (0.1 m/s–0.4 m/s) to investigate the effect of subcooled water flow rate on the CIWH phenomenon in the tube. A total of five velocities are numerically calculated for the model with the modulus of elasticity of the pipe material set to 206 GPa, the model discretization are 10 nodes, the velocity boundary condition at the left inlet and the pressure boundary condition at the right outlet (Figs. 8, 9 and Table 1).

Table 1. Parameters used in the analysis of inlet velocity

Parameter	Value
Tube diameter	73 mm
Saturated steam temperature	470 K
Initial pressure	1450 kPa
Inlet water temperature	295 K
Elastic modulus	206 GPa
Inlet velocity	0.1–0.4 m/s

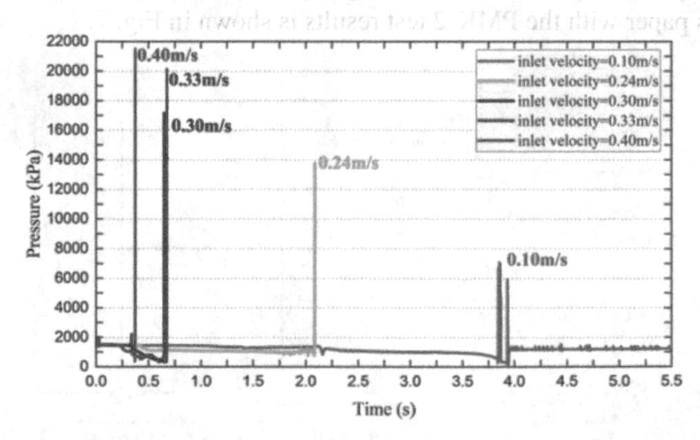

Fig. 8. Effect of subcooled water inlet velocity on CIWH

With the increasing flow rate of subcooled water (0.1 m/s to 0.4 m/s), the peak pressure of condensate water hammer is growing, but the growth rate is gradually slowing

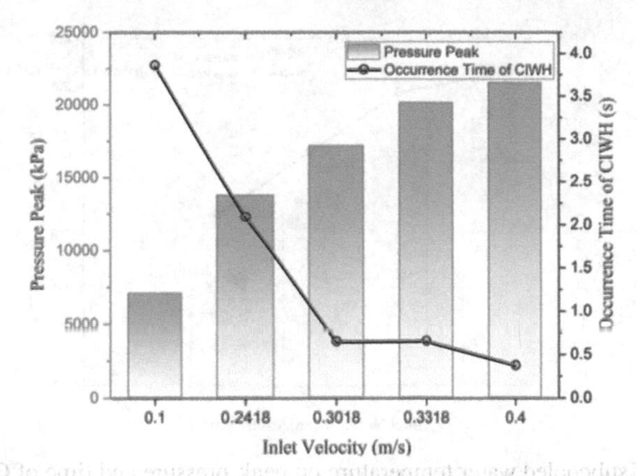

Fig. 9. Effect of subcooled water inlet velocity on peak pressure and time of CIWH occurrence

down; the occurrence time of water hammer is constantly in advance, but when the flow rate is greater than 0.3 m/s, the change of the occurrence time is no longer obvious and is maintained at 0.5 s or less.

4.4 Analysis of the Effect of Inlet Water Temperature on CIWH

Following the calculation conditions in Sect. 4.3, the injection velocity of subcooled water remains unchanged at 0.24 m/s, and only the liquid phase temperature is changed from 285 K to 325 K, every 10 K for a grade, for a total of 5 water temperature conditions (Figs. 10 and 11).

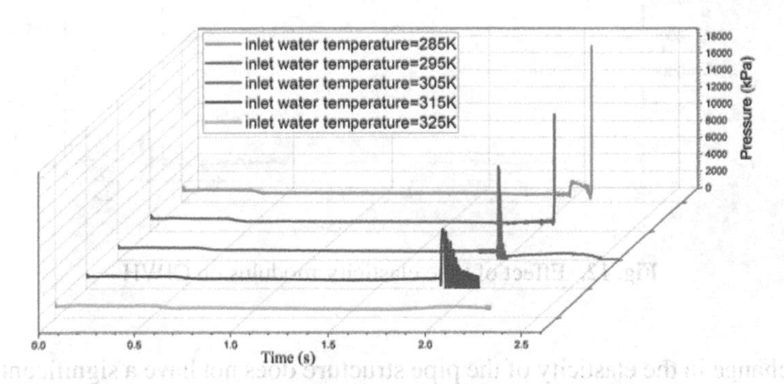

Fig. 10. Effect of subcooled water temperature on CIWH

Except for the liquid phase temperature of 325 K, the sudden pressure change occurs about 2 s after the cold water starting inject. The condensate water hammer pressure peak is the largest at 285 K, reaching 18.5 MPa, and the lowest at 315 K, 7.0 MPa, and enter a state of continuous oscillation after the water hammer occurs. The obvious pressure pulse phenomenon is no longer observed when the temperature increased to 325 K.

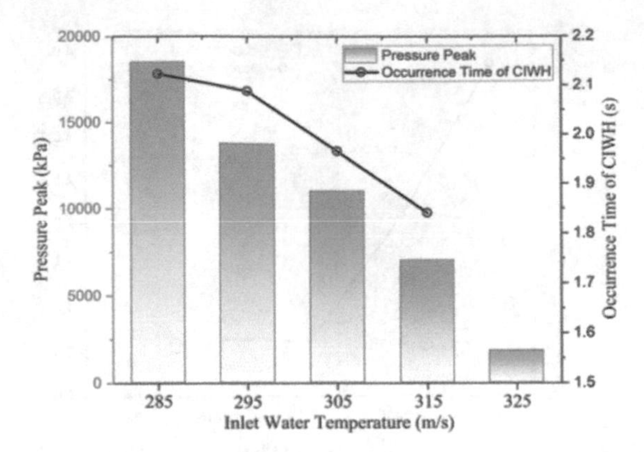

Fig. 11. Effect of subcooled water temperature on peak pressure and time of CIWH occurrence

4.5 Analysis of the Effect of Elastic Modulus on CIWH

Using the numerical algorithm of this paper, a computational study of CIWH phenomenon occurring in pipes with different modulus of elasticity (E) is done. The injection velocity and temperature of the supercooled water are kept constant at 0.24 m/s and 295 K.

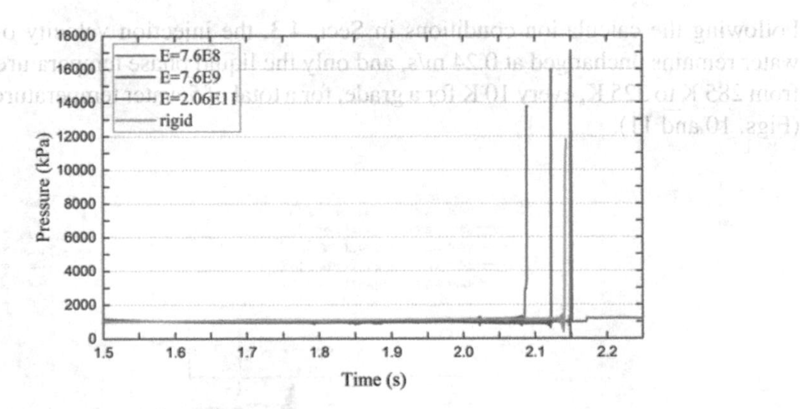

Fig. 12. Effect of tube elasticity modulus on CIWH

The change in the elasticity of the pipe structure does not have a significant effect on the time of occurrence of CIWH, and the peak pressure occurs within 2.1 ± 0.05 s for several conditions in Fig. 12. The change of the pressure peak is mainly related to the deformation of the pipe cross-section. In the pipe with greater stiffness, the smaller the growth of the cross-sectional area caused by the pressure pulse, the smaller the cavitation area formed after gas condensation, and the smaller the stroke of the bubble collapse effect, which shows a lower pressure peak.

As shown in Fig. 13 and Fig. 14, comparing the cross-sectional area before and after the water hammer, we can see the extent of the deformation of the pipe caused by the

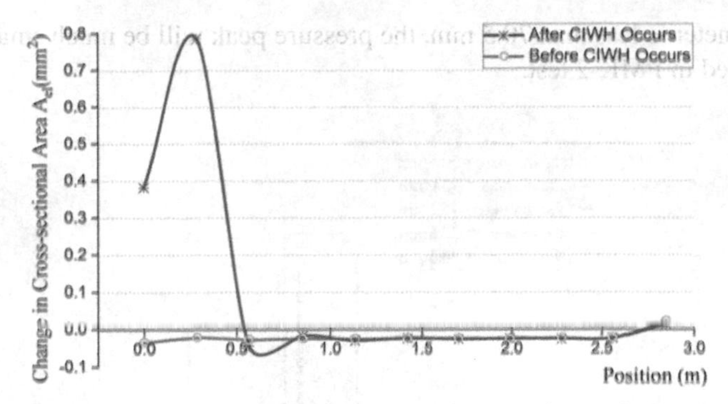

Fig. 13. Comparison of changes in tube cross-sectional area before and after the occurrence of CIWH (E = 760 MPa)

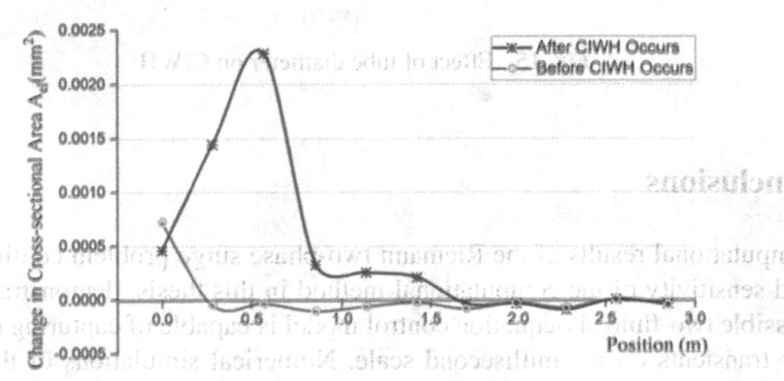

Fig. 14. Comparison of changes in tube cross-sectional area before and after the occurrence of CIWH (E = 206 GPa)

pressure pulse. For the pipe with E = 206 GPa, although the deformation of the pipe cross-section is much smaller than the former, the pressure pulse has a broader impact on the pipe deformation.

4.6 Analysis of the Effect of Tube Diameter on CIWH

The tube diameter is directly related to the mass flow rate of subcooled water injection and the contact area between the two phases, and also has an impact on the peak pressure and occurrence time of condensate hammer. The numerical model of this paper is used to calculate the pressure time history of the same temperature of subcooled water with the same flow rate into different diameter pipes, the modulus of elasticity is taken as 206 GPa, and the diameters are taken as 63 mm, 68 mm, 70.5 mm, 73 mm, 75.5 mm, 78 mm and 80.5 mm.

As shown in Fig. 15, in the diameter range of 70.5 to 80.5 mm, with the increase of pipe diameter, the peak pressure of condensate hammer increases up to 20.3 MPa, but the time of pressure pulse occurs backward, the peak pressure occurs at 1.96 s under 70.5 mm pipe diameter, and this moment is delayed to 2.44 s under 80.5 mm pipe diameter. When the pipe diameter is less than 70.5 mm, the peak pressure will be When the pipe diameter is less than 70.5 mm, the pressure peak will be much smaller than the result recorded in PMK-2 test.

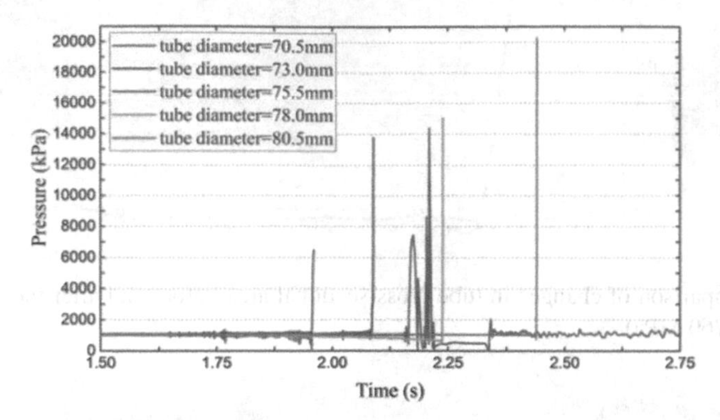

Fig. 15. Effect of tube diameter on CIWH

5 Conclusions

The computational results of the Riemann two-phase surge problem confirm the accuracy and sensitivity of the computational method in this thesis, demonstrating that the compressible two-fluid six-equation control model is capable of capturing pressure and velocity transients on the millisecond scale. Numerical simulations of the important experimental PMK-2 for condensation-induced water hammer are also performed on top of the benchmark problem, and a good match is achieved in terms of pressure pulse amplitude prediction and location of occurrence.

Based on the test equipment conditions, this paper also calculates the subcooled water flow rate and subcooled water temperature factors, which are of interest in this research area. The results show that as the inlet flow rate increases, the peak pressure pulse rises and CIWH occurs earlier; as the subcooled water temperature increases, the peak pressure falls and no longer triggers CIWH if the subcooled water temperature decreases to a certain extent. The pressure peak of CIWH is largely affected by the tube diameter. When the pipe cross-sectional area increases by 7.2% from an initial diameter of 70.5 mm, the pressure peak increases up to 2.1 times if the original, while CIWH is not be observed when the pipe diameter turns to be small.

Acknowledgements. This work was sponsored by the National Natural Science Foundation of China (52271285), Young Talent Project of China National Nuclear Corporation (CNNC-YT-2021), LingChuang Research Project of China National Nuclear Corporation (CNNC-LC-2020), the Oceanic Interdisciplinary Program of Shanghai Jiao Tong University (SL2022PT201) and CSSC-SJTU Marine Equipment Prospective Innovation Joint Fund (1-B1).

References

1. Zouhri, L., Smaoui, H., Carlier, E., et al.: Modelling of hydrodispersive processes in the fissured media by flux limiters schemes (Chalk aquifer, France). Math. Comput. Model. **50**(3–4), 516–525 (2009)
2. Strubelj, L., Ezsöl, G., Tiselj, I.: Direct contact condensation induced transition from stratified to slug flow. Nucl. Eng. Des. **240**(2), 266–74 (2010)
3. Ceuca, S.C., Macián-Juan, R. : CFD simulation of direct contact condensation with ansys cfx using locally defined heat transfer coefficients. In: Proceedings of the International Conference on Nuclear Engineering, Proceedings, ICONE, F (2012)
4. Datta, P., Chakravarty, A., Ghosh, K., et al.: Modeling of steam–water direct contact condensation using volume of fluid approach. Numer. Heat Transf. Part A: Appl. **73**(1), 17–33 (2018)
5. Pham, T.Q.D., Choi, Sanghun: Numerical analysis of direct contact condensation-induced water hammering effect using OpenFOAM in realistic steam pipes. Int. J. Heat Mass Transf. **171**, 121099 (2021). https://doi.org/10.1016/j.ijheatmasstransfer.2021.121099
6. Höhne, Thomas, Gasiunas, Stasys, Šeporaitis, Marijus: Numerical modelling of a direct contact condensation experiment using the AIAD framework. Int. J. Heat Mass Transf, **111**, 211–222 (2017). https://doi.org/10.1016/j.ijheatmasstransfer.2017.03.104
7. Li, S.Q., Wang, P., Lu, T.: CFD based approach for modeling steam–water direct contact condensation in subcooled water flow in a tee junction. Prog. Nucl. Energy **85**, 729–746 (2015)
8. Barna, I.F., Imre, A.R., Baranyai, G., et al.: Experimental and theoretical study of steam condensation induced water hammer phenomena. Nucl. Eng. Des. **240**(1), 146–150 (2010)
9. Tiselj, I., Petelin, S.: Modelling of two-phase flow with second-order accurate scheme. J. Comput. Phys. **136**(2), 503–521 (1997)
10. Ceuca, S.C., Laurinavicius, D. : Experimental and numerical investigations on the direct contact condensation phenomenon in horizontal flow channels and its implications in nuclear safety. In: Proceedings of the Kerntechnik (2016)
11. Hibiki, T., Rassame, S., Liu, W., et al.: Modeling and simulation of onset of condensation-induced water hammer. Prog. Nucl. Energy **130**, 103555 (2020)
12. Milivojevic, S., Stevanovicv, D., Maslovaric, B.: Condensation induced water hammer: numerical prediction. J. Fluids Struct. **50**, 416–436 (2014)
13. Wylie, E.B., Streeter, V.L., Wiggert, D.C.: Fluid transients. J. Fluids Eng. **102**(3), 384–385 (1980)
14. Brucker, G.G., Sparrow, E.M.: Direct contact condensation of steam bubbles in water at high pressure. Int. J. Heat Mass Transf. **20**(4), 371–381 (1977)
15. Tiselj, I., Horvat, A., Černe, G., et al.: WAHALoads-two-phase flow water hammer transients and induced loads on materials and structures of nuclear power plants: WAHA3 code manual (2004)
16. Drew, D., Cheng, L., Lahey, R.T.: The analysis of virtual mass effects in two-phase flow. Int. J. Multiph. Flow **5**(4), 233–242 (1979)
17. Glaister, P.: Flux difference splitting for the Euler equations in one spatial co-ordinate with area variation. Int. J. Numer. Meth. Fluids **8**(1), 97–119 (1988)
18. Datta, P., Chakravarty, A., Ghosh, K., et al.: Modeling and analysis of condensation induced water hammer. Numer. Heat Transf. Part A: Appl. **74**(2), 975–1000 (2018)

Study on Coupling Effect and Dynamic Behavior of Double Bubbles Rising Process

Lanxin Gong, Changhong Peng(✉), and Zhenze Zhang

School of Nuclear Science and Technology, University of Science and Technology of China, Hefei, Anhui, China

{lenovomax,desolate}@mail.ustc.edu.cn, pengch@ustc.edu.cn

Abstract. Gas-liquid two-phase flow widely exists in nuclear energy engineering, in which bubble movement and deformation are critical problems. Because the activity of bubbles in the fluid is a very complex physical process, and the movement process is a flow field-bubble coupling process, which has strong nonlinearity and unsteady, the relevant research is usually based on experiments and simulation.

We built a medium-sized experimental device to generate double bubbles with different sizes and characteristic numbers and recorded the motion trajectory with a high-speed camera. We developed and improved the image processing method to obtain high-quality bubble motion information and realized a good capture of bubble shape and rotation.

The experimental results show that in the two bubbles rising successively, the trailing bubble is affected by the trailing field of the leading bubble, and the bubble velocity, relative distance, deformation rate, and other parameters change accordingly. In addition, through simulation, we get the interaction mechanism of the bubbles under experimental conditions. The results show that the coupling leads to flow field velocity and pressure changes, which explains the experimental results. The research results are helpful for a thorough understanding of the law of bubble movement and provide empirical data support for developing a thermal-hydraulic model.

Keywords: Bubble rising experiment · Bubble coupling · Gas-liquid two-phase flow

1 Introduction

Since the last century, researchers have carried out a lot of research on bubble motion from the aspects of theory, experiment, and numerical simulation [1–3].

There are various forms of bubble movement. For the free bubble, if it is affected by the rigid wall, it will not only deform the bubble interface but also change its original motion state, mainly including the movement of the bubble away from the wall, the movement of the bubble close to the wall, and the bounce movement of the bubble along the wall. For free-space bubbles, it is relatively simple, mainly showing a zigzag and spiral rise. Predecessors have also carefully studied the rise of single bubble and bubble

C. Liu (Ed.): PBNC 2022, SPPHY 283, pp. 973–984, 2023.
https://doi.org/10.1007/978-981-99-1023-6_82

chains and obtained the laws of bubble terminal velocity, bubble deformation, and other motion parameters in some cases.

Generally speaking, in the presence of a wall, bubbles are affected laterally by the wall attraction (lift pointing towards the wall) and the wall repulsion (lift pointing away from the wall). When bubbles are affected by repulsive force and attraction, they appear in a zigzag motion. In the longitudinal direction, bubbles are usually affected by lift, drag, additional mass force, and film-induced force, regardless of whether the wall exists. Because the growth and movement of bubbles are very complex, involving the mass conversion between two phases and the energy transfer between three phases, the growth and movement mechanism of bubbles have not been fully understood. In addition, as a unique flow field boundary, bubbles have an important influence on the dynamics of other bubbles around them, making the problem more complex. When two bubbles rise in parallel, the smaller spacing will lead to bubble fusion; When bubbles rise one after another, the wake of the leading bubble will cause the bubbles that follow to rise faster. These coupling effects significantly change the bubble distribution and two-phase contact area and affect the heat and mass transfer performance.

Clift et al. [1] studied the change of bubble shape and drew the bubble phase diagram. It was found that bubbles with a diameter of less than 1.3 mm remained spherical, and the shape of large bubbles would be the oval and spherical cap. Duineveld [4] studied the floating characteristics of bubbles with a diameter of 0.33–1 mm in purified water, explored the equilibrium velocity and shape coefficient obtained by bubbles and the relationship between Weber number *We*, and found that the maximum Weber number that can float in a stable shape and velocity does not exceed 3.2. Raymond and Rosant [5], Zenit [6], and Magnaudet [6] use different liquids or add different proportions of chemicals to the water to change the density, viscosity, surface tension coefficient, and other parameters of the fluid and explore the changing laws and internal relations of physical quantities such as bubble resistance coefficient, shape, and buoyancy under different *We* and Re. Wu and Gharib [7] found that the bubbles with a diameter of 0.1–0.2 cm were spherical or ellipsoidal in the floating process. When the bubble volume was constant, the bubbles with high rising velocity were generally ellipsoidal; In addition, the upward floating path of bubbles with a diameter less than 1.5 mm is a straight line. When the bubble size is larger, the upward floating path of bubbles will be Z shaped or spiral-shaped.

As for the interaction between two bubbles, Duineveld's experimental research shows that if the horizontal approach velocity of bubbles is the characteristic velocity when *We* are less than 0.18, the two bubbles will become a single bubble. When the bubble radius is less than 0.7 mm, the impact velocity of the two bubbles is always tiny, so the possibility of bubbles fusion is high. When We are less than the critical value, the two bubbles after springing could collide again. Sanada et al. [8] conducted an experimental study on the interaction between two horizontal bubbles and found that when re exceeds a specific range, the two bubbles attract each other. After the collision, there may be two cases of fusion or bounce-off. If the bounce-off occurs, the bubble floating speed will be reduced by about 50% due to vortex shedding and other reasons. The occurrence of different phenomena is related to Re and *We*. It is pointed out that the critical *We* are about 2, and the critical Re is related to the Morton number Mo.

As for the terminal velocity, it depends on the bubble shape and is related to the Eotvos number, Reynolds number, and Morton number. Wallis [9], Grace [10], Jamialahmadi [11], Bozzano [12], Sung-Hoon Park [13], and others proposed a series of correlations based on experiments. As for the application of final velocity, it is widely used in system analysis programs (such as RELAP5 [14]) to determine the selection of flow patterns, heat transfer correlations, etc.

2 Experimental Equipment

The central part of the experimental device is a 30 × 30 × 40 cm acrylic water tank. In the tank, the rubber tube and Ruhr joint are connected with the flat head stainless steel needle tube; Outside the tank, the micro air pump and the gas pipe are connected with the flow regulating valve and the micro syringe. When bubbles need to be generated, open the air pump, control the gas flow rate by adjusting the flow regulating valve, and change the bubble size by changing the diameter of the syringe needle in the cylinder. As shown in Fig. 1, when recording bubbles from the front, a strip-shaped parallel light source is set at the back of the cylinder; When shooting bubbles on the left side, place a rectangular light source on the right side of the cylinder, as shown in Fig. 1. The high-speed camera is supported by stable support and is connected to the computer, which can process the captured images in real-time. To simplify the subsequent image processing, we record the bubbles in the dark environment and keep the white LED light source so that the captured image has a white background and the bubble itself is black. In the experiment, we set the camera to shoot at the frame rate of 500 fps.

Table 1. Experimental group and needle diameter

Serial number	1	2	3	4	5	6
OD/mm	2.42	1.86	1.29	3.46	2.78	4.00
Number of experiments	21	23	17	14	16	12

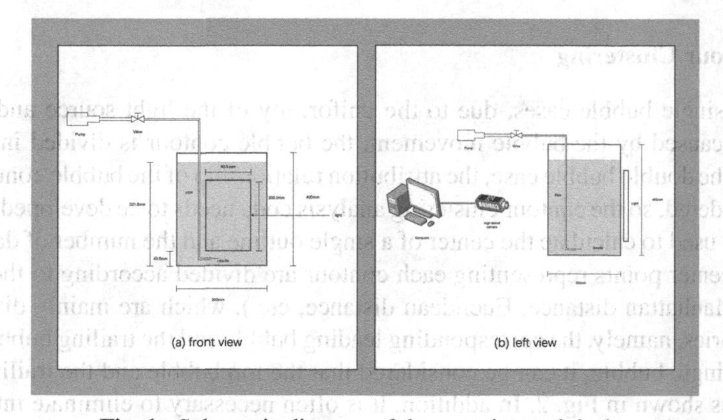

Fig. 1. Schematic diagram of the experimental device

Under the above experimental conditions, we generated a free-rising single bubble and a continuous rising double bubble. It is noted that it is difficult for parallel bubbles to combine into a single bubble due to the influence of initial needle diameter and spacing. Several experiments were carried out under each working condition to generate single and double bubbles repeatedly, and the rising motion of bubbles was photographed with a camera. The bubble video with high contrast, slight noise, and within the width of the light source is selected to study the law of motion and deformation. We used the following diameter needles to generate bubbles of different sizes (Table 1).

3 Image Processing

We use mature image processing technology, and the Python program and OpenCV library are selected to develop the corresponding image processing program. Image processing mainly includes four steps: preprocessing, image segmentation, contour clustering, and feature extraction.

3.1 Preprocessing

Preprocessing includes image frame clipping, keyframe interval selection, gray level processing, binarization processing, Gaussian filtering, etc. The image frame is cut to retain the image of the experimental section, and the width is slightly wider than the width of the light source. Keyframe interval selection mainly selects the video interval of the bubble rising process. Gray processing prepares for binarization, and binarization processes the image into a two-color image according to the threshold. Gaussian filtering can filter out part of the image noise.

3.2 Image Segmentation

Image segmentation is mainly to identify bubbles and the surrounding environment. The Canny operator processes the image to obtain the bubble boundary, and then the contour data is obtained using the *findcontours* function in OpenCV.

3.3 Contour Clustering

For some single bubble cases, due to the uniformity of the light source and the light refraction caused by the bubble movement, the bubble contour is divided into several parts. For the double bubble case, the attribution relationship of the bubble contour needs to be considered, so the contour clustering analysis code needs to be developed. The sub-function is used to calculate the center of a single outline and the number of data points. Then the center points representing each contour are divided according to the distance (such as Manhattan distance, Euclidean distance, etc.), which are mainly divided into two categories, namely, the corresponding leading bubble and the trailing bubble (for the case of a single bubble, it can be considered that the top bubble and the trailing bubble overlap), as shown in Fig. 2. In addition, it is often necessary to eliminate interference profiles. The elimination of the interference profile depends on the shape factor, which is defined as follows.

$$K = \frac{P^2}{4\pi S} \tag{1.1}$$

where P is the contour perimeter and S is the contour area.

3.4 Feature Extraction

According to the results of segmentation and clustering, the feature parameters such as bubble centroid are extracted, and the saved data are output for subsequent data processing. The pictures processed in four steps are shown in Fig. 2.

We define the center of mass and velocity of the bubble as follows,

$$X = \frac{\sum_{i=1}^{i=N} x_i}{N} \tag{1.2}$$

$$Y = \frac{\sum_{i=1}^{i=N} y_i}{N} \tag{1.3}$$

$$v_x^t = \frac{X^{t+1} - X^t}{\Delta t} s \tag{1.4}$$

$$v_y^t = \frac{Y^{t+1} - Y^t}{\Delta t} s \tag{1.5}$$

where x_i, y_i are coordinates of all pixels within the bubble contour. X and Y are obtained bubble centroid coordinates. s is the ratio of true distance to unit pixel. Δt is the time difference between two frames.

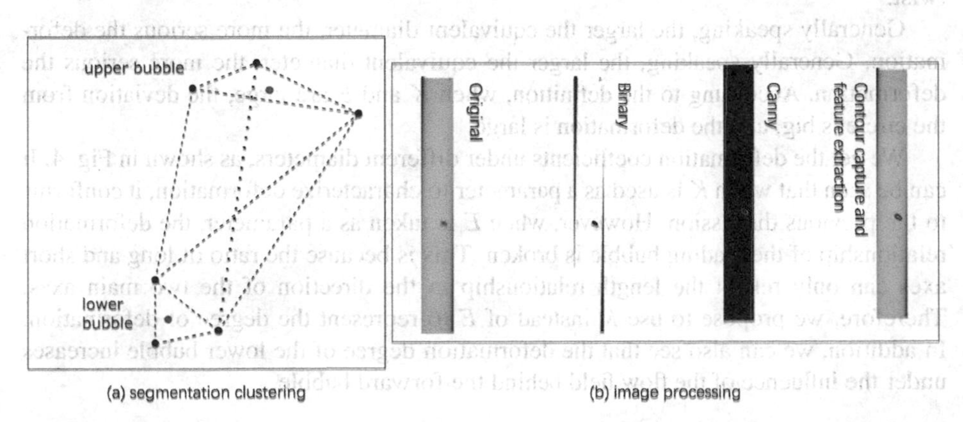

Fig. 2. Schematic diagram of segmentation clustering algorithm and image processing flow

The deformation degree and equivalent volume diameter are related to the long-axis b and short-axis a (assuming the bubble shape is elliptical). The deformation degree of the bubble can also be expressed by the aspect ratio, while the volume diameter could be defined as,

$$E = \frac{b}{a} \tag{1.6}$$

$$d_e = \sqrt[3]{b^2 a} \tag{1.7}$$

During the experiment, we also extracted dimensionless parameters, which are defined as follows,

$$\mathrm{Re} = \frac{\rho v d_e}{\mu} \tag{1.8}$$

$$Eo = \frac{d_e^2(\rho_l - \rho_g)g}{\sigma} \tag{1.9}$$

$$We = \frac{v^2 d_e \rho_l}{\sigma} \tag{1.10}$$

where ρ is the density, v is the velocity of the bubble, g is the gravitational acceleration, σ is the surface tension.

4 Result and Analysis

4.1 Trajectory and Deformation

The trajectory of bubble motion is different under different equivalent volume diameters. As shown in Fig. 3, the captured trajectory shows that almost all bubbles are zigzagging or spiraling. When the equivalent diameter is less than 5 mm, the bubble is easier to twist.

Generally speaking, the larger the equivalent diameter, the more serious the deformation. Generally speaking, the larger the equivalent diameter, the more serious the deformation. According to the definition, when K and E are large, the deviation from the circle is big, and the deformation is large.

We get the deformation coefficients under different diameters, as shown in Fig. 4. It can be seen that when K is used as a parameter to characterize deformation, it conforms to the previous discussion. However, when E is taken as a parameter, the deformation relationship of the leading bubble is broken. This is because the ratio of long and short axes can only reflect the length relationship in the direction of the two main axes. Therefore, we propose to use K instead of E to represent the degree of deformation. In addition, we can also see that the deformation degree of the lower bubble increases under the influence of the flow field behind the forward bubble.

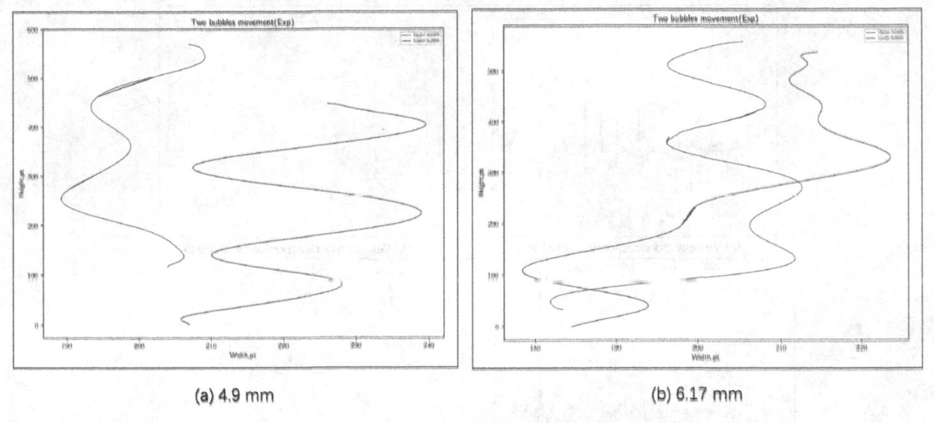

(a) 4.9 mm (b) 6.17 mm

Fig. 3. Centroid locus (red line: top bubble blue line: bottom bubble)

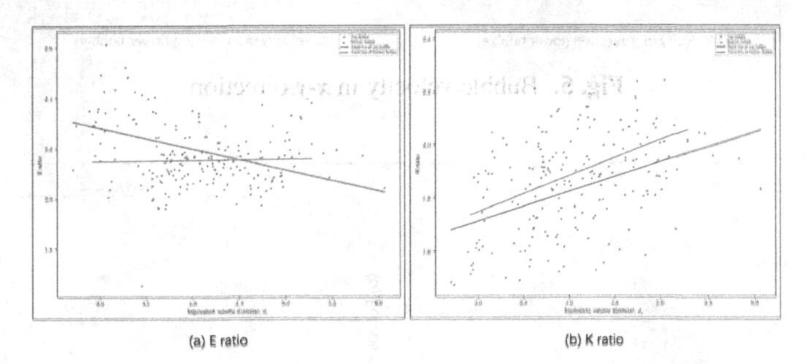

(a) E ratio (b) K ratio

Fig. 4. K and E under different equivalent diameters(E on the left, K on the right)

4.2 Instantaneous Parameters

We place the light source in the middle and rear part of the bubble movement and record the changes in velocity, relative distance, and dimensionless number with time. As shown in Fig. 5 and Fig. 6, the speed in the *x*-direction under different equivalent diameters presents periodic variation characteristics. The speed in the *y*-direction fluctuates up and down in the mean value.

It is observed that the *x*-direction distance between the two bubbles fluctuates around the mean value, while the *y*-direction distance may fluctuate or decrease. This is related to the relative velocity of the two bubbles. The dimensionless parameters also present periodic fluctuations, mainly related to the changes in equivalent diameter and speed.

4.3 Terminal Velocity

In the field of nuclear engineering, people are concerned about the change of terminal velocity with equivalent diameter.

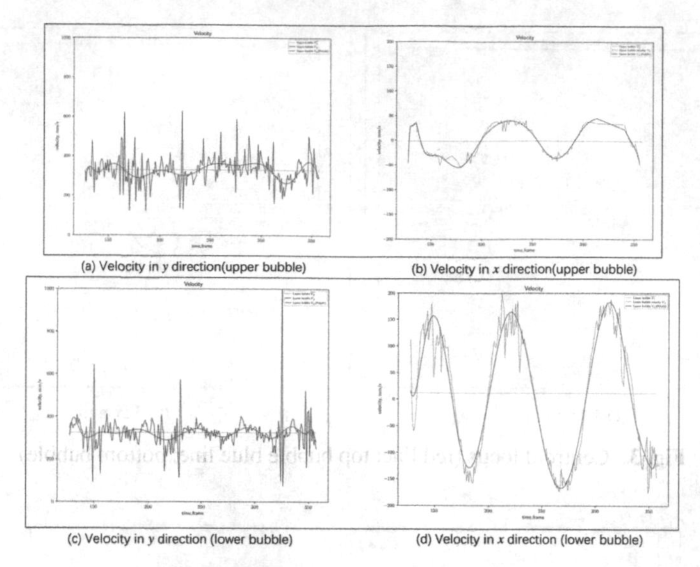

(a) Velocity in *y* direction(upper bubble) (b) Velocity in *x* direction(upper bubble)

(c) Velocity in *y* direction (lower bubble) (d) Velocity in *x* direction (lower bubble)

Fig. 5. Bubble velocity in ***x-y*** direction

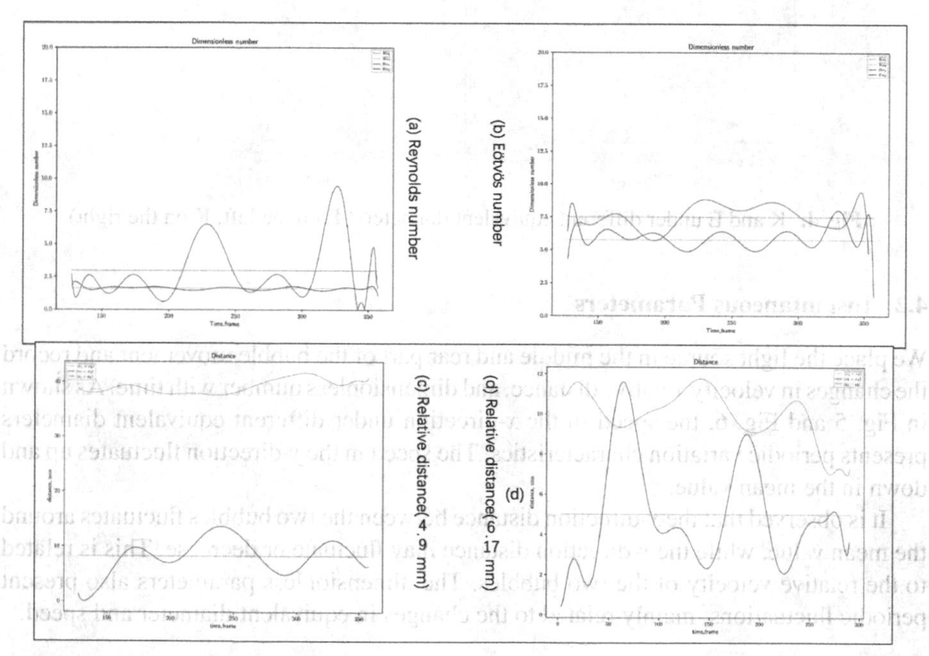

(a) Reynolds number (b) Eötvös number (c) Relative distance(4 .9 mm) (d) Relative distance(6 .17 mm)

Fig. 6. Schematic diagram of time variation of dimensionless parameters and relative distance under different equivalent diameters

As shown in Fig. 7, we used a series of correlations for prediction and found that the Wallis correlation had a significant error, Davis and Taylor[15, 16] correlation were slightly better, and Park correlation and Clift[1] correlation were close in trend. We

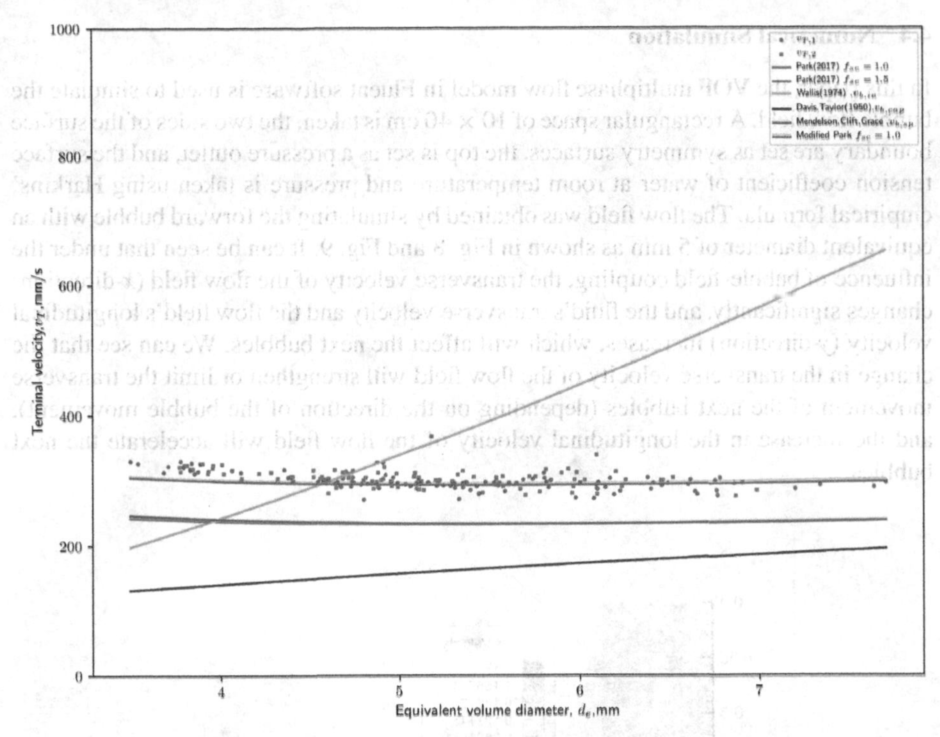

Fig. 7. Bubble terminal velocity under different equivalent diameters and fitting correlations

corrected the Park[13] correlation by adding a velocity offset term $v_{offset} = 0.06\,\text{m/s}$, and found that the predicted results were in good agreement with the experiment.

$$v_{\text{b,in}} = 0.14425 g^{5/6} \left(\frac{\rho_{\text{L}}}{\mu_{\text{L}}} \right)^{2/3} d_{\text{e}}^{3/2} \tag{1.11}$$

$$v_{\text{b, sp}} = \sqrt{\frac{2.14\sigma_{\text{L}}}{\rho_{\text{L}} d_{\text{e}}} + 0.505 g d_{\text{e}}} \tag{1.12}$$

$$v_{\text{b, cap}} = 0.721 \sqrt{g d_{\text{e}}} \tag{1.13}$$

$$v_{\text{b,park}} = \frac{1}{\sqrt{\frac{1}{v_{\text{b,sp}}^2} + \frac{1}{v_{\text{b, non - sp}}^2}}} = \frac{1}{\sqrt{f_{\text{sc}}^2 \left(\frac{144 \mu_{\text{L}}^2}{g^2 \rho_{\text{L}}^2 d_{\text{e}}^4} + \frac{\mu_{\text{L}}^{4/3}}{0.14425^2 g^{5/3} \rho_{\text{L}}^{4/3} d_{\text{e}}^3} \right) + \frac{1}{\frac{2.14\sigma_{\text{L}}}{\rho_{\text{L}} d_{\text{e}}} + 0.505 g d_{\text{e}}}}} \tag{1.14}$$

4.4 Numerical Simulation

In this paper, the VOF multiphase flow model in Fluent software is used to simulate the bubble flow field. A rectangular space of 10 × 40 cm is taken, the two sides of the surface boundary are set as symmetry surfaces, the top is set as a pressure outlet, and the surface tension coefficient of water at room temperature and pressure is taken using Harkins' empirical formula. The flow field was obtained by simulating the forward bubble with an equivalent diameter of 5 mm as shown in Fig. 8 and Fig. 9. It can be seen that under the influence of bubble-field coupling, the transverse velocity of the flow field (*x*-direction) changes significantly, and the fluid's transverse velocity and the flow field's longitudinal velocity (*y*-direction) increases, which will affect the next bubbles. We can see that the change in the transverse velocity of the flow field will strengthen or limit the transverse movement of the next bubbles (depending on the direction of the bubble movement), and the increase in the longitudinal velocity of the flow field will accelerate the next bubbles.

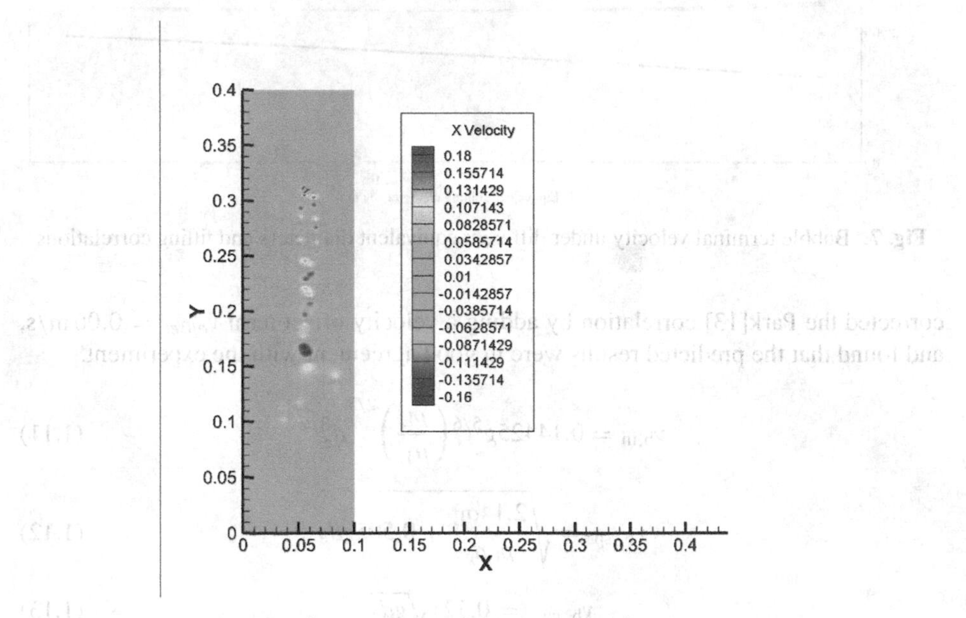

Fig. 8. Bubble flow field velocity (*x* direction)

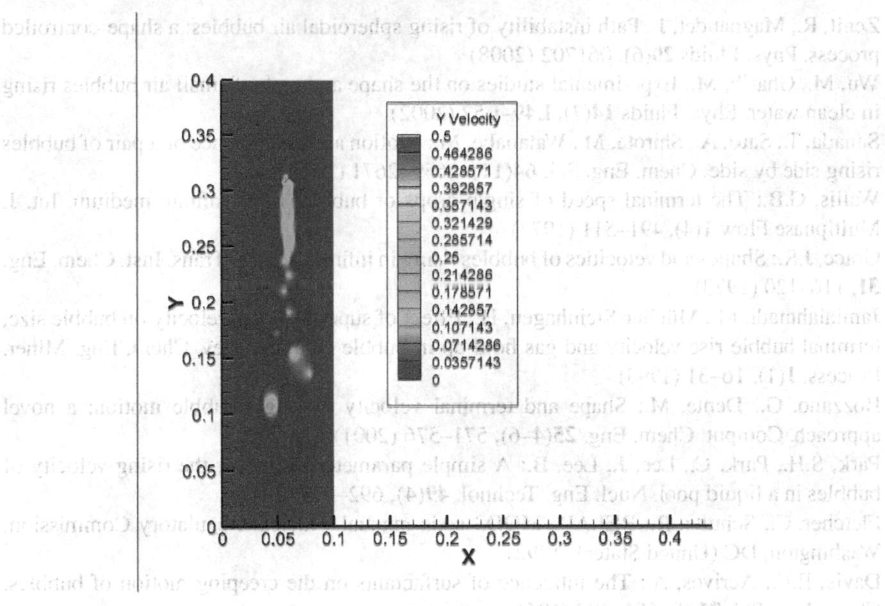

Fig. 9. Bubble flow field velocity (*y* direction)

5 Conclusions

To sum up, by generating double bubbles in different states, we developed an image processing program based on OpenCV to study the motion and deformation law of double bubbles with varying diameters of equivalent. We get the following conclusions: 1) the shape coefficient K can better reflect the shape change during bubble rising than the aspect ratio E. 2) The instantaneous parameters of two bubbles show periodic characteristics. 3) Both simulations and experiments show that the change of the tail flow field caused by the movement and deformation of the first bubble will affect the subsequent bubble and then change its movement. 4) The bottom bubbles were slightly accelerated by the flow field behind the top bubbles. We modified the Park correlation and achieved a good prediction of the terminal velocity of the experiment.

References

1. Clift, R., Grace, J.R., Weber, M.E.: Bubbles, Drops, and Particles. Dover Publications (2005)
2. Feng, Z.G., Michaelides, E.E.: Interparticle forces and lift on a particle attached to a solid boundary in suspension flow. Phys. Fluids **14**(1), 49–60 (2002)
3. Zhang, A.M., Cui, P., Cui, J., Wang, Q.X.: Experimental study on bubble dynamics subject to buoyancy. J. Fluid Mech. **776**, 137–160 (2015)
4. Duineveld, P.C.: The rise velocity and shape of bubbles in pure water at high Reynolds number. J. Fluid Mech. **292**, 325–332 (1995)
5. Raymond, F., Rosant, J.M.: A numerical and experimental study of the terminal velocity and shape of bubbles in viscous liquids. Chem. Eng. Sci. **55**(5), 943–955 (2000)

6. Zenit, R., Magnaudet, J.: Path instability of rising spheroidal air bubbles: a shape-controlled process. Phys. Fluids **20**(6), 061702 (2008)
7. Wu, M., Gharib, M.: Experimental studies on the shape and path of small air bubbles rising in clean water. Phys. Fluids **14**(7), L49–L52 (2002)
8. Sanada, T., Sato, A., Shirota, M., Watanabe, M.: Motion and coalescence of a pair of bubbles rising side by side. Chem. Eng. Sci. **64**(11), 2659–2671 (2009)
9. Wallis, G.B.: The terminal speed of single drops or bubbles in an infinite medium. Int. J. Multiphase Flow **1**(4), 491–511 (1974)
10. Grace, J.R.: Shapes and velocities of bubbles rising in infinite liquids. Trans. Inst. Chem. Eng. **51**, 116–120 (1973)
11. Jamialahmadi, M., Müuller-Steinhagen, H.: Effect of superficial gas velocity on bubble size, terminal bubble rise velocity and gas hold-up in bubble columns. Dev. Chem. Eng. Miner. Process. **1**(1), 16–31 (1993)
12. Bozzano, G., Dente, M.: Shape and terminal velocity of single bubble motion: a novel approach. Comput. Chem. Eng. **25**(4–6), 571–576 (2001)
13. Park, S.H., Park, C., Lee, J., Lee, B.: A simple parameterization for the rising velocity of bubbles in a liquid pool. Nucl. Eng. Technol. **49**(4), 692–699 (2017)
14. Fletcher, C., Schultz, R.: RELAP5/MOD3 code manual. Nuclear Regulatory Commission, Washington, DC (United States) (1992)
15. Davis, R.E., Acrivos, A.: The influence of surfactants on the creeping motion of bubbles. Chem. Eng. Sci. **21**(8), 681–685 (1966)
16. Davies, R.M., Taylor, G.I.: The mechanics of large bubbles rising through extended liquids and through liquids in tubes. Proc. Roy. Soc. Lond. Ser. A Math. Phys. Sci. **200**(1062), 375–390 (1950)

Monte Carlo Simulation and Analysis of Specified Element Samples by Nuclear Resonance Fluorescence Detection

Chen Zhang, Yu-Lai Zheng(✉), Qiang Wang, Yong Li, and Zi-Han Li

China Institute of Atomic Energy, P.O. Box 275-3, Beijing 102413, China
1059463405@qq.com, mczyl@sina.com

Abstract. The detection of explosives and drugs in large cargo distribution centers such as customs and logistics stations has a great effect on preventing smuggling crimes and terrorist incidents. However, the relatively thick shielding of container cargo makes the material composition information obtained by conventional detection methods such as X-ray transmission detection and imaging technology very limited. Nuclear Resonance Fluorescence (NRF) is an emerging nondestructive assay technology that uses the specific resonance energy of nuclides to identify unknown nuclides, which can be used to detect and analyze the isotopic composition of the inspected cargo. In this paper, according to the theoretical analysis of NRF, Geant4 is used to build the NRF backscatter detection model, the collimator structure of the electron accelerator and the background shield of the NRF signal are optimized and calculated, and the NRF process with ^{12}C as the target element is simulated and calculated. The results show that the simulated characteristic energy spectrum of NRF signal is consistent with the theory, the designed background shielding scheme meets the needs of NRF signal identification and detection, and the simulated signal-to-noise ratio data provides the basis for the experiment.

Keywords: Nondestructive Assay · Backscatter · Nuclear Resonance Fluorescence · Genat4 · Nuclear Safeguards

1 Introduction

In recent years, the transportation of cargo through customs containers and land trucks has become a major part of world trade. At the same time, illegal event such as smuggling commercial contraband, explosives, and special nuclear materials (SNM) through borders and customs are also rampant. According to the International Atomic Energy Agency's Illicit Trafficking Database (ITDB), between 1993 and 2021, countries reported a total of 3,928 nuclear safety incidents. In 2021 alone, 32 countries reported 120 incidents to the ITDB [1].

NRF has developed rapidly in the fields of explosives detection, container security inspection and nuclear weapons inspection in recent years. It describes that the nucleus absorbs a photon through resonance and is excited to a specific excited state, and then

C. Liu (Ed.): PBNC 2022, SPPHY 283, pp. 985–993, 2023.
https://doi.org/10.1007/978-981-99-1023-6_83

de-excited by emitting one or more photons, as shown in Fig. 1. For each isotope (Z > 2), the gamma-ray energy spectrum produced by NRF is different. If the gamma-ray beam is adjusted to a specific energy, specific nucleons can be detected by nuclear resonance fluorescence, and the isotopic composition of the cargo can be analyzed. In addition, the emitted NRF photon energy is several MeV, which can penetrate most materials and boxes.

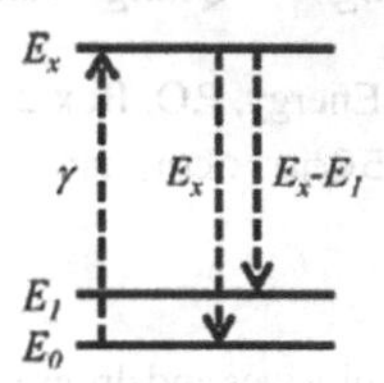

Fig. 1. NRF excitation-de-excitation process

In this paper, an optimization scheme is designed for NRF detection based on an electron accelerator. The influence of different shield thicknesses and different target thicknesses on the NRF count rate is studied from theoretical calculations and Monte Carlo analysis, and a target with ^{12}C as the target element is simulated and calculated. The optimized simulation scheme and the characteristic energy spectrum of the NRF process of the target were obtained.

2 Physical Background

NRF is a typical X(γ, γ')X reaction, which includes two basic processes: transition by absorbing energy; decay by releasing energy. At absolute zero T = 0K, when a photon with energy E is incident, the target atom absorbs energy through isolated resonance to reach the energy level E_r, and then decays to the energy level E_j, and its cross section follows the single-level Breit-Wigner profile [2]

$$\sigma^{NRF}(E) = \pi \frac{2J_1 + 1}{2(2J_0 + 1)} (\frac{\hbar c}{E})^2 \frac{\Gamma_{r,0}\Gamma_{r,j}}{(E - E_r)^2 + (\Gamma_r/2)^2} \tag{1}$$

where J_0 and J_r are the nuclear spins at the ground state and resonance level, respectively. The terms $\Gamma_{r,0}$ and $\Gamma_{r,j}$ denote the partial widths of decay from E_r to ground state and from E_r to E_j, while Γ_r is the total width of the energy level transition. At non-absolute zero degrees (T $\neq$ 0K), the energy level broadening of the NRF reaction cross-section will decrease the NRF cross-section, but compared with the cross-section values of other electromagnetic interaction processes such as photoelectric effect, Compton scattering and electron pair effect, the NRF cross-section still comparable to or even exceeding [3], so even a small amount of special nuclear material can be detected with high sensitivity of the detection system. Due to the conservation of energy and momentum, free nuclei undergoing NRF will recoil with kinetic energy. The recoil energy E_{rec} is determined by the following Compton-like formula [4]

$$E_{rec} = E[1 - \frac{1}{1 + E(1 - cos\chi)/Mc^2}] \cong \frac{E^2}{Mc^2}(1 - cos\chi) \tag{2}$$

where χ is the photon scattering angle relative to the incident direction. For bound nuclei in the atomic lattice, E_{rec} may be large enough to overcome the lattice displacement energy E_d, in which case the kinetic energy transfer is $E_{rec} - E_d$. For unbound nuclei $E_{rec}E_{rec}$ less than E_d, the recoil is transferred across the entire lattice and recoilless NRF is achieved. For outgoing photons, the energy decreases accordingly by $E_{rec} \gg \Delta E_{rec}$, but the energy of photons emitted in the backward direction will be much lower than the resonance energy E_r and it is highly unlikely that another NRF interaction will occur in the same target atom. According to the definition of the angle integral, the differential angular cross-section of the NRF outgoing photon is defined as

$$\frac{d\sigma_r^{NRF}(E)}{d\Omega} = \frac{W(\chi)}{4\pi}\sigma_r^{NRF}(E) \tag{3}$$

The angular correlation function [5] $W(\chi)$ is symmetric around $\chi = \pi/2$, so $W(\chi) = W(\theta)$, where the emission angle θ is relative to the back-beam direction. Then, the angular correlation function $W(\theta)$ is shown in formula (4).

$$W(\theta) = 1 + (R/Q)cos^2\theta + (S/Q)cos^4\theta \tag{4}$$

The constants (R/Q) and (S/Q) represent the contributions of dipole and quadrupole transitions, respectively, and are determined by the sequence of spins $J_0 \rightarrow J_r \rightarrow J_j$. In most experiments, it was roughly isotropically distributed. According to formula (4), the scattering angle of the NRF of ^{12}C is calculated in the order of $0^+ \rightarrow 2^+ \rightarrow 0^+$. The parameters and calculation results are shown in Table 1.

Table 1. Resonance excitation-de-excitation parameters, NRF cross section and scattering angle of ^{12}C

Isotope	$E_r[MeV]$	$\Gamma_r[MeV]$	$\int \sigma_r^{NRF}(E)[eV \cdot b]dE$	(R/Q) (S/Q)	θ
^{12}C	4.438	10.8	6.939	0,4	90°

3 Simulation Setup

Figure 2 is a schematic of the constructed Geant4 simulation model. The model is as follows: the backscattering method is adopted as a whole, a high-energy electron beam is generated by an electron accelerator, electron beam produces X-rays via bremsstrahlung, the collimator confines X-rays to a certain angle range, then high-energy X-rays bombard the target to generate an NRF reaction. The HPGe detector located at 110° relative to the main beam direction receives the NRF photon signal. The detector is provided with a low-energy background shielding lead on the axial end face, and a thick shielding lead facing the accelerator to block the X-Rays from the accelerator.

Figure 3 shows the design structure of the collimator. The overall design is a cylinder. The radius of the inlet end and the outlet end of the cylinder are different, so that the

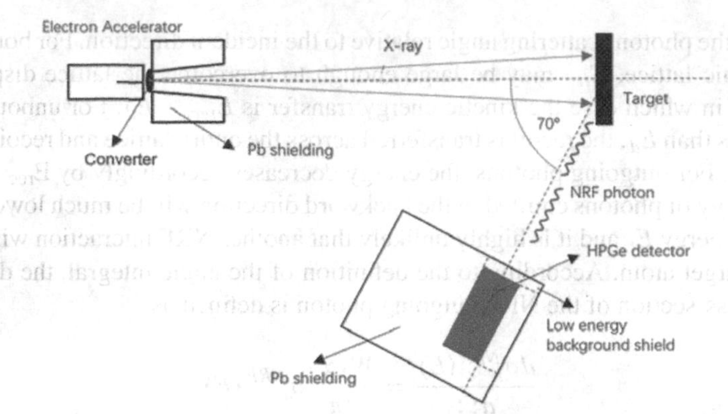

Fig. 2. NRF detection design structure

X-ray with the second highest dose at the scattering angle of 3° –7° is not shielded, and the lifting device overall efficiency. The tungsten alloy converter is placed at the exit of the accelerator beam and at the entrance of the collimator. The inner diameter of the inlet end of the collimator is 0.5 cm, the outer diameter is 5 cm, the inner diameter of the outlet end is 1.2 cm, the outer diameter is 5 cm, and the overall axial length is 10 cm. The diameter of the inlet end of the collimator completely fits the diameter of the converter.

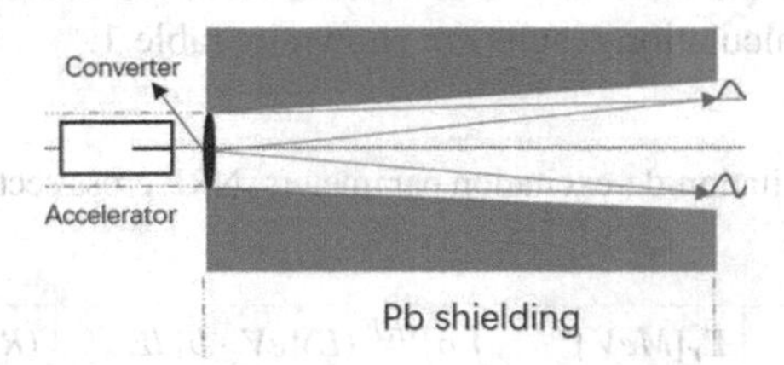

Fig. 3. Collimator structure diagram

The target material is graphite with ^{12}C as the target element. The target shape is a square thick sheet, the width is 10.2 cm × 10.2 cm, and the target thickness is 1.25 cm. The count rate of the NRF peak corresponds to the mass thickness ρD of the measurement object. As thickness increases, NRF photon yield increases according to a power function variant relationship; the NRF photons emitted forward in the thick target have no recoil energy reduction, and the emitted photons can also cause NRF reactions. The thin lead shielding layer of the detector entrance window is 1.25 cm, which can reduce the low-energy background from the target and improve the NRF count rate of the detector. The detector faces the accelerator lead shield of 25 cm, which shields the background photons from the accelerator and other scatterers.

According to the NRF photon exit angle of ^{12}C, it can be seen that the optimal setting angle of the detector is 90°. However, the interrogation system needs to inspect various explosives and nuclear materials. Combined with the 125° scattering angle set for the nuclear material ^{238}U and the approximate isotropic NRF scattering angle characteristics, this design adopts a 110° scattering angle design.

The isotope contained in nuclear materials include ^{238}U, ^{235}U, and the isotope contained in explosives include ^{12}C, ^{14}N, ^{16}O, etc. The NRF energy levels of ^{238}U and ^{235}U are concentrated in 1.5 MeV–2.5 MeV, while the NRF energy levels of ^{12}C, ^{14}N, and ^{16}O are concentrated in 4 MeV–7.5 MeV, the photons in this energy range can easily penetrate the planar and coaxial HPGe detectors with low relative efficiency, and NRF requires excitation at higher incident energy. Therefore, this paper chooses the coaxial HPGe detector with a relative efficiency of 95%, the axial length of the crystal is 80.5 mm, and the radius is 79.5 mm.

4 Simulation Results

The simulation setup for the electron accelerator is as follows: the angular distribution is a Gaussian distribution with a one-dimensional 5° broadening, the energy is a Gaussian distribution with a center value of 10 MeV and a 0.001 MeV broadening, and the number of incident γ particles is 2×10^7. Figure 4 is a thermal diagram of the X-ray count rate at a distance of 20 cm from the converter without a collimator, bright colors indicate high counts, cold colors indicate low counts, and the horizontal and vertical coordinates indicate distances in millimeters. It can be seen that the X-ray count at the central position is higher, and the count rate gradually decreases with the increase of the distance from the central axis.

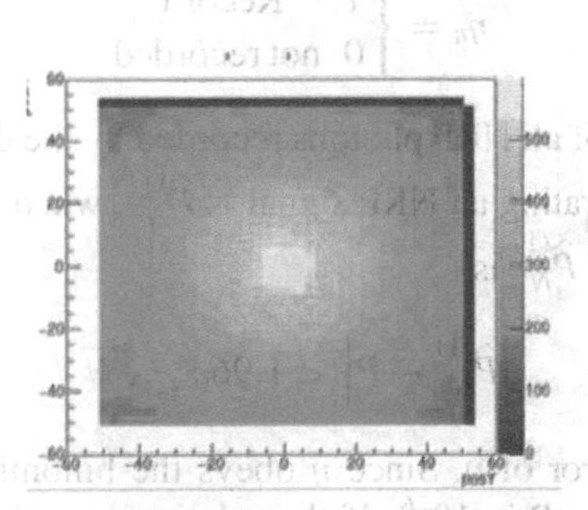

Fig. 4. X-ray energy spectrum generated by converter and count distribution at 20 cm

The selection of the distance between the target and the collimator is constrained by the beam diffusion angle of the collimator and the limitation of the target area; at the same time, the target and the detector have a linkage relationship, that is, if the target is too close to the collimator, the detector will move, increasing its background value. In order to make the X-rays completely incident on the target, the X-ray count rate distributions on the target were measured at distances of 40 cm and 50 cm from the target and the collimator, respectively, as shown in Fig. 5.

It can be seen from the Fig. 5 that as the distance increases, the count rate distribution area on the target increases. When the distance is 50 cm, the counted area ratio can reach 65.6%, and no X-rays escape from the target, which can provide the highest NRF signal

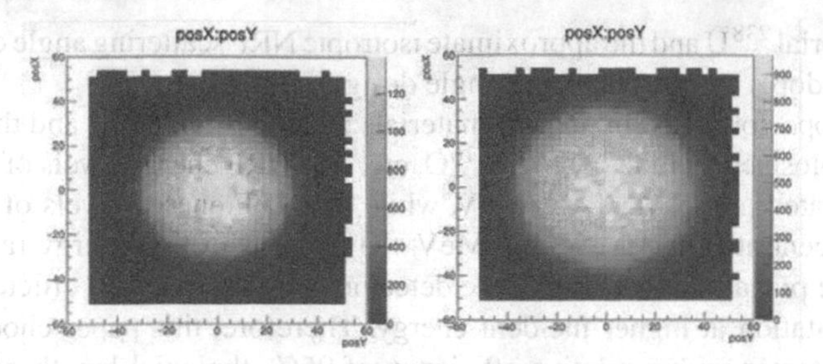

Fig. 5. Count distribution on target Left: Distance 40 cm; Right: Distance 50 cm

rate, which meets the experimental requirements. The purpose of masking is to obtain a clearer NRF peak, and the evaluation factor is introduced accordingly

$$Y_0 = n_{E_r}/n_{off}$$

where n_{E_r} is the NRF photon count generated at a specific resonance energy level when the incident energy is E, n_{off} is the total count of scattered X-rays (background part) received on the detector from the electron accelerator when the source term is unchanged, Its significance is to test the ability of lead shielding to block the corresponding background when generating quantitative target counts (NRF photons). The NRF photons is recorded by the detector, contributing η_n. Then

$$\eta_n = \begin{cases} 1 & \text{Record} \\ 0 & \text{not recorded} \end{cases}$$

To count the number of all NRF photons recorded by the detector as N_1, the approximate probability of generating an NRF signal is $\widehat{P}_N^{(1)}$, when the confidence coefficient is 1-α = 0.95, the error of $\widehat{P}_N^{(1)}$ is

$$\left|\widehat{P}_N^{(1)} - P\right| < 1.96\sigma_\eta/\sqrt{N}$$

σ_η is the mean square error of η. Since η obeys the binomial distribution, according to the simulation analysis, P is 10^{-6}. If the relative error is required to be less than 5%, The magnitude of N is 10^9. A bias factor of 10 is applied to the physical process of bremsstrahlung of electrons, and the simulation result with a detector shielding thickness of 25 cm is counted as 3. This count can be considered to be no statistically. At this time, the evaluation factor Y_0 can meet the design requirements.

In addition, comparing the X-ray energy spectrum before and after adding the shield on the side of the detector facing the accelerator is helpful to evaluate the performance of the shield. Figure 6 shows the X-ray energy spectrum of the shielded entrance end of the detector when the incident particle is 2×10^7. The number of X-ray particles can be statistically 1.5×10^5, while the statistical particle number on the target is 3.3×10^5. According to the calculation of the shielding effect, when the incident particle is 10^9, the shielding ability of X-rays is 7.5×10^6, and it is 1.6×10^7 on the main beam, so the shielding effect of this thickness can at least reach 1.6×10^7 particles (the shielding

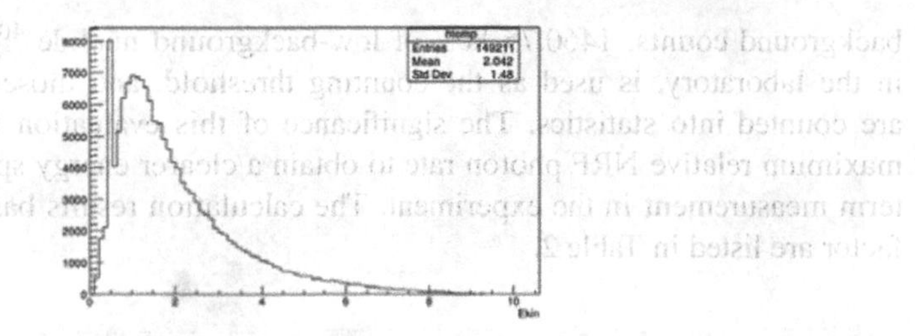

Fig. 6. Detector shield entrance window energy spectrum

effect is at least 6.3×10^{-8}) on the main beam, the detector will receive the background from the electron accelerator.

The source term of the NRF simulation adopts a Gaussian distribution with an energy center of 4.438 MeV and a spread of 0.001 MeV. The surface source is 0.5 mm × 0.5 mm, the length and width are spread by 0.1 mm, and the angular distribution is spread by 0.5° along the axial direction facing the target. The incident particle is γ, and the number is 2 × 10^7. When the distance between the target and the detector is 20 cm, the thickness of the target is 1.25 cm, 2.5 cm, 3.75 cm, 5.0 cm respectively, and the simulation is carried out without adding a thin lead shielding layer on the axial plane of the detector, as shown in Fig. 7 for the detector energy spectrum.

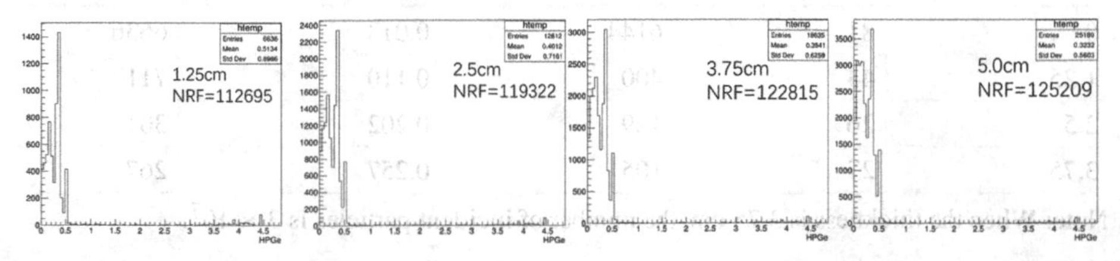

Fig. 7. Detector energy spectrum when the target thickness is 1.25 cm, 2.5 cm, 3.75 cm and 5.0 cm

It can be seen from Fig. 7 that the change of target thickness has no significant effect in the energy spectrum of the signal generated by the target, and the shape of the energy spectrum after normalization has no significant difference. Secondly, in the experimental measurement, the thickness of the sample is expected to be as thin as possible. The purpose of thinning is to reduce the absorption of the γ-ray by the sample material when it is transported in the sample, which is commonly referred to as "self-absorption" [6], thin sample thickness can weaken the count stacking of the detector.

Adding a thin lead shield to the detector entrance window can reduce the low-energy background from the target, thereby increasing the detector's NRF count rate. When the distance between the target and the detector is 20 cm, and the source term is set as above, compare the NRF counts before and after adding a thin lead layer and with different lead layer thicknesses (1.25 cm, 2.5 cm, 3.75 cm lead layer thickness), as shown in the Fig. 8. It can be known that adding a thin lead layer to the detector entrance window, with the increase of lead layer thickness, the shielding effect on the low-energy background from the target becomes more and more significant. For low background counts, 1460.75 keV of low-background nuclide ^{40}K, which is common in the laboratory, is used as the counting threshold, and those below this threshold are counted into statistics. The significance of this evaluation factor is to select the maximum relative NRF photon rate to obtain a clearer energy spectrum through long-term measurement in the experiment. The calculation results based on this evaluation factor are listed in Table 2.

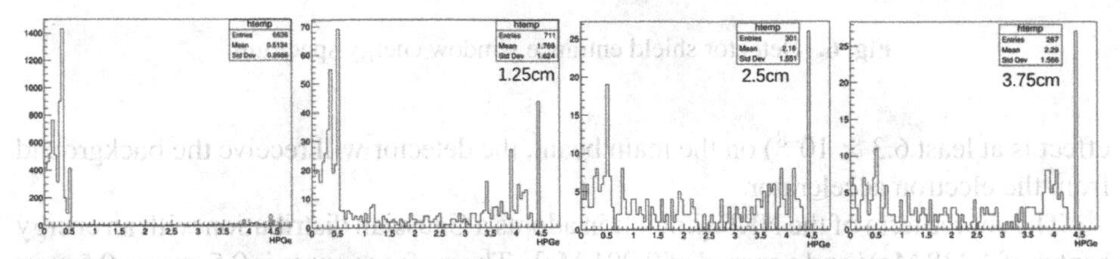

Fig. 8. The order from top to bottom and left to right is lead-free layer, 1.25 cm, 2.5 cm, 3.75 cm lead layer thickness

Table 2. Thin lead layer thickness and count

Target thickness [cm]	NRF count rate	Low background count	Evaluation factor Y	Total count
0	81	6144	0.013	6636
1.25	44	400	0.110	711
2.5	26	129	0.202	301
3.75	27	105	0.257	267

Note: When the thickness is 3.75 cm, the number of incident particles is 3×10^7.

It can be seen from the Table 2 that as the thickness of the thin lead layer increases, the low background count decreases rapidly, and the low background count changes weakly after the lead shielding thickness reaches 2.5 cm; considering that the resonance energy peak region of ^{238}U and ^{235}U is 1.5 MeV–2.5 MeV, the attenuation amplitude should not be too large, and the thickness of the thin lead layer of 2.5cm is suitable.

5 Conclusion

In this paper, by analyzing the relationship between the physical process of X-ray excited NRF process and setting different target parameters, shield thickness and detection efficiency, the following conclusions can be drawn: (1) The yield of NRF photons is ~ 10^6/An incident particle under the calculation conditions in this paper, which requires long-time detection. Optimizing the parameters of the shielding material can improve the de-spectral efficiency; (2) Optimized lead shield thickness for X-rays from accelerators. At a thickness of 25 cm, the shielding effect can be improved. The main beam/background is ~ 6.3×10^8; (3) The thickness of the target has a weak influence on the detection efficiency, and the computational performance can be reduced in order to reduce "self-absorption" and prevent wasted computing performance of emitted recoilless NRF photons.

This paper provides a design method of an NRF backscatter detection scheme. The advantages of this structure design are: on the basis of ensuring the NRF detection efficiency, the design reduces the thickness of the shield, reduces the overall weight of the equipment, and improves the feasibility and practicality of technology application.

Acknowledgements. This work was funded by Continuous Basic Scientific Research Project (WDJC-2019-09).

References

1. Incident and Trafficking Database (ITDB). 2022 Factsheet. https://www.iaea.org/resources/databases/itdb,2022-01
2. Jordan, D.V., Warren, G. A.: Simulation of nuclear resonance fluorescence in Geant4. In: 2007 IEEE Nuclear Science Symposium Conference Record, pp. 1185–1190 (2007)
3. Huang, W., Yang, Y., Li, Y., et al.: Research on SNM detection technology based on LINAC. In: The fifteenth Proceedings of the National Academic Annual Conference on Nuclear Electronics and Nuclear Detection Technology, pp. 505–511 (2010)
4. Vavrek, J.R., Henderson, B.S., Danagoulian, A.: High-accuracy Geant4 simulation and semi-analytical modeling of nuclear resonance fluorescence. Nucl. Instrum. Methods Phys. Res. Sect. B **433**, 34–42 (2018)
5. Hamilton, D.R.: On directional correlation of successive quanta. Phys. Rev. **58**, 121–131 (1940)
6. Zhu, C., Chen, Y., Guo, H., et al.: Research on detection efficiency of high-purity germanium detectors. Nucl. Electron. Detect. Technol. **26**(2), 191–194 (2006)

Research on Public Communication of Small Reactor

Rongxu Zhu(✉), Feng Zhao, Xiaofeng Zhang, Jiandi Guo, and Meng Zhang

Suzhou Nuclear Power Research Institute, Suzhou, Jiangsu, China
zrx441@163.com

Abstract. Based on the conclusions of small reactor environmental impact assessment and social stability risk assessment of small reactors, this paper conducts a special study on the public communication of small reactors. This article discusses the work content, scope and form of public communication of small reactor, and give suggestions on the scope and form of public publicity, public participation and information disclosure. It is matched with small reactor public communication theoretical system and operation manual.

The research on the public communication work of small reactor can provide necessary theoretical basis and technical support for the specific public communication work of small reactor. Provide decision support for local policies and construction departments to implement public communication. It is hoped that this study can provide some reference for the public communication of small reactors in the exploratory stage.

Keywords: small reactor · public communication · NIMBY · EIA · SSRA

Nomenclature

EIA	Environmental Impact Assessment
SSRA	Society Stability Risk Assessment
MEE	Ministry of ecological environment of the people's Republic of China
NDRC	National Development and Reform Commission of the People's Republic of China
PWR	Pressurized Water Reactor
PAZ	Precautionary Action Zone

1 Introduction

In order to expand the application scope of nuclear energy, ensure energy security and meet the needs of low-carbon energy development, China has invested a lot of human and material resources in the development of small reactors in recent years, and has achieved a series of results. The natural avoidance effect of nuclear related projects and the particularity of the utilization of small reactors, such as being close to cities and residents, bring new challenges to public communication. If we simply copy the guidelines

C. Liu (Ed.): PBNC 2022, SPPHY 283, pp. 994–1001, 2023.
https://doi.org/10.1007/978-981-99-1023-6_84

for public communication of nuclear power projects to carry out public communication of small reactors, it is difficult for enterprises and local governments to bear such a huge workload and resource investment, Therefore, it is necessary to study the public communication of small reactors.

2 Relevant Background

In 2012, in order to promote scientific decision-making, democratic decision-making and legal decision-making, prevent and resolve social contradictions, NDRC issued the notice of the national development and Reform Commission on the Interim Measures for social stability, risk assessment of major fixed asset investment projects, requiring NDRC to approve or report to the State Council for approval. Fixed asset investment projects constructed and implemented within the territory of the people's Republic of China shall be subject to SSRA.

The measures for public participation in environmental impact assessment issued by MEE stipulates that the construction departments of reactor facilities and commercial spent fuel reprocessing plants with a core thermal power of more than 300 MWth shall listen to the opinions of citizens, legal persons and other organizations within a radius of 15 km of the facilities or reprocessing plants; The construction departments of other nuclear facilities and uranium mining and metallurgy facilities shall listen to the opinions of citizens, legal persons and other organizations within a certain range according to the specific conditions of environmental impact assessment[1].

The guidelines for public communication of nuclear power projects issued by MEE stipulates that public communication of nuclear power projects should pay attention to the public within a certain range around the plant site (usually 30 km radius of the plant site) that may be directly or indirectly affected by the project construction and operation, and focus on the public within 5 km radius of the plant site [2].

In 2017, the national nuclear accident emergency office issued the guidance on nuclear emergency work of onshore small PWR, which clearly put forward that "the recommended range of small PWR emergency planning area shall not be greater than 3 km, and the specific range shall be proposed by the operating unit after systematic demonstration and scientific calculation, and determined according to the specified procedure".

In the draft for comments on the principles and requirements for the division of nonresidential areas and planning restricted areas of small nuclear power plants, for light water reactors with a single reactor thermal power of less than 300 MWth, the boundary of nonresidential areas can generally be consistent with the plant boundary of nuclear facilities, and the distance from the reactor is generally not less than 100 m; The distance between the boundary of the planned restricted area and the reactor shall not be less than 1 km.

Arrangements for preparedness for a nuclear or radiological emergency (IAEA GS-G-2.1) recommends that the PAZ with reactor thermal power of 100–1000 MWth is 0.5–3 km away from the plant site [3].

3 Small Reactor SSRA

The 6 × 200 MWth small reactor project is planned and arranged at one time and implemented by stages. Phase I project construction 2 × 200 MWth small reactor. The social stability risk investigation scope of the project includes the interest related groups directly and indirectly affected by the project, mainly including: residents, heating enterprises, relevant heat users and other social organizations directly affected by the siting, construction and operation of the project within 5 km of the project site, sensitive objects around the project, including planters, farms, industrial and mining enterprises, schools, hospitals, etc. And government departments in the project location and village committees in the area around the plant site; Mass media and online new media in the project site; Pay due attention to the opinions, suggestions and demands of non-local residents on the project.

The personal questionnaire survey is mainly organized within 5 km around the plant site and the urban area where the project is located. A total of 720 questionnaires were distributed and 720 questionnaires were collected, of which 687 were valid questionnaires (invalid questionnaires were mainly due to lack of ID number, incorrect contact information or incorrect information, etc.). The respondents are mainly men. The main reason is that the household survey is generally filled in by men. The age distribution is mainly middle-aged and young people, taking into account the actual situation of many lefts behind elderly people in rural areas. The occupation is mainly farmers, taking into account the distribution of enterprise and institution employees, individual industrial and commercial households and other professionals. The education level is mainly from junior high school to senior high school, which can effectively reflect the real opinions and demands of the respondents. At the same time, there are a small number of highly intelligent elements, which are representative and extensive. The main findings are as follows [4] (Table 1).

Table 1. Classified statistics outcome of public's attitude to project

Question	Outcome	Proportion	Remarks
Do you understand that this project is a nuclear project	YES	62.6%	1 person did not fill in
	NO	37.3%	
How well do you know the project	Very well	8.2%	2 persons did not fill in
	Basic	28.4%	
	A little	38.4%	
	NO	24.7%	

(*continued*)

Table 1. (*continued*)

Question	Outcome	Proportion	Remarks
Channels of understanding of the project	meeting	4.7%	This topic is multiple choice, and 3 persons did not fill in
	Notice	4.8%	
	Public discussion	76.6%	
	TV	0.9%	
	website	1.3%	
	Nuclear science popularization activities	8.2%	
	micro-blog	0.6%	
	WeChat	11.6%	
	other	1.6%	
Issues most concerned about project construction	environmental effect	39.7%	This topic is multiple choice
	Land requisition compensation	56.6%	
	Providing employment opportunities	57.2%	
Issues most concerned about project construction	conserve energy, reduce emissions	3.5%	This topic is multiple choice
	Nuclear safety issues	40.5%	
	Economic drive	23.0%	
	Safety and quality during construction	11.8%	
	Stability of heating	24.0%	
	other	0.4%	

The construction departments has carried out a small amount of popular science publicity before this questionnaire survey, but the coverage is limited. According to the questionnaire survey, more than half of the respondents understand that the project is a nuclear power project, most of the respondents have a certain understanding of the project, and a few respondents do not understand the project. The respondents understand the project mainly through public discussion, WeChat and nuclear science popularization activities; Respondents focused on land acquisition compensation, employment opportunities, environmental impact, nuclear safety issues, stability of heating, etc.

4 Small Reactor EIA

The environmental impact assessment of the small reactor shows that the radioactive waste gas generated during normal operation is discharged into the atmospheric environment through the chimney after being treated to meet the annual total emission limit specified in GB6249. The radioactive waste liquid generated is reused after being treated to meet the radioactivity level and annual total emission limit specified in GB6249. If it cannot be reused completely, it is discharged into the atmospheric environment through carrier evaporation. The maximum individual effective dose caused by the discharge of airborne effluent is less than 0.25 mSv, which meets the provisions of GB6249 and the safety review principles of small pressurized water reactor nuclear power plant. At the same time, it meets the public individual effective dose constraint value (0.02 mSv/a) caused by two small reactors, and its radiation impact on the environment is acceptable [5].

Within the duration of the site selection accident, the effective dose of public individuals (adults) at the site boundary is less than 10 mSv, and the thyroid equivalent dose is less than 100 mSv, which meets the safety review principles of small PWR nuclear power plant.

The emergency planning area is preliminarily divided into a small reactor centered area with a radius of 1 km of the plant site. The feasibility analysis results of implementing emergency plan at the plant site show that there are no insurmountable difficulties in implementing off-site emergency plan for nuclear accidents.

5 Small Reactor Public Communication

According to the conclusions of EIA and SSRA of small reactor, special research on public communication of small reactor is carried out, discussing the work content, scope and form of public communication of small reactor, and suggesting on the scope and form of public publicity, public participation and information disclosure. So Finally, a complete theoretical system of small pile public communication is developed..

5.1 Work Scope

The public communication scope of small reactor includes the public, enterprises, institutions and social organizations that may be directly or indirectly affected by the project construction and operation within 5 km of the plant site, focusing on the public, enterprises and institutions within 3 km. For nuclear heating reactors, attention should also be paid to the residents around the heating pipeline and end users (relevant residents and enterprises, etc.).

5.2 Work Content

The contents of public communication include public publicity, public participation, information disclosure and public opinion response. The public communication work should be guided by the local government, the construction company shall provide relevant resource guarantee, and the professional technical company shall be responsible for the whole process of technical consultation.

The public publicity shall be based on increasing the public's understanding of nuclear energy and the project, with the purpose of improving the public's acceptability. A combination of visits, exhibition halls, expert lectures and other means shall be adopted to ensure the full coverage of the public publicity objects. Before the publicity work is carried out, the publicity objects should be reasonably and scientifically classified, so as to make the publicity work more targeted, so as to achieve twice the result with half the effort. At the same time, in combination with the social conditions and historical contradictions of the project location, the local public's acceptance of publicity contents and methods should be fully considered, so as to achieve the purpose of both publicity and risk control.

In principle, public participation should include questionnaire survey and symposium. The questionnaire shall be designed according to the project impact and public concerns to ensure easy to understand and comprehensive content. The questionnaire survey shall be determined according to the distribution of the resident population of the public near the plant site and the impact of stakeholders. Generally, the individual questionnaire shall be distributed by household. The questionnaire survey shall include stakeholders, especially direct stakeholders (residents of land acquisition and demolition, enterprises and institutions affected by the project construction), and shall include expert representatives of relevant government departments and professional institutions. When selecting the respondents, we should consider the selection of individuals and organizations with certain professional knowledge background and social level, and also pay attention to the wide representation of the respondents in terms of age, gender, educational background, occupation, etc. The specific number of individual questionnaires shall be decided based on the number of resident households around the project site. Generally, it should not be less than 300.

Information disclosure shall be implemented with reference to the measures for public participation in environmental impact assessment, guidelines for public communication of nuclear power projects and other relevant documents. It needs to be carried out in the form of newspapers, websites and paper posts. The scope of paper posting suggestions is the scope of public communication of the project. The posting place is the bulletin board of relevant administrative village committee and township government, and the postings last for 10 working days.

The construction company and local government shall establish a stability maintenance linkage mechanism and a public opinion response system. Carry out daily public opinion monitoring, conduct 24-h supervision during the information announcement, study and judge the public opinion of the project as soon as possible, and report the negative public opinion as soon as possible. During the publicity period, increasing the update cycle and amount of information of government official websites, microblogs, forums and popular science news can strengthen the effect of online popular science publicity.

5.3 Work Suggestion

(1) Strengthen public publicity

Strengthen the popularization and publicity of the basic knowledge of nuclear energy utilization, reduce and eliminate the public's anxiety and fear about nuclear safety. In view of the public's doubts and concerns about nuclear safety and the impact of radiation on the environment, take the government as the leading role and cooperate with enterprises, do a good job in the publicity, education and public opinion guidance on nuclear safety and the impact of nuclear radiation, and constantly strengthen the communication with the residents near the plant site. Carry out popular science publicity in a way acceptable to the local public. Pay attention to further strengthen popular science publicity for heat users, carefully listen to the opinions of the public, answer questions or questions in time, and improve the public's understanding of nuclear heating reactor. The public's awareness and acceptability of nuclear heating will be improved through visits and field visits. Strengthen information disclosure, ensure the public's right to know and participate in the safe production of nuclear heating reactors, and strive for the public's understanding and trust.

(2) Strengthen public opinion information monitoring

Strengthen the monitoring of public opinion information, especially pay attention to the possible cross regional avoidance effect and a wide range of public opinion risks; Reduce the dissemination and diffusion of negative public opinion of relevant projects through the supervision and guidance of media public opinion.

(3) Mutual trust and benefit sharing among the three parties

Strengthen public relations management. Strengthen mutual communication and consultation with the government, groups affected by the project construction and surrounding residents. Strengthen the publicity of nuclear safety knowledge, optimize the ways of information disclosure, improve the efficiency of information disclosure, and enhance the public's cognitive trust in the project. The construction company shall actively fulfill its corporate social responsibility, strengthen the development assistance of villages near the plant site, and try to localize the employment as much as possible, so as to make the development benefits more benefit the surrounding people. On the premise of law and regulation, local governments take the initiative to communicate and exchange with construction units, timely inform public needs, and promote the construction of benefit sharing mechanism among the government, enterprises and the public; We will promote the reemployment of landless farmers and ensure that the quality of life of landless farmers does not decline.

6 Conclusion

Through the research on the public communication work of small reactors, we can provide necessary theoretical basis and technical support for the specific public communication work of small reactors, also provide specific support for the implementation of public communication work by local policies and construction company, and give opinions and suggestions for the follow-up public communication work of small reactors.

References

1. MEE. Measures for public participation in environmental impact assessment (2018). (in Chinese)
2. National Nuclear Safety Administration. Guideline of public communication for nuclear power projects (2015). (In Chinese)
3. Arrangements for preparedness for a nuclear or radiological emergency ,IAEA GS-G-2.1
4. Project social stability risk assessment report. SNPI (2021)
5. Regulations for environmental radiation protection of nuclear power plant. GB6249 (2011)

Study on a Non-collecting Atmospheric Radon Concentration Measurement System

Chuanfeng Tang[1], Liangquan Ge[2], Shengliang Guo[3(✉)], Zhipeng Deng[1], and Jin Li[1]

[1] College of Nuclear Technology and Automation Engineering, Chengdu University of Technology, Chengdu, Sichuan, China

[2] College of Nuclear Technology and Automation Engineering, Applied Nuclear Technology in Geosciences Key Laboratory of Sichuan Province, Chengdu University of Technology, Chengdu, Sichuan, China

[3] Chengdu Newray Technology Technology Co., Ltd., Chengdu, Sichuan, China

gs133@qq.com

Abstract. Radon in the atmosphere is an important tracer in meteorology and geology and an important index of environmental radioactivity level evaluation. In this paper, NaI (Tl) scintillator detector was developed to directly measure radon concentration in the atmosphere, and a mathematical model of atmospheric radon gamma measurement was proposed, which solved the technical problem of online real-time monitoring of atmospheric radon concentration. It has important scientific and practical value.

In this paper, the characteristic gamma peaks of radon daughters ^{214}Bi (609.31 keV) and ^{214}Pb (351.92 keV) are respectively selected to calculate the radon concentration in the atmosphere. During the measurement period, the variation trend of radon concentration is the same as the theory, which suggests a higher concentration in the morning and a lower in the evening. Finally, the experimental measurement results were compared with the RAD7 radon measuring instrument. The error range of this detection system is 79.73% smaller than that of RAD7 on average, and its detection limit reaches 0.29 Bq/m^3 with a 30-min-measurement at room temperature. This paper proves that it is feasible to directly measure the activity concentration of radon in the atmosphere. The atmospheric radon measurement method proposed in this paper can accurately obtain the concentration of atmospheric radon and has the advantages of convenience, large measuring range, low detection limit, and online measurement.

Keywords: Radon concentration · Online measurement · Gamma-ray measurement · Radioactivity of the atmosphere · Radon-in-air

1 Introduction

Radon and its daughters are members of the natural radioactive uranium and thorium series and are one of the main sources of human natural radiation. According to a report by the United Nations Scientific Committee on the Effects of Radiation, exposure doses to radon and its daughters are about 1.3 mSv per year and are 54% of natural radiation. The U.S. Environmental Protection Agency (EPA) lists radon and its daughters as the

C. Liu (Ed.): PBNC 2022, SPPHY 283, pp. 1002–1014, 2023.
https://doi.org/10.1007/978-981-99-1023-6_85

only major causes of lung cancer other than smoking. In 2005, R. Williamsfield et al., and David Hilld et al., reported that radon inhalation is directly related to lung cancer [1, 2]. In recent years, with the improvement of people's working and living environment, the harm of radon and its daughters has naturally become a hot issue, which is concerned by all sectors of society. At the same time, radon in the atmosphere is also an important tracer in many related research fields such as meteorology and geology. In particular in seismology, earthquakes and similar geological activities such as volcanic eruptions can affect the concentration of radon in the atmosphere, and the concentration of it is commonly used for earthquake prediction [3]. The time scale and scope of these effects are often unclear, so rapid responses and long-term continuous measuring of radon concentration in the atmosphere are also required. There are many kinds of measurement methods of radon in the atmosphere, which can be classified into continuous sampling method, instantaneous sampling method, and passive sampling method by sampling method. For example, passive cumulative sampling is generally used for long-term measurement of radon gas, whereas the active continuous sampling method is used for continuous monitoring of radon concentration changes. And the measurement methods also can be divided into thermoluminescence method, solid nuclear track method, electret method, scintillation chamber method, electrostatic collection method, and activated carbon box method by measuring principle.

In terms of the general situation, the thermoluminescence measurement method has the advantages of being cheap, small-in-volume, no-radiation-source, and it's easy to read data, but the result obtained by this method is greatly affected by ambient temperature and wind speed. The attenuation of thermoluminescence over time and the trace radionuclides contained in the encapsulated or fluorescent material will further reduce the accuracy and reliability of the measurement results [4].

The solid track method has the advantages of low cost, small volume, and no-radiation-source. It is generally suitable for large-scale radon measurement. However, the disadvantage is also very obvious. In a low concentration area, the tracking number will be easier to generate errors because of the small detecting area [5, 6].

The electret method has the advantages of low cost, small volume, and weight, reusable and wide application range [6]. However, special care should be taken in the storage and use of electret, and the natural background should be corrected before measurement.

The scintillation chamber method has the advantages of a low lower limit of detection, ease to use, and high accuracy [7]. However, the detector is large in size and heavy in mass, and it is difficult to remove radon daughters deposited in the scintillation chamber. Therefore, it must be corrected before carrying out a continuous measurement.

The electrostatic collection method has a low lower limit of detection and it's easy to integrate [8]. However, its collection efficiency is greatly affected by air humidity, it must be equipped with a drying tube during measurement, and the drying tube must be replaced frequently in the humid air. It is not suitable for long-term continuous and unattended measurement.

The advantages of the activated carbon box method are that it is easy to use, can be continued after cleansing, and can be used for large area radon measurement in batches. The disadvantage is that the adsorption of radon by activated carbon is greatly affected by atmospheric conditions, and the radon concentration before the end of collection contributes more to the measurement results by this measurement method [9].

The above measurement methods cannot meet the requirements of outdoor continuous measurement in situ because of the response time, measurement residues, or measurement environmental conditions. This paper introduces a new field measurement system of radon concentration in the atmosphere, which consists of a NaI (Tl) detector and directional detector, and can realize the field and continuous measurement of radon concentration in the atmosphere. Calibration experiment and field measurement are also carried out. In this paper, the working principle and performance of the system are introduced in detail, and some field measurement results are given.

2 Materials and Methods

2.1 Measurement System

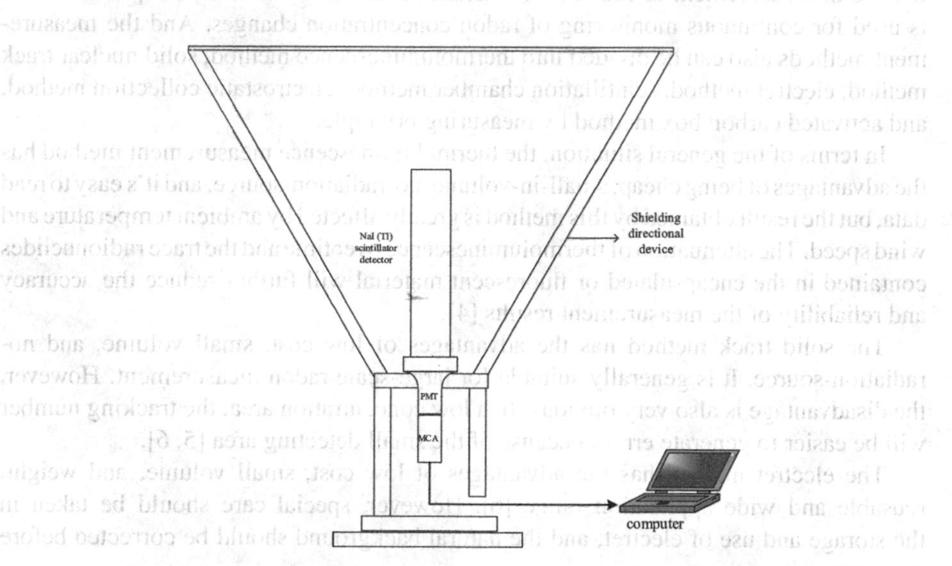

Fig. 1. Radon concentration measurement system for the atmosphere

As shown in Fig. 1, the whole measurement system is roughly composed of a detection part and a directional shielding part. The detection part uses a NaI(Tl) scintillator detector, which is generally composed of three main components: NaI(Tl) scintillator, photomultiplier tube, and pre-amplifier Circuit. NaI(Tl) scintillation detector workflow is as follows:

(1) Produce fluorescence. After gamma-ray enter NaI(Tl) scintillator, it will generate the photoelectric effect, Compton scattering, and electron pair effect with NaI(Tl) scintillator, resulting in secondary charged particles. These charged particles can ionize the atoms of the NaI(Tl) scintillator. When the excited atoms deactivate, they will emit fluorescent photons.

Fig. 2. Photos of measurement system entities

(2) Photoelectric conversion. The photoconductive materials and optical coupling agents are used to transfer the fluorescent photons to the photocathode of the photomultiplier tube, and then the photocathode will emit photoelectrons when the photoelectric effect is generated.
(3) Electron multiplication. After passing through the photomultiplier tube, the number of photoelectrons is changed from 1 to 10^4–10^9, so that a huge flow of electrons is collected at the anode of the photomultiplier tube.
(4) Pulse shaping. The flow of electrons at the anode further forms an electrical pulse signal on the load, which is then output through a pre-circuit.

Then the resulting pulse signal is analyzed by a single-channel pulse analyzer or multi-channel pulse analyzer to form a gamma-ray spectrum.

Another part is the directional shielding device, the function of it is to shield and exclude gamma-rays from the ground and surrounding building materials. Figure 3 shows the comparison between the spectrums after shielding and those without shielding, it can be seen that the directional shielding device is effective in shielding gamma-rays from most of the ground and surrounding building materials, such as the characteristic peak of 40K shown in Fig. 1 can be blocked 75.78%, so the shielding effect of the directional shielding is very obvious.

2.2 Principle

The most important step in calculating the detecting volume is to determine the 95% attenuation thickness of the characteristic energy selected for calculating radionuclide in air. Manjunatha et al. gave the following empirical formula for the mass attenuation

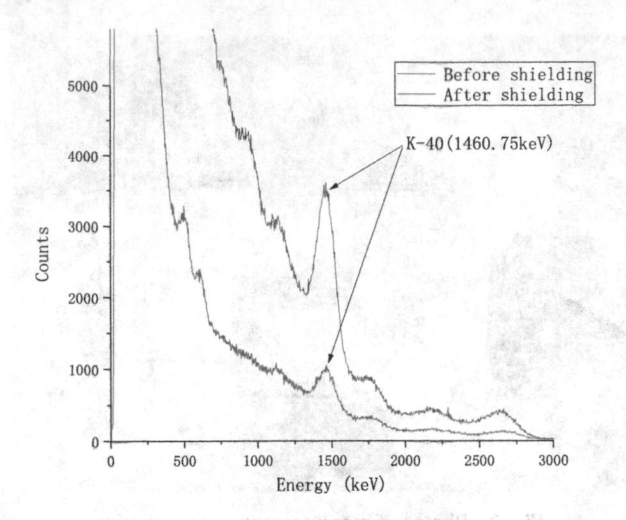

Fig. 3. Comparison of gamma spectrums between the spectrums after shielding and that without shielding with 8000 s measurement

coefficients of the gamma-ray with the energy of 0.1 meV to 20 meV penetrating different substances [10]:

$$\frac{\mu}{\rho_{0.1-20MeV}}=\begin{cases}\frac{1}{\Delta_1 E^{\Delta_2}+\Delta_3}(1<Z<17)\\(\rho_1 E+\rho_2)\rho_3(18<Z<24)\\\frac{1}{\xi_1\ln(E)+\xi_2}(25<Z<30)\\\frac{\eta_1}{E+\mu_2}+\mu_3(31<Z<37)\\\psi_1 E^{\frac{\psi_2}{E}}(38<Z<53)\\\chi_1\exp(\frac{\chi_2}{E+\chi_3})(54<Z<92)\end{cases}\tag{1}$$

In Formula 1, Z is the atomic number of the materials penetrated by gamma photon. Δ, ρ, ξ, η, Ψ, and χ are nonlinear fitting parameters corresponding to the penetrated substances with different atomic numbers, and some values of them can be found in Table 1.

Shao Qiwei gave the transformation formula of the mass attenuation coefficient of different media when studying the variation rule of point source gamma spectrum in the air [11]:

$$\mu=(\frac{\rho}{\rho_i})\cdot[\mu_i+(C^3-1)\cdot\tau_i+(C-1)\cdot\kappa_i]\tag{2}$$

In the type $C=\frac{z}{z_i}$, ρ_i is the known density of the elemental substance of an element; μ_i, τ_i, κ_i is the total attenuation coefficient, photoelectric effect attenuation coefficient,

Table 1. Nonlinear fitting parameters of mass attenuation coefficients with the energy of 0.1–20 meV and atomic number of 1–17

Z	Δ_1	Δ_2	Δ_3	R_2
1	0.047982133	0.691880705	2.238642389	0.99
2	0.107417393	0.675949755	4.312357789	0.99
3	0.140243972	0.659953864	4.814931496	0.99
4	0.154706411	0.643613157	4.514027117	0.99
5	0.169836414	0.626296457	4.128023198	0.99
6	0.182339125	0.607519221	3.591093182	0.99
7	0.2128311	0.588115345	3.324194141	0.99
8	0.254511573	0.566099072	2.980792844	0.99
9	0.328662042	0.541807911	2.694584435	0.99
10	0.391679834	0.515722504	2.045656896	0.99
11	0.516585414	0.487557871	1.453406302	0.99
12	0.649524818	0.457794714	0.618606405	0.99
13	0.887895087	0.42549067	−0.380992808	0.99
14	1.169847578	0.391521387	−1.589755847	0.99
15	1.678263885	0.356046899	−3.219101428	0.99
16	2.311611291	0.319734571	−5.060422121	0.99
17	3.459415953	0.283349367	−7.809427731	0.99

and electron pair effect attenuation coefficient of an element elemental with the incident gamma-ray of a particular energy, If $K = (\frac{\rho}{\rho_i}) \cdot (C^3 - 1)\tau_i + (C - 1) \cdot \kappa_i$ $K = (\frac{\rho}{\rho_i}) \cdot (C^3 - 1)\tau_i + (C - 1) \cdot \kappa_i$, it can be formulated as:

$$\mu = K \cdot \frac{\rho}{\rho_i} \cdot \mu_i \tag{3}$$

When calculating the air attenuation coefficient in this method, the atomic number of the converted elemental substance must be as small as possible, so this paper adopts the mass attenuation coefficient of Al ($Z = 13$) for conversion calculation.

In the case of the same elemental substance, the attenuation coefficient μ of the material has the following relationship with 95% attenuation thickness d:

$$\mu = \frac{\ln 20}{d} \tag{4}$$

According to the above formula, the 95% attenuation thickness of gamma rays of given energy in air can be calculated.

Because the 95% attenuation thickness of gamma rays in the air is much larger than the detector size, the detector can be regarded as a point detector. The detection volume of the detector is the volume of a spherical cone with a height of 95% attenuation thickness and the apex angle of the directional shielding device is taken as the apex angle of the detection volume. Thus, the characteristic energy air attenuation coefficient and 95% attenuation thickness of the nuclides that may be detected are calculated, as shown in Table 2.

Table 2. Air attenuation coefficient and 95% attenuation thickness of the corresponding radionuclide characteristic energy

Radionuclide species	Characteristic energy (keV)	Attenuation coefficient of air (μ, × 10^{-5})	Attenuate 95% thickness (d, cm)
^{214}Bi	609.31	9.67	33175.64
^{214}Pb	351.92	12.30	26065.98
^{7}Be	477.59	10.80	29815.23
^{208}Tl	2614.53	5.13	62483.13
^{40}K	1460.75	6.60	48559.32

MC method is used to simulate the detection efficiency of each characteristic energy detector, as shown in Table 3:

Table 3. The detection efficiency of each characteristic energy detector is obtained by MC method

Radionuclide species	Characteristic energy (keV)	Detection efficiency of the detector (ε, × 10^{-9})
^{214}Pb	351.92	7.52
^{214}Bi	609.31	4.27
	1764.49	1.85
^{40}K	1460.75	2.24
^{7}Be	477.59	6.29
^{208}Tl	2614.53	1.52

The detected effective volume can be regarded as a spherical cone, and the formula of gamma-ray measurement for atmospheric radon is established as follow:

$$AC = \frac{S}{\varepsilon \eta V t} \tag{5}$$

In formula 5, AC is the activity concentration of ^{214}Bi or ^{214}Pb in the atmosphere, and its value is equal to the activity concentration of ^{222}Rn in the atmosphere after equilibrium. S is a net area of the selected characteristic peak; ε is the detection efficiency of the detector for the selected characteristic energy; η is the branching ratio of characteristic gamma rays; V is the detecting volume of the detector; t is the measurement time. In formula 5, V can be obtained from formula 6:

$$V = \frac{2\pi(1 - \sin\theta)d^3}{3} \tag{6}$$

In formula 6, θ is half of the apex angle of the directional shielding device. d is the thickness of the characteristic gamma-ray exposure rate in the air attenuates to 5% of the incidence.

According to formulas 5 and 6, the activity concentration of radon and its daughters in the atmosphere can be calculated by measuring the gamma spectrum.

2.3 In Situ Measurements

First, a fully enclosed lead chamber was built to measure the background gamma spectrum, and the lower limit of detection of the system was calculated. When the measured radionuclide radioactivity level is close to the background, and the confidence level is 95%, the probability of the first- and second-class errors is 5% ($K_\alpha = K_\beta = 1.645$), and the background and measurement time are the same, the lower limit of detection of radionuclide activity concentration measured by gamma energy spectrum, L_D, can be approximated as:

$$L_D = \frac{4.65}{\varepsilon\eta V}\sqrt{\frac{n_b}{t}} \tag{7}$$

In formula 7, η is the branching ratio of characteristic gamma rays, V is the volume of the sample to be analyzed, ε is the detection efficiency of the gamma-ray omnipotent peak, n_b is the background counting rate in the selected peak region measured within the measurement time t, It includes the unshielded nuclides in the measurement system and its surroundings, the counting rate of interference peak caused by high-energy cosmic rays and the contribution of the continuous spectrum of other high-energy gamma rays in the sample [12].

Subsequently, on-site measurement was carried out. The site selection of this experiment needed to eliminate the influence of tall buildings around the measuring device as far as possible, so the open-air parking lot of a university in Chengdu was selected as the experimental measurement site. There are no tall buildings around the parking lot, and the surrounding environment is empty. In this experiment, the activity concentration of ^{222}Rn in the atmosphere, the atmospheric conditions, temperature, and humidity were measured continuously for 10 days.

Preliminary analysis of the measured spectral line is mainly to check the nuclide belonging to the obvious peak that can be detected in the spectral line, and the results are shown in Fig. 4. From the gamma energy spectrum detected, it can detect ^{214}Pb (351.92 keV), ^{214}Bi (609.31 keV, 1764.49 keV), ^{40}K (1460.75 keV), ^{208}Tl (2614.53 keV), However, the content of ^{7}Be (477.59 keV) is low and cannot be clearly displayed on the spectrum line due to the limitation of NaI(Tl) detector's energy resolution. The measured radionuclides include ^{222}Rn and its daughters to be measured in this paper, as well as uranium-series, thorium-series, and non-series radionuclides ^{40}K that enter the atmosphere in the form of aerosols due to human production and life, land dust suspension, etc. As can be seen from Fig. 4, in addition to the characteristic energy peaks of ^{214}Pb and ^{214}Bi of ^{222}Rn's daughters, the characteristic energy peaks of 1460.75 keV of nonseries radionuclide ^{40}K and 2614.53 keV of thorium daughter ^{208}Tl can be obtained by this spectrum line measurement.

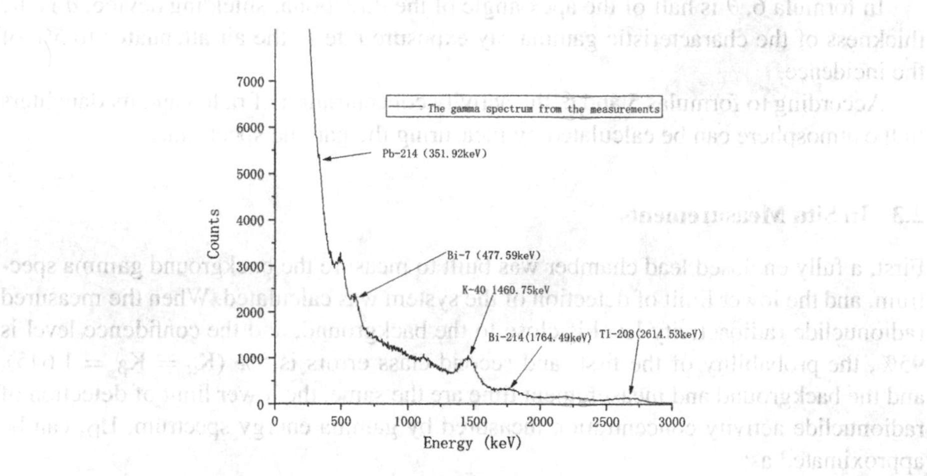

Fig. 4. The measured gamma spectrum line and the characteristic energy peaks of the main nuclides in 1 h

Then the spectra were solved and the activity concentration of ^{222}Rn in the atmosphere was calculated. The RAD7 was used to measure the activity concentration of ^{222}Rn at the same time to compare the results.

The source terms of the uncertainty of the measurement system mainly come from the uncertainty of the counts of gamma photons, the uncertainty of the detecting efficiency, the uncertainty of the detecting volume, the uncertainty of the branching ratio of gamma rays and the uncertainty of the measurement time. As shown in Table 4:

Table 4. The source terms of the uncertainty of the measurement system

The source term of the uncertainty	Counts	Detecting efficiency	Detecting volume	Branching ratio	Measurement time
The value of the uncertainty	$1/\sqrt{S}$	0.005	0.005	0.013	0.00001

The uncertainty formula of radon activity concentration in the atmosphere of this measurement system can be calculated according to the error transfer formula, as shown in formula (8):

$$\mu = \sqrt{(\frac{\partial AC}{\partial S})^2\delta_S^2 + (\frac{\partial AC}{\partial \varepsilon})^2\delta_\varepsilon^2 + (\frac{\partial AC}{\partial \eta})^2\delta_\eta^2 + (\frac{\partial AC}{\partial V})^2\delta_V^2 + (\frac{\partial AC}{\partial t})^2\delta_t^2} \quad (8)$$

3 Results and Discussion

The lower limit of detection of the measurement system is 0.29Bq/m^3 with a measurement time of 30 min, and it can be lower as the measurement time increases. At different measuring times, the results of the measurement system and RAD7 radon detector are drawn into a change curve, as shown in Fig. 5.

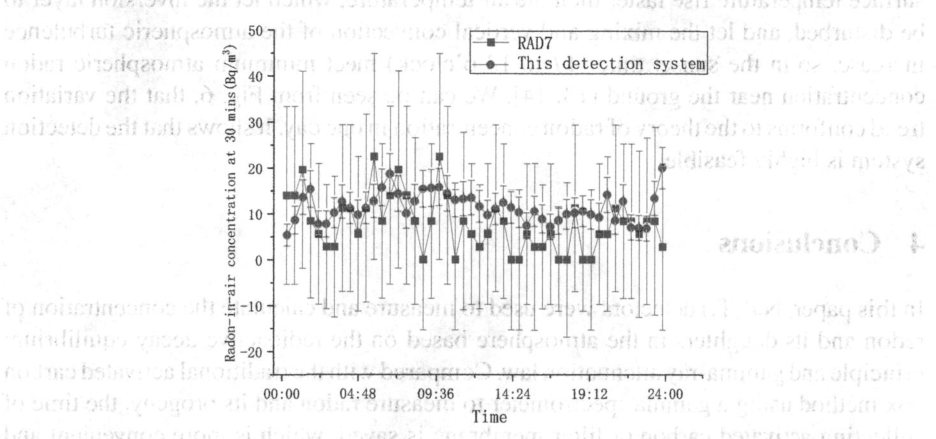

Fig. 5. The measurement data of RAD7 and this measurement system with a measurement time of 30 min

As can be seen from the comparison results in Fig. 5, the radon concentrations measured by this measurement system and the RAD7 have the same variation trend, showing a high concentration at sunrise and a low concentration at sunset. The error range of measurement results is 79.73% smaller on average than that of RAD7. From the above results, it can be seen that this measurement system can realize online measurement and has the advantages of being more convenient and faster.

Draw a line chart of the measurement data obtained by using this measurement system for 3 days, and Fig. 6 is obtained:

In The Sichuan Basin, the inversion layer will form after sunset, and its thickness will continue to accumulate before dawn. When the accumulation of the inversion layer reaches the maximum, atmospheric turbulence will be inhibited, which weakens the mixing effect of radon on the surface in the vertical direction, so the radon concentration reaches the highest point at 6 to 8 o'clock. After sunrise, the sun makes the earth's

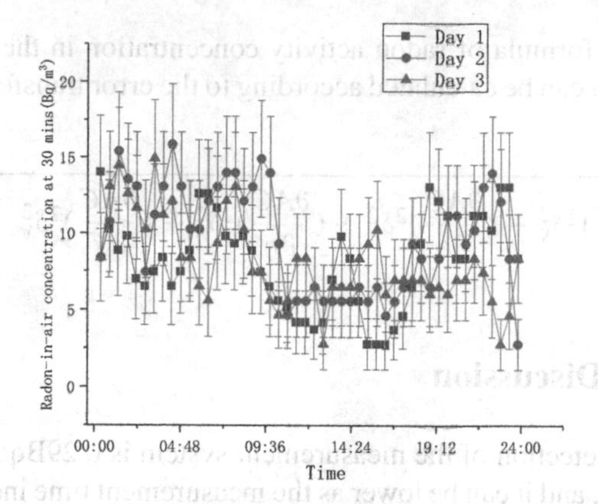

Fig. 6. Radon concentration change curve during 3 days

surface temperature rise faster than the air temperature, which let the inversion layer to be disturbed, and let the mixing and vertical convection of the atmospheric turbulence increase, so in the sunset time (17 to 18 o'clock) meet minimum atmospheric radon concentration near the ground [13, 14]. We can be seen from Fig. 6, that the variation trend conforms to the theory of radon concentration in one day. It shows that the detection system is highly feasible.

4 Conclusions

In this paper, NaI(Tl) detectors were used to measure and calculate the concentration of radon and its daughters in the atmosphere based on the radioactive decay equilibrium principle and gamma-ray attenuation law. Compared with the traditional activated carbon box method using a gamma spectrometer to measure radon and its progeny, the time of collecting activated carbon or filter membrane is saved, which is more convenient and faster.

In order to shield the gamma-rays from the surrounding soil and buildings, we developed a custom lead shielding device with a rotatable direction. Then the effective atmospheric volume of the detector is estimated, and the MC method is used to simulate the detector to obtain the characteristic gamma detection efficiency of the detector for the selected radionuclides. Finally, the atmospheric radon concentration was measured in an open-air parking lot of a university in Chengdu, and the atmospheric radon concentration was calculated according to the established mathematical model of atmospheric radon measurement. The daily variation of the measured radon concentration was preliminarily analyzed, and the theoretical variation trend was consistent with the measured data, which proved the feasibility of the measurement system.

The measured values of low radon concentration in the environment obtained by this measurement system are compared with those obtained by the RAD7 detector. The variation trend of the measured data is basically consistent, and the error range of this method is 79.73% smaller than that of the RAD7 detector on average, which further confirms the feasibility of this measurement system. This measurement system can be used for on-line monitoring of atmospheric radon concentration. Compared with other measurement methods, its detection limit is lower, its measurement process is shorter, more convenient, and faster.

Acknowledgement. This work was funded by Sichuan Science and Technology Program (No. 2020YJ0334).

References

1. Field, R.W., Steck, D.J., Smith, B.J.: Residential radon gas exposure and lung cancer the iowa radon lung cancer study. Am. J. Epidemiol. (2005)
2. Darby, S., Hill, D., Auvinen, A., Barros-Dios, J.M., Baysson, H., Bochicchio, F., et al.: Radon in homes and risk of lung cancer: collaborative analysis of individual data from 13 European case-control studies. BMJ **330**, 223 (2005). https://doi.org/10.1136/bmj.38308.477650.63
3. Iwata, D., Nagahama, H., Muto, J., Yasuoka, Y.: Non-parametric detection of atmospheric radon concentration anomalies related to earthquakes. Sci. Rep. **8**(1) (2018). Accessed 1 July 2022
4. Quanlu, G.: A brief introduction to measurement methods of radon and its daughters. Radiat. Prot. Bull., 35–41 (1994)
5. Ranjbar, A.H., Durrani, S.A.: Scintillator-filled etch-pit method of counting radon-decay alpha tracks, and calibration in a diffusion chamber. Radiat. Meas. **25**(1–4), 757–760 (1995)
6. Nikezic, D., Krstic, D., Savovic, S.: Response of diffusion chamber with LR115 detector and electret to radon and progeny. Radiat. Meas. **44**(9–10), 783–786 (2009)
7. Maozhi, W., Jianliang, Z., Xiaoping, Q., Tao, Y.: The relationship between detection efficiency and setting of instruments for 222Rn, 220Rn measurements with scintillation cell. Nucl. Electron. Detect. Technol., 514–516 (2008)
8. Tan, Y., Tokonami, S., Hosoda, M.: On the calibration of a radon exhalation monitor based on the electrostatic collection method and accumulation chamber. J. Environ. Radioact. **144**, 9–14 (2015)
9. Lian, F., Meng, J.: Research on affecting factors of measuring radon by active carbon box method. Guangdong Trace Elem. Sci. **17**(8), 51–54 (2010). https://doi.org/10.3969/j.issn.1006-446X.2010.08.008
10. Manjunatha, H.C., Seenappa, L., Sridhar, K.N., Sowmya, N., Hanumantharayappa, C.: Empirical formulae for mass attenuation and energy absorption coefficients from 1 keV to 20 MeV. Eur. Phys. J. D **71**(9), 1–22 (2017). https://doi.org/10.1140/epjd/e2017-70679-7
11. Qiwei, S.: Study on the variation of point source γ spectrum in the air (2009)
12. Du, Y.-w., Xiao-dong, X., Qian, W.: Sensitivity analysis of detection limit of atmospheric aerosol nuclide activity concentration by gamma ray spectrum. Sichuan Environ. **35**, 32–37 (2016)
13. Li, L., Ge, L., Cheng, F., Tian, L., Jing, B.: Study on the temporal change of radon concentration. Radiat. Prot., 13–18 (2007)
14. Singh, K., et al.: Variation of radon (222Rn) progeny concentrations in outdoor air as a function of time, temperature and relative humidity. Radiat. Meas. **39**(2), 213–217 (2005)

Numerical Simulation of Flow Boiling Heat Transfer in Helical Tubes Under Marine Conditions

Leqi Yuan[1], Kun Cheng[2(✉)], Haozhi Bian[1], Yaping Liao[1], and Chenxi Jiang[1]

[1] Fundamental Science on Nuclear Safety and Simulation Technology Laboratory, Harbin Engineering University, Harbin, Heilongjiang, China
[2] Science and Technology on Reactor System Design Technology Laboratory, Nuclear Power Institute of China, Chengdu, Sichuan, China
chengkunhrbeu@sina.com

Abstract. Lead-based cooled reactors in most countries and some small reactors at sea use helical tube steam generators. Compared with U-tubes, the convection heat transfer coefficient in the spiral tube is higher, the structure is more compact, and the secondary flow is generated under the action of centrifugal force and gravity, which can achieve the effect of wetting the inner wall of the tube. However, due to the importance of the steam generator in the reactor and the complexity of the flow and boiling in the helical tube, the aggregation behavior of bubbles, the distribution of the two-phase interface and the secondary flow in the tube will significantly affect the heat transfer characteristics, so the gas-liquid phase in the tube is studied. Distribution, changes in heat transfer coefficients, and fluid flow characteristics are very important.

In order to study the boiling heat transfer characteristics of helical once-through steam generators under static and marine conditions to provide safe and reliable energy supply for offshore facilities such as marine floating, this study uses STAR-CCM+ software, VOF method and Rohsenow boiling model to study the heat transfer capacity and flow characteristics of flow boiling in a helical tube under swaying and tilting conditions. The gas-liquid phase distribution characteristics, secondary flow variation characteristics and convective heat transfer coefficient of the fluid under different swing functions and inclined positions are obtained by numerical calculation, and the law of physical parameters changing with the cycle is found. The research results show that the secondary flow and heat transfer capacity in the tube change with the cycle, and the change is most obvious at the tube length of 0.8m. 5% of the normal condition; when the inclination angle is 45°, the maximum increase of the convection heat transfer coefficient is 16.8%, and the maximum decrease is 6.6%.

Keywords: Helical tubes · Ocean conditions · Flow boiling · Secondary flow · Computational fluid dynamics

1 Introduction

The once-through steam generator (OTSG) is a bridge for heat transfer in the primary and secondary circuits of the reactor, which can generate superheated steam, and the

C. Liu (Ed.): PBNC 2022, SPPHY 283, pp. 1015–1030, 2023.
https://doi.org/10.1007/978-981-99-1023-6_86

pressure stabilization does not require dehumidification. And compared with the natural circulation steam generator, the structure is more compact, suitable for small spaces such as ships, and has higher maneuverability to achieve rapid power change. Helical tube once-through steam generators are used in many lead-based cooling reactors. Compared with U-shaped tubes, the convection heat transfer coefficient in the spiral tube is higher and the structure is more compact. More importantly, the fluid in the tube produces secondary flow under the action of centrifugal force and gravity, which can achieve the effect of wetting the inner wall of the tube [1–3].

In recent years, with the continuous development of two-phase computational fluid dynamics (CFD) technology, there are more and more studies on the heat transfer model of helical tube once-through steam generators. Kumar V et al. [4] studied the laminar flow of the fluid in the helical tube, and found the non-uniform distribution of the flow velocity in the tube, and obtained its influence on the heat transfer characteristics by changing the placement method of the helical tube. Niu X J et al. [5] used the VOF model to study the fluid flow and heat transfer characteristics of the helical tube under full-side and half-side heating conditions.Chung Y J et al. [6] studied the relationship between the pressure in the helical tube and the dryout, and found that the centrifugal force caused the liquid film to move to the outside of the tube wall and the secondary flow effect was obvious at higher mass flow rates. And the heat transfer coefficient increases with the increase of pressure, but it is not obvious at low mass flow rate. The experimental results are consistent with the experimental results obtained by Styrikovich et al.Yang Yupeng et al. [7] used the fluid-structure interaction model of computational fluid dynamics to simulate the flow and heat transfer characteristics of the helical tube once-through steam generator. The obtained results are compared with Bartolomei's straight-tube boiling experiment and Santini's spiral-tube boiling experiment, and the errors of both are within 25%. And the numerical simulation of the liquid metal on the shell side is verified with the lead-bismuth liquid metal heat transfer experiment of Xi'an Jiaotong University, the Kalish-Dwyer relation and the Schad relation.

Due to the importance of the steam generator in the reactor and the complexity of flow boiling in the helical tube, the aggregation behavior of bubbles, the distribution of the two-phase interface, and the secondary flow in the tube can significantly affect the heat transfer characteristics. So, it is very important to study the gas-liquid phase distribution in the tube, the change of the heat transfer coefficient and the fluid flow characteristics. Under ocean conditions, the force of the fluid in the pipe will change significantly, which may have a significant impact on the heat transfer characteristics. At present, there is a lack of research on the law of fluid flow and boiling in the spiral pipe under ocean conditions. Therefore, based on the CFD method, this paper simulates the heat transfer characteristics of the uniformly heated helical tube under the conditions of swaying and tilting, and studies the variation law of the secondary flow and convection heat transfer coefficient by changing the swaying function and the inclination angle of the helical tube. The parameters such as flow and convective heat transfer coefficients are analyzed to provide a reference for the subsequent research on the heat transfer of spiral tubes under ocean conditions.

2 Numerical Computation Model

2.1 VOF Model

In this paper, the VOF model is used to simulate the fluid. The VOF method is an interface tracking method based on Euler grid. The immiscible fluids use the same set of governing equations. By defining the phase volume fraction α, two or more Simulation of a variety of immiscible fluids, tracking interphase interfaces. Typical applications include the movement of large air bubbles in liquids, jet breakup phenomena, the flow of liquids in the event of a dam break, and transient and steady-state simulations of any gas-liquid interface [8]. The main assumptions of the model are:

(1) Each component fluid is immiscible, incompressible and does not undergo chemical reaction;
(2) Each component fluid flows at the same speed, ignoring interphase slip;
(3) Each component fluid is in a state of thermal equilibrium.

For the VOF model, the fluid flow and energy transfer process follow specific governing equations. These equations are mathematical expressions of the physical properties of the fluid, which determine the simulation process of the CFD software. The following governing equations are used:

(1) Continuity Equation:

$$\frac{\partial\left(\alpha_q\rho_q\right)}{\partial t}+\nabla\cdot\left(\alpha_q\rho_q\vec{V}_q\right)=S_\alpha \tag{1}$$

S_α—quality source term.

(2) Momentum Equation:

$$\frac{\partial(\rho\vec{V})}{\partial t}+\nabla\cdot(\rho\vec{V}\vec{V})=-\nabla p+\nabla\cdot(\overline{\overline{\tau}})+\rho\vec{g}+\vec{F} \tag{2}$$

where: $\vec{V}$ is the mass average velocity, which satisfies the following relationship:

$$\vec{V}=\frac{\alpha_l\rho_l\vec{V}_l+\alpha_v\rho_v\vec{V}_v}{\rho} \tag{3}$$

(3) Energy Equation:

$$\begin{aligned}\frac{\partial(\rho h)}{\partial t}+\nabla\cdot(\rho h\vec{V})=\frac{\partial p}{\partial t}+\nabla\cdot(k_t\nabla T)\\+(\overline{\overline{\tau}}\cdot\nabla)\vec{V}+S_h\end{aligned} \tag{4}$$

The physical parameters such as density, viscosity and specific heat at constant pressure used in the process of solving the steam-water interface in the VOF model should be the volume-weighted average value, which can be calculated by the following relationship:

$$\rho = \sum_i \rho_i \alpha_i \tag{5}$$

$$\mu = \sum_i \mu_i \alpha_i \tag{6}$$

$$C_p = \sum_i (C_p)_i \alpha_i \rho_i / \rho \tag{7}$$

2.2 Turbulence Model

Considering that the flow in the helical tube is turbulent with high Reynolds number and the possible secondary flow, the realizable K-Epsilon two-layer model is adopted in this paper. The Realizable K-Epsilon Two-Layer Model is a combination of the Realizable K-Epsilon model and the two-layer method, which improves the calculation of swirl and separation flows compared to the standard K-Epsilon model, using two layers The method can significantly improve the calculation accuracy in the low Reynolds number region, and reduce the calculation error caused by the large difference between the viscous bottom layer and the mainstream physical properties in the standard K-Epsilon model [9].

The turbulent kinetic energy and turbulent dissipation rate in this model can be calculated by:

The formula for calculating turbulent kinetic energy:

$$\begin{aligned} \frac{d}{dt}\int_V \rho k dV + \int_A \rho k(\nu)\cdot da = \int_A \left(\mu + \frac{\mu_t}{\sigma_k}\right)\nabla k \cdot da \\ + \int_V \left[G_k + G_b - \rho(\varepsilon - \varepsilon_0 + \gamma_M) + S_k\right] dV \end{aligned} \tag{8}$$

Calculation formula of turbulent dissipation rate:

$$\begin{aligned} \frac{d}{dt}\int_V \rho \varepsilon dV + \int_A \rho \varepsilon(\nu)\cdot da = \int_A \left(\mu + \frac{\mu_t}{\sigma_\varepsilon}\right)\nabla \varepsilon \cdot da \\ + \int_V \left[C_{\varepsilon 1} S \varepsilon + \frac{\varepsilon}{k}(C_{\varepsilon 1} C_{\varepsilon 3} G_b) - \frac{\varepsilon}{k + \sqrt{\nu \varepsilon}} + S_\varepsilon\right] dV \end{aligned} \tag{9}$$

2.3 Boiling Model

In this paper, the Rohsenow boiling model is used to simulate the flow boiling in the tube. The Rohsenow boiling model in STAR CCM+ is combined with the liquid film boiling model, so that the model is suitable for the nucleate boiling stage where the wall temperature is only slightly higher than the liquid saturation temperature. It is also suitable for The high wall temperature results in a liquid film boiling stage in which a continuous film of vapor covers the heated surface. Its empirical relationship is as follows [10]:

$$q_{bw} = \mu_l h_{lat} \sqrt{\frac{g(\rho_l - \rho_v)}{\sigma}} \left(\frac{C_{pl}(T_w - T_{sat})}{C_{qw} h_{lat} Pr_l^{np}}\right)^{3.03} \tag{10}$$

q_{bw} is the wall heat flux in boiling heat transfer, w/m^2;
μ_l is the hydrodynamic viscosity, $Pa \cdot s$;
h_{lat} is the latent heat of vaporization, J/kg;
ρ_l is the liquid density, kg/m^3;
ρ_v is the steam density, kg/m^3;
σ is the liquid surface tension, N/m, The size of the surface tension will affect the shape of the bubble. In the three-dimensional space, the bubble is not strictly a sphere, and the surface tension will reduce the area of the bubble and become closer to a sphere. In this paper, the following relationship is used to calculate the surface tension [10]:

$$\sigma = 0.09537 - 2.24 \times 10^{-6}T - 2.56 \times 10^{-7}T^2 \tag{11}$$

The total correlation coefficient of this formula is 0.9999, which has good correlation with the data and is applicable in a wide temperature range.

C_{pl} is the specific heat of liquid, $J/(kg \cdot K)$;
T_w is the wall temperature, K;
T_{sat} is the Saturation temperature, K;
C_{qw} is the The empirical coefficient determined by the combination of liquid type and wall surface, this paper takes 0.008;
Pr_l is the liquid Prandtl number;
n_p is the Prandtl number index, depends on the working fluid, this paper takes 1.

The steam mass production rate $\dot{m}_{ew}$ at the vaporization core is calculated by the following relation:

$$\dot{m}_{ew} = \frac{C_{ew} q_{bw}}{h_{lat}} \tag{12}$$

C_{ew} is the Model constant for the amount of heat flux required to generate bubbles.

Since the Rohsenow relation does not depend on the fluid temperature and the heat in the computational domain is independent of the fluid temperature, exceeding its applicable range may result in an unrealistically high heat flux, causing the fluid temperature to be higher than the wall temperature. To prevent this, multiply the calculated heat flux by:

$$\max\left[0, \min\left(\frac{T_w - T}{T_w - T_{sat}}\right), 1\right] \tag{13}$$

T is the Fluid temperature near the heated wall.

2.4 Swaying Model Under Ocean Conditions

The equations of motion and dynamics for swaying in ocean conditions are as follows:

$$\begin{aligned} x &= \theta_0 + \theta_m \sin\left(\tfrac{2\pi t}{T}\right) \\ \omega &= \theta_m \tfrac{2\pi}{T} \cos\left(\tfrac{2\pi t}{T}\right) \\ \beta &= -\theta_m \tfrac{4\pi^2}{T^2} \sin\left(\tfrac{2\pi t}{T}\right) \end{aligned} \tag{14}$$

where x, ω, β is the distance, angular velocity and angular acceleration, respectively. The centrifugal acceleration of the fluid in the tube is determined by the following relationship:

$$a = \beta \cdot h \tag{15}$$

Among them, *a* and *h* are the additional centrifugal acceleration of the fluid and the height of the fluid element from the rotation axis, respectively.

Figure 1 is a schematic diagram of the swing motion under ocean conditions. The main parameters affecting the swing of the spiral tube are the swing cycle, swing height and maximum swing angle.

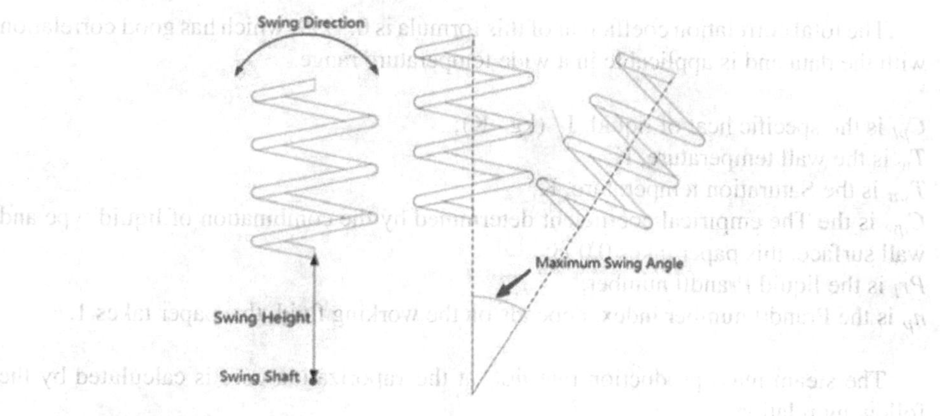

Fig. 1. Swing Motion Model Diagram

3 Validation of Numerical Models

3.1 Boiling Model Validation

Bartolomei (1980) used a vertical circular tube to conduct in-tube flow boiling experiments, and there are perfect experimental data [11], which are easy to use for the verification of the flow boiling model. As shown in the figure below, a model with a pipe diameter of 12.03 mm and a height of 1m is established. The cold water flows through the round pipe from bottom to top and is heated by the pipe wall. The feasibility of the boiling model is verified by comparing the simulated data and experimental data of this model (Fig. 2).

In this paper, two experimental conditions are selected from Bartolomei's experiments, and thermal parameters such as working pressure, inlet temperature and inlet flow rate are determined for simulation to ensure the accuracy of the boiling model. The parameters are shown in the table below.

After verifying the grid independence of the simulated pipe section, the working conditions A1 and A2 in Table 1 are simulated, and the obtained data curve is compared

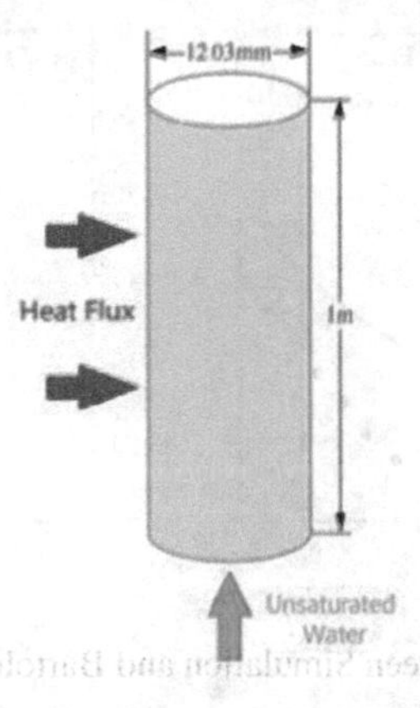

Fig. 2. Schematic Diagram of Test Piece Geometry

Table 1. Bartolomei's Experimental Parameter Table

Condition	P (MPa)	T_{in} (K)	q (MW/m^2)	G_{in} [kg/(m^2·s)]
A1	6.89	495	1.2	1500
A2	6.89	519	0.8	1500

with the circular pipe boiling experiment conducted by Bartolomei. The results are shown in Fig. 3.

The cross-sectional cavitation fraction along the height direction is selected as the monitoring quantity. From the results, it can be seen that the error between the simulated data and the experimental data is larger at 0.65m under the A1 working condition, and the errors at the other positions are smaller; under the A2 working condition, the error before 0.7m is small, and there is a certain error after 0.86m. In general, considering that the current boiling flow formulas in the tube have large errors and cannot obtain very accurate simulation results, it can be considered that the boiling model is reliable for simulating the flow boiling in the spiral tube.

3.2 Helical Tube Mesh Independence Verification

In order to make the used helical tube model have good accuracy and fast calculation speed, the grid-independent verification must be carried out. Taking the operating conditions of a small lead-bismuth fast reactor helical once-through steam generator as the calculation condition [12], and taking it as a modeling reference, the specific thermal parameters are shown in Table 2 below.

The cross-sectional mesh of the spiral tube model used for verification is shown in Fig. 4. The polyhedral mesh generator and the prism layer mesh generator are used. The thickness of the prism layer is 1 mm, the extension of the prism layer is set to 1.5, and the number of prism layers is set to 7, change the base size to change the number of

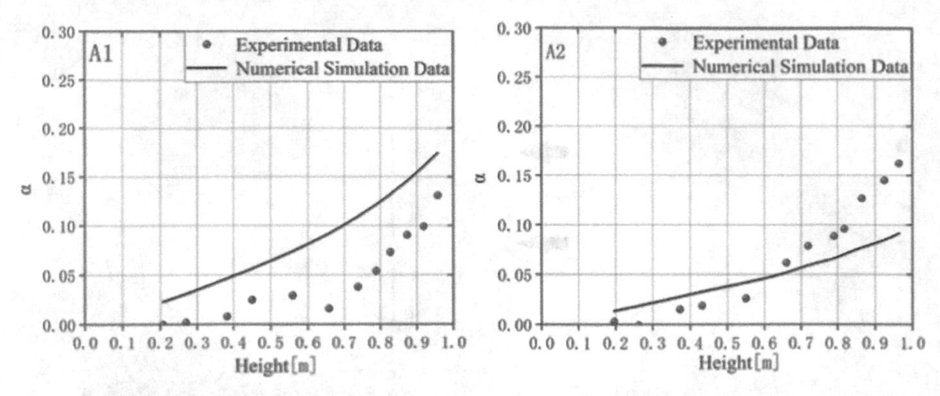

Fig. 3. Comparison between Simulation and Bartolomei's Experimental Data

Table 2. Thermal Boundary Conditions used in the Simulation

Physical Parameters	Symbol	Unit	Value
System Pressure	P	MPa	5.0
Import Temperature	T_{in}	°C	210
Import Speed	v_{in}	m/s	0.665
Heat Flux	q	MW/m^2	0.5

meshes. Monitor the convective heat transfer coefficient at 0.8 m and the cross-section gas content α at the cross-section, and draw the curve of the two with the number of grids, as shown in Fig. 5.

It can be found that when the number of grids is greater than 3×10^6, more stable results can be obtained, and the basic grid size is 0.7 mm.

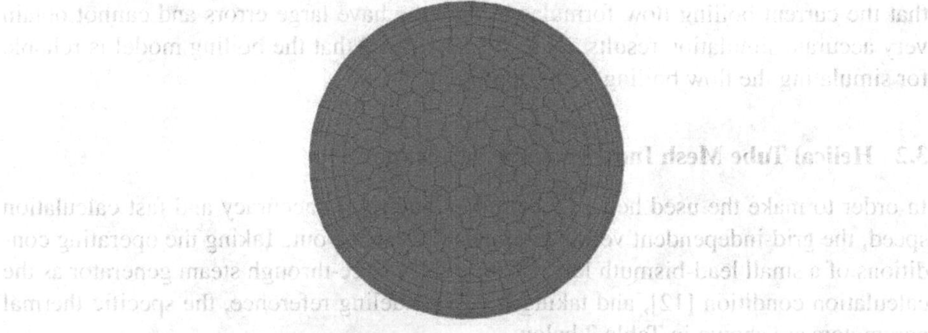

Fig. 4. Grid Schematic

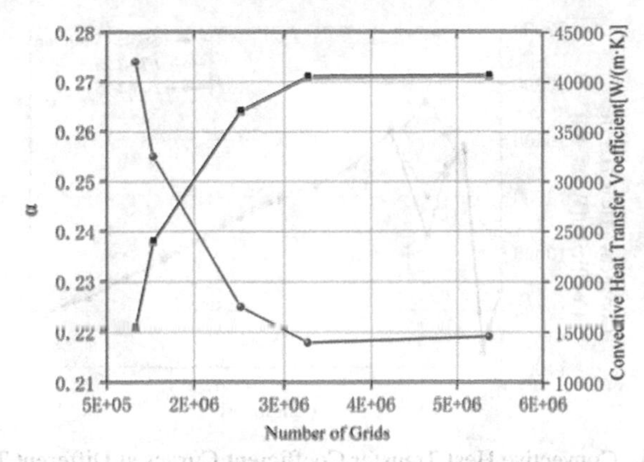

Fig. 5. Mesh Independence Verification of Helical Tubes

4 Analysis of Calculation Results

4.1 Analysis of the Change of Flow Heat Transfer Characteristics with Time

Taking the operating conditions of a small lead-bismuth fast reactor spiral-tube once-through steam generator as the calculation condition, the specific thermal parameters are shown in Table 2 above. The specific motion parameters of the simulated swing motion are shown in Table 3.

Table 3. Swing Motion Parameter Table

Physical Parameters	Symbol	Unit	Value
Swing cycle	T	s	7.0
Maximum Swing Angle	ϕ_m	deg	30
Swing Height	H	m	5.0

In order to avoid the possible instability of the calculation results of the first cycle, take the time cycle from the swing motion to the second cycle to study. Draw the change of the convection heat transfer coefficient of the spiral tube with time under ocean conditions, as shown in the Fig. 6.

It can be found that the three convective heat transfer coefficients change most obviously at about 0.8 m, and the fluctuation range is large. The convective heat transfer coefficient here shows a significant increase and decrease with time. This is because the displacement effect of the liquid phase on the gas phase under the action of the secondary flow is not obvious when the gas content is small, and the secondary flow is less affected by the swing motion when the gas content is large. The gas content here is relatively moderate, which can make the secondary flow change more violently, so the influence of the secondary flow is obviously different.

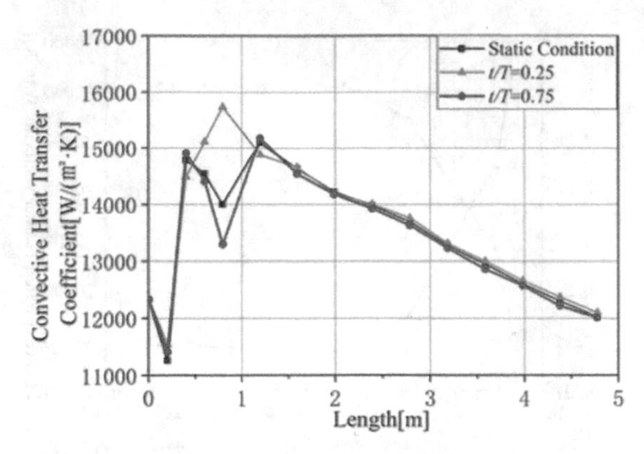

Fig. 6. Convective Heat Transfer Coefficient Curves at Different Times

When t/T is 0.25, the secondary flow is strengthened under the action of swing acceleration, so the convective heat transfer coefficient increases significantly compared with the static state, and the increase range is 12.4%; similarly, when t/T is 0.75, the secondary flow weakened, the convective heat transfer coefficient decreased by 5.0%. At the same time, it can be seen that the convective heat transfer coefficient does not change significantly after the tube length is 2 m, which is because the gas content is higher after 2 m, and the impact of the swing acceleration on the secondary flow is small.

Several circular sections were taken at a pipe length of 0.8 m to observe the change of the secondary flow with the swing motion. As shown in Fig. 7 below, the ratio of the motion time t to the swing cycle T in the second cycle is 0, Secondary flow cloud images at 0.25, 0.5, 0.75 and 1.0:

It can be found from the figure that the secondary flow is the strongest when t/T is 0.25, and the weakest when t/T is 0.75. This is because the acceleration generated by the former's swing motion is in the same direction as the acceleration generated by the fluid's circular motion in the tube, and the combined action of the accelerations strengthens the secondary flow.

In order to further illustrate that the convective heat transfer coefficient and the secondary flow are affected by the ocean conditions, the time-dependent curves of the two at a pipe length of 0.8 m are drawn, as shown in Fig. 8:

In order to observe the effect of the swing motion on the temperature at different times, the pipe wall surface and circular section near the pipe length of 0.8 m were taken. The temperature distributions of the wall temperature and fluid were observed when t/T was 0.25 and 0.75, respectively, as shown in Fig. 9 and Fig. 10 show.

It can be found that the position of the maximum wall temperature at different times has a certain deviation, and the peak position is closer to the upper part when t/T is 0.75. However, the maximum temperature values of the two are basically the same. The maximum temperature of the pipe wall at a pipe length of 0.8 m is 637 °C when the t/T is 0.25 and 0.75, while the minimum temperature of the wall is significantly different, which are 560 °C and 570 °C, respectively.

The temperature cloud map of the fluid circular section shows that the temperature of the fluid does not change much in value, but its spatial distribution changes significantly.

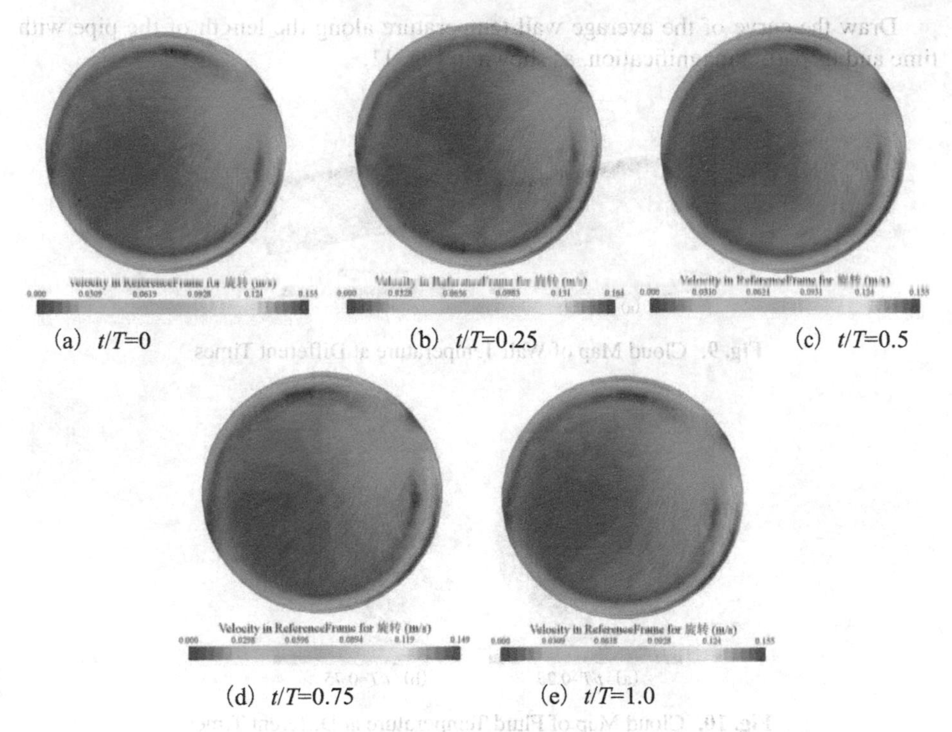

Fig. 7. Cloud Map of Secondary Flow with Time

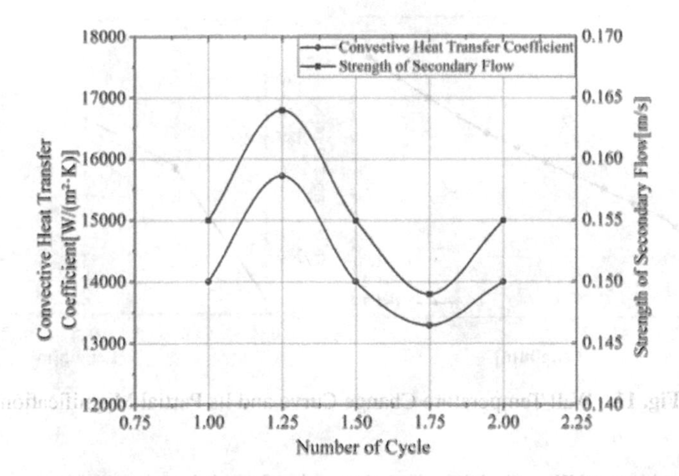

Fig. 8. Convective Heat Transfer Coefficient and Secondary Flow Curve with Time at Pipe Length of 0.8 m

When t/T is 0.75, it is closer to the upper part. This law is consistent with the distribution law of the secondary flow and the peak temperature of the wall.

Draw the curve of the average wall temperature along the length of the pipe with time and its partial magnification, as shown in Fig. 11.

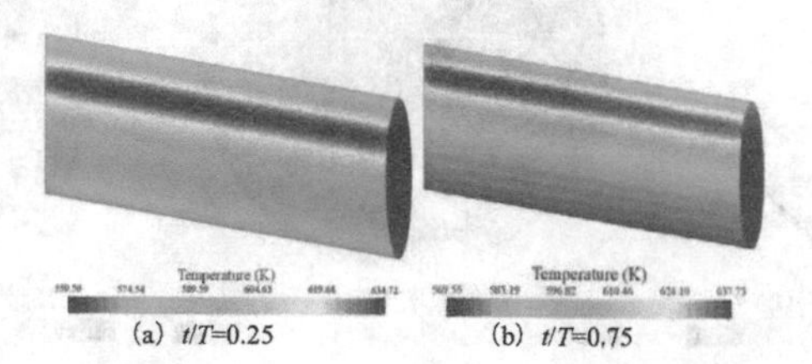

Fig. 9. Cloud Map of Wall Temperature at Different Times

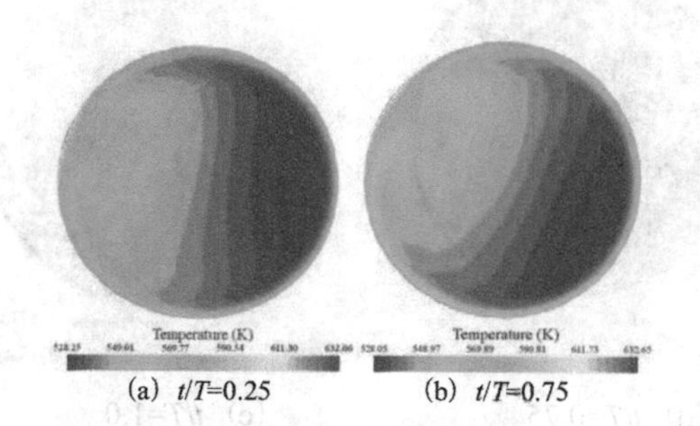

Fig. 10. Cloud Map of Fluid Temperature at Different Times

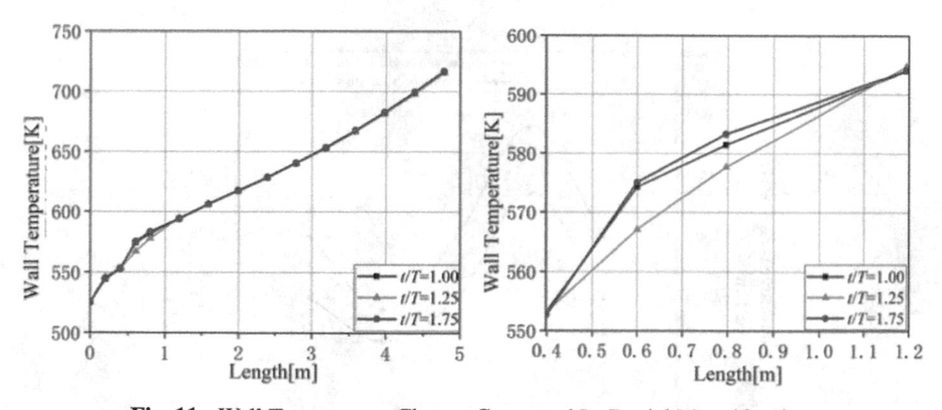

Fig. 11. Wall Temperature Change Curve and Its Partial Magnification

From the laws of Figs. 9, 10 and 11, it can be found that the maximum value of the wall temperature does not change much with time, but the average temperature fluctuates to a certain extent with time. This is because the highest temperature of the pipe wall is located in the secondary flow stagnation zone, and the secondary flow changes have little effect on it. The wall at the lower wall temperature is under the scouring of the secondary flow, and the change of the secondary flow has a significant impact on it, and the temperature change in the lower temperature region is also more obvious. In the range of tube length from 0.4 m to 1.2 m, when the secondary flow is strong, the wall temperature is lower, and the average temperature drops by 5 °C; when the secondary flow is weak, the average wall temperature is higher, and the average temperature increases by 1 °C.

4.2 Influence of Ocean's Inclination Condition on Heat Transfer Characteristics

The helical tube may be inclined for a long time in marine conditions, and when the spiral tube is inclined, the centrifugal force in some positions will be strengthened by the gravitational component, and some positions will be weakened. This will lead to changes in the secondary flow, which in turn changes the heat transfer performance. As shown in Fig. 12 below, the centrifugal force on the left is weakened by gravity, while the right is enhanced.

The spiral tube was simulated at an inclination angle of 20° and 45°, and the spatial distribution of gas and liquid phases was observed along the circular section of the tube under some working conditions. As shown in Fig. 13 below.

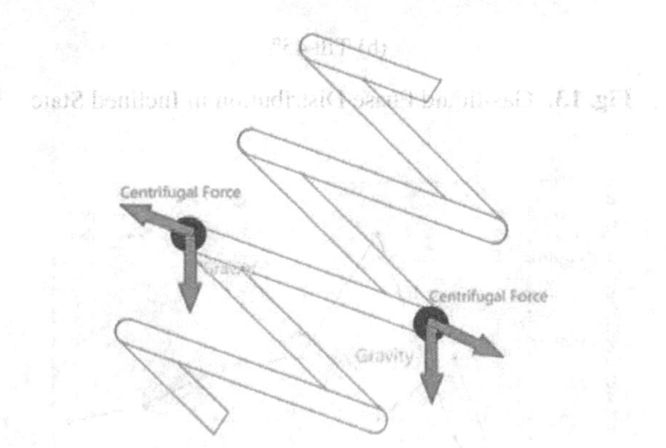

Fig. 12. Force Analysis of Fluid in Spiral Tube in Inclined State

It can be found from the figure that the gas phase of the first section in the state of tilting 45° is carried by the secondary flow, and the gas phase accumulation area is closer to the horizontal axis of symmetry of the tube than the first section without tilt. This is because the swing acceleration at this position is strengthened by the gravitational acceleration component, so the secondary flow is stronger, and the gas phase is carried by the secondary flow to approach the horizontal axis of symmetry.

In the same way, the swing acceleration is weakened by the gravitational acceleration component at the second section position in the state of tilting 45°, and the gas phase is closer to the upper part of the circular section. On the third and fourth sections, the effect of the secondary flow on the distribution of the gas and liquid phases is not obvious due to the high content of the gas phase. The convective heat transfer coefficient distribution curve along the tube length in the inclined state is made, as shown in Fig. 14:

It can be found that at the tube length of 0.8 m, the convective heat transfer coefficient is larger than that of the no-tilt condition due to the enhanced secondary flow due to the

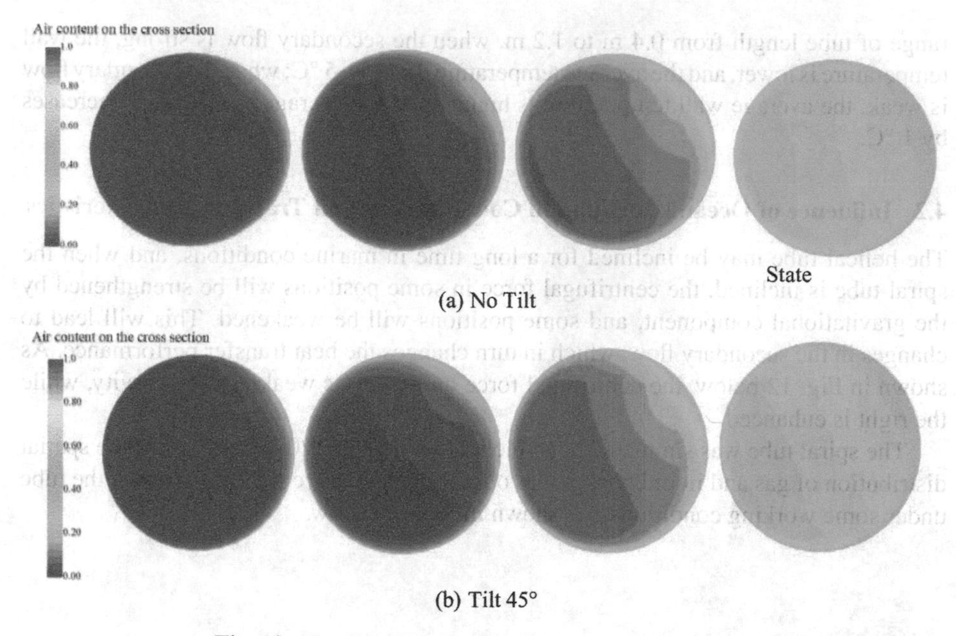

(a) No Tilt

(b) Tilt 45°

Fig. 13. Gas-liquid Phase Distribution in Inclined State

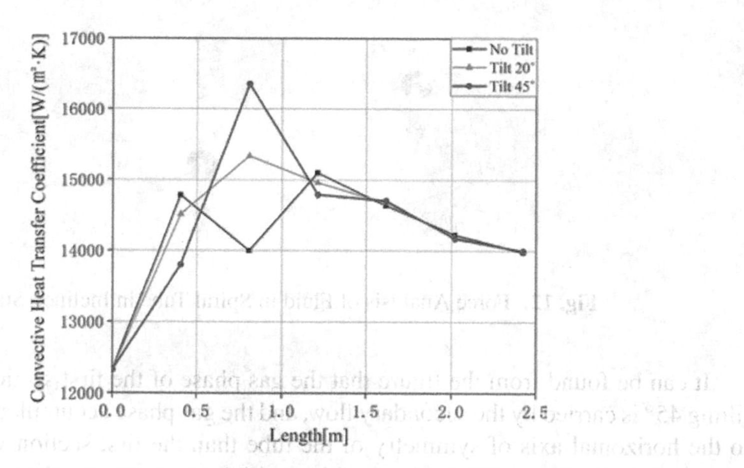

Fig. 14. Convective Heat Transfer Coefficient Curve of Spiral Tube in Inclined State

gravitational acceleration component. Similarly, the reduction of convection heat transfer coefficient can be seen at the tube lengths of 0.4 m and 1.2 m.

And it can be clearly found that the increase or decrease of the convective heat transfer coefficient increases with the increase of the inclination angle. When the inclination angle is 20°, the maximum increase of convection heat transfer coefficient is 9.5%, and the maximum decrease is 1.8%. When the inclination angle is 45°, the maximum increase of convection heat transfer coefficient is 16.8%, and the maximum decrease is 6.6%.

5 Conclusions

In order to study the specific influencing factors of the swing motion on the heat transfer characteristics of the helical tube, simulations were carried out under different swing cycles, maximum swing angles and swing heights, and the exact effects of the three on the heat transfer of the helical tube were verified.

For the inclined working condition, this paper mainly carries out the simulation under the inclined angle of 20° and 45°, and draws certain conclusions according to the distribution of cross-section gas content and the change of convective heat transfer coefficient.

The main conclusions are as follows:

(1) When the swing motion is a sinusoidal function, the curve of the secondary flow intensity at the local position in the pipe also presents a sinusoidal relationship, reaching a maximum value in 1/4 cycle and a minimum value in 3/4 cycle. At the same time, the distribution of gas content in the cross-section also shifted to a certain extent.
(2) The convective heat transfer coefficient of the spiral tube under the swing motion fluctuates with time, and the change is most obvious at the position where the tube length is 0.8 m. However, the convective heat transfer coefficient of the middle and rear pipe sections has little change. This is because the secondary flow is not significantly affected by the swing acceleration when the gas holdup is high, and the convective heat transfer coefficient at the rear position changes less than 1%.
(3) The fluctuation of the secondary flow intensity and the convective heat transfer coefficient with time under the swing motion shows a high degree of consistency, and the increase of the convective heat transfer coefficient at 0.8 m can reach 12.4% of the normal condition, and the reduction is 5% of the normal condition.
(4) The peak value of the wall temperature under the swing motion basically does not fluctuate, but the value at the lower wall temperature fluctuates greatly. When the secondary flow is strong, the average wall temperature is low, and conversely, the wall temperature is high. The main change positions of the wall temperature are concentrated in the tube length of 0.4 m to 1.2 m, and the maximum change temperature reaches 10 °C. At the same time, the position of the wall temperature peak will shift cycleically with the swing motion.
(5) Under the inclined condition, the secondary flow at different positions of the spiral tube is strengthened and weakened respectively, and the convective heat transfer coefficient also changes accordingly. When the inclination angle is 20°, the maximum increase of convective heat transfer coefficient is 9.5%, and the maximum decrease is 1.8%; when the inclination angle is 45°, the maximum increase of convective heat transfer coefficient is 16.8%, and the maximum decrease is 6.6%. When the inclination direction of the spiral tube changes, the change of the convection heat transfer coefficient may also change accordingly.

Acknowledgments. This work is financially supported by National Natural Science Foundation of China (12005215).

References

1. Wang, Z.: IEA releases new edition of world energy outlook report.Foreign Nucl. News (12), 2 (2013)
2. Chen, Z.: Thermal-hydraulics Design and Safety Analysis of a 100MWth Small Natural Circulation Lead Cooled Fast Reactor SNCLFR-100. University of Science and Technology of China (2015)
3. Li, Z.Y.: Development and military application of lead-bismuth reactors abroad. Foreign Nucl. News (07), 29–31 (2020)
4. Kumar, V., Nigam, K.D.P.: Numerical simulation of steady flow fields in coiled flow inverter. Int. J. Heat Mass Transf. **48**(23–24), 4811–4828 (2005)
5. Niu, X., Luo, S., Fan, L.L., et al.: Numerical simulation on the flow and heat transfer characteristics in the one-side heating helically coiled tubes. Appl. Therm. Eng. **106**, 579–587 (2016)
6. Chung, Y.J., Bae, K.H., Kim, K.K., et al.: Boiling heat transfer and dryout in helically coiled tubes under different pressure conditions. Ann. Nucl. Energy **71**, 298–303 (2014)
7. Yang, Y.P.: Numerical study of liquid metal helical coil once-through tube steam generator. Atomic Energy Sci. Technol. **55**(07), 1288–1295 (2021)
8. Wu, T.T.: Numerical Simulation of Cross Flow in Microstructure of Gas Diffusion Layer in Proton Exchange Membrane Fuel Cell. TianJin University (2014)
9. Shih, T.H., Liou, W.W., Shabbir, A., et al.: A new k-ϵ eddy viscosity model for high reynolds number turbulent flows. Comput. Fluids **24**(3), 227–238 (1995)
10. Rohsenow, W.M.: A method of correlating heat transfer data for surface boiling of liquids. MIT Division of Industrial Cooporation, Cambridge (1951)
11. Ustinenko, V., Samigulin, M., Ioilev, A., et al.: Validation of CFD-BWR, a new two-phase computational fluid dynamics model for boiling water reactor analysis (2008)
12. Ding, X.Y.: Athermal Hydraulic Model for a Helical Coiled Tube Once Through Steam Generator of Lead Cooled Fast Reactor, pp. 196–206. China Academic Journal Electronic Publishing House (2019). https://doi.org/10.26914/c.cnkihy.2019.047767
13. Wongwises, S., Polsongkram, M.: Evaporation heat transfer and pressure drop of HFC-134a in a helically coiled concentric tube-in-tube heat exchanger. Int. J. Heat Mass Transf. **49**(3–4), 658–670 (2006)
14. Yang, Y., Wang, C., Zhang, D., et al.: Numerical study of liquid metal helical coil once-through tube steam generator. Atomic Energy Sci. Technol. **55**(7), 1288 (2021)
15. Zaman, F.U., Qureshi, K., Haq, I., et al.: Thermal hydraulics analysis of a helical coil steam generator of a small modular reactor. Ann. Nucl. Energy **109**, 705–711 (2017)

Numerical Simulation of the Transient Flow Characteristics and Thermal Stratification Phenomena in the Passive Residual Heat Removal System of NHR-200-II

Yiwa Geng and Xiongbin Liu(✉)

Institute of Nuclear and New Energy Technology (INET), Tsinghua University, Collaborative Innovation Center of Advanced Nuclear Energy Technology, Key Laboratory of Advanced Reactor Engineering and Safety of Ministry of Education, Beijing, China

lxb@mail.tsinghua.edu.cn

Abstract. The NHR-200-II nuclear heating reactor is a multi-purpose small integral pressurized water reactor (iPWR) developed by the Institute of Nuclear and New Energy Technology (INET) of Tsinghua University. The design of NHR-200-II features a reactor core with thermal power of 200MW, in-vessel hydraulically-driven control rods and passive residual heat removal (PRHR) systems, et.al. Passive residual heat removal experiments were conducted in a scaled integral test facility for NHR-200-II. The PRHR experiments in the scaled facility were simulated by a layered RELAP5 system model to study the flow characteristics of the PRHR system in different primary fluid temperatures and different valve states. The phenomenon of reversed flow occurred in some primary heat exchangers in the numerical simulations when the primary fluid temperature was higher than certain level, which was consistent to the experiments. The simulated uneven outlet temperature distribution of the primary heat exchangers was also consistent with the experimental data when the isolation valves for the steam generator was kept open. Thermal stratification effect in the headers of the PRHR system played an important role in the phenomenon of uneven outlet temperature distributions, and the layered RELAP5 model was proven to be an efficient method for preliminary estimation of thermal stratification effect in the headers.

Keywords: NHR-200-II · Passive Residual Heat Removal System (PRHRS) · RELAP5 · Natural Circulation

1 Introduction

Various advanced small modular reactors (SMRs) are currently under development in the worldwide, including the Westinghouse Small Modular Reactor (W-SMR) [1], NuScale [2], SMART (System-integrated Modular Advanced Reactor) [3], mPower [4] and IRIS [5]. NHR-200-II is also a new type of advanced SMR designed by the Institute of Nuclear and New Energy Technology (INET) of Tsinghua University, the NHR-200-II reactor can be a safe, clean, affordable, and less carbon-footprint choice of nuclear power generation

C. Liu (Ed.): PBNC 2022, SPPHY 283, pp. 1031–1045, 2023.
https://doi.org/10.1007/978-981-99-1023-6_87

[6, 7], and the reactor can be used for district heating, power generation, process heat, desalination, et al.

NHR-200-II has several engineered safety systems [7]. The passive residual heat removal (PRHR) systems are key part of the safety features of the reactor. A series of passive residual heat removal experiments have been conducted in a scaled integrated test facility for NHR-200-II by INET [8, 9]. To explain the flow phenomenon in the experiments, it was necessary to establish a detailed numerical model to analysis the transient characteristics of the PRHR system in the scaled integral test facility of NHR-200-II.

In this paper, a layered RELAP5 model of the PRHR system was established, numerical simulations with different primary fluid temperatures were carried out, and the simulation results were discussed and compared to the experiments.

2 Design of the PRHRS Test Facility of NHR-200-II

The NHR-200-II reactor has two parallel intermediate circuits, and two parallel PRHR columns are connected to each intermediate circuit. For the sake of simplicity, only one of the PRHR columns was chosen for modeling. The schematic of the PRHR column was shown in Fig. 1. The PRHR column consists of seven primary heat exchanger (PHE) branches, a residual heat exchanger (RHE) branch and a steam generator (SG) branch. A hot header and a cold header that connecting all the branches are arranged above the top of reactor pressure vessel. The seven PHEs are placed inside the reactor pressure vessel, and the RHE is a finned tube heat exchanger that is installed in the air-cooling tower of NHR-200-II. PHEs are connected asymmetrically to the hot header and the cold header by T-shaped junctions, and the steam generator (SG) is connected to the headers by low resistance Y-shaped junctions. The RHE is connected between the hot leg and cold leg of the SG. The pressurizer (PRZ) is placed at the highest location of the loop.

The PRHR column removes heat from the core by three coupled natural circulation loops. The first natural circulation loop is the natural circulation of the primary fluid inside the reactor pressure vessel, the second natural circulation loop is the fluid circulation in the PRHR loop (Fig. 1), and the third loop is the natural circulation of air in the air-cooling tower.

An integral test facility was built to study the characteristics of the PRHR system of NHR-200-II. Design of the test facility was similar to the original PRHR system of NHR-200-II with some exceptions due to site and equipment limitations. The main differences between PRHR system in the integral test facility and the PRHR system of NHR-200-II were:

(1) The PRHR system in the integral test facility was a scaled model, the height ratio is about 1:5, the hydraulic diameter ratio is about 1:3 for most pipes, the system volume ratio is about 1:50, and the thermal power ratio of the integral test facility is about 1:20 to that of the prototype.
(2) Temperature and pressure of the PRHRS of the integral test facility were slightly lower than those of the prototype due to site limitations.
(3) The heat sink of the residual heat exchanger (RHE) in the test facility was water, instead of air in the prototype.

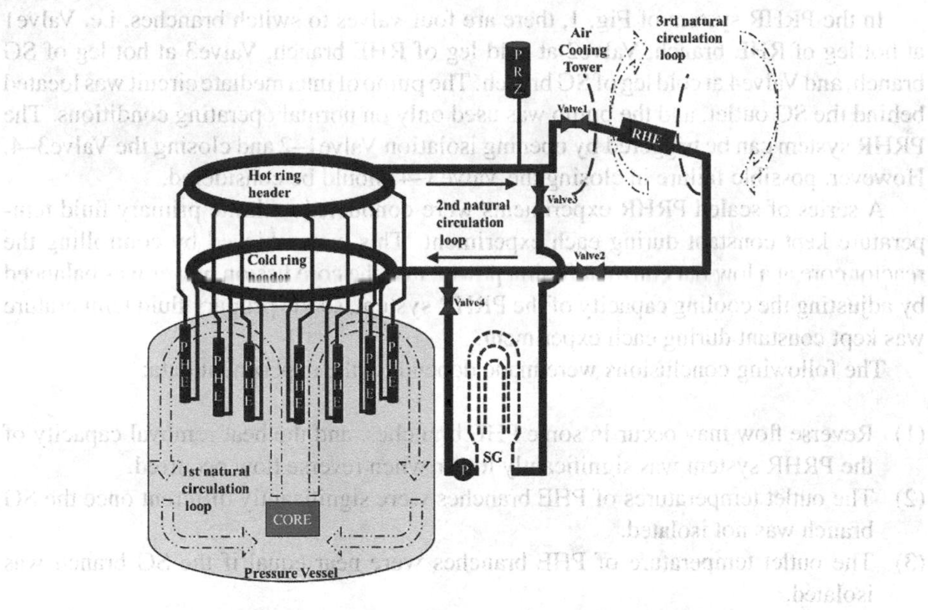

Fig. 1. Schematic Diagram of a PRHR system.

(4) The RHE was a smooth tube-array heat exchanger in the test facility, while the RHE was a finned-tube air cooler in the prototype.
(5) The number of PHE was six for the test facility due to space constraints, while there are seven PHE in the prototype.

The geometry information of key components of the integral test facility is summarized in Table 1.

Table 1. Geometry Information of Key Components in the integral Test Facility of NHR-200-II

Components name	SG	PHE	RHE	Cold ring header	Hot ring header
Elevation (m)	3.05 (center) (951P-953P)	0 (center)	1.427 (center)	2.198	2.432
Hydrulic diameter (mm)	118.0	60.0	76.0	92.0	92.0
Heat transfer area (m2)	9360	39.8	8.09	/	/
Tube/Pipe length (m)	3.79785 (length of heat transfer tubes)	1.6 (length of heat transfer tubes)	2.76 (length of heat transfer tubes)	8.69 (length of the circular pipe)	8.69 (length of the circular pipe)

In the PRHR system of Fig. 1, there are four valves to switch branches, i.e. Valve1 at hot leg of RHE branch, Valve2 at cold leg of RHE branch, Valve3 at hot leg of SG branch, and Valve4 at cold leg of SG branch. The pump of intermediate circuit was located behind the SG outlet, and the pump was used only on normal operating conditions. The PRHR system can be triggered by opening isolation Valve1–2 and closing the Valve3–4. However, possible failure in closing the Valve3–4 should be considered.

A series of scaled PRHR experiments were conducted with the primary fluid temperature kept constant during each experiment. This was achieved by controlling the reactor core at a low but constant fission power, and the core fission power was balanced by adjusting the cooling capacity of the PRHR system, so the primary fluid temperature was kept constant during each experiment.

The following conclusions were made depend on the experiment data:

(1) Reverse flow may occur in some PHE branches, and the heat removal capacity of the PRHR system was significantly lower when reverse flow occurred.
(2) The outlet temperatures of PHE branches were significantly different once the SG branch was not isolated.
(3) The outlet temperature of PHE branches were near equal if the SG branch was isolated.

To explain those observations, transient flow characteristics of the PRHR system was simulated with different combination of valve states (opening or closing) and different primary fluid temperatures.

3 Numerical Model of the PRHRS Test Facility

A REALP5 model was setup according to the geometry and hydraulic parameters of the PRHR system of the scaled integral test facility. The node diagram of the RELAP5 model was shown in Fig. 2. In this diagram, the six primary heat exchangers (PHEs) were named with starting numbers of 1–6, and the cold header and hot header were named starting with numbers 7 and 8 correspondingly.

Unlike other pipe components, the hot and cold headers in the test facility were closed circular pipes, and the pipe diameter of the headers was significantly larger than other pipe components, so possible local recirculation and thermal stratification may occur in the hot and cold headers. To capture the secondary flow phenomena in the headers, a layered model was used for the headers. The layered header model composed of three layers that arranged vertically, and the center layer was connected to its neighborhoods to simulate vertical flow mixing at low flow rates. The schematic diagram of layered model for the headers was shown in Fig. 3.

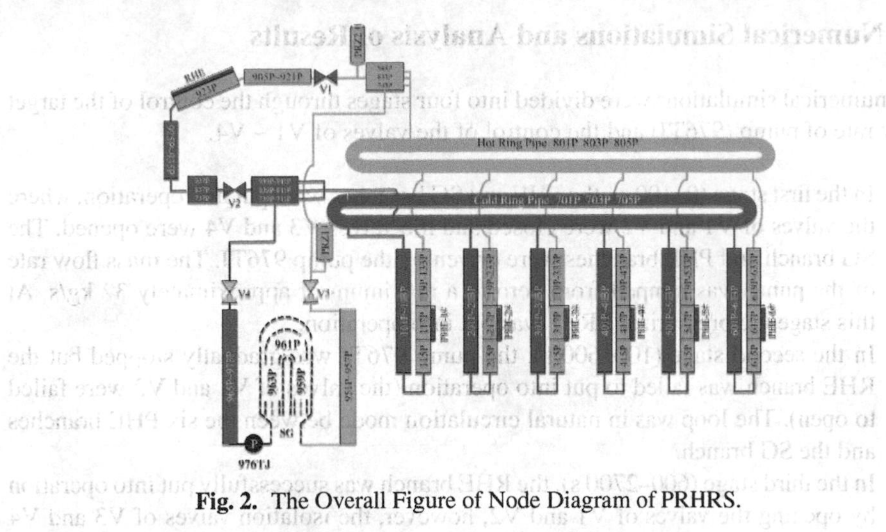

Fig. 2. The Overall Figure of Node Diagram of PRHRS.

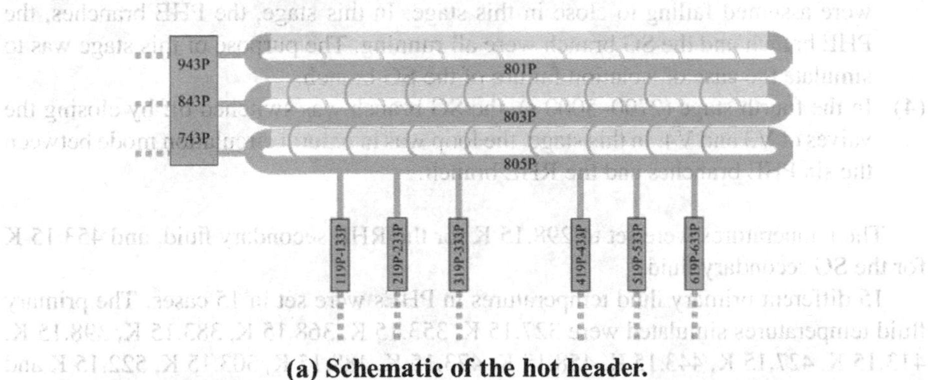

(a) Schematic of the hot header.

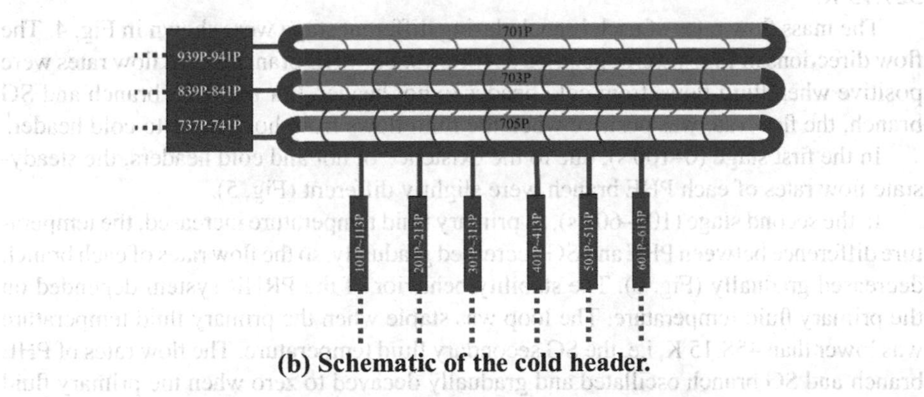

(b) Schematic of the cold header.

Fig. 3. Node Diagram of Hot and Cold Header with Layered Model.

4 Numerical Simulations and Analysis of Results

All numerical simulations were divided into four stages through the control of the target flow rate of pump (976TJ) and the control of the valves of V1 ~ V4.

(1) In the first stage (0–100 s), the PHE and SG branches were put into operation, where the valves of V1 and V2 were closed and the valves V3 and V4 were opened. The SG branch and PHE branches were driven by the pump 976TJ. The mass flow rate of the pump was ramped from zero to a maximum of approximately 32 kg/s. At this stage the pressurizer PRZ1 was put into operation.
(2) In the second stage (100–600 s), the pump 976TJ was gradually stopped but the RHE branch was failed to put into operation (the valves of V1 and V2 were failed to open). The loop was in natural circulation mode between the six PHE branches and the SG branch.
(3) In the third stage (600–2700 s), the RHE branch was successfully put into operation by opening the valves of V1 and V2, however, the isolation valves of V3 and V4 were assumed failing to close in this stage. In this stage, the PHE branches, the PHE branch and the SG branch were all running. The purpose of this stage was to simulate the case of isolation failure of the SG branch.
(4) In the fourth stage (2700–5000 s), the SG branch was switched off by closing the valves of V3 and V4. In this stage, the loop was in natural circulation mode between the six PHE branches and the RHE branch.

The temperatures were set to 298.15 K for the RHE secondary fluid, and 453.15 K for the SG secondary fluid.

15 different primary fluid temperatures in PHEs were set in 15 cases. The primary fluid temperatures simulated were 327.15 K, 353.15 K, 368.15 K, 383.15 K, 398.15 K, 413.15 K, 427.15 K, 443.15 K, 458.15 K, 473.15 K, 488.15 K, 503.15 K, 522.15 K and 527.15 K.

The mass flow rates of each branch during different stage were shown in Fig. 4. The flow directions in Fig. 4 were specified as follows. For PHE branches, the flow rates were positive when fluid flows from cold header to hot header. For the RHE branch and SG branch, the flow rate was positive when the fluid flows from hot header to cold header.

In the first stage (0–100 s), due to the existence of hot and cold headers, the steady-state flow rates of each PHE branch were slightly different (Fig. 5).

In the second stage (100–600 s), as primary fluid temperature increased, the temperature difference between PHE and SG decreased gradually, so the flow rates of each branch decreased gradually (Fig. 6). The stability behavior of the PRHR system depended on the primary fluid temperature. The loop was stable when the primary fluid temperature was lower than 458.15 K, i.e. the SG secondary fluid temperature. The flow rates of PHE branch and SG branch oscillated and gradually decayed to zero when the primary fluid temperature was higher than 458.15 K.

In the third stage (600–2700 s), all the branches of the PRHR system were running, including the SG branch, the RHE branch and the PHE branches.

The PRHR system in this stage was found to be stable as the primary fluid temperature were between 327.15 K and 383.15 K. Steady flow rates of each branch under different primary fluid temperatures were shown in Fig. 7.

The flow rates in each branch of the PRHR system oscillated with decreasing magnitude when the primary fluid temperature is 458.15 K.

The flow rates of each branch of the PRHR system became steady after an initial unstable period for the primary fluid temperature range from 473.15 K to 533.15, and the flow rates of 2#PHE, 3#PHE, 4#PHE and RHE branches were positive, while flow rates of other PHE branches, i.e. 1#PHE, 5#PHE, 6#PHE, were negative (Fig. 8).

In the fourth stage (2700–5000 s), the steam generator was isolated from the PRHR system. The PRHR system became stagnated when the primary fluid temperature was in the range of 327.15 K to 458.15 K. In the primary fluid temperature of 473.15 K to 527.15 K, the flow in the PRHR system became steady-state with reverse flow in some PHE branches (Fig. 9), and the reverse flow occurred in the 1#PHE, 4#PHE, 5#PHE and 6#PHE. The flow rates of each branch increased with primary fluid temperature. The flow rates in the reversed PHE branches were basically the same value, while the flow rates in the positive PHE branches were different (Fig. 9).

Thermal stratification effect in the hot and cold ring headers and corresponding Y-junctions was shown in Fig. 10 and Fig. 11, where the temperature contours of the headers and the Y-junctions were plotted for primary fluid temperature of 522.15 K. It was clearly shown in Fig. 10 and Fig. 11 that the layered header model can capture some detail of local flow phenomena in the headers and Y-junctions, such as thermal stratification or local recirculation.

It was observed in the scaled PRHR experiments that the outlet temperatures of the six PHE branches were significantly uneven if the SG branch was not isolated, while the outlet temperatures were nearly equal if the SG branch was isolated. In the numerical simulations, the phenomenon of uneven outlet temperatures of PHE branches were successfully captured in the third stage (600–2700 s in Fig. 12), and the outlet temperatures of PHE branches were nearly equal in the fourth stage (2700–5000 s in Fig. 12). The numerical simulations agreed the experimental data quantitively.

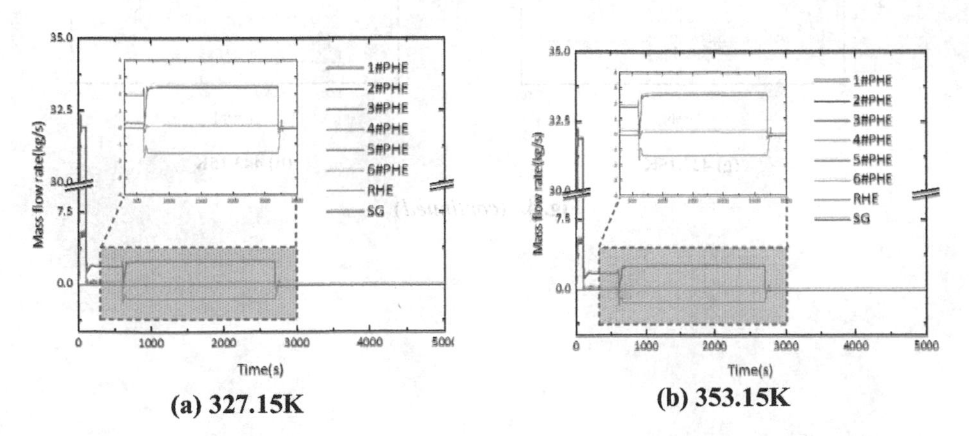

(a) 327.15K **(b) 353.15K**

Fig. 4. Mass Flow Rate of Each Branch at Different Primary Fluid Temperature.

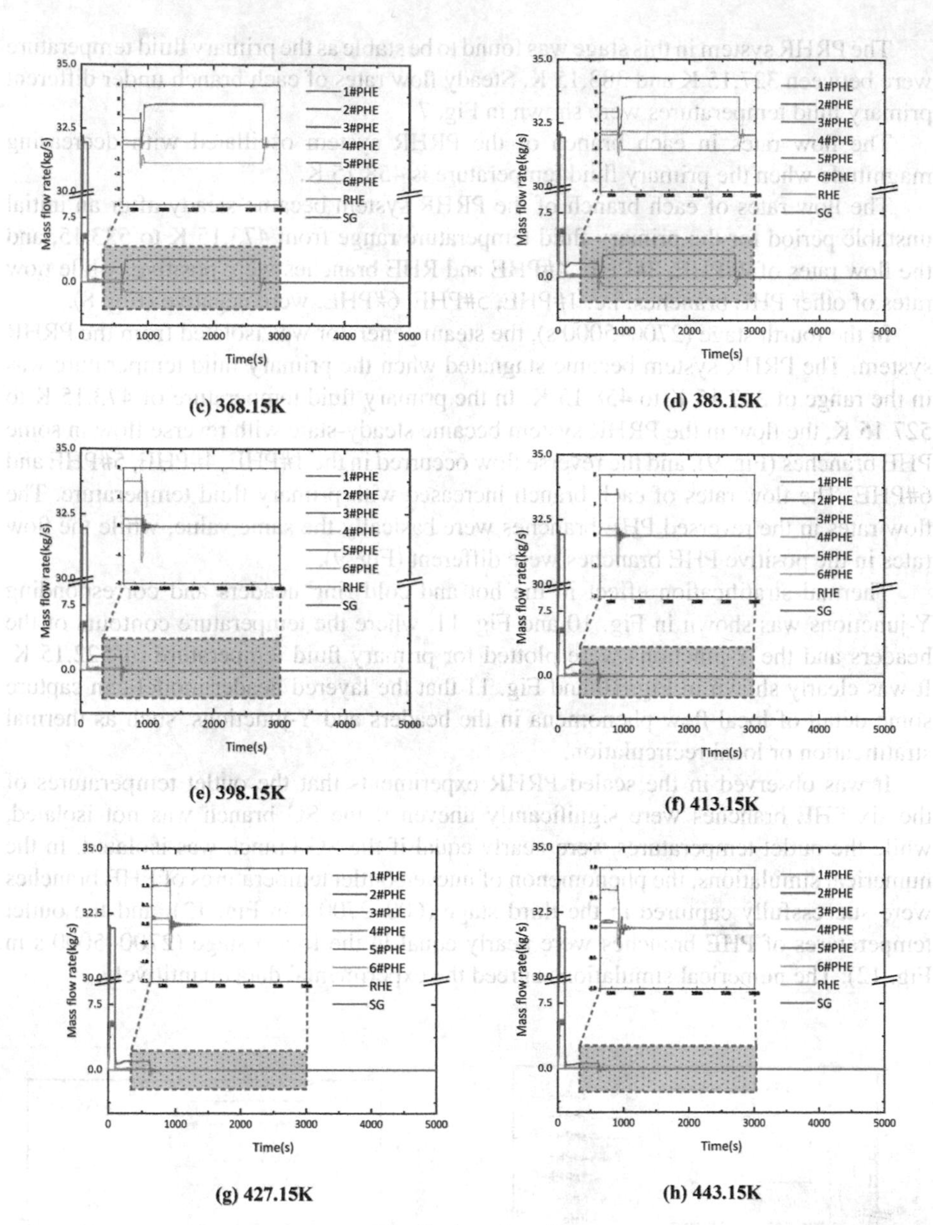

Fig. 4. (*continued*)

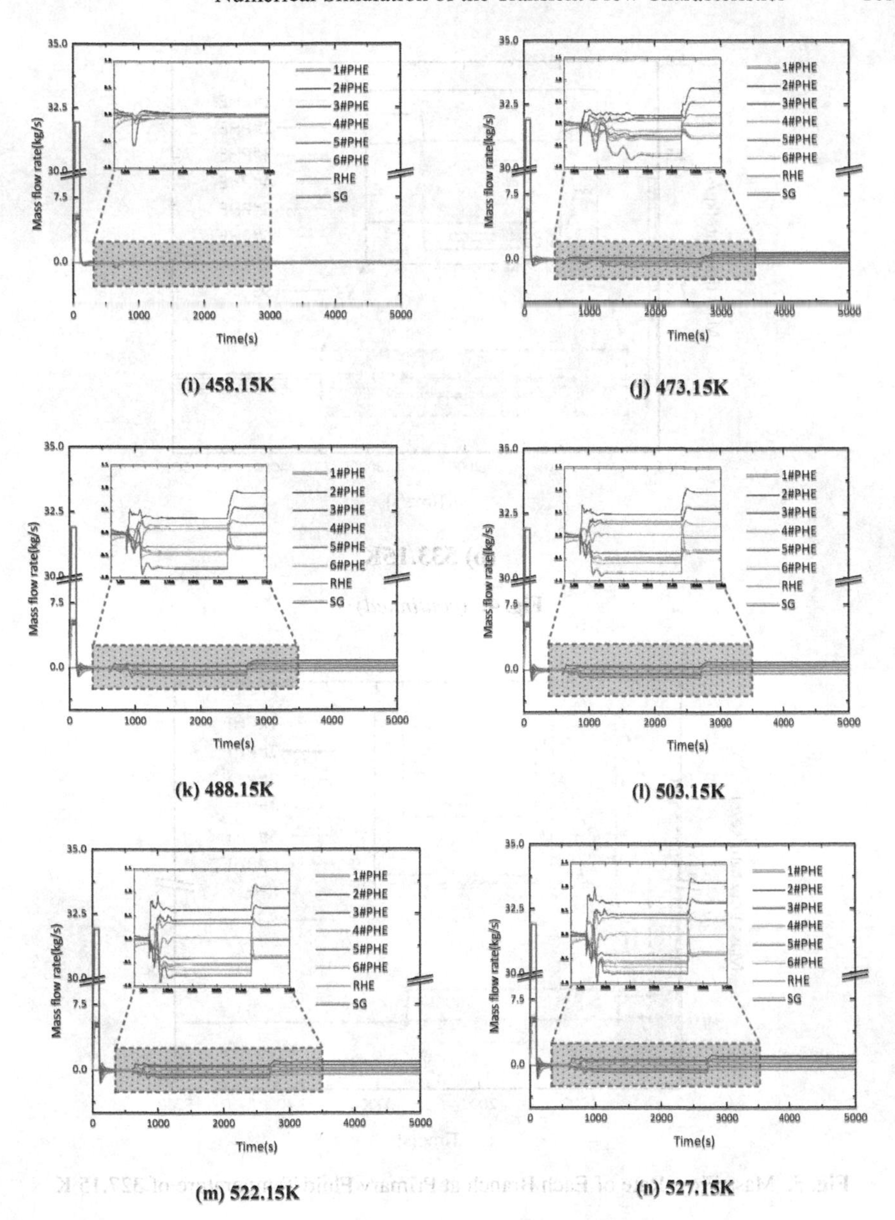

Fig. 4. (*continued*)

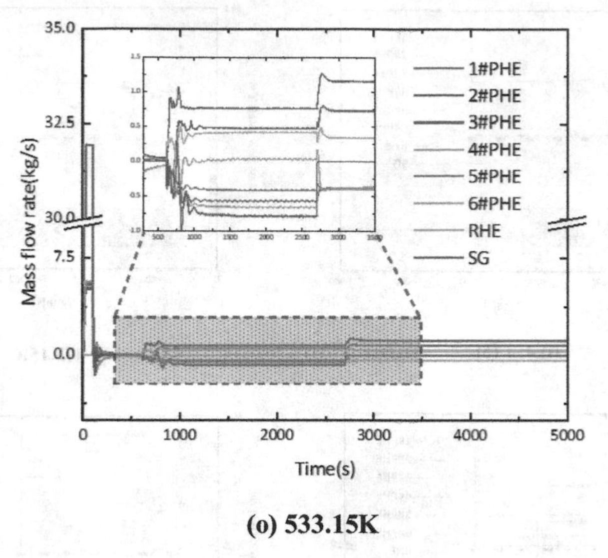

(o) 533.15K

Fig. 4. (*continued*)

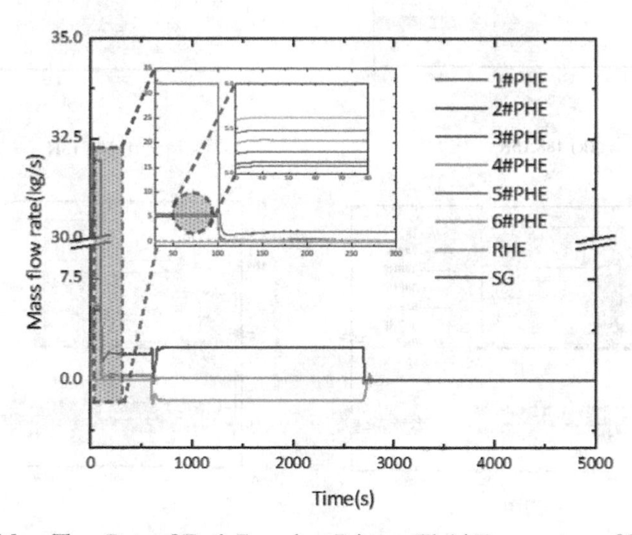

Fig. 5. Mass Flow Rate of Each Branch at Primary Fluid Temperature of 327.15 K.

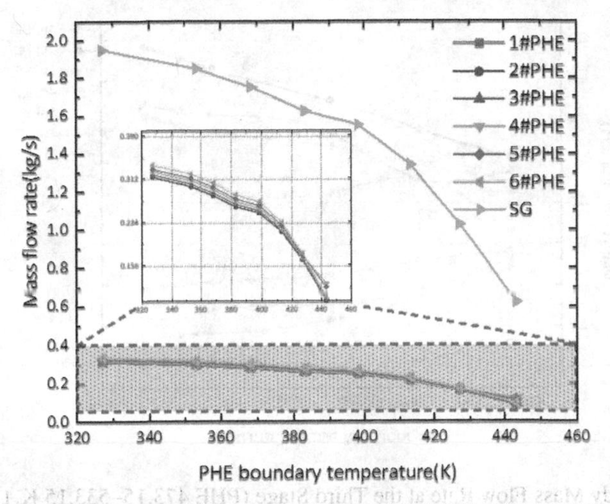

Fig. 6. Steady Mass Flow Rate at the Second Stage (PHE 327.15–443.15K, t = 500.0 s).

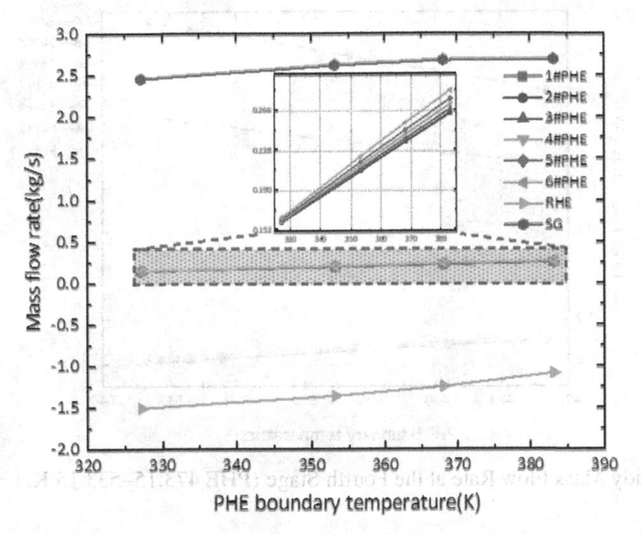

Fig. 7. Steady Mass Flow Rate at the Third Stage (PHE 327.15–383.15 K, t = 2000.0 s).

The uneven outlet temperatures of the PHE branches in the third stage can be explained by the reverse flow in some PHE branches and the thermal stratification effects in the headers. Significant thermal stratification occurred in the third stage, since the fluid from SG branch can flow into the lower part of the cold header or hot header through the low-resistance Y-junctions in the headers. In the fourth stage, since the valves of the SG branch closed, the thermal stratification phenomenon in the headers became insignificant, and the outlet temperatures of PHE branches became the same value.

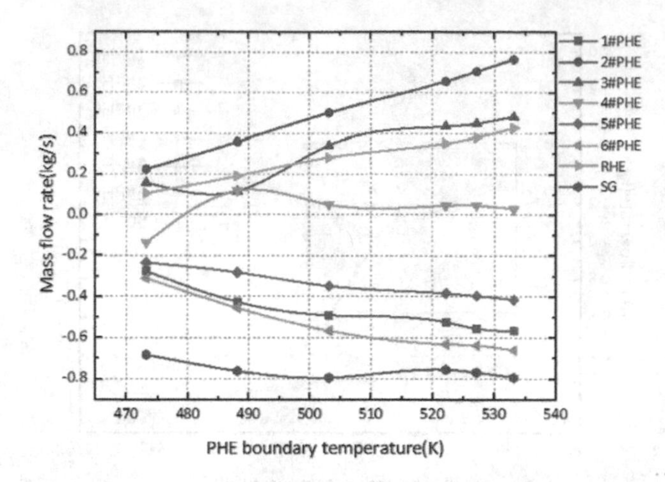

Fig. 8. Steady Mass Flow Rate at the Third Stage (PHE 473.15–533.15 K, t = 2000.0 s).

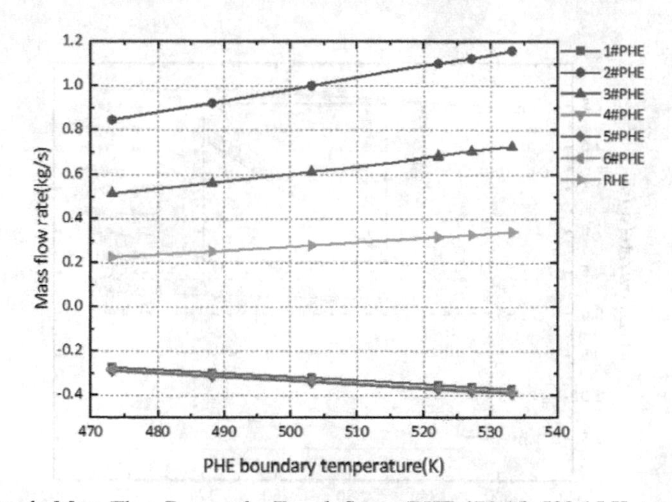

Fig. 9. Steady Mass Flow Rate at the Fourth Stage (PHE 473.15–533.15 K, t = 4000.0 s).

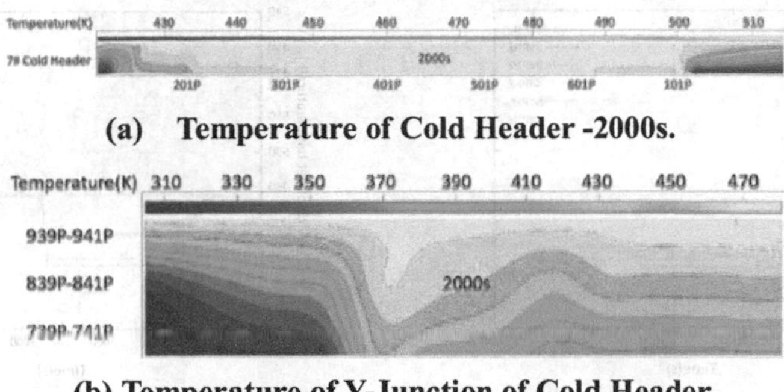

(a) Temperature of Cold Header -2000s.

(b) Temperature of Y-Junction of Cold Header (939P-941P/839P-841P/739P-741P) -2000s.

Fig. 10. Thermal Stratification of Cold Header and Y-Junction.

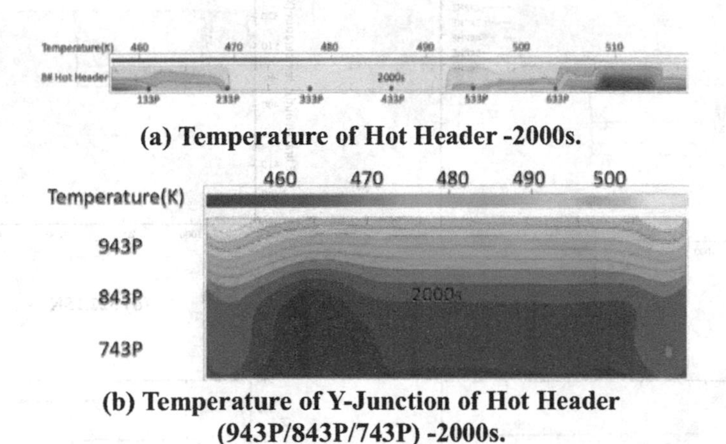

(a) Temperature of Hot Header -2000s.

(b) Temperature of Y-Junction of Hot Header (943P/843P/743P) -2000s.

Fig. 11. Thermal Stratification of Hot Header and Y-Junction.

5 Conclusions

In this paper, a layered RELAP5 model was developed for the PRHR system of a scaled test facility for the NHR-200-II reactor. Numerical simulations were carried for both the PRHR scenario of SG branch isolated and the PRHR scenario of SG branch not isolated. The main conclusions include:

(1) For the case of SG branch not isolated, the numerical simulations shown that the outlet temperatures of all PHE branches were significantly uneven (600–2700 s in Fig. 12), and for the case of SG branch isolated, the outlet temperatures of PHE branches were nearly equal (2700–5000 s in Fig. 12). In both cases, the numerical results were qualitatively consistent with the experimental observations.

(2) The primary fluid temperature has a significant impact on the PRHR system. The phenomenon of reversed flow occurred in some PHE branches when the primary

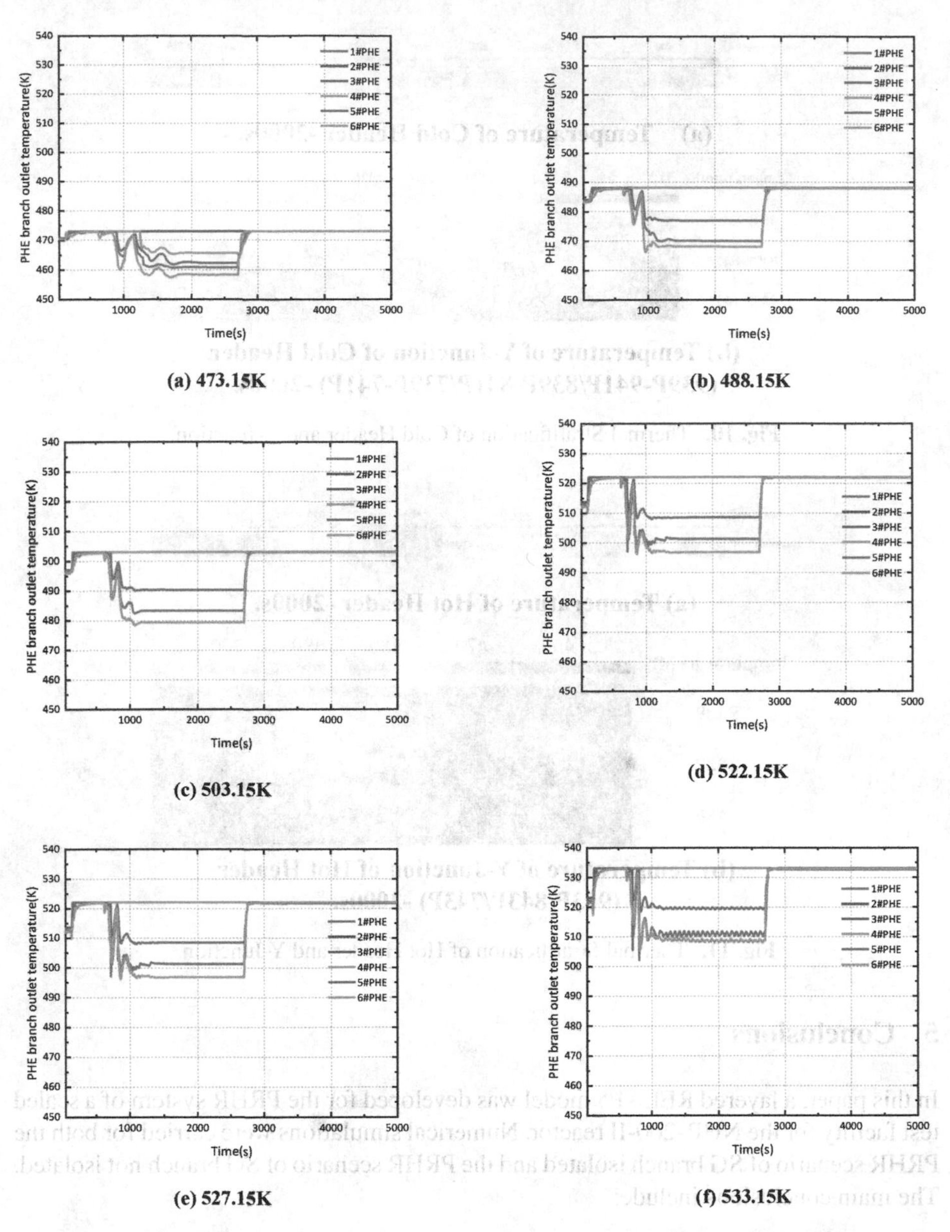

(a) 473.15K (b) 488.15K

(c) 503.15K (d) 522.15K

(e) 527.15K (f) 533.15K

Fig. 12. PHE Branch Outlet Temperature Dispersion (PHE 473.15–533.15 K).

fluid temperature was higher than 473.15 K (Fig. 8 and Fig. 9), which was also consistent with the experiment observations.

(3) The uneven outlet temperatures of the PHE branches when the SG branch was not isolated can be explained by the reverse flow in some PHE branches and the thermal stratification effect in the headers. The layered RELAP5 nodalization for

the hot and cold headers can be used as a preliminary estimation method of the thermal stratification effect, however, detailed 3-dimensional CFD methodology is still necessary to accurately capture the phenomenon of thermal stratification in the hot and cold headers of the PRHR system.

References

1. W (2013) http://www.westinghousenuclear.com/SMR/index.htm
2. IAEA.: Status of Small Medium Sized Reactor Designs. IAEA (2011a)
3. Park, K.B.: SMART design and technology features. In: Interregional Workshop on Advanced Nuclear Reactor Technology for Near Term Deployment, Austria (2011c)
4. IAEA (2011b) http://www.iaea.org/NuclearPower/Downloads/Technology/meetings/2011-Jul4-8-ANRT-WS/3_USA_mPOWER_BABCOCK_DELee.pdf
5. Carelli, M.D., Conway, L.E., Oriani, L.: The design and safety features of the IRIS reactor. In: ICONE11e36564, 11th International Conference on Nuclear Engineering, Tokyo, Japan (2003b)
6. Dazhong, W.A.N.G., Jiagui, L.I.N., Changwen, M.A., et al.: Design of 200MW nuclear heating station. Nucl. Power. Eng. **14**(4), 289–295 (1993)
7. Zhang, Z., Gao, Z., Wang, Y., et al.: Inherent safety of 200MW nuclear heating reactor. Nucl. Power. Eng. **03**, 227–231+255 (1993)
8. Test of Passive Residual Heat Removal System for Low-Temperature Reactor (Internal Technical Report), Institute of Nuclear and New Energy Technology (INET), Tsinghua University, Beijing, China (2018)
9. Xu, Z., Wu, X.: Dynamic analysis of passive residual heat removal system of 200 MW low-temperature nuclear heating reactor. Nucl. Power. Eng. **02**, 61–65 (2008)

Fatigue Analysis Method of Steel Containment of Floating Nuclear Power Plant

Mingxuan Liu[1], Xinyang Fan[1], Danrong Song[2], Bin Zheng[2], and Meng Zhang[1](✉)

[1] Yantai Research Institute and Graduate School, Harbin Engineering University, Yantai, Shandong, China
zhangmeng@hrbeu.edu.cn
[2] Nuclear Power Institute of China, Chengdu, Sichuan, China

Abstract. Floating nuclear power plant (FNPP) is a movable nuclear power plant built on the floating platform, which can provide clean and stable power for remote coastal areas, and are currently a hot research topic in the field of nuclear power. The steel containment is located in the reactor compartment of the FNPP and it is an important safety guarantee structure. Fatigue and fracture have been an important issue for ship and offshore structures for a long time. Fatigue failure of containment will have serious consequences.

In order to research the fatigue life analysis method of steel containment of the first FNPP in China, the paper adopts miner linear cumulative damage theory and spectral analysis method, based on the American Society of Mechanical Engineers (ASME) standards and relevant standards of China Classification Society (CCS), and uses AQWA to analyze Wave load of FNPP. The hydrodynamic calculation results are imported into finite element model to analyze the structural response of each point of containment, and calibrate the transfer function data of each key point by using the linear system theory and regular wave periodic evaluation method. The fatigue analysis of each point is carried out according to the transfer function and the wave dispersion diagram drawn by the forty years monitoring sea conditions of the working sea area of the FNPP. The result shows that the fatigue life of steel containment is superior and meets the service requirements.

Keywords: Floating nuclear power plant · Containment vessel · Fatigue analysis · Transfer function · Regular wave simulation method

1 Introduction

With the continuous adjustment and optimization of China's energy structure and the continuous promotion of the strategy of strengthening the country through the sea, it is increasingly difficult for traditional fossil energy sources as well as emerging energy sources such as wind, wave and solar energy to meet the energy demand brought by the development of coastal oil and gas resources and islands. Offshore floating nuclear power plant refers to a movable floating marine platform equipped with nuclear reactor and power generation system, which is a product of the organic combination of mobile small nuclear power plant technology and ship and marine engineering technology. As

C. Liu (Ed.): PBNC 2022, SPPHY 283, pp. 1046–1059, 2023.
https://doi.org/10.1007/978-981-99-1023-6_88

early as in the 1970 s, researchers in the United States proposed the idea of floating nuclear power plants [1], and the world's first floating nuclear power plant 'Akademik Lomonossov' was also launched in Russia in 2016 [2]. In floating nuclear power plants, a sealed steel containment structure is usually installed to wrap around the reactor and other auxiliary power generation equipment structures to protect the reactor from normal operation as well as to protect the external environment. Compared to traditional onshore containment, the environment and loads on the small steel containment and support structure of an offshore floating nuclear power plant are very different, especially because the complexity of the marine environment leads to more complex loads on the containment and support structure.

The alternating loads caused by these complex sea conditions may cause fatigue damage to the structure and generate cracks, which in turn threaten the safety of the floating nuclear power plant structure and cause the structure to fracture when the cracks expand to a certain extent, resulting in serious accidents. Floating nuclear power plants. However, there are few studies on the fatigue of floating nuclear power plant containment. Therefore, in order to ensure the operational safety of floating nuclear power plants and protect the surrounding personnel and external environment from nuclear radiation, it is important to carry out research on the fatigue assessment method of floating nuclear power plant steel containment in marine environment to ensure the safe operation of floating nuclear power plants during the design life and reduce economic losses.

The current fatigue assessment methods for marine structures can be divided into, simplified algorithm [4], design wave method [5] and direct calculation of spectral analysis method, among which spectral analysis method has the advantages of high accuracy and can reflect the specific structural details of the ship is widely used in the field of marine engineering, previously Hadi and Yang et al. used spectral analysis method for fatigue reliability analysis of marine platforms [6, 7], Zhang et al. used the spectral analysis method to evaluate the fatigue strength of a small waterline surface catamaran [8]. In contrast, in the study of nuclear power system pressure-bearing equipment, the transient method is usually used to assess its fatigue damage because its stress response time course is easily accessible [9, 10].

In this paper, based on the above research, the fatigue reliability study of a type of floating nuclear power plant containment is carried out by combining the spectral analysis method commonly used in marine engineering with the fatigue strength analysis of floating nuclear power plant containment, while referring to ship-related codes and ASME-related codes [11], using the regular wave simulation method.

2 Fatigue Strength Spectrum Analysis Method

2.1 Spectrum Analysis Method

The FNPP has a complex working environment and is subjected to the combined action of wind, wave and current. The main part of the fatigue load that causes the fatigue failure of the containment structure of the FNPP is the wave load. The key points of evaluating the fatigue strength of the containment of FNPP are the selection of wave spectrum, the calculation of transfer function and the calculation of fatigue damage.

Spectrum analysis method is a commonly used method in ship and ocean engineering to study load and structural response. Its theoretical basis is the linear system transformation in random process theory. The method firstly obtains the power spectral density function (PSD) of the structural stress response, after that establishes the relationship between the stress response power spectral density function and the rain flow stress range distribution, and then selects a suitable S-N curve and Miner cumulative damage theory to calculate the fatigue damage of the structure (Fig. 1).

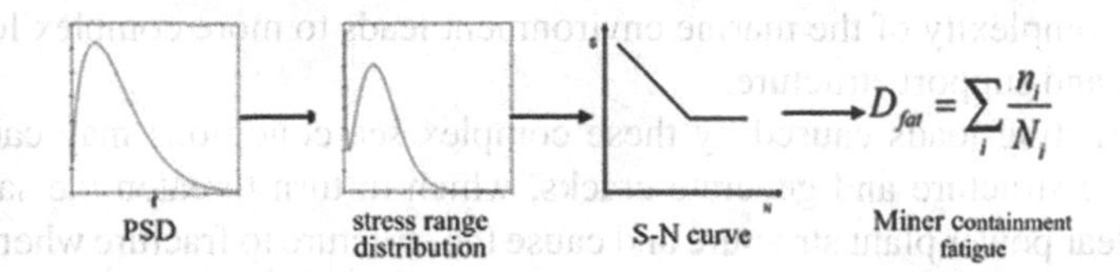

Fig. 1. Fatigue analysis flow

The FNPP can be regarded as a typical dynamic system in ships and offshore engineering structures. The wave process a acting on the hull is the input of the system, and the alternating stress B caused by the wave action in the structure is the response of the containment. In general, the relationship between the response process of the containment and the wave load input process of the FNPP can be written as:

$$X(t) = L[\eta(t)] \tag{1}$$

where, B represents the operator that transforms C into D. When e is a linear operator, the system is linear.

In the fatigue analysis of ship structure, the calculation of wave load and structural response are based on linear theory. Under this condition, if the wave is a stationary random process, the alternating stress obtained by transformation is also a stationary random process. According to the random process theory, there is the following relationship between the power spectral density functions of two stationary random processes:

$$G_X(\omega) = |H(\omega)|^2 \cdot G_\eta(\omega) \tag{2}$$

In function (2), $H(\omega)$ is a transfer function or frequency response function of linear dynamical system. $|H(\omega)|^2$ is response amplitude operator (RAO).

The physical meaning of $H(\omega)$ is the ratio of the amplitude of the response process to the amplitude of the input process when the linear dynamic system vibrates with a circular frequency of ω.

2.2 Wave Spectrum

For the spectral analysis method for fatigue assessment of ship structures, since the FNPP is located in a shallow water depth and the wave dispersion diagram is the joint distribution of meaningful wave height and spectral peak period, the improved JONSWAP spectrum is selected for analysis, and its expression is as follows:

$$S(f) = \beta_J H_{\frac{1}{3}}^2 T_P^{-4} f^{-5} \exp[-\frac{5}{4}(T_P f)^{-4}]\gamma^{\exp[-(\frac{f}{f_P}-1)^2/2\sigma^2]} \tag{3}$$

In function (3):

$$B_J = \frac{0.06238}{0.23 + 0.0336\gamma - 0.185(1.9 + \gamma)^{-1}}[1.094 - 0.01915 \ln \gamma]$$

γ is the crest factor, mean value is 3.3.

σ is the peak shape parameter. When the frequency is on the left side of the maximum value point, it is taken as 0.07, and when the frequency is on the right side of the maximum value point, it is taken as 0.09.

2.3 Regular Wave Simulation Method

The transfer function is determined by the system through experiments under the action of rule input or random input, or by the system's theoretical analysis of rule input. Based on the regular wave test method in the pool test, we propose the regular wave simulation method, that is, using the wave load calculation program to obtain the response of the ship motion and external hydrodynamic pressure of a series of regular waves arranged according to a certain initial phase interval under the heading angle and circular frequency. The external hydrodynamic pressure and various inertial forces related to the motion of the hull are applied to the finite element model of the hull structure to obtain the stress response. For the stress response of a series of regular waves under the heading angle and circular frequency, the maximum stress response at the point is fitted by Fourier transform, and the stress amplitude of the heading angle and circular frequency can be obtained. The value of the transfer function under the heading angle and circular frequency can be obtained by comparing the obtained stress amplitude with the wave amplitude.

2.4 Fatigue Cumulative Damage Calculation and S-N Curve

After obtaining the damage caused by each cycle, the selection of a suitable fatigue accumulation damage theory is also one of the core elements of fatigue calculation. The S-N curve and the fatigue cumulative damage analysis method of linear cumulative damage theory are commonly used to evaluate the fatigue of structures in the codes of classification societies of various countries.

When the fatigue load spectrum is expressed as a continuous probability density function corresponding to a certain period of time, the fatigue cumulative damage degree can be expressed as

$$D = \int_L \frac{dn}{N} = \int_0^\infty \frac{N_L f_S(S) dS}{N} = N_L \int_0^\infty \frac{f_S(S) dS}{N} \tag{4}$$

S is the stress range, is the probability density function of the stress range distribution, is the number of cycles required to achieve fatigue failure under a single cyclic load with a stress range of, is the total number of cycles of the internal stress range during the whole time period considered, is the number of cycles of the included stress range, and represents the integral of the whole time interval considered. According to the cumulative damage theory, when the damage degree is accumulated, the fatigue failure of the structure will occur.

The S-N curve is often used to reflect the relationship between the stress range S and the number of cycles required for the structure to achieve fatigue failure under a single cyclic load at the level of the stress range, i.e. the fatigue life n. It is generally obtained by fitting the fatigue test results. A large number of research results show that under a given stress range s, the discrete type of the parameter m is small and can be regarded as a certain value, the fatigue life n and parameter a should be treated as random variables, and it is generally considered that n obeys lognormal distribution. Expressed as

$$NS^m = A \tag{5}$$

Take logarithm on both sides of the equation

$$\lg N + m \lg S = \lg A \tag{6}$$

Equation (6) is a commonly used double log-linear model of the S-N curve. The small steel containment material studied in this paper is Steel-SA-738Gr.b. Therefore, the parameters are referred to the appendix of ASME BPVC Volume III [11].

The tensile strength of Steel-SA-738Gr.b is 585-705mpa. According to the S-N curves, the S-N curve of the material can be obtained by interpolation. The curve is transformed into a double logarithmic linear form, taking m = 3 and a = 11.464.

3 Containment Fatigue Strength Analysis

3.1 Hydrodynamic Analysis

The structural hydrodynamic model is shown in Fig. 2. The right-hand rectangular coordinate system is used. The origin is taken at the intersection of the intersection line of the longitudinal section and the middle transverse section of the platform and the base plane. The X axis is the longitudinal axis, and the point from the tail to the head is positive; Y-axis is the transverse axis, and it is positive from the centerline to the port; The z-axis is the vertical axis, and upward from the base is positive.

For model hydrodynamic analysis, AQWA software is used for hydrodynamic analysis of FNPP. During frequency domain hydrodynamic calculation, the minimum frequency of each wave direction is set at 0.01592 Hz and the maximum frequency is set at 0.27 Hz, with a total of 50 frequency points.

The frequency response curve of the longitudinal bending moment of the middle hull cross-section (x = 0.328 m cross-section) of the reactor bay at 0° wave incidence angle is shown in the following Fig. 3:

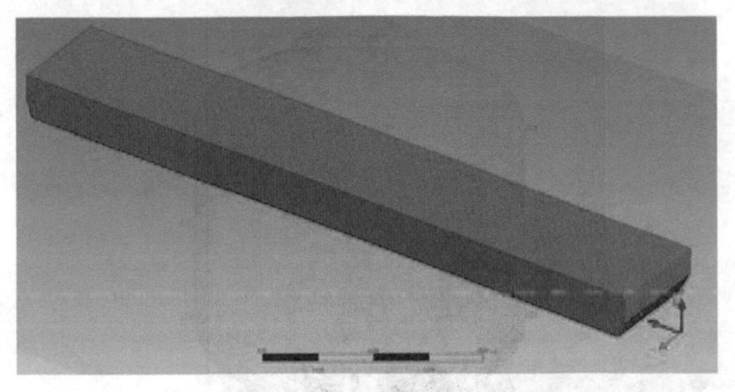

Fig. 2. Hydrodynamic model of FNPP.

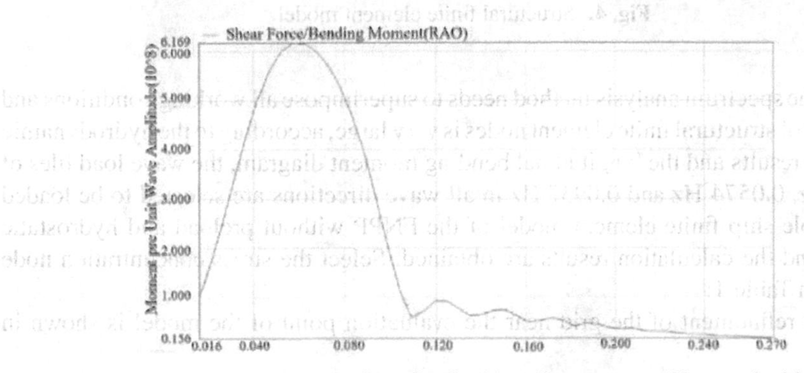

Fig. 3. Frequency response curve of longitudinal bending moment (0° wave incidence angle)

Taking 0° wave incidence angle as an example, it can be seen from the calculation results that the peak longitudinal bending moment in the transverse section of the hull in the middle of the stack is about $6.169*10^8$ N*m under 0° wave incidence angle and unit wave amplitude, and the corresponding wave frequency is 0.0574 Hz. Another longitudinal bending moment value at the waistline is $3.425*10^8$ N*m corresponding to the wave frequency of 0.03147 Hz, and the longitudinal bending moment value is $3.035*10^8$ N*m corresponds to a wave frequency of 0.0937 Hz.

3.2 Selection of Fatigue Damage Assessment Points

The finite element model of the FNPP structure is constructed with shell181 and beam188 elements. The mesh size of the bottom and supporting parts of the containment is 0.1 M, the mesh size of the upper part of the containment is 0.2 m, and the mesh size of the rest parts is 0.8 m. The total number of elements on the ship is about 1.56 million. The finite element model of the containment is shown in Fig. 4.

Fig. 4. Structural finite element model.

Since the spectrum analysis method needs to superimpose all working conditions and the number of structural finite element nodes is very large, according to the hydrodynamic calculation results and the longitudinal bending moment diagram, the wave load files of 0.03147 Hz, 0.0574 Hz and 0.0937 Hz in all wave directions are selected to be loaded on the whole ship finite element model of the FNPP without preload and hydrostatic pressure, and the calculation results are obtained. Select the stress concentration node as shown in Table 1:

A local refinement of the grid near the evaluation point of the model is shown in Fig. 5.

3.3 Fatigue Life of Containment

Wave scatter diagram is a common method to describe the marine environment in ship and ocean engineering. Table 2 shows the monitoring data of the nearby platform in the sea area where the floating nuclear power plant works. In the table, H_s denotes the meaningful wave height and T_p denotes the spectral peak period.

The long-term distribution of the stress range within the design life of the FNPP containment can be obtained from the short-term distribution combined with the distribution of various sea conditions that may be encountered in operation. In a given sea state, the ship may sail in any course. In the calculation, several courses are divided, and it is assumed that the probability of each course is equal.

The FNPP can set a course every 15° from 0° to 360° in the marine environment. There are 24 courses in total, and the probability of each course angle is 1/24. In order to simplify the calculation, the FNPP, as a symmetrical structure, can simplify the structural response caused by the symmetrical course. Therefore, in the actual calculation, take a course every 15° from 0° to 180°, a total of 13 courses, of which the probability of 0° and 180° is 1/24, and the probability of other courses is 1/12 (Fig. 6).

Therefore, in the regular wave experimental simulation method, AQWA software is used to calculate the response of hull motion and external hydrodynamic pressure

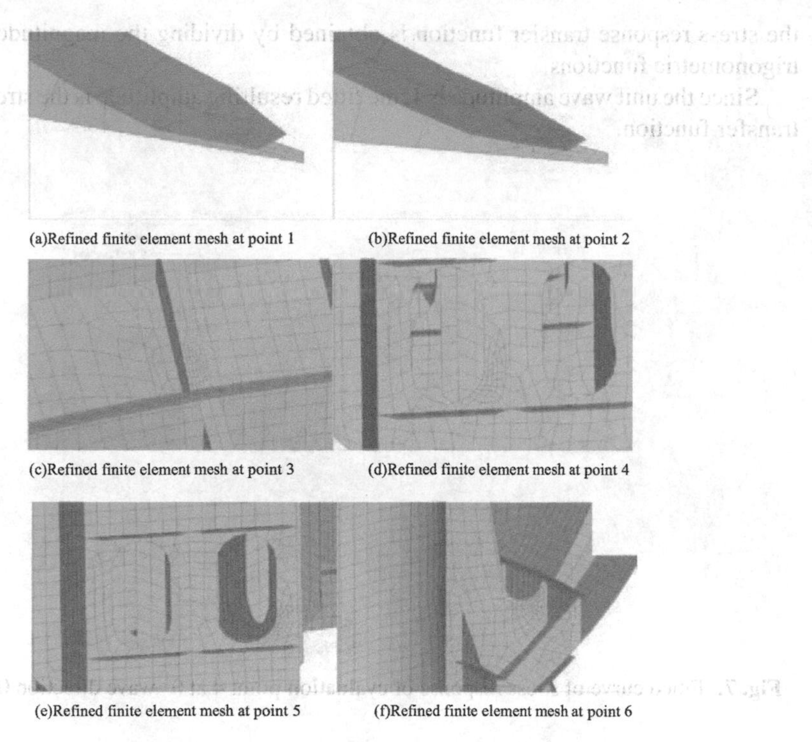

Fig. 5. Finite element refinement mesh

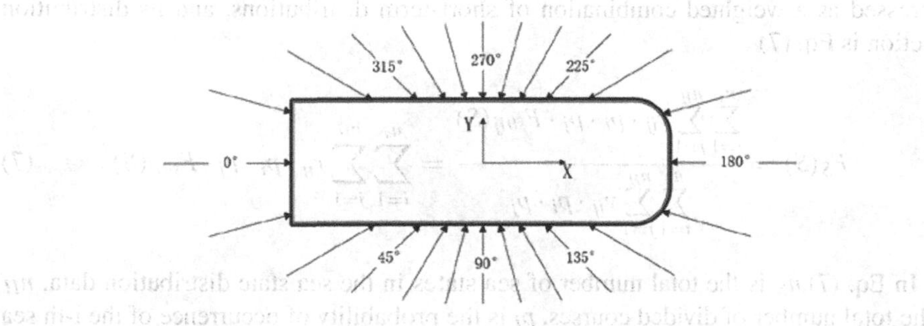

Fig. 6. Schematic diagram of wave incidence angle.

of a series of regular waves with unit wave amplitude of 1 at each heading angular circular frequency arranged at a certain initial phase interval, the phase is taken as 0° to 360° with 45° interval, 8 regular waves at each wave direction frequency, and the wave load file is extracted. The wave load is the wave surface pressure, and the wave surface pressure is mapped to the wet surface of the hull to calculate the structural response. The structural response is the response of the structure under the action of unit wave amplitude. The maximum stress value at each calculation point acting on the top of the FNPP containment is extracted and fitted using the fast Fourier transform (Fig. 7), and the stress response transfer function is obtained by dividing the magnitude of the two trigonometric functions.

Since the unit wave amplitude is 1, the fitted resulting amplitude is the stress response transfer function.

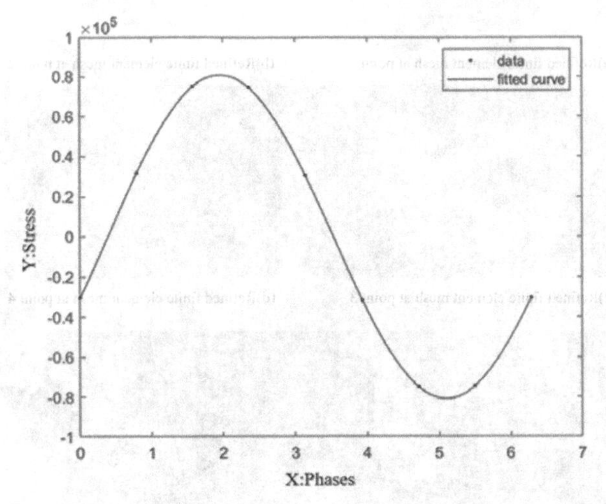

Fig. 7. Fitted curve of stress response of evaluation point 4 at 0° wave direction 0.26451 Hz.

Under the above conditions, the long-term distribution of the stress range can be expressed as a weighted combination of short-term distributions, and its distribution function is Eq. (7)

$$F_S(S) = \frac{\sum_{i=1}^{n_s}\sum_{j=1}^{n_H} v_{ij}\cdot p_i\cdot p_j\cdot F_{s\theta ij}(S)}{\sum_{i=1}^{n_s}\sum_{j=1}^{n_H} v_{ij}\cdot p_i\cdot p_j} = \sum_{i=1}^{n_s}\sum_{j=1}^{n_H} r_{ij}\cdot p_i\cdot p_j\cdot F_{s\theta ij}(S) \tag{7}$$

In Eq. (7),n_S is the total number of sea states in the sea state distribution data, n_H is the total number of divided courses, p_i is the probability of occurrence of the i-th sea state, which can be obtained according to the frequency of occurrence of each sea state in Table 3;p_j is the frequency of occurrence of the j-th heading. v_{ij} is the average zero crossing rate of stress alternating response under the i-th sea state and the j-th heading.v_0 is the total average zero crossing rate of stress response considering all sea conditions and heading.

$$v_0 = \sum_{i=1}^{n_s}\sum_{j=1}^{n_H} v_{ij}\cdot p_i\cdot p_j \tag{8}$$

Table 1. Fatigue damage point calculation number and location.

Evaluation point	Location
1	Containment bottom support
2	Containment bottom support
3	T-section at bottom of containment
4	Containment bottom support
5	Containment bottom support
6	Containment pressurizer reinforcing rib

3.4 Containment Fatigue Life Correction

In addition to the influence of marine environmental load on the structure, the marine environmental conditions also have a great impact on the fatigue performance of materials, mainly in the form of corrosion. *Fatigue strength guide for hull structures (2021)* of CCS [12] stipulates that for the normal bending stress of hull girder during simplified stress analysis and the hot spot stress under overall load conditions during finite element stress analysis, the corrosion correction factor $f_{cl} = 1.05$;

For the bending normal stress under lateral load in simplified stress analysis and the hot spot stress under local load in finite element stress analysis, the corrosion correction factor $f_{cl} = 1.1$.

In the direct calculation method of fatigue assessment, the fatigue safety factor needs to be superimposed for calculation. In this regard, *GUIDELINES FOR FATIGUE STRENGTH ASSESSMENT OF OFFSHORE ENGINEERING STRUCTURES (2013)* of CCS provides relevant provisions [13].

Fatigue failure criteria can be based on fatigue damage or fatigue life. When based on fatigue damage, the fatigue strength of the calculated point shall meet Eq. (9)

$$D \leq \frac{1.0}{S_{ftg}} \tag{9}$$

D-- Fatigue damage degree;

S_{ftg}-- Fatigue strength safety factor.

The fatigue safety factor of the small steel containment of the FNPP is selected by reference to the fixed floating structure. The fatigue damage assessment location is accessible for inspection and maintenance in a dry environment, and the failure consequences are serious. Considering the special nature of the small steel containment of FNPP, we select 5 as the fatigue safety factor. Therefore $D \leq 0.2$.

See Table 4 for the cumulative fatigue damage degree and the corrected fatigue life of the final six fatigue assessment points of the containment.

T_1 is fatigue life considering corrosion correction factor/year.

T_2 is the fatigue life considering the fatigue safety factor and corrosion.

Table 2. Wave Dispersion

Spectral Peak period – Tp(s)															
Hs(m)	0–1	1–2	2–3	3–4	4–5	5–6	6–7	7–8	8–9	9–10	10–11	11–12	12–13	13–14	14–15
0.0–0.5	0.094 (1)	1.836(2)	6.383(3)	11.998(4)	1.508(5)	0.665(6)	0.290(7)	0.018(8)	0.005(9)	0.002(10)	0.005(11)	0.012(12)	0.014(13)	0.018(14)	0.026(15)
0.5–1.0	0	0	0	7.584(16)	27.675(17)	4.405(18)	1.220(19)	0.144(20)	0.003(21)	0	0	0	0	0	0
1.0–1.5	0	0	0	0.193(22)	6.793(23)	10.673(24)	2.290(25)	0.043(26)	0.008(27)	0.001(28)	0	0	0	0	0
1.5–2.0	0	0	0	0.005(29)	0.119(30)	3.580(31)	4.138(32)	0.248(33)	0.003(34)	0	0	0	0	0	0
2.0–2.5	0	0	0	0	0.020(35)	0.112(36)	2.874(37)	0.933(38)	0.002(39)	0	0	0	0	0	0
2.5–3.0	0	0	0	0	0	0.021(40)	0.452(41)	1.566(42)	0.028(43)	0	0	0	0	0	0
3.0–3.5	0	0	0	0	0	0	0.012(44)	0.740(45)	0.253(46)	0	0	0	0	0	0
3.5–4.0	0	0	0	0	0	0	0	0.132(47)	0.426(48)	0.012(49)	0	0	0	0	0
4.0–4.5	0	0	0	0	0	0	0	0.004(50)	0.164(51)	0.135(52)	0	0	0	0	0
4.5–5.0	0	0	0	0	0	0	0	0	0.008(53)	0.054(54)	0.004(55)	0	0	0	0
5.0–5.5	0	0	0	0	0	0	0	0	0	0.013(56)	0.024(57)	0	0	0	0
5.5–6.0	0	0	0	0	0	0	0	0	0	0.002(58)	0.009(59)	0	0	0	0
6.0–6.5	0	0	0	0	0	0	0	0	0	0	0.001(60)	0.001 (61)	0	0	0
6.5–7.0	0	0	0	0	0	0	0	0	0	0	0	0	0	0	0

Table 3. Fatigue cumulative damage results.

Evaluation point	Cumulative fatigue damage	Fatigue life/years
1	0.009708	4120
2	0.003918	10209
3	0.02289	1747
4	0.08070	496
5	0.01025	3903
6	0.05275	758

Table 4. Fatigue cumulative damage results after correction.

evaluation point	Cumulative fatigue damage	T_1/years	T_2/years
1	0.01292	4120	619
2	0.005215	7670	1534
3	0.03047	1747	263
4	0.1074	496	74
5	0.01364	3903	587
6	0.07021	758	114

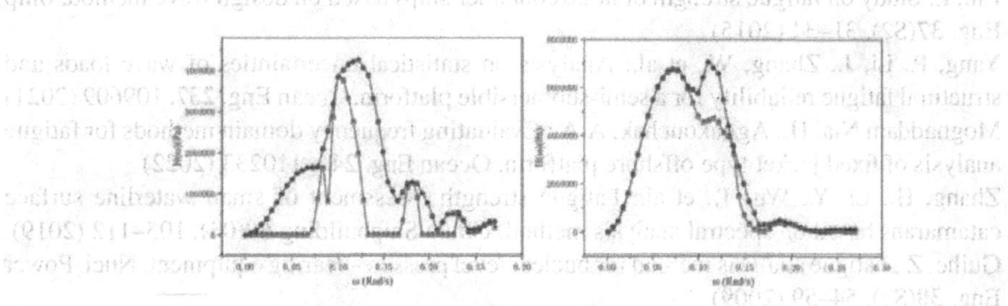

(a) evaluation point 4 (b) evaluation point 6

Fig. 8. Comparison of transfer function results

4 Conclusion

Based on the spectral analysis method and regular wave simulation method, this paper analyzes the small steel containment vessel of FNPP. It is concluded that the maximum fatigue cumulative damage is at No. 8 calculation point, the fatigue cumulative damage degree is 0.1074, and the fatigue life is 74 years. It is located at the bottom support, which meets the design requirements of FNPP. At the same time, the parameters are conservative and the fatigue life is short.

For the regular wave simulation method, take the evaluation point 4 and evaluation point 6 with short fatigue life as an example. Figure 8 shows the transfer function plots of evaluation point 4 and evaluation point 6 at 45° wave direction. The black curve is the simplified theoretical analysis method, and the red curve is the regular wave experimental simulation method. It can be concluded from the figure that the regular wave simulation method has a smoother and more accurate curve, although the calculation is more complex.

References

1. Thangam Babu P.V., Reddy, D.V.: Existing methodologies in the design and analysis of offshore floating nuclear power plants. Nucl. Eng. Des. **48**(1), 167–205 (1978)
2. Arctic's first offshore floating nuclear power plant on the verge of completion. Ship Eng. **39**(04), 9 (2017)
3. Guo, X., Kong, F., Zhu, C., Zhu, G.: Study on safety guidelines in the design of floating nuclear power plants. Nucl. Power Eng. (2021)
4. Liu, Y., Ren, H., Feng, G., et al.: Simplified calculation method for spectral fatigue analysis of hull structure. Ocean Eng. **243**, 110204 (2022)
5. Lin, I.: Study on fatigue strength of large container ships based on design wave method. Ship Eng. **37**(S2), 31–34 (2015)
6. Yang, P., Li, J., Zhang, W., et al.: Analysis on statistical uncertainties of wave loads and structural fatigue reliability for a semi-submersible platform. Ocean Eng. **237**, 109609 (2021)
7. Moghaddam Nia, H., Aghakouchak, A.A.: Evaluating frequency domain methods for fatigue analysis of fixed jacket type offshore platform. Ocean Eng. **246**, 110233 (2022)
8. Zhang, H., Li, Y., Wu, T., et al.: Fatigue strength assessment of small waterline surface catamarans based on spectral analysis method. China Shipbuilding **60**(03), 105–112 (2019)
9. Guihe, Z.: Fatigue analysis method for nuclear level pressure-bearing equipment. Nucl. Power Eng. **30**(S2), 54–59 (2009)
10. Wen, J., Fang, Y., Lu, Y., Zou, M., Zhang, Y., Sun, Z.: Methods and steps for fatigue analysis of nuclear safety equipment. At. Energy Sci. Technol. **48**(01), 121–126 (2014)
11. ASME Boiler and Pressure Vessel Code: SECTION III (2004)
12. CCS. Guidelines for fatigue strength assessment of offshore engineering structures (2013)
13. CCS. Guidelines for fatigue strength of ship structure (2021)

Correlating IASCC Growth Rate Data to Some Key Parameters for Austenitic Stainless Steels in High Temperature Water

Caibo Xie, Songhan Nie, Yiqi Tao, and Zhanpeng Lu(✉)

Shanghai University, Shanghai, China

zplu@t.shu.edu.cn

Abstract. Austenitic stainless steels have been widely used for fabricating reactor core-internal components in PWRs due to its high strength, ductility and fracture toughness. The accelerated failure or degradation of austenitic stainless steel represented by IASCC has become one of the key problems affecting the safe and efficient operation of reaction core-internal in PWR nuclear power plants. IASCC is generally divided into three stages: crack initiation, crack propagation and instable fracture. Among the three stages, the crack initiation stage would occupy the major service time, the crack growth stage is featured by quasi-steady crack propagation at a certain rate, and the instable fracture stage should be avoided. Stress intensity factor K at the crack tip is often used to represent the mechanical driving force for SCC as well as IASCC.

In this paper, SCC crack growth rate (CGR) data of austenitic stainless steels irradiated in high temperature water were compiled and reanalyzed to evaluate the influence of key parameters such as radiation dose and mechanical properties on IASCC sensitivity and crack growth rate of these materials in PWR nuclear power plant environment. The CGR-K curves of the irradiated materials were also analyzed. The effects of low, medium and high doses of neutron irradiation are compared, and the analysis process is illustrated with examples. In the research process, abnormal CGR and K of materials under a specific irradiation dose was found, so this phenomenon was analyzed. The CGR data and irradiation dose of austenitic stainless steel in different K range were analyzed. And proposed a way to judge the type of change:type I, type II and type III. Finally, the yield strength of the material under the same irradiation dose was found, and combined with other research data, it was further demonstrated that the neutron irradiation dose had a significant effect on the crack growth rate.

Keywords: Austenitic stainless steel · Irradiation assisted stress corrosion cracking · Crack growth rate · High temperature water

1 Introduction

There are pressurized water reactor (PWR), boiling water reactor, heavy water reactor and other types of commercial nuclear power plants. PWR is the most widely used because

C. Liu (Ed.): PBNC 2022, SPPHY 283, pp. 1060–1072, 2023.
https://doi.org/10.1007/978-981-99-1023-6_89

of its mature technology and rich operation experience, accounting for more than half of the operating nuclear power plants in the world. Prolonged neutron irradiation leads to the changes of mechanical properties due to irradiation induced hardening effect, the changed of local chemical compositions due to radiation-induced segregation effect, the increase of physical defects such as dislocation loops, and the water decomposition due to radiolysis, and finally affects the SCC in high temperature water.

IASCC is generally divided into three stages: crack initiation, crack propagation and instable fracture [1]. At the initial stage of crack initiation, the surface of sensitive materials begins to produce microcracks under the coupling effect of environmental and mechanical factors, and these microcracks are invisible under the light microscope. Over time, the microcracks merge with each other to form an initial crack with a length of 10 μm. In the crack growth stage, the crack expands at a certain rate, which is affected by environmental and mechanical factors. The laboratory usually studies the process by introducing the stress intensity factor K at the crack tip. In the instable fracture stage, the crack expands rapidly until the material fracture.

It is well known that SCC of materials in high temperature and high pressure water depends on three factors: materials, environment and relatively high stress is shown in Fig. 1. The key parameters affecting IASCC mainly include the material itself (such as microstructure, microchemistry and yield strength, etc.) and environmental parameters (such as hydrochemistry, irradiation temperature and irradiation dose, etc.). The key parameters such as irradiation temperature and irradiation dose have great influence on the crack growth rate of austenitic stainless steel IASCC. There are also many studies on the influence of material factors such as microstructure, microchemistry and yield strength on IASCC sensitivity of austenitic stainless steel.

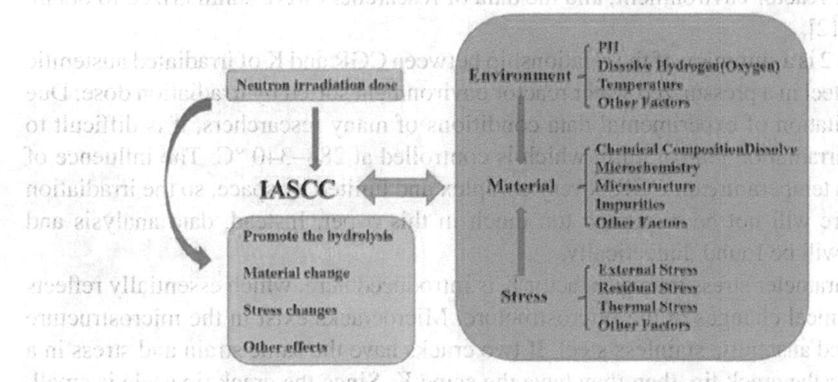

Fig. 1. The main influencing factors of IASCC

The study of IASCC behavior of austenitic stainless steel needs to pay attention to three stages: crack initiation, crack propagation and instable fracture. The crack propagation stage is the most important. Stress corrosion cracking promoted by irradiation is affected by material itself, medium environment, irradiation temperature and irradiation dose, etc. The research process is complicated and uncertain. The relationship between CGR and K is the key parameter in IASCC process of austenitic stainless steel.

SCC or IASCC is used for the study of materials. There are many types of specimens, such as compact tension (CT) specimen [2], four-point bending specimen [3], round compact tension (RCT) specimen [4] and so on. In this paper, the advantages and disadvantages of sample types are not discussed too much, and only the experimental results are concerned.

2 Compiling IASCC Data of Austenitic Stainless Steels in Simulated PWR Primary Water

Austenitic stainless steels can be divided into many types, among which 304 and 316 based austenitic stainless steel is widely studied, so here only around 304 and 316 based austenitic stainless steel is discussed.

The stress corrosion cracking experiment promoted by neutron irradiation needs special equipment, and the experimental difficulty exists objectively due to the limitation of equipment and other conditions. Radiation doses can range from a few percent of dPa to more than 100 dPa. Experimental data on crack growth rate of 0.06–47.0 dPa austenitic stainless steel were collected in this paper.

The CGR data of irradiated austenitic stainless steel in simulated BWR or PWR environments were obtained from CT samples, RCT samples or four-point bending samples. In the experiment, factors such as irradiation temperature, material type and water environment should be controlled, and the irradiation dose should be reasonably controlled. The CGR data of the sample should be matched with K one by one, and the relation data between CGR and K should be obtained.

Based on 304 and 316 austenitic stainless steel, the materials were irradiated in the light water reactor environment, and the data of researchers were summarized to obtain Fig. 2 [4–12].

Figure 2 is a summary of the relationship between CGR and K of irradiated austenitic stainless steel in a pressurized water reactor environment sorted by irradiation dose. Due to the limitation of experimental data conditions of many researchers, it is difficult to unify the irradiation temperature, which is controlled at 288–340 °C. The influence of irradiation temperature on CGR is very complex and limited by space, so the irradiation temperature will not be discussed too much in this paper. Instead, data analysis and problems will be found dialectically.

The parameter stress intensity factor K is introduced here, which essentially reflects the mechanical changes in the microstructure. Microcracks exist in the microstructure of irradiated austenitic stainless steel. If two cracks have the same strain and stress in a region near the crack tip, then they have the same K. Since the crack tip scale is small, K represents the stress and strain at the crack tip.

As shown in Fig. 2, as a whole from the reference curve, CGR generally shows an increasing trend with the increase of K. However, we found that after data integration, each data was not strictly linear and was greatly affected by the radiation dose.

The CGRs of SCC of unirradiated austenitic stainless steel is lower than that of IASCC of irradiated austenitic stainless steel under the same conditions, and the difference may be several times. Taking the CGRs data of austenitic stainless steel under 3 dpa irradiation dose as an example, the CGR of unirradiated austenitic stainless steel is about 3×10^{-10} m/s [2–5] while the CGR of irradiated austenitic stainless steel can reach 9×10^{-10} m/s or 1×10^{-9} m/s. The difference is about three times. This indicates that neutron irradiation can promote the crack growth rate (CGR) of austenitic stainless steel IASCC to a certain extent. The effect of irradiation dose on CGR of austenitic stainless steel IASCC in PWR is complicated. From the figure, we can roughly divide the radiation dose into low dose, medium dose and high dose. Low radiation dose mainly refers to 0–9 dpa. Under this condition, the change of CGR is not obvious compared with that without radiation. K is mainly between 12 and 20 MPa $m^{0.5}$. Based on literature review and PWR experience, 3 dpa is the threshold of material irradiation. In the case of 0 3 dpa irradiation dose, the CGR of the material will not change with the increase of the irradiation dose, but still maintain the original growth rate. Under the irradiation dose of 3–9 dpa, irradiation promoted the CGR of austenitic stainless steel. For example, the CGR of 3 dpa was higher than that of 6.3 dpa and 8.0 dpa. K remained in a certain range, while CGR changed abnormally. It is speculated that the CGR is affected by the changes of microstructure defects, dislocation loops and radiation dose.

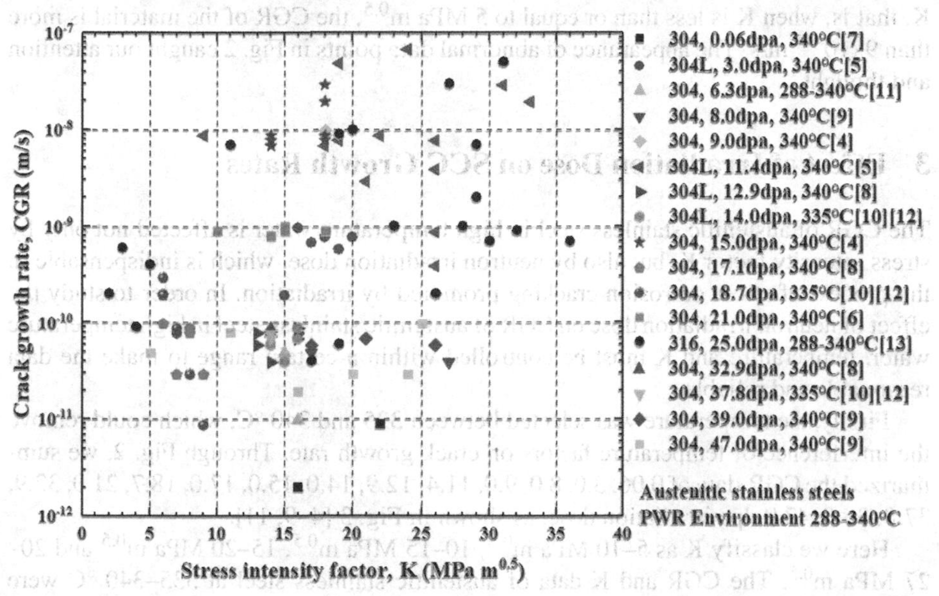

Fig. 2. Relationship between CGR and K of austenitic stainless steel at 0.6–47.0 dpa in pressurized water reactor [4–12]

The radiation dose here refers to the radiation dose of approximately 11.4–25.0 dpa. In the case of medium irradiation dose, CGR and K maintain a positive correlation. At 17.1 and 18.7 dpa, the data are mainly concentrated at the lower left of the figure near the reference curve. At 21.0, 12.9 and 14.0 dpa, the data mainly concentrated in the center of the graph near the curve. Although the data at 11.4, 15.0, and 25.0 dpa are mainly distributed in the upper right of the graph, they generally follow the trend of increasing CGR as K increases. In these data, the phenomenon of 25.0 dpa is worth separate discussion and will be carried out in the future.

High radiation dose refers to radiation dose above approximately 25.0 dpa. As can be seen from the figure, although the CGR of austenitic stainless steel under high irradiation dose is still higher than that of non-irradiated stainless steel, compared with medium irradiation dose, the CGR of austenitic stainless steel is abnormally reduced, mainly between $1\text{x}10^{-11}$ m/s and $1\text{x}10^{-10}$ m/s. These results indicate that high irradiation dose is not conducive to the acceleration of the growth rate of austenitic stainless steel IASCC in PWR, but has a certain inhibitory effect.

The relationship between CGR and K of austenitic stainless steel in pressurized water reactor remains relevant, and CGR increases with K in most cases. However, with the increase of irradiation dose, CGR and K of austenitic stainless steel do not increase simultaneously, but atrophy occurs. It may be caused by the excessive damage to the material caused by neutron irradiation.

It is worth noting that although the CGR and K relationship data of irradiated austenitic stainless steel were partially dispersed, CGR and K still showed a positive correlation. It should be pointed out that the 316 based austenitic stainless steel material under the neutron irradiation dose of 25 dpa in Fig. 2 has abnormal conditions. At lower K, that is, when K is less than or equal to 5 MPa $m^{0.5}$, the CGR of the material is more than $9\text{x}10^{-11}$ m/s. The appearance of abnormal data points in Fig. 2 caught our attention and thought.

3 Effect of Irradiation Dose on SCC Growth Rates

The CGR of austenitic stainless steel in high temperature water is affected not only by stress intensity factor K, but also by neutron irradiation dose, which is indispensable in the process of stress corrosion cracking promoted by irradiation. In order to study the effect of neutron irradiation dose on CGR of austenitic stainless steel in high temperature water, temperature and K must be controlled within a certain range to make the data reasonable and reliable.

Firstly, the temperature was selected between 335 and 340 °C, which could remove the interference of temperature factors on crack growth rate. Through Fig. 2, we summarized the CGR data of 0.06, 3.0, 8.0, 9.0, 11.4, 12.9, 14.0, 15.0, 17.0, 18.7, 21.0, 32.9, 37.8, 39.0, 47.0 dpa irradiation dose, as shown in Fig. 3 [4–9, 11].

Here we classify K as 5–10 MPa $m^{0.5}$, 10–15 MPa $m^{0.5}$, 15–20 MPa $m^{0.5}$ and 20–27 MPa $m^{0.5}$. The CGR and K data of austenitic stainless steel at 335–340 °C were processed to obtain the CGR and neutron irradiation dose of austenitic stainless steel under different normalized K conditions.

When K is normalized to 5–10 MPa $m^{0.5}$, as shown in Fig. 4 [5–9, 11], the CGR of the material is the highest at 11.4 dpa, close to $1\text{x}10^{-8}$ m/s. When K is 5–10 MPa $m^{0.5}$, CGR of all irradiation dose data is greater than $9\text{x}10^{-12}$ m/s. At 17.1 dpa and 18.7 dpa, the CGR values of the materials are close to each other, basically within the range of 10^{-11} m/s and 10^{-9} m/s. At 21.0 dpa, the CGR of the material increased slightly to about 10^{-9} m/s. CGR at 47 dpa was basically within the data range of 17.1 dpa and 18.7 dpa irradiation dose.

As shown in Fig. 5 [4–9, 11], when K is in the range of 10–15 MPa $m^{0.5}$, CGR of all data in the figure is greater than $4\text{x}10^{-11}$ m/s. CGR data under 3.0 dpa were higher than

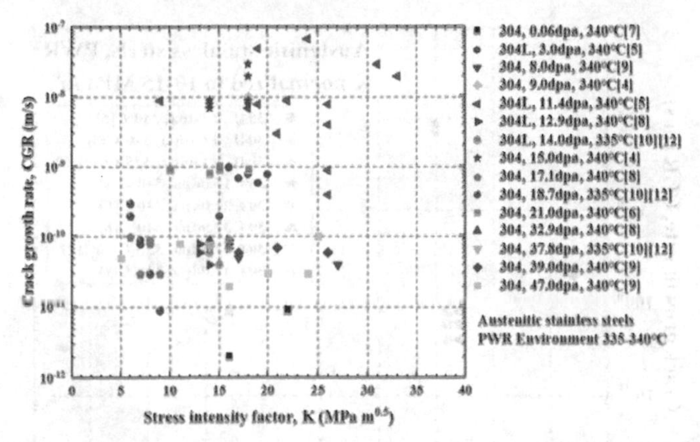

Fig. 3. CGR and K of materials with different neutron irradiation doses at 335–340 °C [4–9, 11]

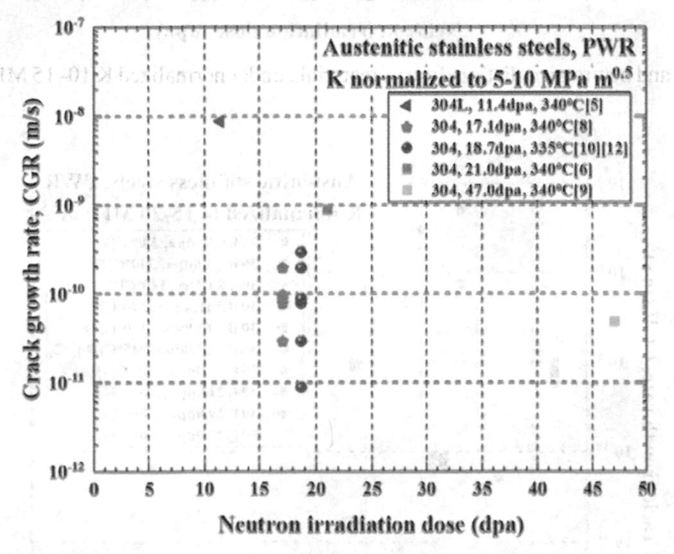

Fig. 4. CGR and neutron irradiation dose of austenitic stainless steel under normalized K 5–10 MPa $m^{0.5}$ [5–9, 11]

10^{-10} m/s. The CGR data range of 12.9 dpa and 14.0 dpa irradiation doses were similar. The CGR at 15.0 dpa was two orders of magnitude higher than that at 12.9 dpa and 14.0 dpa. The CGR data at 21.0 dpa irradiation dose were about 1 order of magnitude higher than those at 12.9 dpaand 14.0 dpa irradiation dose. The CGR data of 21.0 dpa irradiation dose was about 0.1 times that of 15.0 dpa irradiation dose. When the neutron irradiation dose increased from 32.9 dpa to 37.8 dpa and from 37.8 dpa to 47.0 dpa, the CGR data at these three irradiation doses were close to 10^{-10} m/s (between 5×10^{-11} m/s and 1×10^{-10} m/s). The CGR data at these three irradiation doses are not sensitive to the neutron irradiation dose.

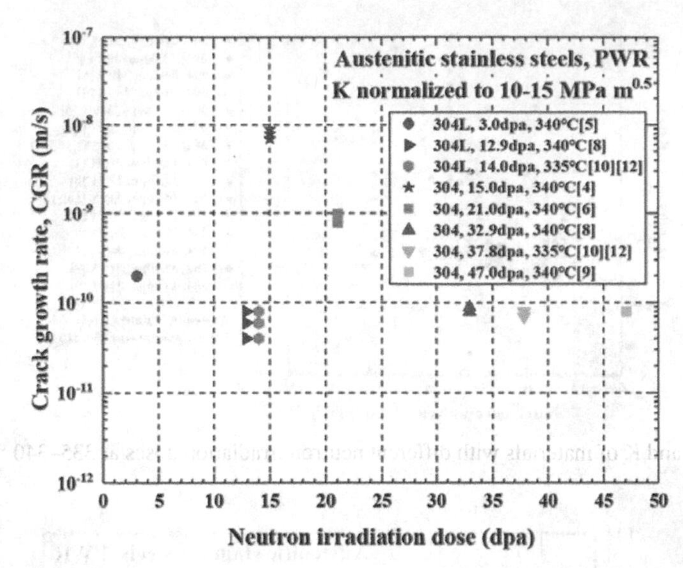

Fig. 5. CGR and neutron irradiation dose of materials under normalized K 10–15 MPa $m^{0.5}$ [4–9, 11]

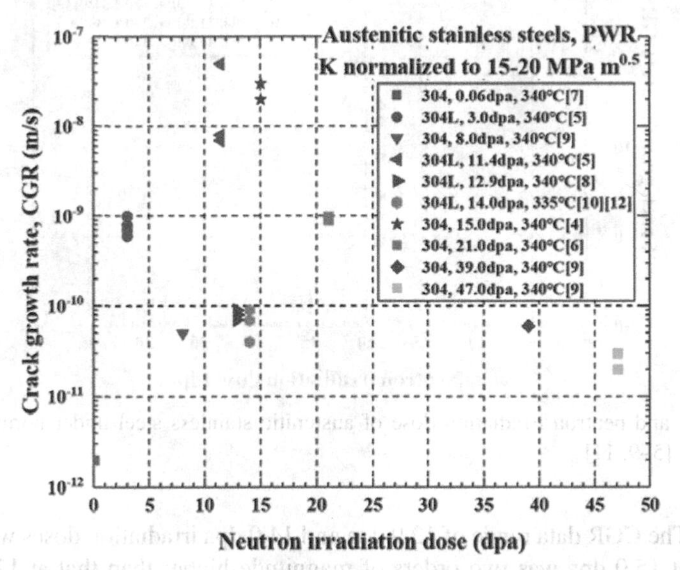

Fig. 6. CGR and neutron irradiation dose of materials under normalized K 15–20 MPa $m^{0.5}$ [4–9, 11]

CGR (3.0 dpa) < CGR(21.0 dpa) < CGR(15.0 dpa) > CGR(12.9 dpa, 14.0 dpa, 32.9 dpa, 37.8 dpa and 47.0 dpa). The peak value of CGR data with neutron irradiation dose of 15 dpa in Fig. 5 has aroused our concern. As for the reason of the peak value? What causes the spike remains to be studied and solved.

As shown in Fig. 6 [4–9, 11], when K is 15–20 MPa $m^{0.5}$, CGR at 0.06 dpa, a very low irradiation dose, is very low, about 2×10^{-12} m/s. However, when the irradiation dose reached 3.0 dpa, the CGR data directly reached nearly 10^{-9}, with a significant increase. When the irradiation dose was higher than 8.0 dpa, the CGR of the material increased significantly, and the CGR was greater than 2×10^{-11} m/s. The CGR data range of 11.4 dpa was close to that of 15.0 dpa, up to about 10^{-8} m/s. CGR data at 12.9 dpa and 14.0 dpa are close in range, but are still two orders of magnitude lower than CGR data at 11.4 dpa and 15.0 dpa. The CGR data at 21.0 dpa was between 12.9 dpa and 14.0 dpa and 11.4 dpa and 15.0 dpa, with a value of about 10^{-9} m/s.

On the whole, it shows that when the neutron irradiation dose is greater than or equal to 15 dpa, the crack growth rate decreases with the increase of the dose. At the same time, the irradiation dose of 11.4 dpa and 15.0 dpa showed two similar peak values. Similar to the situation where K is 10^{-15} m/s as shown in Fig. 5, the phenomenon of peak value arouses concern. The two can be compared and further studied.

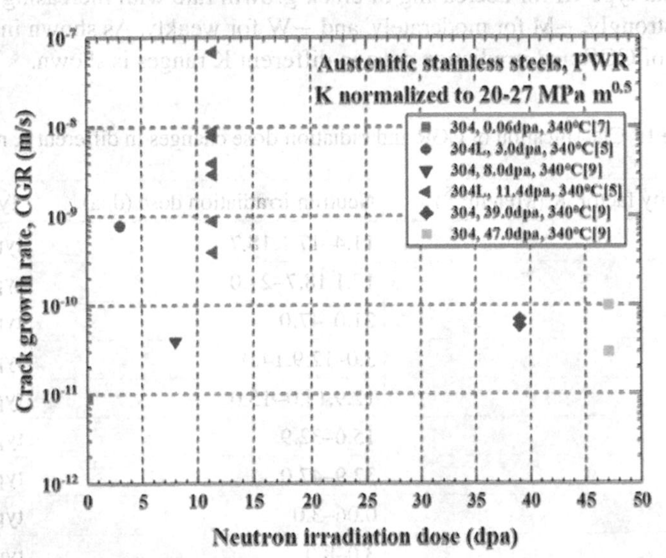

Fig. 7. CGR and neutron irradiation dose of materials under normalized K 20–27 MPa $m^{0.5}$ [5, 6, 8]

As shown in Fig. 7, when K is in the range of 20–27 MPa $m^{0.5}$, the range of CGR data fluctuates greatly. The main reason is that the CGR data of 0.06 dpa is small and that of 11.4 dpa is large. When the irradiation dose increased from 0.06 dpa to 3.0 dpa, the CGR increased by about 2 steps. From 3.0 dpa to 8.0 dpa, its CGR decreased by more than one order of magnitude, but it was still about four times the CGR data at 0.06 dpa. CGR data at 11.4 dpa fluctuated from 4×10^{-10} m/s to nearly 10^{-7} m/s, with a fluctuation range of more than 2 orders of magnitude. Compared with the CGR data at 8.0 dpa, the CGR data at 39.0 dpa was 1.5 or 1.75 times higher, and the CGR data at 47.0 dpa was 2.5 or 0.75 times higher.

Among them, CGR data at 11.4 dpa fluctuated from 4×10^{-10} m/s to nearly 10^{-7} m/s, with a fluctuation range of more than 2 orders of magnitude. In the same range of K, why is there such a high CGR data, and why is there such a large fluctuation range of CGR data in the same radiation dose? These questions remain to be explored.

Based on the above data, we propose a method to judge the phase of CGR data varying with neutron irradiation dose.

$$\frac{\delta CGR}{\delta Dose} > 0\text{: type I,}$$

$$\frac{\delta CGR}{\delta Dose} = 0\text{: type II}$$

$$\frac{\delta CGR}{\delta Dose} < 0\text{: type III}$$

Each type is divided into three categories: type I for increasing, type II for nearly not changing, and type III for decreasing of crack growth rate with increasing dose. There are –S for strongly, –M for moderately, and –W for weakly. As shown in Table 1, the change rate of CGR and irradiation dose in different K ranges is shown.

Table 1. Classification of CGR and radiation dose changes in different K ranges

Stress intensity factor, K (MPa m$^{0.5}$)	Neutron irradiation dose (dpa)	Type
5–10	11.4–17.1,18.7	type III-S
5–10	17.1,18.7–21.0	type I-S
5–10	21.0–47.0	type III-S
10–15	3.0–12.9,14.0	type III-M
10–15	12.9,14.0–15.0	type I-S
10–15	15.0–32.9	type III-S
10–15	32.9–47.0	type II
15–20	0.06–3.0	type I-S
15–20	3.0–8.0	type III-S
15–20	8.0–11.4	type I-S
15–20	11.4–12.9,14.0	type III-S
15–20	12.9,14.0–15.0	type I-S
15–20	15.0–47.0	type III-S
20–27	0.06–3.0	type I-S
20–27	3.0–8.0	type III-S
20–27	8.0–11.4	type I-S
20–27	11.4–39.0	type III-S
20–27	39.0–47.0	type II

4 Verification of Effect of Irradiation Doses on SCC Growth Rate

Bosch [13] et al. recently studied the correlation between neutron irradiation and mechanical properties of 316 material cold worked processed in pressurized water reactor environment, which provided reference value for our paper. He studied the relationship between neutron irradiation and stress-strain, tensile strength, yield strength and other mechanical properties, and we only take the yield strength related content.

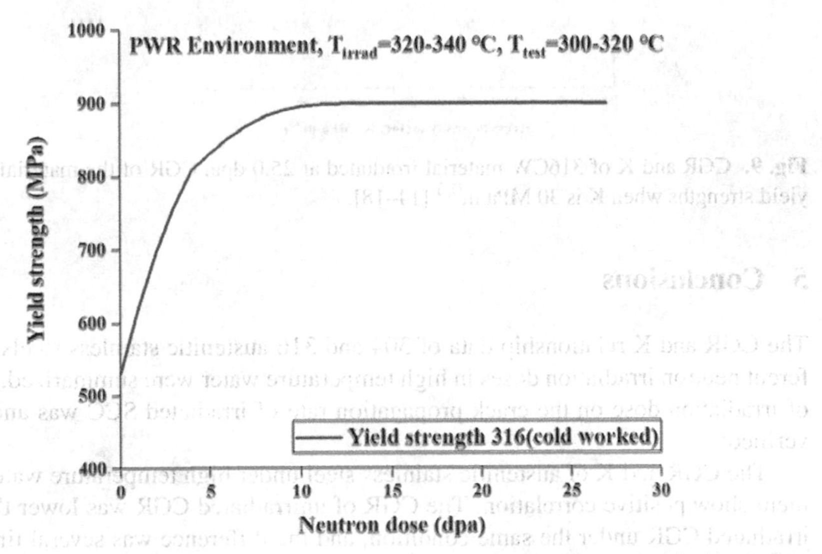

Fig. 8. Variation of yield strength of cold worked 316 material with neutron irradiation dose in pressurized water reactor [13]

Figure 8 shows the change of yield strength of 316CW material in PWR environment with neutron irradiation dose under the irradiation temperature of 320–340 °C and the test temperature of yield strength of 300–320 °C.

In Fig. 9 we can see the CGR distribution of 316CW material at 25 dpa. Through the study of Terachi [14] et al., Castano [15] et al., Shoji [16] et al., Toloczko [17] et al. and Donghai Du [18] et al., we can find the CGR situation of 316CW materials without irradiation under different yield strengths of K is 30 MPa $m^{0.5}$. It can be clearly seen that the CGR of about 800 MPa yield strength is between 10^{-10} m/s and 10^{-9} m/s, which is nearly 2 orders of magnitude lower than the CGR after irradiation at the same K. CGR with yield strengths of 500 MPa and more than 200 MPa is between 10^{-11} m/s and 10^{-10} m/s, which is nearly 3 orders of magnitude lower than CGR after irradiation.

We can obviously score that the CGR of the cold worked 316 material after neutron precipitation is much higher than that of the cold worked 316 material without neutron irradiation. This shows that neutron irradiation has a great influence on the change of material CGR. In addition, the microstructure defects and dislocations of materials may be increased due to the influence of neutron irradiation on the microstructure and mechanical properties of materials.

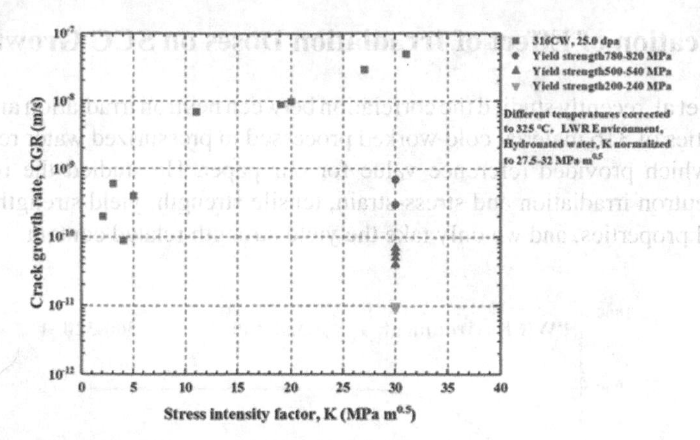

Fig. 9. CGR and K of 316CW material irradiated at 25.0 dpa, CGR of the material at different yield strengths when K is 30 MPa $m^{0.5}$ [14–18].

5 Conclusions

The CGR and K relationship data of 304 and 316 austenitic stainless steels under different neutron irradiation doses in high temperature water were summarized. The effect of irradiation dose on the crack propagation rate of irradiated SCC was analyzed and verified:

The CGR and K of austenitic stainless steel under high temperature water environment show positive correlation. The CGR of unirradiated CGR was lower than that of irradiated CGR under the same condition, and the difference was several times. There is a certain degree of dispersion in the relation data between CGR and K.

In the range of different K, the relationship between CGR data and irradiation dose of austenitic stainless steel is different. Peak data points of CGR appear at some irradiation doses, and the specific reasons need to be studied. We propose a classification method for CGR and radiation dose variation trend and apply it to the data presented in this paper.

Finally, the significant effect of neutron irradiation on CGR of austenitic stainless steel was verified based on yield strength.

Acknowledgments. This work has been supported by National Science and Technology Major Projects 2019YFB1900905, Natural Science Foundation of China (NSFC No. 51771107), and Independent Research and Development Project of State Key Laboratory of Advanced Special Steel, Shanghai University (SKLASS 2020- Z00).

References

1. Chen, W.: Modeling and prediction of stress corrosion cracking of pipeline steels. Trends Oil Gas Corros. Res. Technol. 707–748 (2017)
2. Du, D., Song, M., Chen, K., et al.: Effect of deformation level and orientation on SCC of 316L stainless steel in simulated light water environments. J. Nucl. Mater. **531**, 152038 (2020)
3. Dd, A., Kai, S.B., Gswa, B.: IASCC of neutron irradiated 316 stainless steel to 125 dpa-ScienceDirect. Mater. Charact. **173**, 110897 (2021)
4. Chen, K., Ickes, M.R., Burke, M.A., et al.: The effect of potassium hydroxide primary water chemistry on the IASCC behavior of 304 stainless steel (2021)
5. Chopra, O.K., Rao, A.S.: A review of irradiation effects on LWR core internalmaterials-ASCC susceptibility and crack growth rates of austenitic stainless steels. J. Nucl. Mater. **409**, 235–256 (2011)
6. Materials reliability program: corrosion testing of decommissioned PWR vessel internals material samples (MRP-222), EPRI, Palo Alto, CA, 1015478 (2007)
7. Materials Reliability Program: A Review of the Cooperative Irradiation Assisted Stress Corrosion Cracking Research Program (MRP-98)-EPRI Report 1002807 (2003)
8. Karlsen, T.M., Espeland, M., Horvath, A.: Summary Report on the PWR Crack Growth Rate Investigation, IFA-657. OECD Halden Reactor Project, Report HWR-773, May 2005
9. Chen, Y., Alexandreanu, B., Natesan, K., Rao, A.S.: Stress corrosion cracking and fracture toughness tests of irradiated Type 304 stainless steel. In: 19th International Conference on Environmental Degradation of Materials in Nuclear Power Systems-Water Reactors, 18–22 August 2019
10. Karlsen, T.M., Espeland, M., Horvath, A.: HWR-773, OECD Halden Reactor Project, May 2005
11. Jenssen, A., Stjarnsater, J., Pathania, R.: Proceedings of 14th International Conference on Environmental Degradation of Materials in Nuclear Power Systems – Water Reactors, American Nuclear Society, Lagrange Park, IL, 2009
12. Nakano, J., Karlsen, T.M., Espeland, M.: HWR-843, OECD Halden Reactor Project, August 2008
13. Davis, R.B., Andresen, P.: Cooperative IASCC Research (CIR) Final Report, EPRI 1007378, October 2002
14. Bosch, R.W., Renterghem, W.V., Dyck, S.V., et al.: Microstructure, mechanical properties and IASCC susceptibility of stainless steel baffle bolts after 30 years of operation in a PWR. J. Nucl. Mater. **543**, 152615 (2021)
15. Terachi, T., Yamada, T., Miyamoto, T., et al.: SCC growth behaviors of austenitic stainless steels in simulated PWR primary water. J. Nucl. Mater. **426**(1–3), 59–70 (2012)
16. Marin, M.C.: Crack Growth Rate of Hardened Austenitic Stainless Steels in BWR and PWR Environments (2003)
17. Shoji, T., Li, G.F., Kwon, J.H., et al.: Quantification of yield strength effects on IGSCC of austenitic stainless steels in high temperature waters. In: Proceedings of 11th International Conference on Environmental Degradation of Materials in Nuclear Power Systems-Water Reactors (2003)
18. Toloczko, M.B., Andresen, P.L., Bruemmer, S.M.: SCC crack growth of cold-worked type 316 SS in simulated bwr oxidizing and hydrogen water chemistry conditions. Proceedings of 13th International Conference on Environmental Degradation of Materials in Nuclear Power Systems-Water Reactors (2007)

Research on the Optimization of Commercial PWR'S Financial Evaluation Model

Meifang Bo[1], Baojun Zheng[2], Shengli Zhang[1], Zhengxu Ren[3], and Bojie Liu[1](✉)

[1] China Nuclear Power Engineering Co., Ltd., Beijing, China
liubja@cnpe.cc

[2] CHINERGY Co., Ltd., Beijing, China

[3] CNNP Rich Energy Co., Ltd., Beijing, China

Abstract. Considering industry standards, policies, market environment, actual costs of nuclear power plants (NPPs) and technical characteristics of nuclear power, and combining with industrial circumstances when these industry standards were published, we analyzed the limitations of current commercial PWR's fianancial evaluation methods. Taking HPR1000 as a case study, we deeply estimated the generation cost and cash flow. Moreover, optimization direction of the financial evaluation model after comparative analysis was proposed. This research aims to provide reference to similar NPPs'financial evaluation and basis of investment decision-making for investors, owners and general contractors.

Keywords: Commercial PWR · Financial Evaluation · Optimization of Model

1 Introduction

In China, the nuclear power plant (NPP) is regarded as an independent legal entity in the economic evaluation of commercial NPPs, and the on-grid electricity price, an important indicator of the economic feasibility of the NPPs, is calculated on the basis of investment in total capital cost, power generation cost and benchmark internal return rate. The economic evaluation is based on the *Economic Evaluation Guidelines of the Nuclear Power Plant Construction Project* (NB/T20048–2011) (hereinafter referred to as the "NB"), which was released in 2011. Nevertheless, significant changes have taken place in domestic power market policy, actual operating cost of NPPs, the business environment of the external nuclear power market, the safety standards and the technical approach of NPPs since its release 10 years ago. Therefore, the gaps open up the room for research on how can current economic evaluation further contribute to the scientific decision-making in investment, rational resource allocation, and sound development in the whole chain of nuclear power industry.

In addition, the industry standard *NB* focuses on guiding studies on the economic feasibility of NPPs. However, with regard to the commercial NPPs, the shareholders,

Financial support was provided by the R&D project of China Nuclear Power Engineering Co., Ltd (KY22246)

C. Liu (Ed.): PBNC 2022, SPPHY 283, pp. 1073–1084, 2023.
https://doi.org/10.1007/978-981-99-1023-6_90

shareholders of listed companies with the long-term development plan at heart in particular, are more concerned about the internal rate of return of all parties involved, while the general contractors pay more attention to the investment of total capital cost per kilowatt and the construction units lay emphasis on whether the project can be approved. The analysis on different schemes can be conducted through scientific economic evaluation, serving as solid data support for all parties, and the results can provide insight in understanding the gap between the theoretical and actual benefits of the project, the allocation options for subsequent development, and the range of maximum total capital cost per kilowatt within the project budget.

2 Limitations of the Current Economic Evaluation Model for Pressurized Water NPPs

On the one hand, *NB* was officially released in 2011, and the calculation parameters, criterion parameter and boundary conditions in its appendix were in consistent with the market environment and industry expectations before finalization. However, the capital cost and expected revenue based on the market environment at that time is vastly different from the current one. On the other hand, *NB* was compiled mainly based on the data of Generation 2+ NPPs that were widely built and operated in China, without taking into account the advanced design, the differences in generation costs between the Generation 2+ and Generation 3 NPPs and the actual cost of NPPs in recent years. Furthermore, the existing *NB* model lacks the in-depth research and analysis on the cash flow planning including financing, loan repayment, and corporate dividend of the whole project, and the evaluation system takes the on-grid electricity price measured by the capital benchmark rate of return as the main indicator, excluding the rate of return of the investment of all parties, which is more closely related to the investors.

2.1 The Background of *NB* Release

China's electricity pricing mechanism has undergone a series of changes since the reform and opening up, evidenced by a number of relevant policies implemented, such as the repayment of capital with interest (RCI pricing), "price pegged to the increase in transportation" and "operating price" since 1985. Before 2013, nuclear power on-grid electricity price varies among different generating sets based on compensation costs and reasonable revenue, and against which background the *NB* was compiled and released.

2.2 The Applicable Analysis of *NB* to the PWR Reactor Types

The *NB* regulates that it is suitable for the economic evaluation of PWRs and the economic evaluation of other nuclear facilities could refer to this standard. However, this standard does not clearly indicate the reactor types which are applicable to this standard and there is also no content stating about the relationship between the evolvement of PWR nuclear technology and its applicability to this standard.

The formulation of *NB* is mainly based on the economic parameters and data of the Generation 2+. The effects of technical innovation on economic evaluation methods and parameters are not considered in current *NB*. Therefore, the methods and parameters in this *NB* could not objectively and comprehensively meet the demands of economic evaluation for current PWRs, especially for the third generation PWRs.

2.3 Analysis of Macro-economic Environment Change for Nuclear Power Enterprises

Compared with the year 2011 when the *NB* was published, the market environment regarding the fund cost and expected returns has changed significantly. From 2010 to 2019, the interest rate under macro-economy had been declined obviously. Under the current economic environment, the internal rate of returns for the capital fund could decrease to 7%–8% while the rate of returns for investors could decrease to 6%–7%.

2.4 Analysis of Business Environment Change for Nuclear Power Enterprises

Compared with the year 2011, the business environment of nuclear enterprises has changed obviously. Due to the higher safety requirement, the project cost of the third generation PWRs has increased. Meanwhile, against the backdrop of benchmark tariff reduction, power market revolution, weakened fiscal and tax support, the nuclear power enterprises are under huge business difficulties. For instance, the power generation capability of some nuclear reactors is compared with their full capacity, posing risks to the business operation of nuclear enterprises. Moreover, with the successive increase of transected electricity, NPPs in certain regions participate in the annual bidding transaction, monthly bidding transaction, straight-powered protocol for large users and trans-provincial and trans-regional power transaction, facing dual pressure from both declined planned electricity and market bidding.

Economic evaluation based on the *NB* would deviate from actual circumstances. The actual operation time of Generation 2+ far exceed 7000 h and operation costs are much higher than the regulations in *NB*, resulting in higher electricity price. Moreover, actual fiscal and tax policies also have minor differences with *NB*. As for the third generation, its economics are facing risks in terms with both power generation and electricity price. Thus, to accurately evaluate the economics of the Generation 3 NPPs, it is imperative to focus on the rationality and accuracy of economic evaluation methods. Under the background of declined interest rate, the return rate could be further decreased. *NB* regulates that the pre-tax benchmark return rate before financing is 7% and the after-tax benchmark return rate after financing is 9%. The above-mentioned return rates could be lowered. Besides, more concentration could be paid to financial internal return rate of investors, which is the core concern of investors. The financial internal return rate of investors is advocated as a core parameter in this study and it could be further optimized to 8%–10%.

2.5 Limitations of Current Financial Evaluation for PWRs

(1) Economic indicators

Based on the financial evaluation of PWRs with different Under the hypothesis of identical internal return rate, the on-grid electricity price of PWRs with different power show large disparity. Moreover, there is no significantly linear relationship between power generation cost and on-grid electricity. Focusing on the on-grid electricity price and capital internal return rate in economic evaluation would result in the overlook of other economic indicators. Therefore, apart from the internal return rate and on-grid electricity price, we should also concentrate on the power generation cost and payment schedule, achieving more precise and comprehensive economic analysis.

(2) Cost control in the whole life period

According to the feedback from NPPs, the actual operation cost has increased, exceeding the parameters regulated in *NB*. Specifically, cost from uranium mining, conversion and separation is higher than the international market, resulting in high cost of nuclear fuels. The increase of nuclear fuel cost and operation and maintenance cost counteract the advantages of investment decline brought by the batch production of Generation 2+ and design maturation of HPR1000. Therefore, apart from the investment control, we should focus on the cost from the period of both construction and operation. Operation and maintenance cost as well as nuclear fuel cost should be supplemented as auxiliary indicators in the future economic evaluation of PWRs.

(3) Indicator discordance with practice

In financial evaluation, the table of capital cash flow is formulated based on designated either on-grid electricity price or capital internal return rate. Common practice is to designate capital internal return rate as 9% and evaluate the on-grid electricity price, thus evaluating the feasibility of the on-grid electricity price of specific NPPs. However, investors pay more attention to dividend distribution and there exists large uncertainty with capital internal return rate as the only economic indicator.

3 Optimization of Economic Evaluation Model

In the future, the financial evaluation of NPPs should be transformed from the single indicator to multi indicators. Currently, focusing only on the on-grid electricity price and capital internal return rate would lead to incomplete analysis of economics of NPPs and simultaneously controlling multiple indicators is conducive to achieving comprehensive economic evaluation of NPPs. For instance, cost control should be expanded from construction to the whole life period of NPPs, including operation and maintenance. Table shows advocated multi indicators in future economic evaluation of NPPs (Table 1).

Table 1. Proposed multiple indicators for optimized model

Indicator type	Indicator name	Indicator standard
Key project cost indicator	Overnight cost per kilowatt	Absolute value
Key project cost indicator	Total capital cost per kilowatt	Absolute value
Auxiliary project cost indicator	Overnight cost	
Auxiliary project cost indicator	Total capital cost	
Key cost indicator	Average power generation cost	Absolute value
Key cost indicator	Operation and maintenance cost	Absolute value
Key cost indicator	Fuel cost	Absolute value
Key return indicator	Financial internal return rate of investors	6%–7%
Auxiliary return indicator	Capital financial internal return rate	
Auxiliary return indicator	Financial internal return rate of total investment (after tax)	
Other auxiliary indicators	Profitability and other indicators	

4 Preliminary Research on the Cost of Power Generation of HPR1000

Cost of power generation is the basis of on-grid electricity price evaluation. Because the FOAK reactor of HPR1000 has only operated for 2 years, the data for the cost of operation and maintenance is not thorough. Thus, this study only conducts preliminary research on the cost of power generation of HPR1000. Differences of operation and maintenance cost between HPR1000 and the Generation 2+ result from their technical differences.

Cost of materials consumed during operation includes expense from nuclear fuels, chemicals, water and electricity. Water consumption difference between HPR1000 and the Generation 2+ is clear but the quantity of other materials consumed in the operation of HPR1000 is still unclear. Standard of material expenses for HPR1000 should be identified based on the practice of the Generation 2+ and the in-service HPR1000. The cost for each staff of HPR1000 and the Generation 2+ should be the same, but the staff quota of HPR1000 should be modified based on actual situation. Maintenance of NPPs includes general overhaul and routine operation and maintenance. General overhaul planning is related with the type of nuclear reactors. Correspondingly, the expense of general overhaul is also different for each type of reactor. With the progress of science and technology as well as the management of power plants, the repair fee rate should be declined. Boasting with higher design standard, the repair fee rater of HPR1000 could be decreased from 1.35% to 1.2%.

Life period is one of major differences between the Generation 2+ and the HPR1000. The life period of the HPR1000 is 60 years, which mainly influences the design of major equipments. Since the capital cost is low in today's market, a shorter payback period is conducive to sustainable bonus distribution. Therefore, adopting 20 years as depreciation period is still feasible in the economic evaluation of HPR1000.

Based on the comparative analysis and advanced design concept, we estimated the power generation cost of HPR1000 after optimization of depreciation period and repair fee rate. We estimated the power generation cost and on-grid electricity price based on the total capital cost of 37.75 billion RMB and the capital internal return rate of 9% (Table 2).

Table 2. Cost composition of HPR1000

	Cost item	Cost value (RMB/MWh)	Percentage(%)
1	**Investment cost**	**106.6**	**45%**
1.1	Depreciation	75.4	71%
1.2	Amortization	4.0	4%
1.3	Financial expense	27.2	25%
2	**Fuel cost**	**68.7**	**30%**
2.1	Nuclear fuel procurement	46.3	67%
2.2	Reprocessing cost	22.4	33%
3	**Operation and maintenance cost**	**52.6**	**22%**
3.1	General overhaul cost	26.5	50%
3.2	Wages and welfare	10.7	20%
3.3	Other expenses	6.6	13%
3.4	Material cost	5.4	10%
3.5	Water cost	0.4	1%
3.6	Medium and low radioactive waste treatment and disposal cost	0.5	1%
3.7	Other operation and maintenance cost	2.6	5%
4	**Decommissioning Funds**	**7.5**	**3%**
	合计	**235.5**	**100%**

5 Financial Evaluation of HPR1000

Based on the above-mentioned optimization clue, we formulated 8 schemes to analyze the economics of HPR1000 under each condition.

5.1 Scheme Estimation

(1) Original Scheme (Option 1)

The power generation cost, on-grid electricity price, and the rate of return for all parties were calculated at the total capital cost of 37.75 billion RMB, 1.35% rate of general overhaul cost, depreciation period of 25 years and capital return rate of 9%. The long-term loan is reimbursed by deducted VAT, refunded VAT, depreciation, amortization, and undistributed profits for repayment in order, and cash for circulation will be used to repay short-term loans annually. Dividend will be distributed under the condition of guaranteed loan repayment in the form of cash. The dividend distribution ratio is as follows (Table 3).

Table 3. Average dividend distribution ratio of option 1

No	Year after operation	Average ratio of cash dividends to distributable net profit
1	1–5	0%
2	6–10	15%
3	11–15	65%
4	16–20	98%
5	After 20	100%

As seen from the table, in the first five years after the unit is put into commercial operation, the ratio of cash dividends to distributable net profit is 0, suggesting that there will be no dividend in the first 5 years, and the trend continue till the 6th year; from the 6th to 10th years after the operation, the average distribution ratio of annual dividend is 15%, which is still far below the value proposed by the listed company (30%); from the 11th to 15th years, the average distribution ratio of annual dividend is 65%, and the full allocation can't be achieved until 17th year. The dividend distribution is so conservative that the listed company's investment in new projects and its rolling development get impeded.

(2) Base Case (Option 2)

The total capital cost, power generation cost, and investment return of all parties were retrodicted with 1.35% general overhaul rate, depreciation period of 25 years, capital return rate of 9%, and the on-grid electricity price, which was calculated according to that of the benchmark local thermal power desulfurization and denitrification. The arrangement of cash flow and dividend is similar to the Option 1 in nature with comparable annual dividend distribution ratio.

(3) General overhaul Scheme (Option 3)

The total capital cost, power generation cost, investment return of all parties were retrodicted with 1.2% general overhaul rate, adjusted from 1.35% according to the estimated quota of annual overhaul expense and fixed assets, depreciation period of 25 years, capital return rate of 9%, and the on-grid electricity price calculated according to that of benchmark local thermal power desulfurization and denitrification. The arrangement of cash flow and dividend is similar to the previous case in nature with comparable annual dividend distribution ratio.

(4) Depreciation Scheme (Option 4)

The total capital cost, power generation cost, investment return of all parties were retrodicted with 1.2% general overhaul rate, depreciation period of 20 years, capital return rate of 9%, and the on-grid electricity price calculated according to that of benchmark local thermal power desulfurization and denitrification. The arrangement of cash flow and dividend is similar to the previous case in nature with comparable annual dividend distribution ratio.

(5) Rate of return Scheme I (Option 5)

The option 5 is to adjust the order of cash flow, the dividend distribution and repayment scheme. The total capital cost and power generation cost were retrodicted with 8% of capital return rate and 9% of return rate of all parties, considering low loan interest rate for the rolling development of the company. Combining the study of the dividend distribution of the listed companies with the research on the dividend of NPPs, the arrangement of cash flow is as follows.

Cash dividends for shareholders are guaranteed at a certain amount, and the rest is repaid to long-term loans in the order of deducted VAT, refunded VAT, depreciation, amortization, and undistributed profits remained after dividend. The short-term loan will be used to compensate for the long-term loan that could not be covered by dividend and repayment. The corresponding dividend distribution plan is that there will be no dividend in the first two years after commercial operation, 860 million RMB per year from the 3rd to 7th year, 1.2 billion RMB per year from the 8th to12th year, and full percent of the profit from the 13th year onwards through the subsequent operation.

(6) Rate of return Scheme II (Option 6)

The dividend distribution and loan repayment scheme is adjusted in option six, while other parameters remain the same as Option 4. The total capital cost and power generation cost were retrodicted with 8% of capital return rate and 10% of return rate of all parties. The slight changes in the cash flow and the corresponding dividend distribution can be found in comparison with Option 5. There will be no dividend in the first two years after commercial operation, 1.08 billion RMB per year from the 3rd to 7th year, 1.66 billion RMB per year from the 8th to 12th year, and full percent of the profit from the 13th year onwards through the subsequent operation.

(7) Rate of Return Scheme III (Option 7)

The dividend distribution and loan repayment scheme is adjusted in option seven, while other parameters remain the same as Option 4. The total capital cost and power generation cost were retrodicted with 7% of capital return rate and 9% of return rate for shareholders. The slight changes in the cash flow and the corresponding dividend distribution can be found in comparison with Option 5. There will be no dividend in the first two years after commercial operation, 1.03 billion RMB per year from the 3rd to 7th year, 1.4 billion RMB per year from the 8th to 12th year, and full percent of the profit from the 13th year onwards through the subsequent operation.

(8) Rate of Return Scheme IV (Option 8)

The dividend distribution and loan repayment scheme is adjusted in option eight, while other parameters remain the same as Option 4. The total capital cost and power generation cost were retrodicted with 7% of capital return rate and 10% of return rate for shareholders. The slight changes in the cash flow and the corresponding dividend distribution can be found in comparison with Option 5. There will be no dividend in the first two years after commercial operation, 1.31 billion RMB per year from the 3rd to 7th year, 1.6 billion RMB per year from the 8th to 12th year, and full percent of the profit from the 13th year onwards through the subsequent operation.

5.2 Results and Comparative Analysis

(1) Estimation results

Table 4 shows the results of power generation cost and its composition for each scheme while Table 5 demonstrates main indicators.

Table 4. Estimation of power generation cost

Item	Option 1	Option 2	Option 3	Option 4	Option 5	Option 6	Option 7	Option 8
Investment cost	**106.6**	**104.2**	**105.3**	**106.3**	**115.9**	**118.4**	**124.1**	**127.5**
Depreciation	75.4	73.7	74.4	75.2	76.8	75.9	79.5	78.3
Amortization	4.0	3.9	3.9	4.0	4.0	4.0	4.2	4.1
Financial expense	27.2	26.6	26.9	27.1	35.1	38.6	40.4	45.0
Fuel cost	**68.7**	**68.7**	**68.7**	**68.7**	**68.7**	**68.7**	**68.7**	**68.7**
Nuclear fuel procurement	46.3	46.3	46.3	46.3	46.3	46.3	46.3	46.3
Reprocessing cost	22.4	22.4	22.4	22.4	22.4	22.4	22.4	22.4
Operation and maintenance cost	**52.6**	**52.0**	**49.4**	**49.1**	**49.7**	**49.4**	**50.6**	**50.2**

(continued)

Table 4. (*continued*)

Item	Option 1	Option 2	Option 3	Option 4	Option 5	Option 6	Option 7	Option 8
General overhaul cost	26.5	25.8	23.2	23.4	23.9	23.7	24.8	24.4
Wages and welfare	10.7	10.7	10.7	10.7	10.7	10.7	10.7	10.7
Other expenses	6.6	6.6	6.6	6.6	6.6	6.6	6.6	6.6
Material cost	5.4	5.4	5.4	5.4	5.4	5.4	5.4	5.4
Water cost	0.4	0.4	0.4	0.4	0.4	0.4	0.4	0.4
Medium and low radioactive waste treatment and disposal cost	0.5	0.5	0.5	0.5	0.5	0.5	0.5	0.5
Other operation and maintenance cost	2.6	2.5	2.6	2.1	2.1	2.1	2.2	2.2
Decommissioning funds	**7.5**	**7.4**	**7.4**	**7.5**	**7.7**	**7.6**	**8.0**	**7.8**

Table 5. Estimation of main indicators

Item	Unit	Option 1	Option 2	Option 3	Option 4	Option 5	Option 6	Option 7	Option 8
On-grid electricity price	RMB/MWh	400.2	393.2	393.2	393.2	393.2	393.2	393.2	393.2
Average power generation cost	RMB/MWh	235.5	232.2	230.8	231.6	242.0	244.1	251.4	254.3
Total capital cost per kilowatt	RMB/kilowatt	15573	15207	15369	15518	15856	15665	16415	16177
Return rate of investors	%	6.9	6.9	6.9	6.9	9.0	10.0	9.0	10.0
Capital return rate	%	9.0	9.0	9.0	9.0	8.0	8.0	7.0	7.0

(2) Comparative Analysis

Based on the results of the above options, it can be concluded that the rate of return of shareholders from all parties and the total capital cost per kilowatt increase by modification of power generation cost of the base case in the model and optimization of cash flow with current low interest rate under the condition that the on-grid electricity price and energy being the same, i.e. the same revenue from electricity sales.

The model enables shareholders to prepare for the rolling development of the company and invest in new projects from a long-term perspective. In addition, as the total capital cost of projects gradually increase, the general contractor can better allocate its resources against background of mounting investment in the Generation 3 NPPs by diverting premium resources to design, construction and procurement, which will in turn benefit the owner. The optimization that attaches more importance to the whole life cost of the project rather than limited to the investment in the total capital cost can be more conducive to the sustainable development of the entire nuclear power industry.

It can be concluded that the yield of the plant can be further increased with the same construction investment. Meanwhile, the project can endure higher construction investment under the condition of the same or even higher yield, urging relevant units to apply higher safety standards to improve the safety and reliability of the plant. Based on the current market environment and operating practice of dividend strategy of groups including China National Nuclear Corporation (CNNC), it is recommended that Option 6 can generate stable shareholder yield and total capital cost per kilowatt. Finally, the risk resisting capability of nuclear units is lifted.

6 Conclusions

Considering the current market and the experience practice of in-service NPPs, we optimized the economic evaluation model of NPPs. Since in listed nuclear power companies, investors pay more attention to dividends and the long-term development of the company, the optimized capital internal return rate, internal return rate of investors and total capital cost could be obtained via optimized capital flow and bonus allocation. This research provides viable investment strategies for decision makers.

The flexible arrangement of cash flows and allocating bonus to investors in priority would significantly enhance the debt risk compared with the baseline scenario. Moreover, if the interest rate increases with the change of financial market, the debt rate of the proposed scenarios would increase, posing debt risks to the project. However, the asset liability under current circumstances could meet the requirements. It is noted that our research is based on the current market environment and the variance of power market and financial market would influence the results of optimization.

References

1. National Development and Reform Commission. Ministry of Housing and Urban-Rural Development of the People Republic of China. Methods and Parameters of economic evaluation for construction projects (3rd edn.). Beijing: China Planning Press (2006)
2. Wei, X.Y.: Problems and suggestions in economic evaluation of infrastructure projects. Constr. Econ. **42**(S1), 74–75 (2021)
3. Qin, H.: Construction of The Project Economic Evaluation Research. Master thesis (2015)
4. Peng, F., Luo, D.K., Yin, C.F., et al.: Bibliometric study on economic evaluation of fossil energy exploration and development articles and its visualization. Pet. Sci. Bull. **5**(1), 132–140 (2020)

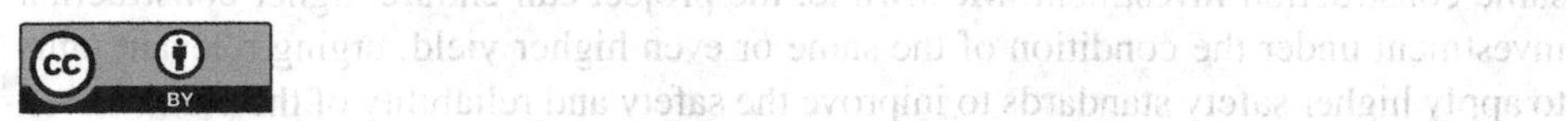
BY

Design of Radioactive Waste Classification and Detection System for Nuclear Facility Decommissioning

Kang Chang[1](✉), Guangming Cheng[2], Jinwei Zeng[2], Wenzhang Xie[1], Chenyu Shan[1], Qingxin Lei[1], Jia Huang[1], Feng Liu[1], Xiajie Liu[1], and Li Li[1]

[1] China Nuclear Power Technology Research Institute Co., Ltd., Shenzhen, China
changkang@cgnpc.com.cn

[2] Guangxi Fangchenggang Nuclear Power Co., Ltd., Fangchenggang, China

Abstract. The most important indicator of the radioactive waste classification and detection system is its identifiable minimum activity of low level radioactive waste, that is, the lower detection limit. In order to design a low level and very low level waste classification and detection system that can perform large-scale and rapid measurement, and explore the influence of various factors on the low limit of detection, this paper establishes the layout model of the detection unit of the classification detection system by Monte Carlo method to simulate the number of NaI crystals, the density of the detected object, the distance between the detection crystal and the detected object, the distance between the adjacent measured waste and the measurement time, etc. Meanwhile this paper analyzes the characteristics of the low limit of detection, and performs operations research analysis based on the principle of being able to detect radioactive waste with specific activity of not less than 0.1 $Bq{\cdot}g^{-1}$, maximizing the shielding effect and optimizing the shielding weight, and designs a low level and very low level waste classification and detection system that can measure radioactive waste which density is not less than 0.25 g/cm^3 and which specific activity is not less than 0.1 $Bq{\cdot}g^{-1}$.

Keywords: Monte carlo · Low limit of detection · Radioactive waste · Classification

1 Introduction

During the decommissioning of nuclear facilities, a large amount of radioactive waste will be generated, covering different types of waste such as medium, low and very low level radioactive and clearance level waste. According to Chinese current radioactive waste classification standard, different types of radioactive waste need to adopt different disposal methods. Timely and accurate classification of decommissioned nuclear facilities is one of the key links in radioactive waste management, and it is also an important means to minimize waste and reduce disposal and supervision costs [1–5].

When using physical measurement techniques such as radiation detection to classify radioactive waste, especially when classifying clearance level waste (EW), very low-level

C. Liu (Ed.): PBNC 2022, SPPHY 283, pp. 1085–1094, 2023.
https://doi.org/10.1007/978-981-99-1023-6_91

waste (VLLW) and low-level waste (LLW), due to the very low level of radioactivity of the first two types of waste, the low limit of detection of instrument used has higher requirements. Therefore, this paper refers to other classification detection devices (e.g. box-type waste detecting device and waste barrel gamma scanning device) and uses Monte Carlo calculation software to simulate the low limit of detection of the module of a low-very low radioactive waste classification detection system, and studies relationship between the different detection crystal volume and crystal arrangement, different measurement time, different shielding thickness around the detection crystal, the distance between the detection crystal and the upper surface of the waste, and the density with the low limit of detection, and optimizes the detection module of the low-very low radioactive waste classification detection system.

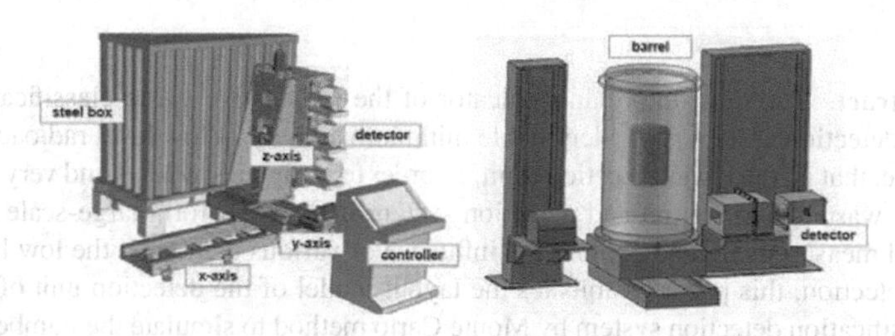

Fig. 1. Box-type waste detecting device and waste barrel gamma scannig device

2 Materials and Methods

2.1 Detector Source Response Coefficient

The detector source response coefficient K of the classification detection system is an important design parameter of the detector, that is, the counting rate generated by the radioactive source per unit specific activity in the measurement chamber of the classification detection system in the detector. In the development phase, we uses the F8 counting card in the MCNP program and SCORE card in the FLUKA program to simulate the calculation to obtain the pulse count rate produced in the detector in the measurement chamber of the specific activity of the waste in the detector. Detector source response coefficient.

$$K = Tally_{FB} \times V \times 1 \quad (1)$$

where K is the detector source response coefficient of the classification detection system, cps·Bq^{-1}·m^{-3}; $Tally_{F8}$ is counting rate (energy range of 100 keV ~ 3 meV), cps; V is the volume of waste to be measured in the simulation calculation, m^3; I is the decay branch ratio of gamma rays produced by radionuclides in the waste to be measured.

2.2 Minimum Detectable Concentration

Minimum detectable concentration ($M_{in}DC$, MDC) [4–6] represents the minimum activity concentration value that the classified detection system can analyze for specific nuclides in the measured waste. It is an important indicator of the sensitivity of the classification detection system when it is applied in a given situation. The premise that the waste to be tested can be reliably detected is that the activity concentration of the waste to be measured is greater than the minimum detectable concentration.

$$M_{in}DC = \frac{2.706/t + 4.65\sqrt{I_b/t}}{K} \tag{2}$$

where: $M_{in}DC$ is the minimum detectable concentration, Bq·m^{-3}; I_b is the environmental background count rate, cps; t is the given measurement time, s.

2.3 Model of Simulation

As shown in Fig. 2, this paper establishes the detection module model in the above-mentioned radioactive waste classification detection system in the Monte Carlo simulation software MCNP and FLUKA. In the calculation model, the size of a single NaI crystal is 40 cm × 10 cm × 5 cm; the size of the waste detection box (stainless steel) is 65 cm × 65 cm × 10 cm; the radioactive source term is divided into two parts. One is the environmental background, and it is considered that the environmental dose rate in the working area of the detection system is the upper limit of the white area's dose rate, and it is equivalent to a uniform spherical shell surrounding the whole detection device. The spherical shell is a source of air uniformly distributed with Cs-137 sources. The other part is the radioactive waste contained in the detection box, which is set as a cuboid source in the simulation.

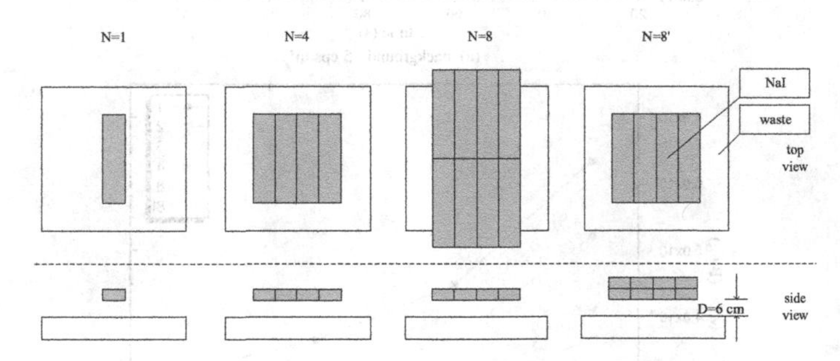

Fig. 2. Lay-out of the detectors and waste box

3 Results and Discussion

3.1 Effect of Crystal Arrangement on the Low Limit of Detection

According to the Nuclear Safety Guidelines HAD401/04 "Radioactive Waste Classification", the level of Cs-137 control is 0.1 Bq·g^{-1}, that is, the classification detection system for requirements of Cs-137.

Usually, the larger the crystal volume or the number of crystals are, the smaller the low limit of detection is. However, due to the high volume NaI crystal price, this paper selects several different NaI detection crystal arrangement and quantity as shown in Fig. 1 to simulate the detection of the lower limit change, in order to choose the most cost-effective crystal arrangement and quantity. As shown in Fig. 3(a), when the background count rate is 5 cps·cm^{-3} and when the detection time is 30 s, no detection scheme can meet 0.1 Bq·g^{-1}; when the detection time is 60 s, at least 4 NaI detection crystals should be selected to meet 0.1 Bq·g^{-1}; as shown in Fig. 3(b), when the background count rate is 0.5 cps·cm^{-3}, there is no limit to the number of NaI detection crystal blocks; in Fig. 3(a)(b), it shows that the detection efficiency of the 8 NaI crystal tiling type is better than the detection efficiency of the two-layer superposition form, mainly because the gamma ray energy is low, and the radiation energy can be deposited by selecting one layer of crystal, and the two-layer superposition form increases the lower limit of background influence detection caused by the crystal itself.

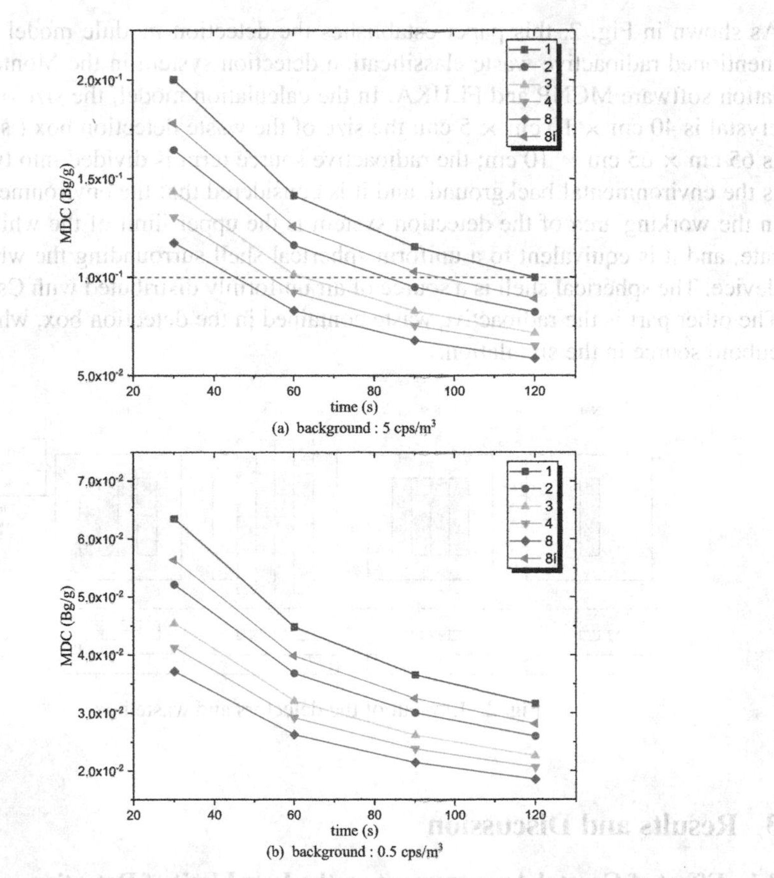

Fig. 3. Difference in the mdc with the volume of NAI crystals in different environment background

3.2 Effect of Average Distance Between Detector and Waste Upper Surface on Low Limit of Detection

In the actual measurement process of the radioactive waste classification detection system, the average distance between the detector and the waste surface is required to be adjustable, so as to avoid the overflow of the waste to be detected and the detector can achieve the best detection effect, so different source distances are used to simulate the change of the low limit of detection. As shown in Fig. 4, it can be seen that the low limit of detection increases as the average distance between the detector and the upper surface of the waste increases. Figure 4(a), when the NaI background count rate is 5 cps·cm^{-3} and when the measurement time is 60 s, the average distance between the NaI detector and the upper surface of the waste should not exceed 10 cm, even if the measurement time is relaxed to 120 s, the average distance should not exceed 20 cm. And when the NaI background count rate is 0.5 cps·cm^{-3}, the average distance between the NaI detector and the upper surface of the waste should not exceed 50 cm.

3.3 Influence of the Density of Waste to be Measured on the Low Limit of Detection

In addition, the types of waste of the decommissioning of nuclear facilities are various and the density varies greatly, including metal materials such as steam generators in nuclear power plant, waste gas masks, and radioactive medical waste. In order to explore the impact of waste density on the low limit of detection and avoid discomfort to the detection system due to excessive or small waste density, different waste densities are used to simulate changes in the low limit of detection.

As shown in Fig. 5, under the same measurement conditions, the low limit of detection decreases with increasing waste density. Figure 5(a) when the measurement time is 60 s, if the NaI background count rate is 5 cps·cm^{-3}, the waste density should not be less than 0.25g·cm^{-3}. Figure 5(b) when NaI background count rate is 0.5 cps·cm^{-3} there are no restrictions on waste density.

3.4 Effect of Shielding on the Low Limit of Detection

The background count rate in the NaI detection crystal mainly comes from the noise of the NaI crystal, the environmental background of the area where the detection system is located, and the adjacent waste box to be tested. From the above analysis, it can be seen that the background count rate of different levels also has an important impact on the detection lower limit, so shielding must be set around the detector. Since the noise of the NaI crystal itself cannot be eliminated by shielding, and the background count rate in the NaI detection crystal with different intervals of the waste box to be detected is not shielded. The result is shown in Table 1. The environmental background is much lower than the influence of the adjacent waste box to be detected on the background of the NaI detection crystal, so the following only simulates the waste box to be detected at different intervals, changes in the lower limit of detection under different shielding thicknesses.

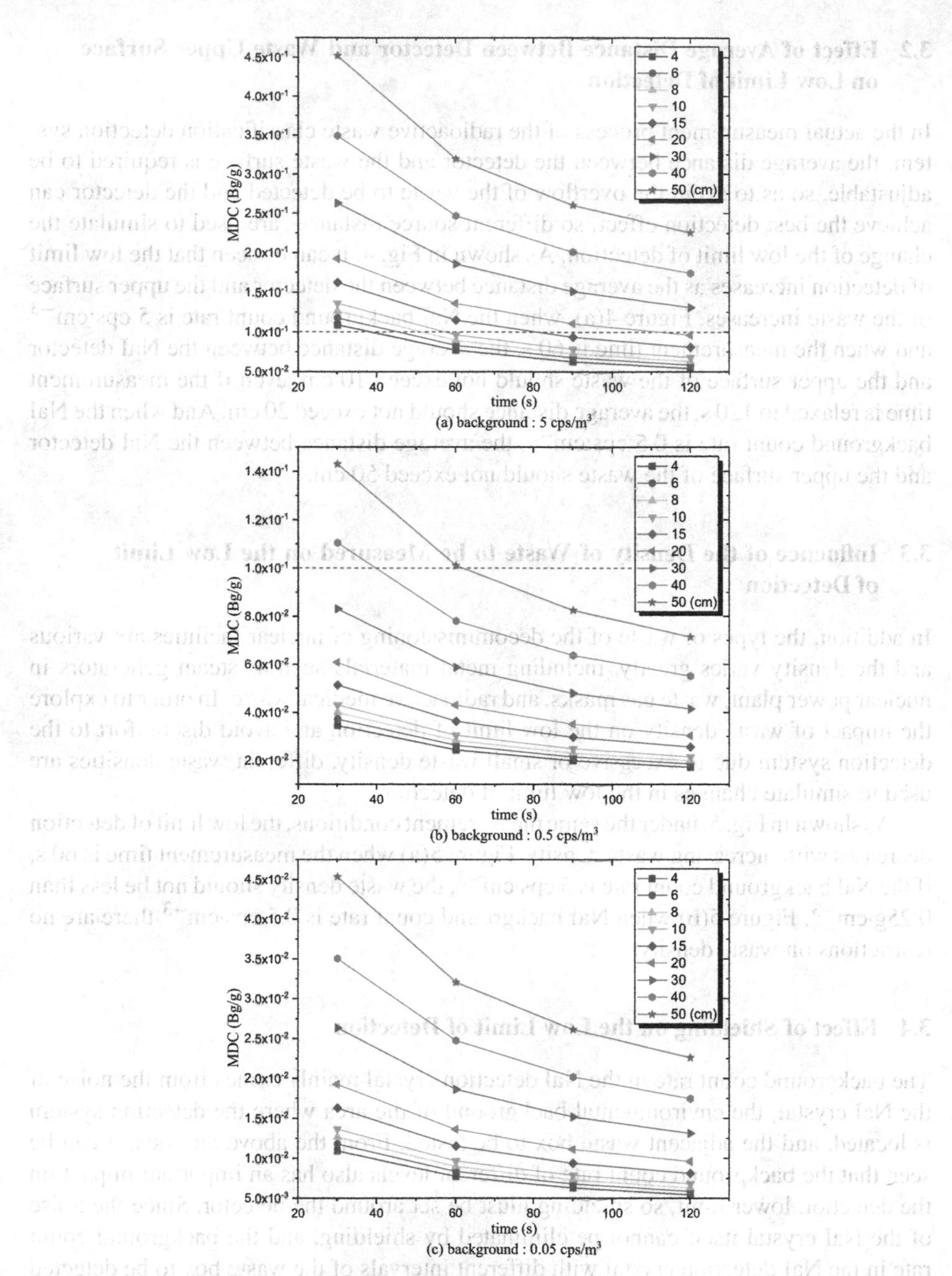

Fig. 4. Difference in the mdc with the average distance between the detector and the upper surface of waste in different environment backgrounds

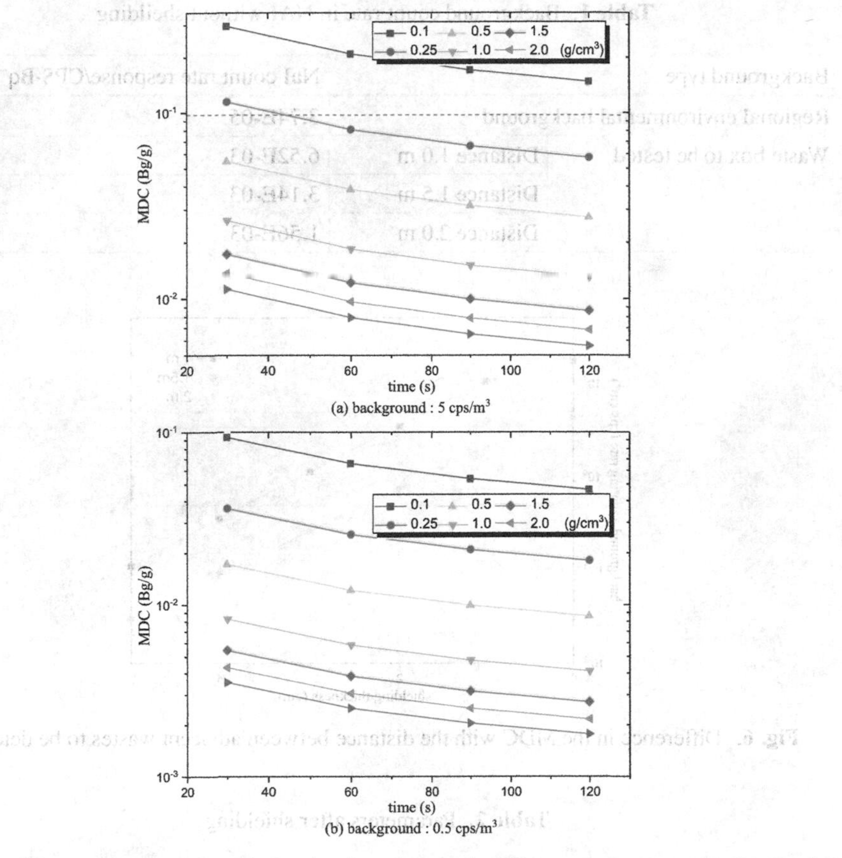

Fig. 5. Difference in the mdc with the density of waste in different environment backgrounds

As shown in Fig. 6(a), when the spacing between adjacent waste boxes is 1 m, the thickness of the shielding lead layer is 2 cm and 4 cm, respectively, the background count rate can be reduced to 5 cps·cm^{-3} and 0.5 cps·cm^{-3} as shown in Fig. 6(a) and (b), when the spacing between adjacent waste boxes is 1.5 m or more, the background count rate can be reduced to 5 cps·cm^{-3} and 0.5 cps·cm^{-3} when the thickness of the shielding lead layer is 1 cm and 3 cm, respectively. As shown in Table 2, for a conservative estimate, when the distance between adjacent waste boxes is selected for 1 m, the background count rate is reduced to 5 cps·cm^{-3} and 0.5 cps·cm^{-3}, respectively. The weight of lead shielding is about 139.3 kg and 303.1 kg, because the aging lead used for shielding is expensive, so it is proposed to use 2 cm thickness lead shielding.

Table 1. Background count rate in NAI without sheilding

Background type		NaI count rate response/CPS·Bq^{-1}
Regional environmental background		3.74E-05
Waste box to be tested	Distance 1.0 m	6.52E-03
	Distance 1.5 m	3.14E-03
	Distance 2.0 m	1.56E-03

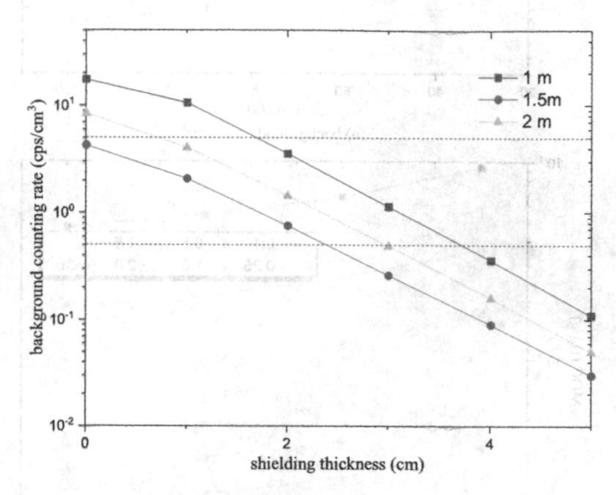

Fig. 6. Difference in the MDC with the distance between adjacent wastes to be detected

Table 2. Parameters after shielding

Shield lead layer thickness/ cm	Background count rate/ cps·Bq^{-1}	Number of crystals	Minimum source distance/ cm	Waste minimum density/ g·cm^{-3}	Lead shield weight/ kg
2	5	4	20	0.25	139.3
4	0.5	Unlimited	50	Unlimited	303.1

4 Conclusions

In this paper, Monte Carlo software MCNP and FLUKA are used to simulate the influence of the number of NaI crystals, the composition of the measured object, the distance between the detection crystal and the measured object, and the measurement time on the lower limit of detection. The results show that:

1) In the case of the same arrangement method, the lower limit of detection varies with the increase in the number of detected crystals; the low limit of detection of the

tiling type is better than the two-layer superposition form; it is expected to use 8 NaI crystal tiling type arrangement method in engineering.

2) The low limit of detection deteriorates as the average distance between the detector and the upper surface of the waste increases;
3) The low limit of detection increases with increasing waste density;

Taking into account the weight and shielding effect of lead shielding, it is proposed to use 2 cm thickness lead shielding to reduce the background count rate to 5 cps·cm^{-3} below, at the same time, the average distance between the NaI detector and the upper surface of the waste should be controlled at 20 cm, and the waste density should be controlled not less than 0.25 g·cm^{-3}.

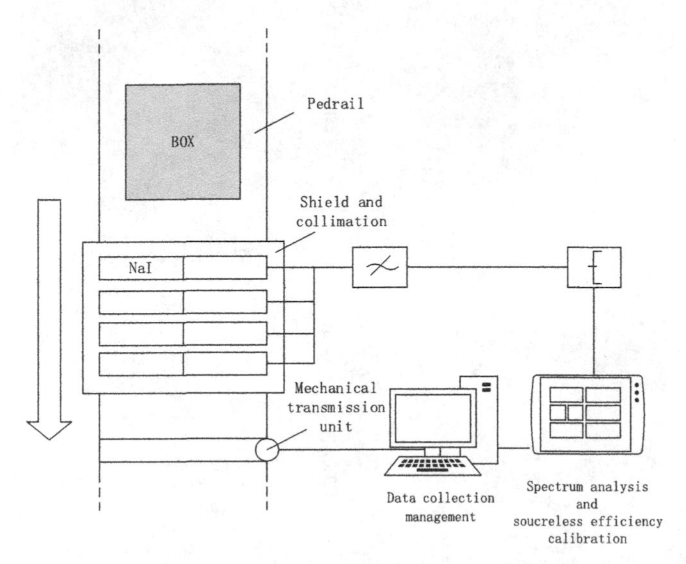

Fig. 7. Device schematic

Based on the above research contents, this paper designs the device as shown in the Fig. 7. The conveyor transports the detection cassettes containing the radioactive waste into the detection chamber with shielding and collimation through the pedrail. The source-less efficiency and energy spectrum analysis program completes the calibration and measurement of the waste, according to which the radioactive of the waste is judged and classified, and the transmission device is controlled to continue to move.

References

1. Ministry of Environmental Protection, Ministry of Industry and Information Technology, National Defense Science and Industry Bureau. Classification of Radioactive Waste (2017)
2. Ren, X., Liu, W.: Radwaste management for nuclear facility decommissioning. Radiat. Prot. Bull. **28**(4), 1–7 (2008)

3. Wang, P., Liao, Y., Wei, G., et al.: The requirement of radioactive waste clearance level. Nucl. Saf. **14**(2), 6–11 (2015)
4. Teng, Y., Zuo, R., Wang, J.: Technical methods of evaluation of nearsurface disposal of very low level radioactive waste. Bull. Mineral. Petrol. Geochem. **30**(1), 59–64 (2011)
5. Xie, Y.: Summarization of safety assessment on disposal of low level radioactive wastes. Sichuan Environ. **30**(5), 54–58 (2011)
6. Hao, R.: Practical formulae for the detection limit in low-level radioactivity measurement. Acta Metrologica Sinica **4**(4), 303–310 (1983)
7. Sha, L., Wei, W., Xuan, Y.: Study on processing method of data near the lower limits of detection for radiation environmental monitoring. Admin. Technique Environ. Monit. **18**(1), 38–43 (2006)
8. Huang, N.: Concept and calculation of detection limit in low level radioactive sample measurement. Radiat. Prot. Bull. **24**(2), 25–32 (2004)

Hot-Pressing Sintering and Analytical Characterization of Coated Particle-Dispersed Fuel Pellets

Wentao Liu[1,2](✉), Zongyi Shao[1,2], Ying Meng[2], Zongshu Li[1,2], and Zheng Sui[1,2]

[1] CNNC Key Laboratory of Fabrication Technology of Reactor Irradiation Special Target, Baotou 014035, China
838280916@qq.com

[2] China North Nuclear Fuel Co., Ltd., Baotou 014035, China

Abstract. Coated particle-dispersed fuel pellets have the characteristics of high thermal conductivity and multilayer protection against fission products, which are an important component of ATF fuel. In this paper, by carrying out research on the dressing process of TRISO particles, it is realized that SiC powder completely encapsulates TRISO particles. The hot-pressing sintering method was used to study the influence of different sintering aids, different hot-pressing sintering holding time, different hot-pressing sintering temperature, different hot-pressing sintering pressure and different powder particle size on the density of pellets. The power with a particle size of 10nm is sintered at 3% sintering aid, 1690 °C, 1.5h holding time, 80 MPa vacuum hot pressing, and a full ceramic micro-encapsulated fuel core with a relative density of 95% or more is prepared, and the internal particle structure is complete.

Keywords: Coated particle-dispersed fuel pellets · TRISO · Hot press sintering

1 Introduction

The purpose of research and development of advanced reactor systems is to meet the economic, environmental and social development needs of the 21st century. Its technical goals involve four aspects: sustainability, economy, safety and reliability, nuclear non-proliferation and physical protection. Specifically, it contains eight technical goals, covering various technical directions related to the design and implementation of reactors, energy conversion systems and fuel cycle facilities.

The nuclear fuel element is the core component of the reactor, and its advanced nature and safety are the important basis for the advanced nature and safety of the reactor. ATF (Accident Tolerant Fuel) fuel that can tolerate accidents to a certain extent and has inherent safety, is an important development direction in the field of nuclear fuel in the world. As an important part of ATF, the coated particle-dispersed fuel obtained from TRISO particle-dispersed silicon carbide matrix is aimed at the weakness of the traditional UO_2-Zr alloy fuel system, and combines the mature TRISO (Tri-Structural Isotropic) particles with the tolerance of fission products. As well as the advantages of

C. Liu (Ed.): PBNC 2022, SPPHY 283, pp. 1095–1103, 2023.
https://doi.org/10.1007/978-981-99-1023-6_92

good thermal performance and thermal stability of SiC matrix, it is a more promising development direction in advanced accident-resistant nuclear fuel.

The coated particle-dispersed fuel is based on a fuel pellet with three-dimensionally isotropic TRISO-coated particles embedded in a SiC matrix. The structure of the pellets and the internal TRISO particles is shown in Fig. 1. Such pellets feature high thermal conductivity, multiple layers of protection against cracking products, and high burnup, which facilitates normal and transient operability of the reactor.

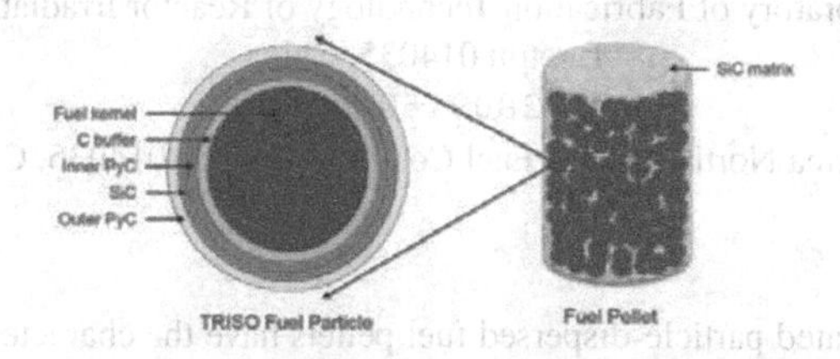

Fig. 1. Schematic diagram of TRISO particle and coated particle dispersion fuel

After the concept of coated particle dispersion fuel in which TRISO particles are dispersed in a silicon carbide matrix was proposed, researchers from the United States, South Korea and other countries have conducted in-depth research in this direction, and prepared products with good thermal conductivity. All-ceramic microencapsulated die with uniform phase distribution.

In this paper, the effect of binder on the dressing of TRISO granules and the effect of pre-compression molding pressure on the green density were studied through mixing, molding and hot-pressing sintering experiments. After the SiC matrix and TRISO particles were uniformly mixed and molded, the effects of different sintering aid contents, different hot-pressing holding times, different hot-pressing sintering temperatures, different hot-pressing sintering pressures, and different powder particle sizes on the properties of pellets were studied.

2 Materials and Methods

2.1 Materials

The TRISO particles used in the study are fuel particles with spherical UO_2 as the core. The outer layer of the core is sequentially coated with loose pyrolytic carbon, inner dense pyrolytic carbon, silicon carbide and outer dense pyrolytic carbon. The SiC powder used is β phase and its purity is greater than 99.9%. The particle size distribution of the raw material powder is relatively uniform, but the deviation between the median value and the average value is small, and the powder particle size as a whole presents a normal distribution centered on the median value.

Due to the poor sintering performance of SiC powder, in the process of preparing by NITE liquid phase sintering method, adding a small amount of sintering aid for co-sintering can improve the sintering performance of the material to a certain extent. The sintering aids used in the test are Al_2O_3 and Y_2O_3 powders with a purity $> 99.9\%$.

2.2 Methods

Put SiC powder with a particle size of 10nm and 3wt% sintering aid (the mass ratio of nano-scale Al_2O_3 and Y_2O_3 is 7:3) into a ball mill jar, with anhydrous ethanol as the dispersant, stainless steel balls as the grinding balls, The ratio is 3:1, and the ball milling time is 4 h. After ball milling, the powder is dried, crushed and sieved; the sieved mixed powder is evenly wrapped on the surface of the TRISO particles; the TRISO particles after uniformly wrapping the base powder and the base powder are mixed uniformly according to a certain volume ratio, and then pre-compressed and formed; finally Vacuum hot pressing sintering, heating rate 10 °C·min-1, sintering temperature 1650 ~ 1750 °C, 1 ~ 2.5 h holding time, 65 ~ 90 MPa.

3 Results and Discussion

3.1 Influence of Binder on Dressing Effect of TRISO Particles

In order to improve the compatibility of SiC and TRISO particles during the green forming process of the coated particle-dispersed fuel pellets, and at the same time prevent the TRISO particles from contacting each other, it is necessary to coat a layer of SiC on the surface of the TRISO particles.

Due to the large difference in particle size between TRISO particles and SiC powder, it is difficult for SiC to directly coat the surface of TRISO particles without binder, as shown in Fig. 2(a). After investigation and test, use glycerol as binder, absolute ethanol as diluent, configure 10% glycerol-absolute ethanol as binder, and use glue tip dropper to evenly wrap the surface of TRISO particles with a layer of adhesive. Then, it is dispersed into the matrix powder (SiC + sintering aid) to realize the bonding of a layer of matrix powder on the surface of the TRISO particles. As shown in Fig. 2(b), the surface of the TRISO particles is completely covered and the dressing effect is good.

Fig. 2. Mixture macro effect diagram:(a) The effect of dressing without binder.(b) The effect of dressing without binder

3.2 Influence of Pre-press Forming Pressure on Green Density

The weight of the TRISO particles and the SiC powder after adding the sintering aid are calculated and weighed according to the volume fraction, and the ingredients are mixed, and then loaded into the pre-compression mold, and the floating female mold is pre-compressed using a micro-controlled pressure testing machine to determine the molding pressure and raw material. The relationship between blank density and TRISO particle integrity is shown in Fig. 3. It can be seen from Fig. 3 that the higher the pressing pressure, the higher the green density of the pellets. The higher the green density, the better the density of pellets after sintering, so the molding pressure should be raised as high as possible. It was observed in the experiment that when the pressure was lower than 2.0 KN, the TRISO particles were not damaged, and when the pressure reached 2.5 KN, the surface layer of some TRISO particles appeared to fall off, and there was abnormal noise during the pressing process.

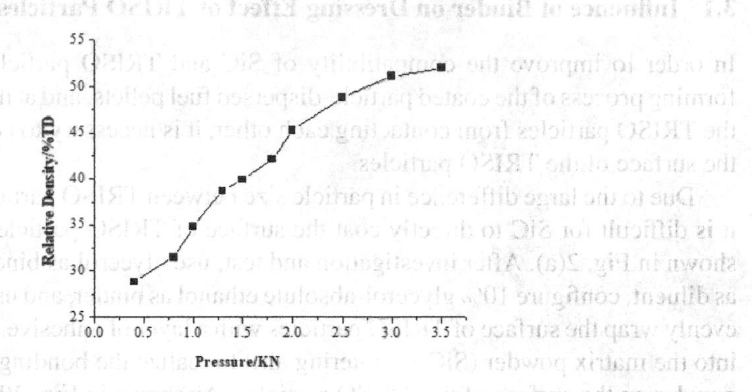

Fig. 3. Variation curve of green density with pressure

3.3 Effects of Different Hot Pressing Sintering Processes on the Properties of Pellets

Table 1 shows the density changes of pellet fuel pellets covered by pressureless sintering under different additions of sintering aids. When the addition of sintering aids is less than 3wt%, the density of pellets increases with the addition of sintering aids., when the addition amount of sintering aid is greater than 3wt%, the increase in the density of pellets is no longer obvious.

Table 2 shows the change results of the density and phase of the pellets at different hot pressing sintering temperatures. It can be seen from the results that as the sintering temperature increases, the density of the pellets increases, and the β-SiC in the pellets will be partially converted into α -SiC, when the sintering temperature is higher than 1710 °C, obvious α-SiC diffraction peaks appear in the pellet phase.

Table 1. Effect of Sintering Aid Additives on the Density of Pellets

Addition of sintering aid	Relative density of pressureless sintered pellets
0wt%	58.32%T.D
1wt%	67.24%T.D
3wt%	78.92%T.D
5wt%	78.95%T.D
7wt%	78.86%T.D
10wt%	78.97%T.D

Table 2. Influence of sintering temperature on the density and phase of pellets

Hot pressing sintering temperature	Pellets relative density	Pellets phase
1660 °C	89.77%T.D	Pure β-phase
1690 °C	92.14%T.D	No obvious α-phase diffraction peaks
1710 °C	95.31%T.D	obvious α-phase diffraction peaks
1760 °C	96.58%T.D	obvious α-phase diffraction peaks

Table 3 shows the change results of the density of pellets under different hot pressing and holding time. With the prolongation of holding time, the diffraction peak of α-SiC phase in the pellets gradually increased; when the holding time was 1h, the inner and outer layers of the pellets appeared Phenomenon, the pellets are not burned through, and after the holding time exceeds 2h, the density of the pellets does not increase significantly.

Table 3. The effect of holding time on the density of pellets

Holding time	Pellets relative density
1 h	The pellets are not fully burned, and the phenomenon of internal and external delamination occurs
1.5 h	95.31%T.D
2 h	95.89%T.D
2.5 h	95.81%T.D

Table 4 is divided into the change results of the density of pellets under different hot-pressing pressures. The analysis results show that with the increase of hot-pressing pressure, the density of pellets gradually increases. When the hot-pressing pressure is in the range of 75 ~ 80 MPa, the The increase in density decreases. When the hot pressing pressure reaches 90 MPa, the punch on the die breaks.

Table 4. The effect of sintering pressure on the density of pellets

Hot pressing pressure	Pellets relative density
65 MPa	92.69%T.D
70 MPa	93.92%T.D
75 MPa	94.89%T.D
80 MPa	95.26%T.D
90 MPa	The punch on the die breaks

Table 5 shows the experimental results of SiC powder with different particle sizes. It can be seen from the results that with the decrease of powder particle size, the density of pellets gradually increases.

Table 5. Effect of Powder Particle Size on Pellets Density

Powder particle size	Sintering parameters	Pellets relative density
500 nm	3wt% sintering aid/1690 °C/80 MPa	94.89%T.D
100 nm		95.34%T.D
10 nm		96.23%T.D

3.4 Performance Characterization of Coated Particle-Dispersed Fuel Pellets

Using SiC matrix powder with a particle size of 10nm, vacuum hot-pressing sintering at 3wt% sintering aid addition, 1690 °C hot-pressing sintering temperature, 1.5 h holding time, and 80 MPa hot-pressing sintering pressure, the obtained all-ceramic micro-encapsulated dispersed fuel pellets were prepared. The real thing is shown in Fig. 4.

Fig. 4. Coated particle dispersion fuel pellets

The SEM of the SiC matrix is shown in Fig. 5, the second phase composed of sintering aids exists inside the SiC matrix, and the distribution is uniform. The metallographic photos of TRISO particles can clearly see that the outer layer of the particles has 4 layers of cladding layers, the particle structure is complete, and there is no damage during the molding and sintering process. After core sintering, the thickness and element distribution of each layer of TRISO particles are still comparable to those of the original particles, and the thickness of the outer dense pyrolytic carbon layer is 19 μm.

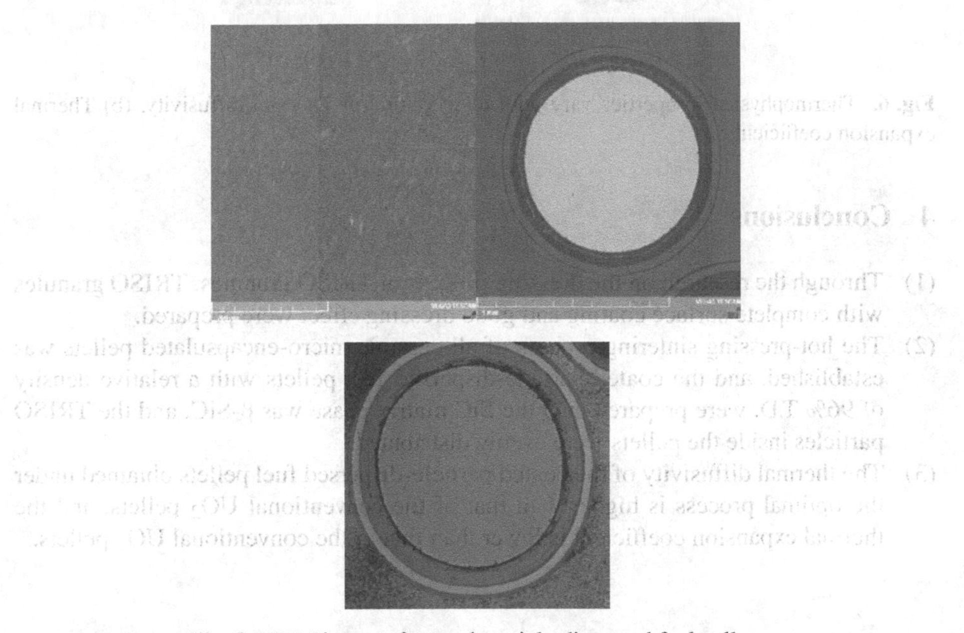

Fig. 5. SEM image of coated particle-dispersed fuel pellets

It can be seen from Fig. 6(a) that the thermal diffusivity of the coated particle-dispersed fuel core gradually decreases with the increase of temperature, and the thermal diffusivity at each temperature point is lower than that of the base SiC. The addition of TRISO particles reduces the core to a certain extent. The thermal diffusivity is still significantly higher than that of the traditional UO_2 fuel core. It can be seen from Fig. 6(b)that the thermal expansion coefficient of the fuel core increases gradually with the increase of temperature, and the change trend is linear. As the content of TRISO particles increases, the thermal expansion coefficient of the core decreases slightly. In the range of 0 – 1000 °C, the thermal expansion coefficient of the coated particle-dispersed fuel core is significantly lower than that of the traditional UO_2 core.

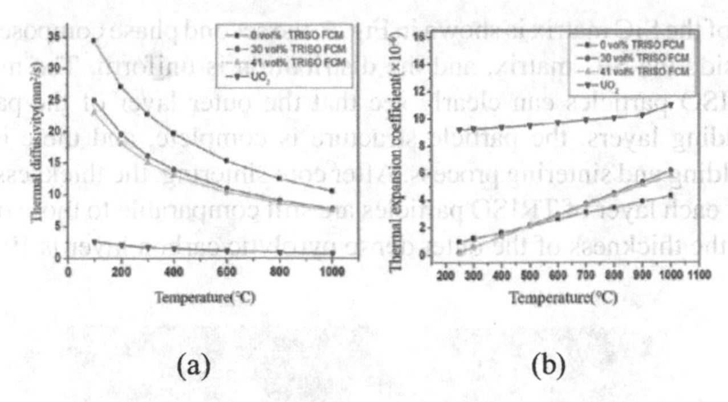

Fig. 6. Thermophysical properties vary with temperature:(a) Thermal diffusivity. (b) Thermal expansion coefficient

4 Conclusions

(1) Through the research on the dressing process of TRISO granules, TRISO granules with complete surface coating and good dressing effect were prepared.
(2) The hot-pressing sintering process of all-ceramic micro-encapsulated pellets was established, and the coated particle-dispersed fuel pellets with a relative density of 96% T.D. were prepared, and the SiC matrix phase was β-SiC, and the TRISO particles inside the pellets were evenly distributed.
(3) The thermal diffusivity of the coated particle-dispersed fuel pellets obtained under the optimal process is higher than that of the conventional UO_2 pellets, and the thermal expansion coefficient is lower than that of the conventional UO_2 pellets.

References

1. Terrani, K.A., Kiggans, J.O., Silva, C.M., et al.: Progress on matrix SiC processing and properties for fully ceramic microencapsulated fuel form. J. Nucl. Mater. **457**, 9–1 (2015)
2. Lee, H.G., Kim, D., Lee, S.J., et al.: Thermal conductivity analysis of SiC ceramics and fully ceramic microencapsulated fuel composites. Nucl. Eng. Des. **311**, 9–15 (2017)
3. Sonat Sen, R., Pope, M.A., Ougouag, A.M., et al.: Assessment of possible cycle lengths for fully encapsulated microstructure fueled light water reactor concepts. Nucl. Eng. Des. **255**, 310–320 (2013)
4. Terrani, K.A., Kiggans, J.O., Katoh, Y., et al.: Fabrication and characterization of fully ceramic microencapsulated fuels **426**, 268–276 (2012)
5. 刘荣正，刘马林，刘兵，等.碳化硅基新型包覆燃料颗粒的设计及制备[J].原子能科学技术 p. 7 (2016)
6. 许多挺，刘彤，任启森，等.事故容错燃料芯块热学性能分析[J].核动力工程 **37**(2), 82–86 (2016)
7. 刘荣正，刘马林，邵友林，等.碳化硅材料在核燃料元件中的应用[J].材料导报**29**(1), 1–5 (2015)
8. 邵友林，朱钧国，杨冰，等.包覆燃料颗粒及应用[J].原子能科学技术**39**, 117–121 (2015)
9. 李满仓，刘仕倡，秦冬，等. 全陶瓷微封装燃料压水堆弥散可燃毒物中子学分析[J].原子能科学技术**53**(7), 1188–1194 (2019)
10. 黄智恒，贾德昌，杨治华，等.碳化硅陶瓷的活化烧结与烧结助剂[J].材料科学与工艺**12**(1), 103–107 (2004)
11. 武安华，曹文斌，李江涛，等. SiC烧结的研究进展[J].粉末冶金工业**12**(3), 28–32 (2002)

12. 王静，张玉军，龚红宇.无压烧结碳化硅研究进展[J].陶瓷 **4**, 17–19 (2008)
13. Liu, F.B.: Properties and manufacturing method of silicon carbide ceramic new materials. Appl. Mech. Mater. **416–417**, 1693–1697 (2013)
14. 王坤.碳化硅技术陶瓷无压烧结工艺研究[J]. 陶瓷 **6**, 43–53 (2019)
15. 陈巍，曹连忠，张向军等. SiC陶瓷无压烧结工艺探讨兵器材料科学与工程**27**(5), 35–37 (2004)

Study on Capiliary Characteristics of Stainless Steel Wire Mesh Wick of Alkali Metal Heat Pipe

Yiru Zhu, Luteng Zhang(✉), Zhiguo Xi, Zaiyong Ma, Wan Sun, Longxiang Zhu, and Liangming Pan

Chongqing University, Chongqing, China
ltzhang@cqu.edu.cn

Abstract. The capillary characteristics of the wicks are of great significance to the normal operation of the heat pipe,and this study carried out the wick experiment of vertical reel stainless steel wire mesh in liquid sodium working medium and the visual observation experiment of sodium film in the wire mesh wick. The experimental results show that when the temperature of liquid sodium is about 400 °C, the capillary phenomenon of stainless steel wire mesh occurs more obviously, and after 450 °C, the evaporation of sodium is gradually obvious and accompanied by the mass fluctuation of the wire mesh wick. The visual observation experiment of sodium liquid film in the wire mesh found that the wetting and transition point of sodium on the surface of the stainless steel wire mesh was about 410 °C, which was a good verification of the occurrence of the more obvious capillary phenomenon of the wick at about 400 °C.

Keywords: Alakali metal heat pipe · Wick · Capillary characteristics · Contact angle

1 Introduction

Heat pipe is an efficient heat transfer component that uses working medium to transfer heat at different parts of the phase change. The heat pipe principle was first proposed by Gaugler [1] in 1944. In 1965, Cotter [2] first proposed a relatively complete theory of thermal management. Because of their superiority, Heat pipes are also used for cooling abyssal sea reactors and abyssal sea reactors. However, the heat transfer capacity of the heat pipe is limited by its own heat transfer limit. When the heat transfer limit of the heat pipe occurs, the heat of the reactor core cannot be exported in time, which leads to a reactor safety accident. The capillary limit is due to the fact that the capillary indenter produced by the evaporation and condensation sections of the heat pipe is not enough to overcome the pressure drop caused by the return of the working medium. The working fluid cannot flow back to the evaporation section normally. It will cause the evaporation section to dry up, and cause the wall temperature of the evaporation section to rise rapidly, and even burnout the heat pipe wall. Since the heat pipes are all operating at high temperatures in the heat pipe cooling reactor. The heat pipes must use high temperature heat pipes. The working medium of high-temperature heat pipes

C. Liu (Ed.): PBNC 2022, SPPHY 283, pp. 1104–1113, 2023.
https://doi.org/10.1007/978-981-99-1023-6_94

is generally lithium, sodium, potassium and other alkali metals, so high-temperature heat pipes are often referred to as alkali metal heat pipes. Because of the high viscosity and density of alkali metals, a greater driving force is required in the reflux of the wick. Therefore, exploring the capillary force of the high-temperature heat pipe wick can guide the selection of the operating conditions of the heat pipe, so as to avoid the occurrence of the capillary limit of the high-temperature heat pipe and ensure the safety of the operation of the space reactor.

The most commonly used methods for testing capillary characteristics in the wick are the bubble method [3] and the capillary rising method. Shufeng Huang [4] tested the capillary characteristics of a new type of stainless steel fiber-powder composite wick with the capillary rise method and compared it with a composite wick with a single structure and other structures. Guanghan Huang [5] used an infrared camera to set up a capillary rise rate test device that measured the capillary characteristics of the axial channel wick of the alkaline corrosion treatment. Yong Tang [6] used an infrared camera combined with capillary rise method to measure the capillary performance of the sintered groove composite wick, sintered wick and groove wick with ethanol as the working medium. The results show that the sintered powder wick has better capillary characteristics than the single structure wick. Heng Tang [7] measured the capillary characteristics of a new micro-V-shaped channel wick using the capillary ascending method and acetone as the working medium. Daxiang Deng [8] used a new infrared thermal imaging method to test the capillary characteristics of the sintered groove wick with ethanol as the working medium. Daxiang Deng [9] also tested the capillary characteristics of the micro-V-channel wick with ethanol and acetone for the working medium. Li [10] calculated the capillary characteristics of different porosity wicks in acetone using the mass change curve assessment recorded by the electronic balance.

Most of the above scholars are experiments on the capillary characteristics of sintered wicks, groove wicks and composite wicks with ethanol and acetone as working medium. Due to the reactive chemical properties of the alkali metal working medium, there are fewer test experiments for the capillary core performance of alkali metal heat pipes. In this paper, 304 stainless steel is used as the material of the wire mesh wick, sodium is used as the working medium, and the capillary characteristics of the wire mesh wick is studied by the capillary rising method, and the capillary characteristics of the sodium heat pipe are preliminarily explored. The sodium film in the screen was photographed, and the spread of liquid sodium at different temperatures on the stainless steel wire mesh was obtained, and the law of the capillary ability of the wire mesh wick changed with temperature was verified.

2 Experimental Methods and Principles

2.1 Experiment on Capillary Ability of the Wire Mesh Wick

The capillary ability experiment of the wick is measured by the capillary rise method and the quality change process of the wick is recorded with an electronic balance. Figure 1 shows the entire experimental setup and schematic. The experimental principle is that when the wire mesh wick sample is extended into a stainless steel test tube containing liquid sodium, the liquid sodium will be sucked into the wick due to the influence of

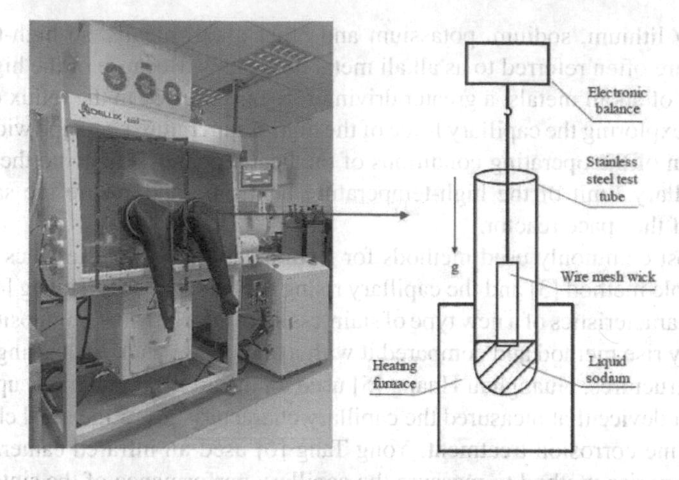

Fig. 1. Capillary ability test equipment diagram

capillary force and can reach a certain height. When the rise of sodium in the wire mesh wick reaches stability, the following relationship is satisfied:

$$h = \frac{2\sigma\cos\theta}{\rho g r_p} = \frac{2\sigma}{\rho g r_{eff}} \tag{1}$$

wherein θ is the contact angle between sodium and the wire mesh wick, the σ is the surface tension of liquid sodium, r_p is the liquid sodium density, r_{eff} is the effective capillary radius of the wick, and r_p is the pore radius. The experimental equipment includes glove box, well-type heating furnace, electronic balance and so on. The glove box provides an inert gas environment with its own dehydration and deaeration functions, which can avoid the contamination of the sodium working medium by oxygen and water vapor during the experiment. Both the well furnace and the electronic balance are placed inside the glove box in the argon atmosphere. The electronic balance has a maximum range of 220 g, an accuracy of 0.2 mg, and can be connected to the computer segment to output real-time quality changes. The test tube contains a K-type thermocouple for internal liquid sodium temperature measurement, and the XSR21A series paperless recorder is used to implement the output temperature data, and the paperless recorder error is 0.2%· F, which F is the set range.

2.2 Sodium Membrane Spreading Observation Experiment

The spreading observation experiment of liquid sodium on the surface of stainless steel wire mesh is based on a heating stage and a microscope, and the experimental equipment diagram is shown in Fig. 2. The hot stage contains platinum electric heating material and crucibles inside and can be heated at a maximum heating temperature of up to 1500 °C. Cold water circulation devices and inert gas runners are provided around the hot stage. Cooling circulation units are used to cool hot stage materials, and inert gases can reduce air pollution to sodium during experiments. The microscope has a maximum magnification of 1000X and can observe samples in the micron range.

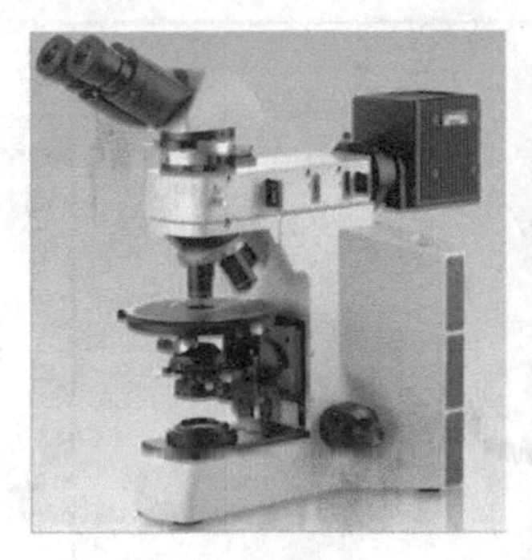

Fig. 2. Hot stage experimental equipment diagram

During the experiment, argon gas is continuously pumped in the crucible of the hot stage, and then the sodium block is placed inside the crucible of the hot stage, and then the cleaned stainless steel wire mesh is covered on the surface of the sodium block. Close the hot stage cover and place under the microscope and select multiples of the microscope objective, focusing until the mesh structure is clearly visible. Then turn on the chiller and turn on the heating to photograph the screen at the set temperature.

3 Analysis of Experimental Processes and Results

The experiment adopted 800 mesh 304 stainless steel wire mesh wick as the material, and it was fixed by three-layer rolling. The wick was cleaned in absolute ethanol and acetone successively using an ultrasonic cleaner. Samples of the cleaned wick was sent to the inside of the glove box through the glove box transition chamber. Removed the sodium from the kerosene in the glove box and cut off the surface oxide layer of sodium with a knife. Weigh 120 g of sodium on a balance and place in a stainless steel test tube. The balance indication was adjusted and zeroed, and the wick was weighed and fixed with a hook under the balance, resulting in an initial mass of 44.847 g. The tube clamping height was adjusted by the motor, and the initial liquid sodium infiltration depth was calculated to be about 1.2 cm. The experimental heating temperature program was set to rise first and then fall. When the water content and oxygen content in the gloves dropped below 0.5 ppm and 1 ppm, respectively, and the balance indication was no longer significantly changed, the well furnace was opened and the experiment began. The experimental results of the 800 mesh 3-layer wire mesh aspiration core are shown in Fig. 3.

As can be seen from the Fig. 3, the mass of the wick remains essentially unchanged when heated to 300 °C at the initial temperature. At 300 °C, the stepper motor is driven to move the wick sample downward so that it comes into contact with the liquid sodium, so there is a turning point of mass degradation at 300 °C. In the temperature range of 300–400 °C, the mass of the wick increases only slightly, and is accompanied by small mass fluctuations, and the analysis may be due to the combined effect of sodium flow and sodium evaporation. After the temperature reaches 400 °C, the mass of the wick as shown in Fig. 3 increases significantly and basically reaches the initial level. When

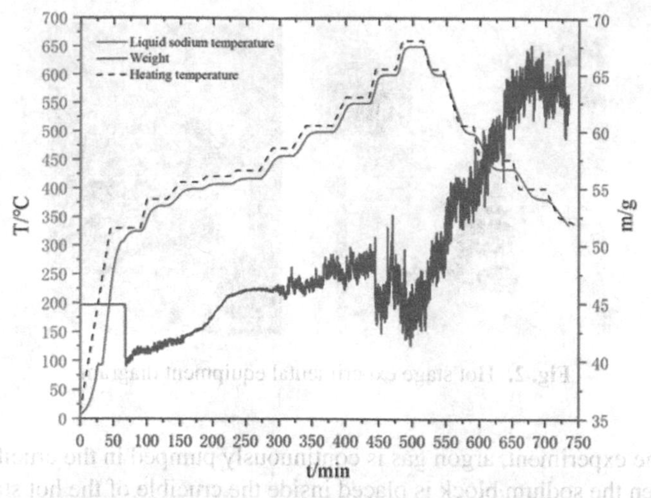

Fig. 3. 800 mesh wick experimental result diagram

heated from 400 °C to 450 °C, the mass of the wick changes are small and there are no significant mass fluctuations. When the liquid sodium temperature reaches 450 °C, the evaporation of metallic sodium is significantly enhanced, which directly leads to significant fluctuations in the quality of the wick in the range of temperature 450 °C to 650 °C. When heated to around 520 °C at 450 °C, the mass of the wick increases slightly. Bader [11] shows that the contact angle of sodium on the 304 stainless steel wire varies with temperature as shown in Fig. 4, which shows that sodium has a wetting transition point near about 400 °C. Therefore, the quality of the wick changes significantly at 400 °C in the experiment, because the contact angle of liquid sodium on the surface of stainless steel is suddenly reduced, so that the capillary ability of the wick is enhanced. The experimental results are basically consistent with the results of the contact angle change given in the literature. And after 500 °C, the liquid sodium and stainless steel have basically been completely wet, the mass change of the wick is not as obvious as when it is 400 °C, and the violent sodium evaporation phenomenon makes the quality of the wick fluctuate greatly.

However, during heating from 520 °C to 650 °C, the quality of the wick decreases slightly. However, in the cooling stage after 650 °C, due to the increase in surface tension of liquid sodium, the capillary effect is enhanced, the suction effect of the wick on sodium is enhanced, the quality of the wick continues to rise, and the quality of the wick reaches stability after the cooling reaches 400 °C. Figure 5 shows the results of the wick, which shows that the sample has been basically completely infiltrated at the end of the experiment, and the highest infiltration height of sodium measured with a ruler is about 15 cm, and the final wick weighing mass has reached 49.378 g.

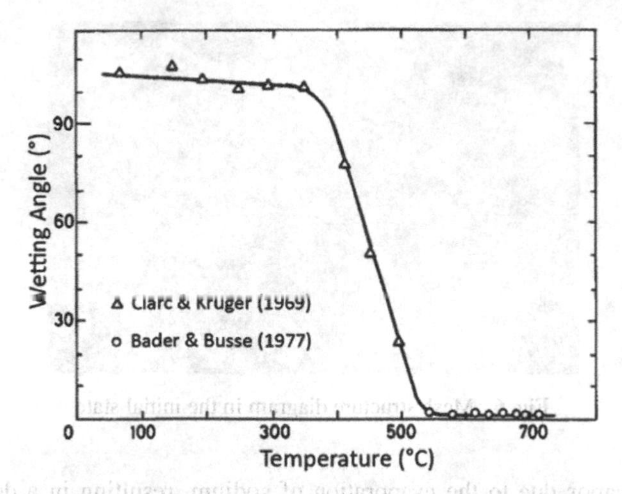

Fig. 4. The contact angle between liquid sodium and stainless steel changes with temperature

Fig. 5. wire mesh wick sodium aspiration experimental result diagram.

The results of the wire mesh wick sodium liquid film observation experiment are shown in the following figure. Figure 6 shows the microstructure diagram of the wick in the initial state, and the surface of the wire mesh shows a silvery-white luster. At a heating temperature of 400 °C, the wire mesh surface loses its original luster, producing yellow and green corrosive products. One of the corrosion products contains a Cr_2O_3 oxide layer on the surface of stainless steel that reacts with sodium to form a $NaCrO_2$ ternary oxide. At this time, a clear metallic luster appears inside the wire mesh, and liquid sodium begins to be gradually sucked onto the surface of the wire mesh wick Fig. 7.

When the heating temperature reaches 500 °C, liquid sodium is sucked onto the surface of the wire mesh due to capillary force, and the experimental results are shown in Fig. 8. When the heating temperature reaches 550 °C, the objective lens is contaminated

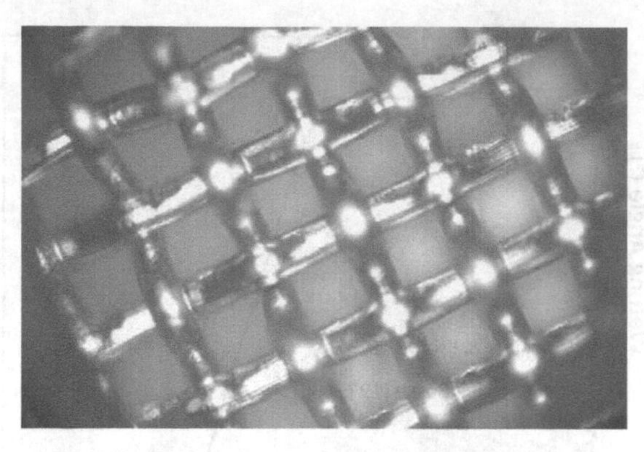

Fig. 6. Mesh structure diagram in the initial state

with sodium vapor due to the evaporation of sodium, resulting in a decrease in the brightness of the picture as shown in Fig. 9. At 600 °C, due to the evaporation of sodium, the reason for the reduction of the sodium film inside the wire mesh and the reason why the sodium vapor covers the lens cannot be observed significantly metallic. The observation experiment of sodium film spreading in the wire mesh is a good verification of the results of the increase of the capillary capacity of the 800 mesh wire mesh wick with the increase of temperature, and the increase of sodium suction of the wick leads to the increase of its own quality. During the experiment, a momentary temperature transition point was captured, it was about 410 °C. Since the argon gas flow has been passed through during the experimental process, the temperature on the surface of the wire mesh will be slightly lower than the temperature of the heating wire, so the temperature transition point should be before 410 °C.

Fig. 7. Spread of sodium film in the wick at 400 °C

Fig. 8. Spread of sodium film in the wick at 500 °C

Fig. 9. Spread of sodium film in the wick at 600 °C

4 Conclusion

In the capillary ability experiment of the wick, when heated from 300 °C to 450 °C, the quality of the wick increased to a certain extent, and the increase of sodium vapor after 450 °C made the quality of the suction core fluctuate significantly. In the temperature range of 520–650 °C, the quality of the wick is slightly reduced. The magnitude of the fluctuations also increases as the temperature increases. And in the process of cooling down by 650 °C for the first time, with the increase of the surface tension of liquid sodium, the wick has a significant increase in mass. Before and after the experiment, the sodium infiltration height of the 800-mesh wick increased from 1.2 cm to 15 cm, and the mass increased from 44.847 g to 49.378 g. The observation experiment of the spread of sodium film in the wire mesh proved that when near 410 °C, there is a transition point of liquid sodium and stainless steel wetting, and the capillary ability increases more obviously, and more sodium is sucked on the suction core, which is mutually verified with a strong growth when the quality of the wick is about 400 °C in the capillary experiment. And at 500 °C, liquid sodium has good wetting properties on the surface of stainless steel wire mesh, and liquid sodium has been better spread on the surface of the wire mesh wick.

Acknowledgement. The authors appreciate the support of National Natural Science Foundation of China (No.11905021), the Project of Science and Technology on Reactor System Design Technology Laboratory (Contract No. HT-KFKT-24–2021002), the Project of Nuclear Power Technology Innovation Center of Science Technology and Industry for National Defense(HDLCXZX-2021-ZH-036), and Operation Fundation of Key Laboratory of Nuclear Reactor System Design Technology of Nuclear Power Institute of China.

References

1. Gaugler, R.S.: Heat transfer device: Google Patents (1944)
2. Cotter, T.: Theory of heat pipes. Los Alamos Scientific Laboratory of the University of California (1965)
3. Adkins, D.R., Dykhuizen, R.C.: Procedures for measuring the properties of heat-pipe wick materials[R]. Sandia National Labs., Albuquerque, NM (United States) (1993)
4. Huang, S., Wan, Z., Zhang, X., et al.: Evaluation of capillary performance of a stainless steel fiber–powder composite wick for stainless steel heat pipe. Appl. Therm. Eng. **148**, 1224–1232 (2019)
5. Huang, G., Yuan, W., Tang, Y., et al.: Enhanced capillary performance in axially grooved aluminium wicks by alkaline corrosion treatment. Exp. Thermal Fluid Sci. **82**, 212–221 (2017)
6. Tang, Y., Deng, D., Lu, L., et al.: Experimental investigation on capillary force of composite wick structure by IR thermal imaging camera. Exp. Thermal Fluid Sci. **34**(2), 190–196 (2010)
7. Tang, H., Tang, Y., Yuan, W., et al.: Fabrication and capillary characterization of axially micro-grooved wicks for aluminium flat-plate heat pipes. Appl. Therm. Eng. **129**, 907–915 (2018)
8. Deng, D., Tang, Y., Huang, G., et al.: Characterization of capillary performance of composite wicks for two-phase heat transfer devices. Int. J. Heat Mass Transf. **56**(1–2), 283–293 (2013)
9. Deng, D., Tang, Y., Zeng, J., et al.: Characterization of capillary rise dynamics in parallel micro V-grooves. Int. J. Heat Mass Transf. **77**, 311–320 (2014)
10. Li, J., Zou, Y., Cheng, L.: Experimental study on capillary pumping performance of porous wicks for loop heat pipe. Exp. Thermal Fluid Sci. **34**(8), 1403–1408 (2010)
11. Bader, M., Busse, C.A.: Wetting by sodium at high temperatures in pure vapour atmosphere. J. Nucl. Mater. **67**(3), 295–300 (1977)

Analytical Method Research of Source Term for Steam Generator Tube Rupture Accident of Small Modular Reactor

Ming-ming Xia(✉), Jun-long Wang, Jian-ping Zhu, Chao Tian, and Jia-jia Liu

Science and Technology on Reactor System Design Technology Laboratory, Nuclear Power Institute of China, Chengdu, Sicuan, China
xiamingmin@gmail.com

Abstract. Based on the full reference to the existing engineering practice and safety review experience, and considering the actual design characteristics of the small modular reactor ACP100, a set of source term analysis method suitable for ACP100 of the steam generator tube rupture (SGTR) accident is proposed, and the source term analysis and consequence evaluation of ACP100 SGTR accident was carried out using this method. The analysis shows that the radiological consequences of the accident source term calculated by this method meet the acceptance criteria of small modular reactor. The analysis results of this article can provide support for the follow-up review of accident source term of ACP100.

Keywords: Small modular reactor · SGTR · Source term analysis method

1 Introduction

The steam generator tube rupture (SGTR) is one of the design basis accidents with high frequency and great impact during the life of the pressurized water reactor nuclear power plant, and it is also the focus of domestic nuclear safety supervision and review agencies. After SGTR, the water and steam in secondary coolant were radioactively polluted due to the release of primary coolant containing radionuclides to the secondary coolant through the break in the tube of steam generator. The radioactively polluted secondary coolant steam is released to the environment through the condenser extraction or the safety valve of the steam generator, resulting in radioactive contamination of the environment [1]. Therefore, it is great significance to carry out the SGTR source term analysis conservatively and reasonably.

This study fully investigates the SGTR source term analysis methods of M310 and Hualong 1 (ACP1000). By comparing the design differences of M310, ACP1000 and ACP100, according to the calculation assumptions and parameters of M310 and ACP1000, combined with the design characteristics of ACP100, proposed the SGTR source term analysis method of ACP100, and an ACP100 nuclear power unit SGTR source term analysis and radiological consequence evaluation are carried out using this method.

C. Liu (Ed.): PBNC 2022, SPPHY 283, pp. 1114–1120, 2023.
https://doi.org/10.1007/978-981-99-1023-6_95

2 Calculation Assumptions and Parameters

In this study, the calculation assumptions and parameters of M310, ACP1000 and ACP100 regarding the primary coolant activity and the iodine carryover coefficient in the steam generator were compared and analyzed.

2.1 Primary Coolant Activity

In tho aooidont oondition, tho primary ooolant aotivity will havo an aotivity poak phonomenon. At this time, the primary coolant activity is in a transient value, and transient assumptions is related with steady state value before the accident.

2.1.1 Primary Coolant Activity in Steady State

SGTR source term analysis of M310, the primary coolant activity in steady state is normalized at 4.44 GBq/t in I-131 equivalent, it is the maximum value of 200 reactor-year operations in a French nuclear power plant [2]. This value is not conservative enough to be lower than the operating limit of primary coolant activity [3]. SGTR source term analysis of ACP1000, the primary coolant activity in steady state is normalized at 37 GBq/t DE I-131, it is assumed that the fuel element cladding failure rate is 0.25%. This value corresponds to the technical specification the maximum radioactivity limit condition is conservative enough [4]. Therefore, ACP100 refers to assumption of ACP1000, and the primary coolant activity in steady stat is assumed that the fuel element cladding failure rate is 0.25%, and the calculation results are normalized at 7.4 GBq/t DE I-131, as shown in Table 1.

2.2 Primary Coolant Activity in Transient State

2.2.1 Primary Coolant Noble-Gas Activity in Transient State

Primary coolant noble-gas activity in transient state of M310 and ACP1000 are the steady state value multiplied by the crest factor, the crest factor is taken from the operating experience of the French nuclear power station. The main factors affecting the crest factor are the reactor power and operating pressure. M310, ACP1000 and ACP100 have different reactor power, but the fuel assemblies used are of the same type, the mechanism of fission products released from damaged fuel rods to the primary coolant is the same, and the primary coolant system pressure is similar, so the primary coolant noble-gas activity in transient state calculation of ACP100 can use the same crest factor as that of M310 and ACP1000. The crest factor is shown in Table 1.

2.2.2 Primary Coolant Iodine Activity in Transient State

For M310, the primary coolant iodine activity in transient state is the steady state value multiplied by the crest factor. ACP1000 considers two case of preaccident iodine spike and concurrent iodine spike, referring to the assumptions in RG1.183 [5]. For preaccident iodine spike case, primary coolant iodine concentration to the maximum value (typically 60 μCi/gm DE I-131) peimitted by the technical specification. For concurrent iodine spike, the increase in primary coolant iodine concentration is estimated using a spiking model that assumes that the iodine release rate from the fuel rods to the primary coolant increases to a value 335 times greater than the release rate corresponding to the iodine concentration at the equilibrium value (typically 1.0 μCi/gm DE I-131) specified in technical specifications, the assumed iodine spike duration should be 8 h. Through comparative analysis, the iodine peak release phenomenon in the SGTR source term analysis of M310 considers the transient state before the accident, which is similar to the preaccident iodine spike case of ACP1000, but the M310 does not consider the peak release phenomenon similar to concurrent iodine spike case of ACP1000.

According to the stipulations in Analysis criterion of the design basis accident source terms for pressurized water reactor nuclear power plant(NB/T 20444-2017RK) [6], the SGTR source term analysis in new pressurized water reactor nuclear power plants should consider the preaccident iodine spike case and concurrent iodine spike case, SGTR preaccident iodine spike case accident is a limit accident, and SGTR concurrent iodine spike case accident is a rare accident. Therefore, the ACP100 SGTR accident source term analysis intends to consider preaccident iodine spike case and concurrent iodine spike case.

1) Primary coolant activity in preaccident iodine spike
   If ACP100 directly refers to the assumption of RG1.183, it is obviously too conservative to increase primary coolant activity in preaccident iodine spike to 2220 GBq/t DE I-131. Therefore, for the SGTR preaccident iodine spike, the primary coolant activity in preaccident iodine spike calculation method is proposed in this paper of ACP100.

Referring to ACP1000, the iodine peak assumes that the primary coolant activity in preaccident iodine spike increases from 37 GBq/t DE I-131 to 2220 GBq/t DE I-131, that is, the primary coolant activity in preaccident iodine spike increases to 60 times steady state value. Therefore, assuming SGTR preaccident iodine spike of ACP100, the primary coolant activity in preaccident iodine spike also increases to 60 times the steady state value(7.4 GBq/t DE I-131), that is, 444 GBq/t DE I-131. Under this assumption, primary coolant activity in preaccident iodine spike of ACP100 is shown in Table 1.

2) Primary coolant activity in concurrent iodine spike

For SGTR concurrent iodine spike accident analysis of ACP100, referring to the assumption of RG1.183, assuming that the iodine release rate from the fuel rods to the primary coolant increases to a value 335 times greater than the release rate corresponding to the iodine concentration at the equilibrium value (7.4 GBq/t DE I-131) specified in technical specifications, that is, the leakage rate from the fuel element into the primary coolant is 335 times the normal leakage rate, the assumed iodine spike duration should be 8 h.

The leakage rate of iodine from the fuel element into the primary coolant in transient state:

$$L_i = \frac{C \cdot R_i}{I_i \cdot f_i \cdot \eta} \quad (1)$$

L_i is the leakage rate of the nuclide from the fuel element into the primary coolant in transient state, 1/s;
C is the iodine concentration increase times in transient state;
R_i is the equilibrium iodine release rate of nuclide, Bq/s;
I_i is the inventory of the nuclide in core, Bq;
f_i is the nuclide fraction in the fuel pellet-cladding gap;
η is the fuel cladding damage fraction, 0.25%.

Table 1. Primary coolant activity

Nuclides	Primary coolant activity in steady state (GBq/t)	Crest factor	primary coolant activity in preaccident iodine spike (GBq/t)	Leakage rate of iodine from the fuel element into the primary coolant in transient state (1/s)
Kr-85m	1.37E + 01	2.38	3.25E + 01	/
Kr-85	4.57E-01	1.00	4.57E-01	/
Kr-87	2.00E + 01	2.34	4.69E + 01	/
Kr-88	3.27E + 01	2.28	7.46E + 01	/
Xe-133m	5.31E + 00	2.22	1.18E + 01	/
Xe-133	1.56E + 02	1.89	2.96E + 02	/
Xe-135	1.13E + 02	1.35	1.52E + 02	/
Xe-138	3.59E + 01	2.84	1.02E + 02	/
I-131	4.94E + 00	/	2.97E + 02	5.63E − 04
I-132	4.74E + 00	/	2.84E + 02	1.60E − 03
I-133	7.47E + 00	/	4.48E + 02	5.59E − 04
I-134	1.44E + 00	/	8.64E + 01	1.01E − 03
I-135	4.04E + 00	/	2.43E + 02	5.86E − 04
DE I-131	7.4	/	444	/

2.3 Iodine Carrying Coefficient in Steam Generator

The design of the once-through steam generator used in ACP100 is quite different from the U-tube steam generator used in M310 and ACP1000 [7]. The U-tube steam generator is a saturated steam generator. The heat transfer tube of the steam generator is completely immersed in water. The iodine of primary coolant leaking to the secondary coolant through the break of the steam generator tube enters the secondary coolant. The iodine of liquid phase is carried into the gas phase on the secondary coolant with the steam, and then released to the environment through the safety valve of the steam generator.

For the once-through steam generator, the secondary coolant is divided into a preheating section, an evaporation section and a superheating section according to the characteristics of the boiling phase change of the secondary coolant. In the superheat section, the iodine in the liquid flashes quickly into the vapor phase and is released to the environment through the steam generator safety valve. Compared with the U-tube steam generator, the iodine in the once-through steam generator is more easily carried by the vapor from the liquid phase to the vapor phase. Therefore, the carry coefficient is obviously not conservative enough if ACP100 directly refers to the iodine of M310 or ACP1000 on the secondary coolant of the steam generator. Since the liquid on the wall of the heat transfer tube will evaporate to dryness in the superheating section of the once-through steam generator, during this process, the iodine in the secondary coolant liquid phase will flash into the secondary coolant vapor phase. So it can be conservatively assumed that iodine carryover factor on the secondary coolant of steam generators is 1.

3 SGTR Source Term Analysis and Radiological Consequence Evaluation of ACP100

3.1 Source Term Analysis Results

Using the calculation assumptions and parameters in Sect. 2, the cumulative source term to the environment after SGTR accident calculated using the ASTA program are shown in Table 2. ASTA is a program independently developed by nuclear power institute of china for calculating the release of radionuclides to the environment under accident conditions.

3.2 Analysis Results of Radiological Consequences

The individual dose limits in the "Principles for Safety Review of Modular Small Pressurized Water Reactor Nuclear Power Plant Demonstration Projects (Trial version) in Chinese" are respectively determined as: The effective dose that the public individual (adult) may receive in each rare accident should be controlled below 5 mSv, and the thyroid equivalent dose should be controlled below 50 mSv; In each extreme accident, the effective dose that the public (adult) may receive should be controlled below 10 mSv, and the thyroid equivalent dose should be controlled below 100 mSv.

Using the site meteorological data of an ACP100 nuclear power plant, the radioactive consequences outside the factory were calculated for the source term of preaccident iodine spike case and concurrent iodine spike case are shown in Table 3, meet the acceptance criteria.

Table 2. Activities released to the environment after SGTR accident

Nuclides	Activities released to the environment after SGTR (Bq)	
	Preaccident iodine spike case	Concurrent iodine spike case
Kr-85m	1.59E + 11	1.59E + 11
Kr-85	2.40E + 09	2.40E + 09
Kr-87	1.95E + 11	1.95E + 11
Kr-88	3.52E + 11	3.52E + 11
Xe-133m	6.18E + 10	6.16E + 10
Xe-133	1.55E + 12	1.54E + 12
Xe-135	7.88E + 11	7.73E + 11
Xe-138	1.87E + 11	1.87E + 11
I-131	1.23E + 12	1.32E + 11
I-132	1.05E + 12	1.27E + 11
I-133	1.83E + 12	2.38E + 11
I-134	2.66E + 11	9.51E + 10
I-135	9.62E + 11	1.67E + 11

Table 3. Consequence of SGTR accident

Consequence	Effective dose (mSv)	Thyroid equivalent dose (mSv)
Preaccident iodine spike case	5.99	80.3
Concurrent iodine spike case	0.7	9.14

4 Conclusions

In this paper, on the basis of fully learning from the existing nuclear power projects, combined with the actual design characteristics of ACP100, a set of source term analysis methods for SGTR preaccident iodine spike case and concurrent iodine spike case suitable for ACP100 are proposed, and this method is used to analyze the SGTR accident source term of an ACP100 nuclear power plant has been identified, and its radiological consequences also meet the radiological consequences acceptance criteria in "Principles for Safety Review of Small Modular Pressurized Water Reactor Nuclear Power Plant Demonstration Projects (trial version) in Chinese".

References

1. Fan, Y., et al.: Calculation and analysis of steam generation tube rupture accident source term in PWR. Nucl. Tech. **43**(06), 31–36 (2020). (in Chinese)

2. Chen, Y., et al.: Analysis of radiological consequence of SGTR accident at PWRS. Radiat. Prot. Bull. **31**(006), 1–5 (2011). (in Chinese)
3. Chen, Y., et al.: Comparison of SGTR accident source terms for AP1000 and CPR1000. Radiat. Prot. **31**(003), 129–133 (2011). (in Chinese)
4. Tao, J., et al.: SGTR accident source term analysis methodology for HPR1000. Nucl. Sci. Eng. **39**(02), 267–273 (2019). (in Chinese)
5. NRC. Alternative radiological source terms for evaluating design basis accidents at nuclear power reactors[R]. RG1.183 (2000)
6. National Energy Administration. Analysis criterion of the design basis accident source terms for pressurized water reactor nuclear power plant. NB/T 20444–2017RK (2017). (in Chinese)
7. Song, D., Qin, Z.: Progress of LING LONG SMR technology and demonstration project. China Nucl. Power **11**(1), 21–25 (2018). (in Chinese)

Research on the Alarm Threshold for Steam Generator Tube Leak Monitoring of Small Modular Reactor

Ming-ming Xia(✉), Jun-long Wang, Tao Xu, Bo-chen Huang, Jian-ping Zhu, Yi-rui Wu, Jian-xin Miao, and Xin Chen

Science and Technology on Reactor System Design Technology Laboratory, Nuclear Power Institute of China, Chengdu, Sicuan, China

xiamingmingmail@163.com

Abstract. On the basis of fully learning from the existing principles of leak monitoring alarm thresholds setting of the steam generator tube in pressurized water reactor nuclear power plants, the principle of setting for N-16 leak monitoring alarm thresholds of steam generator tube in small modular reactors is proposed, and the N-16 monitoring alarm thresholds are given. The calculation method for total gamma count rate of steam generator tube leak monitoring alarm thresholds of the small modular reactor is established, and the total gamma count rate under the condition of the leak corresponding to the N-16 monitoring alarm thresholds is calculated to guide the leak monitoring total gamma alarm thresholds set.

Keywords: Small modular reactor · Steam generator tube leak monitoring · Alarm thresholds

1 Introduction

The steam generator tube is the weakest part of the pressure boundary of the primary circuit [1]. If the tube is ruptured, the radioactive nuclides will leak from the primary circuit to the secondary circuit and cause pollution of the secondary circuit, and the leak of the secondary circuit itself will lead to the release of radioactivity into the environment, which will seriously affect the safe operation of the pressurized water reactor nuclear power plant and pollute the environment, causing harm to equipment and personal safety. In the design of the leak monitoring system for the steam generator tube of the nuclear power plant in my country, two strategies are usually used to monitor the total gamma count rate and N-16 activity in the secondary circuit [2]. The two strategies are redundant with each other and are suitable for different working conditions: the N-16 activity monitor is the main measurement and supplemented by the total gamma count rate when the reactor power is higher than 20%; the N-16 activity monitor is no longer representative when the nuclear power level is lower than 20% [3]. And the rationality of the monitoring alarm threshold directly affects whether the monitoring system can run stably and effectively, so choosing a reasonable alarm threshold is the key to the design [4].

C. Liu (Ed.): PBNC 2022, SPPHY 283, pp. 1121–1126, 2023.
https://doi.org/10.1007/978-981-99-1023-6_96

At present, VVER, M310 and ACP1000 units are typically set for leak monitoring alarm thresholds of steam generator tube in domestic nuclear power plants. Small modular reactors are different from the steam generator types of VVER, M310 and ACP1000 units, and the reactor power is also quite different, so the alarm threshold settings for steam generator tube leak monitoring of VVER, M310 and ACP1000 units cannot be directly used. Therefore, in this study, on the basis the principle of setting the leak monitoring alarm threshold for the steam generator tube of the VVER, M310 and ACP1000 units, and according to the design characteristics of the small modular reactor, the leak monitoring alarm threshold of the small modular reactor steam generator tube is carried out.

2 Setting of the Steam Generator Tube Leak Monitoring Alarm Thresholds in Domestic Nuclear Power Plants

2.1 VVER Unit

VVER determines that 1 kg/h is the normal design basis leak rate according to a large number of operating data of the units, and the N-16 first-level alarm threshold of steam generator tube leak monitoring is 2 kg/h, which is twice the normal design basis leak rate. The second-level alarm threshold of 5 kg/h is the safe operation limit of the primary side to the secondary side leak rate. The Russian side did not give the specific calculation process in the setting value report.

The total gamma first and second level alarm thresholds of the steam generator tube leak monitoring of VVER units are 1×10^{-6} Gy/h and 2×10^{-6} Gy/h respectively. There is no specific calculation method in the alarm threshold list provided by the Russian side. The speculation should be mainly based on the following aspects: the primary coolant source term, the secondary coolant radioactivity limit and the damage of the tube.

2.2 M310 Unit

For M310, the N-16 first-level alarm threshold of steam generator tube leak monitoring is 5 L/h, and the second-level alarm threshold is 70 L/h. The method does not give the basis for selecting the alarm threshold. It is speculated that the first-level alarm threshold is about 3 times the design basis leak rate under normal operating conditions, and the second-level alarm threshold of 70 L/h is derived from the operating regulations. The M310 steam generator tube leak monitoring total gamma first and second level alarm thresholds are 200 cps and 300 cps respectively. The French side does not give a specific calculation method, and it is speculated that the design idea is similar to that of the VVER.

3 Leak Monitoring Alarm Thresholds for Steam Generator Tube of Small Modular Reactor

3.1 Small Modular Reactor Design Features

The small modular reactor adopts once-through steam generator, a total of 16 units [5], of which 4 units are in a group, and each group is connected together and shares a main steam pipeline. In the small modular reactor, one N-16 monitor and one total gamma monitor are respectively set on each main steam pipeline, and the two devices are mutually redundant. Under the normal operating conditions of the small modular reactor, the total leak rate of the primary coolant to the secondary coolant is 3.6 kg/h, so the total leak rate of the steam generator corresponding to one main steam pipeline is 1.8 kg /h under normal operating conditions.

3.2 Leak Monitoring N-16 Alarm Thresholds for Steam Generators Heat Transfer Tubes

Due to the N-16 alarm threshold for the leak monitoring design for the steam generator tube of the VVER and M310 units, a consensus has been reached between the review, design department and the owner. In the absence of extensive operational data to support, the practice of linking normal leak rates to operational thresholds is easier to interpret and sufficiently conservative. Therefore, this study attempts to determine the N-16 first and second level alarm thresholds from the following two aspects:

(1) Determine the alarm threshold based on the review of similar nuclear power plants, the consensus reached between the design department and the owner on the monitoring alarm threshold.
(2) The threshold setting is linked to the normal leak rate. Therefore, the first-level alarm threshold of the steam generator leak monitoring of the small modular reactor is to be considered to be twice the normal design basis leak rate, which is 3.6 kg/h; the second-level alarm threshold is 9 kg/h that set to be 5 times the normal design basis leak rate.

3.3 Total Gamma Alarm Threshold Setting for Leak Monitoring of Steam Generator Tube

Referring to the VVER and M310 units, the setting of total gamma alarm threshold setting for leak monitoring of steam generator tube should be determined by the primary coolant source term, the secondary coolant radioactivity limit and the size of the damage to the tube. Since there is currently no relevant specification for the secondary coolant radioactivity limit, this study only considers two factors, the primary coolant source term and the size of the damage to the tube. Since the N-16 monitor and the total gamma monitor are redundant with each other, the first and second level alarm thresholds of the two should correspond to each other, indicating the same leak level of the steam generator tube under different working conditions. Therefore, according to the N-16 alarm threshold value of steam generator tube leak monitoring, the total gamma of leak monitoring under 20% power level can be calculated through theoretical analysis, so as to guide the setting of the total gamma alarm threshold value of steam generator tube leak monitoring.

(1) Total gamma count rate calculation

The calculation formula of the count rate is as follows:

$$C = \sum_i k_i \times \eta_i \times A_{vi} \tag{1}$$

$$A_{vi} = \frac{A_{pi} \times q \times \rho_v \times e^{-\lambda_i t}}{Q \times \rho_p} \tag{2}$$

where C is the total gamma count rate, cps;
k_i is the detection efficiency of the nuclide i in the Monte Carlo model, cps/(Bq/m^3);
η_i is the carrying coefficient of the nuclide i;
A_{vi} is the radioactive concentration of the nuclide i in the steam at the probe, (Bq/m^3);
A_{pi} is the radioactive concentration of the nuclide i in the primary coolant, Bq/kg;
q is the leak rate, L/h;
ρ_v is the steam density of the main steam pipeline at 20% power level, kg/m3;
λ_i is the decay constant of the nuclide i, 1/s;
t is the transit times between leak location and probe at the 20% power level, s;
Q is the steam flow rate at the 20% power level, L/h;
ρ_p is the average density of the primary coolant, kg/m^3.

(2) Calculation of detection efficiency factor

The efficiency factor of the detector is related to the structure of the detector, the structure of the pipeline, and the relative position of the detector and the pipeline. Small modular reactor steam generator monitoring uses NaI detector. The steam pipe is filled with secondary side steam, the steam density is 19.07 kg/m^3, and the steam pipe is wrapped with a layer of thermal insulation material. The MCNP calculation result is a normalized result, and the result needs to be processed:

$$k_i = \mathrm{F8}_i \times V_{source} \times I_i \tag{3}$$

where, $\mathrm{F8}_i$ is the output result of pulses number of the MCNP that emits gamma-rays by the nuclide i, cps/γ;

V_{source} is the volume of the steam pipe, m^3;
I_i is the probability that the nuclide i emits gamma-rays.

(3) Analysis of the effect of N-16 on the total gamma count rate

For M310 and VVER units, the transit time from the leak location and probe at low power is longer than that at full power, the N-16 decay share is large, and the contribution to the total gamma count rate is small. However, the transit time from the leak location to the outlet of the once-through steam generator tube at low power is still shorter than the half-life of N-16, and the transit time is shown in Table 1. The total gamma count rate still has a large contribution.

Table 1. N-16 transit times between leak location and outlet

Transit times between leak location and outlet (s)		Half-life of N-16(s)
20% Nuclear power level	100% Nuclear power level	
2.14	0.53	7.13

(4) Retention of radionuclides on the secondary side of steam generator

For radioactive gas nuclides, such as N-16 and nobla-gas, due to their insolubility in water, 100% of the gas nuclides leaking from the primary coolant to the secondary coolant are instantly released into the vapor phase of the secondary side. For iodine and particle-type fission product nuclides, after leaking to the secondary side of the steam generator, they will be retained by the water on the secondary side. The water on the secondary side of the steam generator evaporates in the form of water vapor and droplets. Iodine and particulate fission product nuclides are entrained into the vapor phase. For iodine, due to the once-through steam generator used in small modular reactor, the once-through steam generator has to go through the process from supercooled water to hot steam in the secondary side working medium. According to the characteristics of the boiling phase change of the working medium on the secondary side, the secondary side is divided into a preheating section, an evaporation section and a superheating section. In the superheating section of the once-through steam generator, the iodine in the water will flash quickly into the gas phase, so it is conservatively assumed that the iodine carrying coefficient on the secondary side of the steam generator is 1. The droplet entrainment fraction for particulate nuclides is equal to the water content level in the steam, i.e. 0.25%.

(5) The total gamma count rate corresponding to the leak rate of the steam generator tube

In summary, according to formula (1), under low power, when the leak rate is 3.6 kg/h, the total gamma count rate is about 35.8 cps; the leak rate is 9 kg/h, and the total gamma count rate is 89.6 cps.

(6) Environmental background

The energy response range of the total gamma monitoring channel for leak of the steam generator tube is wide, and it is easily disturbed by the environmental background. The selection of the threshold value needs to consider this aspect. In this study, the environmental background count rate was considered to be 5 cps

(7) The total gamma alarm threshold for leak monitoring of the steam generator tube.

Considering the environmental background, and considering the design margin of −20% from an engineering point of view, it is recommended that the total gamma count rate is 50 cps as the first-level alarm threshold and 110 cps as the second-level alarm threshold.

4 Conclusions

In this study, referring to the setting of the leak monitoring alarm threshold of the VVER and M310 steam generator tube, and combined with the design value of the leak rate of the small modular reactor steam generator tube, the leak monitoring of the small modular reactor steam generator tube is proposed. Based on the principle of threshold setting, the N-16 alarm threshold for leak monitoring of the steam generator tube is given. The first-level alarm threshold is 3.6 kg/h, and the second-level alarm threshold is 9 kg/h. The calculation method of the total gamma count rate in the case of the small modular reactor for steam generator tube leak monitoring is established. According to the leak rate level of the leak monitoring N-16 alarm threshold, the total gamma alarm threshold of leak monitoring is set through theoretical analysis and calculation. The first-level alarm threshold is 50 cps, and the second-level alarm threshold is 110 cps.

References

1. Li, R., Ling, Q.: ^{16}N transmission analysis for steam generator tube leak monitoring in nuclear plant. J. Univ. South China (Sci. Technol.) **24**(03), 6–9 (2010). (in Chinese)
2. Xing, H., et al.: The study of online ^{16}N detecting system for marine nuclear power plant. Nucl. Electron. Detect. Technol. (01), 48–51+66 (2005). (in Chinese)
3. Xu, H., et al.: Research of Alarm Threshold Value of Steam Generator Tube Leak Total Gamma Monitoring Channel of Nuclear Power Plant. Zhongguo Fushe Weisheng **26**(03), 351–356 (2017). (In Chinese)
4. Jia, J., et al.: Alarm threshold value of SG sewerage radiation monitoring channel in nuclear power plant. J. Wuhan Univ. Technol. (Inf. Manag. Eng.) **35**(01), 52–55 (2013). (in Chinese)
5. Song, D., et al.: Researsh and development for ACP100 small modular reactor in china. China Nucl. Power **10**(02), 172–177+187 (2017). (in Chinese)

Thermodynamic Equilibrium Analysis of Steam Reforming Reaction of Radioactive Waste Oil

Xuan Wu(✉), Wenyu Li, Li Lin, Yi Liang, Jiaheng Zhang, Wenlu Gu, Jiheng Fan, EnWei Shen, and KouHong Xiong

Nuclear Power Institute of China Chengdu, Sichuan, China
wx139x@139.com

Abstract. At present, there are much radioactive waste oil temporarily stored in nuclear fuel processing plants, nuclear industry research institutes and operating nuclear power plants in China, which brings great storage pressure and safety risks to the operating nuclear facilities. In this paper, the components analysis of 40# waste oil used in nuclear facilities was carried out, and the elemental composition and chemical composition of the waste oil were obtained. The analysis showed that the main elements in the waste oil were C and H, and the main chemical components were alkanes, alkenes, aromatic hydrocarbons and alcohols with carbon chain length of 10–40. Using Aspen Plus software, the process flow model of waste oil's steam reforming treatment was established. Based on the components analysis results of the waste oil, organic mixtures such as ethanol, ethane and propane were selected as the model components, and the element composition close to waste oil was obtained by adjusting the proportion of each component. The mixture was used as the source input of Aspen Plus to achieve good simulation results. The experimental results obtained under Pt catalyst at 400 °C were in good agreement with the simulation results, which confirmed the validity of the model. The thermodynamic equilibrium analysis of waste oil steam reforming reaction was carried out by using the verified model. The influence of reaction temperature (350–1150 °C), pressure (0.01–100bar) and water to carbon ratio (0.01–100) on reforming reaction and off gas components in balanced state was studied. The conclusions are as follows: (1) The steam reforming reaction of waste oil has no obvious inhibition when the reaction pressure is less than 1bar, so the reforming reaction should be carried out under the condition of negative pressure less than 1bar; (2) The temperature should be maintained above 750 °C to ensure the complete steam reforming reaction; (3) Carbon deposition can be completely eliminated when the water/carbon ratio is higher than 1, and when the water/carbon ratio is higher than 10, the product components do not change with the water/carbon ratio.

Keyword: WASTE OIL STEAM REFORMING ASPEN

C. Liu (Ed.): PBNC 2022, SPPHY 283, pp. 1127–1133, 2023.
https://doi.org/10.1007/978-981-99-1023-6_97

1 Introduction

In the process of operation, maintenance and decommissioning of nuclear facilities, radioactive nuclides such as ^{137}Cs, ^{90}Sr and ^{60}Co will be mixed into industrial oil to form radioactive waste oil [1]. The steam reforming treatment process developed by Studsvik is suitable for a variety of radioactive organic wastes, including waste oil, and has many advantages including high volume reduction ratio, stable solid products, and less environmental pollution [2]. Previously, Lin Li et al. from Nuclear Power Institute of China carried out a simulation on the steam reforming of waste resin and analyzed the balance products. They obtained the reaction parameters and operating gas velocity suitable for the treatment of radioactive waste resin [3–5]. However, there are few researches on steam reforming treatment of radioactive waste oil.

In the steam reforming process of waste oil, more hydrogen production is preferred, which is conducive to the subsequent off gas oxidation. Pressure, temperature and water-carbon ratio are the key parameters affecting the balanced off gas composition. A lot of experiments are needed to determine the appropriate reaction pressure, temperature and water-to-carbon ratio. But we can save a lot of effort by simulating the reaction process on the computer. Some researchers have carried out thermodynamic analysis on steam reforming of methanol [6, 7], ethanol [8–10], glycerol [11, 12] and other organic compounds, and obtained reforming reaction temperature, pressure and other operating parameters.

In this paper, Aspen Plus, a chemical process simulation software, was used to build a process model for oil steam reforming. In order to select the most suitable operating parameters for the oil steam reforming test, and to provide reference for the future engineering application of steam reforming process in the treatment of other radioactive wastes, the thermodynamic analysis of the oil steam reforming reaction was carried out and the effects of reaction temperature, pressure and water-carbon ratio on reforming reaction and off gas components was studied.

2 Oil Characterization

The feed composition is needed for thermodynamic analysis using Aspen Plus. 40# lubricating oil is widely used in various domestic nuclear facilities, thus we carried out the analysis of 40# oil from the aspects of elemental composition and chemical composition to provide input for Aspen Plus.

2.1 Elemental Composition

X-ray Fluorescence Spectrometer, Inductively Coupled Plasma Emission Spectrometer and electronic balance were used elemental composition analysis of 40# oil, and the mass fractions of carbon, hydrogen, oxygen, nitrogen, phosphorus and sulfur in the oil were measured. The results were shown in Table 1.

According to the analysis results, the elemental composition of the oil is approximately $C_{35}H_{66.8}$, omitting the very low content of N, O, P and S.

Table 1. Elemental composition of 40# oil.

Elements	Content/wt%	Measurement Method
C	85.06	NB/SH/T 0656–2017
H	13.53	NB/SH/T 0656–2017
N	0.033	SH/T 0657–2007
O	1.179	Substraction
P	0.021	GB/T 17476–1998
S	0.198	GB/T 11140–2008

2.2 Chemical Composition

In this study, Agilent 7890–5977 GC/MSD was used to analyze the chemical composition of the oil. The chromatographic peak signal was compared with the NIST17 spectral database by computer to obtain the chemical composition. The results indicate that the oil is a mixture of various organic compounds, and its main chemical components are alkanes, alkenes, aromatic hydrocarbons and alcohols containing 10–40 carbon atoms. The main elements in waste oil are C and H, indicating that alkanes, alkenes, aromatic hydrocarbons and other hydrocarbons account for a relatively high proportion.

According to the above analysis results, the mixture of ethanol, ethane, propane, ethylene, benzene and toluene was selected as the feedstock, and the ratio (mole ratio) of each component of the mixture was 1:20:25:10:10:10. The mixture has an elemental and chemical composition similar to waste oil, and is suitable for the analysis of waste oil steam reforming reaction.

3 Simulation Modeling

The Aspen Plus model for waste oil steam reforming, as shown in Fig. 1, consists of three main operating units, PYRCT, CARBFLIT, ONSHT and RFRCT, along with some mixers and separators.

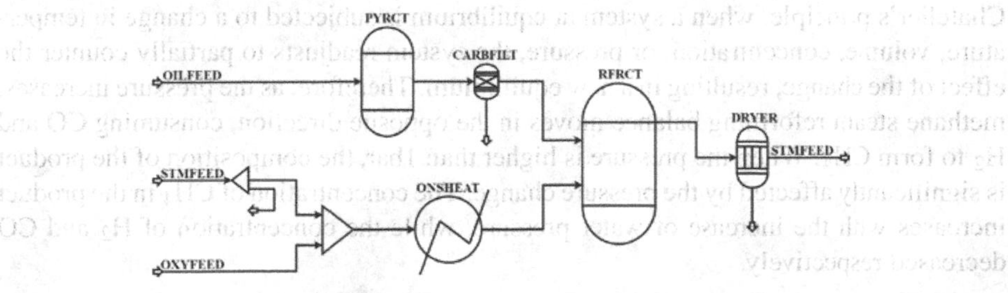

Fig. 1. Simulation scheme of oil steam reforming.

ONSHT is used to simulate steam generator and superheater. It preheats water vapor to reaction temperature and feeds the vapor into the reforming reactor RFRCT. RFRCT adopts the RGibbs reactor as the reactor model, which can calculate the product composition when the system reaches chemical equilibrium and phase equilibrium. PYRCT reactor is used to simulate the pyrolysis process of waste oil. The waste oil reaches pyrolysis equilibrium in the reactor to generate small-molecular-weight organic matter, hydrogen and carbon.

4 Results and Discussion

4.1 Effect of Reaction Pressure

Figure 2 presents the variations of product concentrations versus reaction pressure. The reaction was carried out in 750 °C and steam/carbon ratio of 1.

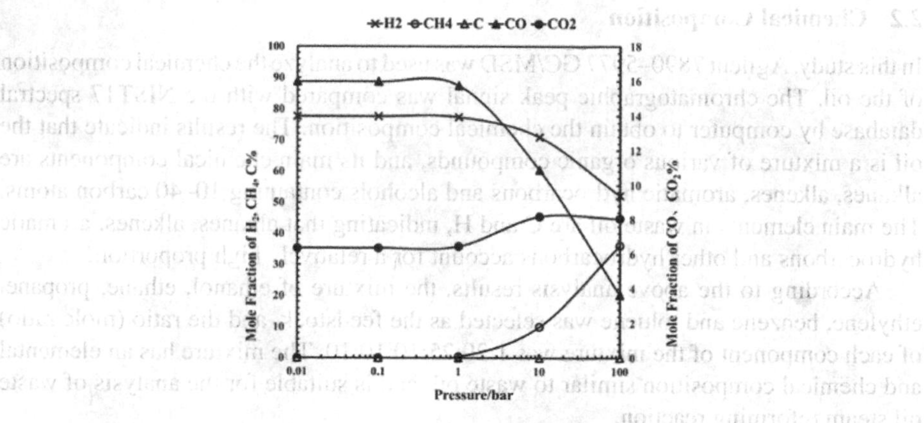

Fig. 2. Effect of pressure at T = 750 °C, S/C = 1.

At T = 750 °C and S/C = 1, methane steam reforming reaction and water gas conversion reaction are the main factors that affect the product composition. The concentration of H_2 decreases with the increase of pressure. The composition of the product remains constant when the pressure is lower than 1 bar. However, methane steam reforming reaction is a reaction in which the number of gas molecules increases, and according to Le Chatelier's principle, when a system at equilibrium is subjected to a change in temperature, volume, concentration, or pressure, the system readjusts to partially counter the effect of the change, resulting in a new equilibrium. Therefore, as the pressure increases, methane steam reforming balance moves in the opposite direction, consuming CO and H_2 to form CH_4. When the pressure is higher than 1bar, the composition of the product is significantly affected by the pressure change. The concentration of CH_4 in the product increases with the increase of water pressure, while the concentration of H_2 and CO decreased respectively.

4.2 Effect of Reaction Temperature

The influence of temperature can be seen from Fig. 3. Simulation is carried out at pressure of 1 bar and steam/carbon ratio of 1. Under these conditions, the mole fraction of hydrogen has significantly increased with increasing temperature up to 750 °C. The increasing temperature shifts the reaction equilibrium toward the products side and thus producing more hydrogen.

The concentration of CO_2 in reforming products tends to increase first and then decrease with the increase of temperature. When the temperature rises to 550 °C, the concentration of CO_2 reaches the peak. This is because the production of CO_2 is mainly affected by the water-gas shift reaction, which is endothermic. Therefore, the rise of temperature has an inhibitory effect on the production of CO_2. At low temperature, the reaction is weakly inhibited, and the mole fraction of CO also increases because the rise of temperature promotes the methane steam reforming reaction, which then shifts the equilibrium of the water-gas shift reaction to the direction of CO_2 production. While above 550 °C, the inhibition of high temperature on the water-gas conversion reaction is more obvious, so the content of CO_2 gradually decreases with the increase of temperature.

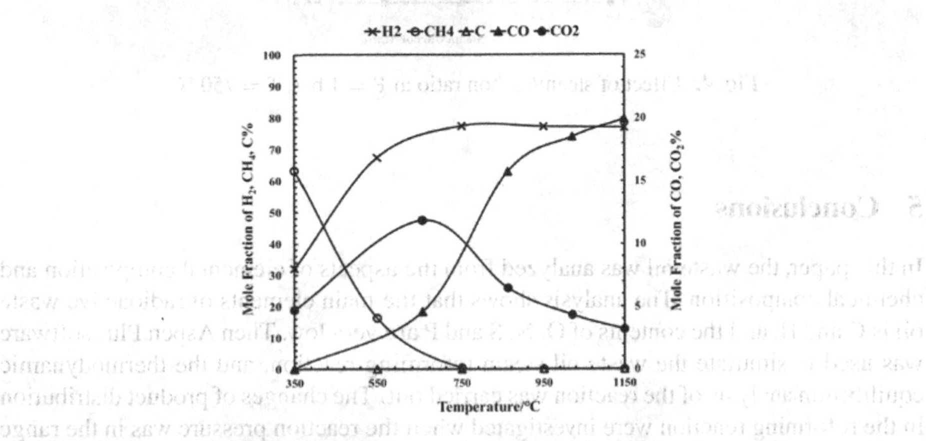

Fig. 3. Effect of temperature at P = 1 bar, S/C = 1.

4.3 Effect of Steam/carbon Ratio

The influence of steam/carbon ratio can be seen from Fig. 4. The operation temperature and pressure are fixed as 750 °C, and 1 bar, respectively. The effect of steam/carbon ratio was investigated in the range of 0.01–100.

There is a large amount of carbon and no methane in the product at low steam/carbon ratio, which is the result of methane decomposition reactions. The water-gas shift reaction mainly occurs when the water-carbon ratio is between 0.01 and 1. Therefore, with the increase of the water-carbon ratio, the carbon decreases continuously until the ratio reaches 1, during which the H_2 content also keeps rising. When the water-carbon ratio is higher than 0.1, the water-gas conversion reaction is encouraged, and the carbon dioxide content keeps rising, while the CO content reaches the peak when the water-carbon ratio is 1. Subsequently, as carbon is completely consumed, the increasing water vapor content will consume CO and generate CO_2, and concentration remains constant when the water-carbon ratio is above 10.

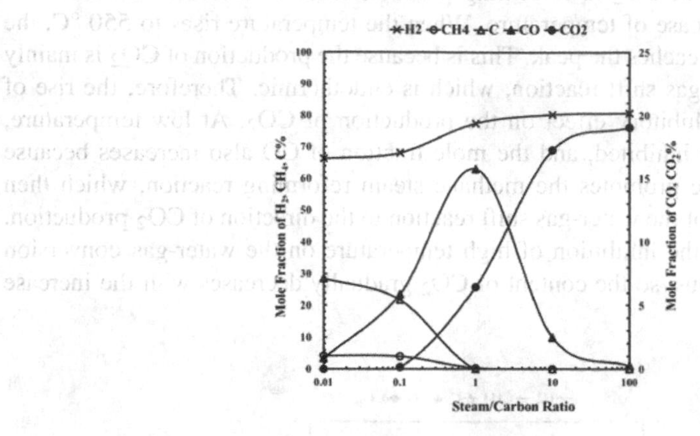

Fig. 4. Effect of steam/carbon ratio at P = 1 bar, T = 750 °C.

5 Conclusions

In this paper, the waste oil was analyzed from the aspects of elemental composition and chemical composition. The analysis shows that the main elements of radioactive waste oil is C and H, and the contents of O, N, S and P are very low. Then Aspen Plus software was used to simulate the waste oil steam reforming reaction, and the thermodynamic equilibrium analysis of the reaction was carried out. The changes of product distribution in the reforming reaction were investigated when the reaction pressure was in the range of 0.01–100 bar, the temperature in 350–1150 °C, and the water carbon ratio in 0.01–100. The following conclusions are drawn:

(1) When the reaction pressure is less than 1bar, there is no obvious inhibition on the steam reforming reaction of waste oil, so the reforming reaction should be carried out under a negative pressure less than 1bar;
(2) The temperature should be maintained above 750 °C to ensure the complete steam reforming reaction;
(3) Carbon can be completely eliminated when the water/carbon ratio is higher than 1, and when the water/carbon ratio is higher than 10, the product components do not change with the water/carbon ratio.

References

1. Zhang, L., Xiong, Y.: Electrochemical advanced oxidation treatment of simulated radioactive waste oil . Southwest University of Science and Technology (2019)

2. Neeway, J.J., Jantzen, C.M., Brown, C.F., et al.: Radionuclide and contaminant immobilization in the fluidized bed steam reforming waste product. INTECH Open Access Publisher (2012)
3. Lin, L., Zhang, H., Li, W., et al.: Numerical simulation analysis on radioactive spent resin steam reforming fluidization based on VOF model. Sci. Technol. Eng. **20**(30), 12657–12663 (2020)
4. Lin, L., Zhang, H., Li, W., et al.: Equilibrium product analysis of steam reforming waste resin based on gibbs free energy minimum principle. Sichuan Environ. **39**(05), 170–174 (2020)
5. Lin, L., Chen, X., Li, W., et al.: Coupled numerical simulation of flow field reaction in a vertical tube for steam reforming of radioactive waste. Sci. Technol. Eng. **16**(04), 200–204 (2016)
6. Amphlett, J., Evans, M., Jones, R., et al.: Hydrogen production by the catalytic steam reforming of methanol part 1: the thermodynamics. The Canadian J. Chem. Eng. **59**(6), 720–727 (1981)
7. Faungnawakij, K., Kikuchi, R., Eguchi, K.: Thermodynamic evaluation of methanol steam reforming for hydrogen production. J. Power Sources **161**(1), 87–94 (2006)
8. Garcia, E., Laborde, M.A.: Hydrogen production by the steam reforming of ethanol: thermodynamic analysis. Int. J. Hydrogen Energy **16**(5), 307–312 (1991)
9. Rossi, C., Alonso, C., Antunes, O., et al.: Thermodynamic analysis of steam reforming of ethanol and glycerine for hydrogen production. Int. Jo.f hydrogen energy, **34**(1), 323–332 (2009)
10. Fishtik, I., Alexander, A., Datta, R., et al.: A thermodynamic analysis of hydrogen production by steam reforming of ethanol via response reactions. Int. J. Hydrogen Energy **25**(1), 31–45 (2000)
11. Chen, H., Zhang, T., Dou, B., et al.: Thermodynamic analyses of adsorption-enhanced steam reforming of glycerol for hydrogen production. Int. J. Hydrogen Energy **34**(17), 7208–7222 (2009)
12. Adhikari, S., Fernando, S., Gwaltney, S.R., et al.: A thermodynamic analysis of hydrogen production by steam reforming of glycerol. Int. J. Hydrogen Energy **32**(14), 2875–2880 (2007)

Simulation Study of the Neutron Scattering Camera Based on Plastic Scintillator and MPPC

Ji Li[1(✉)] and Qing Shan[2]

[1] Nuclear Power Institute of China, Chengdu, Sichuan, People's Republic of China
lijі2615@163.com
[2] Department of Nuclear Science and Engineering, College of Material Science and Engineering, Nanjing University of Aeronautics and Astronautics, Nanjing, Jiangsu, People's Republic of China

Abstract. Nuclear safety has always been the lifeline of the development of the nuclear industry, and the supervision and management of special nuclear materials is a very important part of nuclear safety. By measuring fission neutron generated by the special nuclear materials, the type and position of special nuclear materials can be determined. In this paper, a neutron scattering camera (NSC) based on plastic scintillator and MPPC was designed to detect the fission neutron in the n-γ mixed field, and then to realize the localization and discrimination of fission materials. The designed NSC contains two layers of detector arrays. The first and second layers are consisted of five and nine detection units, respectively. In order to discriminate neutrons and gamma-rays, EJ-276 plastic scintillator is chosen as the detection medium because of its PSD performance. At the same time, MPPC was used to collect the fluorescence generated in the scintillator. For optimizing the NSC, the Geant4 Monte Carlo simulation toolkit is used to study the whole detection process of the NSC. In the simulation, the factors affecting image reconstruction in neutron source image reconstruction have studied by simulation. The influences of the thickness and radius of the detection units in two layers, the distance between two layers on the image reconstruction were studied in detail. According to the simulation results, the thickness of front detector unit, radius of detector unit, thickness of back detector unit and distance between two layers were determined to be 3 cm, 5 cm, 8 cm and 50 cm, respectively.

Keywords: Neutron Scattering Camera · Plastic Scintillator · Image Reconstruction · MPPC · TOF

1 Introduction

Special nuclear measurement is a very important part of nuclear material supervision and management. In order to safely use and transport these special nuclear materials, it is usually necessary to add corresponding shielding structures on the outside. The energies of characteristic gamma-rays emitted from the radionuclides in special nuclear materials are rather low. For example, the energy of characteristic gamma-rays of ^{235}U is 185.7 keV, which has limited penetration in high Z materials. So, it is difficult to

C. Liu (Ed.): PBNC 2022, SPPHY 283, pp. 1134–1146, 2023.
https://doi.org/10.1007/978-981-99-1023-6_98

measure special nuclear materials by detecting gamma-rays directly. Since uranium and transuranic elements can release fission neutrons through spontaneous fission or induced fission, it is possible to measure special nuclear materials in the shielding structure by detecting fission neutrons [1].

Neutron scattering camera (NSC) can detect the fission neutron, combined with the image reconstruction algorithm, the localization and discrimination of special nuclear materials can be realized [2–5]. It can provide an important means for the supervision of special nuclear materials and plays a very important role in nuclear security, non-proliferation, customs inspection and counter terrorism. At present, it is mainly used liquid scintillator coupled photomultiplier tube (PMT) as detection unit to detect neutrons in NSC. Liquid scintillator has certain toxicity and is not easy to package. Besides, the use of PMT also makes the NSC has larger volume, which is not conducive to the portable improvement of NSC system.

EJ-276 plastic scintillator not only has the same n-γ discrimination ability as liquid scintillator, but also has the advantages of non-toxic and not easy to leak. Compared to the liquid scintillator, EJ-276 has many advantages, such as more convenient to use, safer operation and more stable physical properties. At the same times, multi-pixel Photon Counter (MPPC) has strong anti-magnetic field interference ability and smaller volume. In this paper, the NSC based on EJ-276 plastic scintillator and MPPC is studied, and the Geant4 [6] Monte Carlo method is used to optimize the NSC.

2 Principle and Performance Index of Neutron Scattering Camera

2.1 Principle of Neutron Scattering Camera

The schematic diagram of NSC was shown in Fig. 1. NSC is consisted of two layers of organic scintillator detector arrays. The incident neutron was firstly interacted with the hydrogen nucleus through the elastic scattering in the front detector array layer, and the recoil proton and the scattered neutron were generated. The scattered neutron entered the back detector array layer and may occur another elastic scattering. If there is only one scattering event occurs in the front and the scattered neutron is detected by the back detector layer, then this is an effective detection event for the neutron scattering camera.

Since the recoil proton has a large mass, its energy can be completely deposited in the front detector, so the recoil proton energy *Ep* can be measured by the response of the front detector layer. The scattered neutron energy *E'n* can be calculated by the time of flight method. Then, the incident neutron energy can be acquired by adding *Ep* and *E'n*. According to the conservation of kinetic energy and momentum in the elastic scattering process, the neutron scattering angle θ and the incident neutron energy *En* can be calculated as follows:

$$\tan^2 \theta = \frac{E_p}{E_n'} \quad (1)$$

$$E_n = E_p + E_n' \quad (2)$$

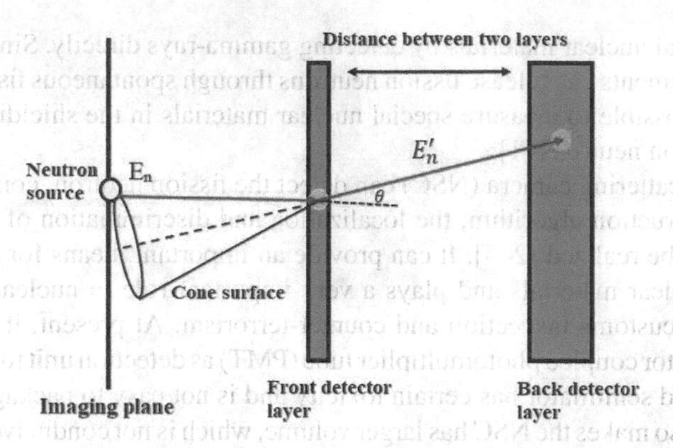

Fig. 1. The schematic diagram of NSC

According to the interaction position of elastic scattering of neutrons in the two detector array layers and neutron scattering angle θ, a cone can be reconstructed, and any position on the conical surface may be the position of the neutron source. But, one cone surface cannot accurately determine the position of the neutron source, the position of the neutron source can be determined by intersection area of multiple cone surfaces.

Simplified Back projection is a common image reconstruction algorithm for neutron scattering camera. The plane where the neutron source is located is called imaging plane X_b. The projection of the reconstructed cone from each neutron scattering event on X_b is an ellipse. X_b is divided into $M \times N$ pixel grids of the same size. When any one, two or three of the four vertices of the grid are fallen into the ellipse, it can be determined that the grid is passed through by the ellipse. The entire imaging plane is traversed to judge the grid passed through by the ellipse, and the pixel value of the grid passed through by the ellipse is added by one until the determination of all conforming events. The region with the largest pixel value is the location of the neutron source.

2.2 Performance Index of Neutron Scattering Camera

The performance indexes of NSC include position resolution of neutron source, primary scattering ratio, coincidence ratio of primary scattering and detection efficiency. The details of them are as follows.

(1) The grid with the largest pixel value in the reconstructed image is the reconstructed position of the point source. By intercepting the pixel value distribution of the point source position along the Y or Z direction, the one dimensional distribution of the neutron source direction can be obtained. Using Gaussian fitting to fit the one dimensional distribution, and the full width at half maximum (FWHM) of the fitting curve is the position resolution of the neutron source.

(2) The primary scattering ratio refers to the ratio of the number of neutrons that have only one elastic scattering in the front detector to the number of neutrons that have elastic scattering in the front detector, and is related to the size of the front detector. The coincidence ratio of primary scattering refers to the ratio of number of effective

detection events to the number of coincidence neutrons detected by the front and back detector layers. It is a performance indicator of the NSC system and is related to the geometric parameters of the NSC.

(3) The detection efficiency is defined as the ratio of detected neutrons to the number of neutrons emitted from neutron source. In Geant4 simulation, in order to improve the simulation efficiency, the cone angle of neutron emission is set as 2θ, and the absolute detection efficiency is given as following

$$\varepsilon = \frac{N_c}{N_{2\theta} \times \frac{4\pi}{\Omega}} \tag{3}$$

where *EC* is the number of effective detection events, *E2θ* is the total number of neutrons emitted from neutron source when the emission cone angle is 2θ, Ω represents the solid angle of cone angle and is calculated as following:

$$\Omega = 2\pi(1 - \cos\theta) \tag{3}$$

3 Simulation

The structural schematic diagram of designed NSC was shown in Fig. 2. There are five and nine detector units in front and back detector layer, respectively. In the front detector array, the distance between adjacent detector units is 20 cm, and the fifth detector unit is located in the center position. In the back detector array, the distance between adjacent detector units is also 20 cm.

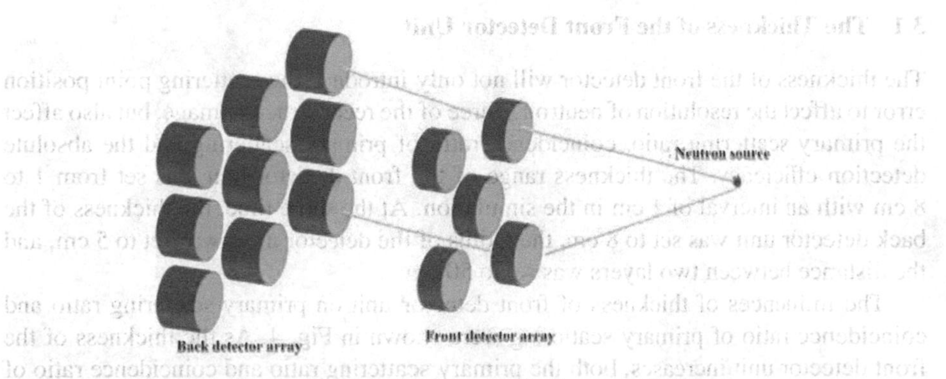

Fig. 2. The structural schematic diagram of NSC

The uranium and plutonium are special nuclear materials, and both of them can generate fission reaction with the neutrons. ^{252}Cf neutron source is spontaneous fission neutron source, and its fission neutrons energy range is similar to that of uranium and plutonium. So, the neutron sources used in Geant4 simulation is chosen to ^{252}Cf neutron sources. In Geant4 simulation, the ^{252}Cf point source was set at the center of the imaging plane which is 2m away from the front detector array, and the cone angle of emission neutrons was 90°. The neutron emission spectrum of the ^{252}Cf neutron source was defined according to the probability density function of the emission neutron energy, as shown in Fig. 3.

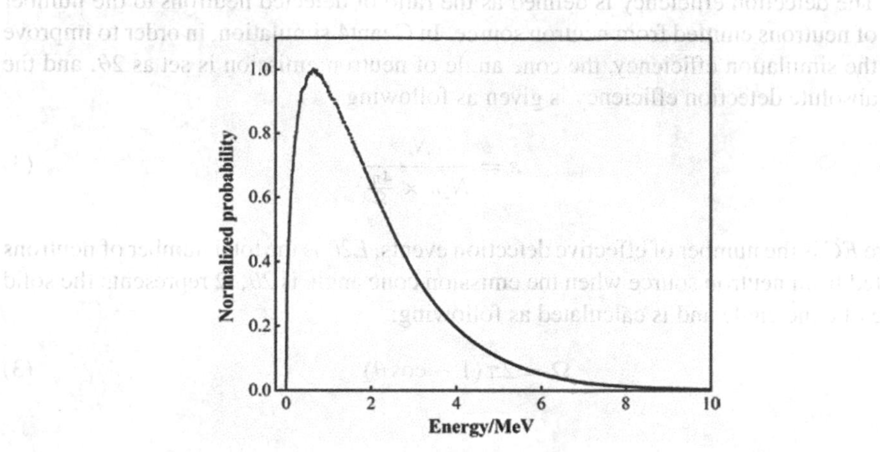

Fig. 3. The spectrum of the ^{252}Cf neutron source [6]

Using Geant4 simulation toolkit, the influences of the thickness and radius of the front and back detector unit and the distance between front and back detector on the image reconstruction were studied. Based on simulation results, the structural parameters of the neutron scattering camera were optimized.

3.1 The Thickness of the Front Detector Unit

The thickness of the front detector will not only introduce the scattering point position error to affect the resolution of neutron source of the reconstructed image, but also affect the primary scattering ratio, coincidence ratio of primary scattering and the absolute detection efficiency. The thickness range of the front detector unit was set from 1 to 8 cm with an interval of 1 cm in the simulation. At the same time, the thickness of the back detector unit was set to 8 cm, the radius of the detector units was set to 5 cm, and the distance between two layers was set to 50 cm.

The influences of thickness of front detector unit on primary scattering ratio and coincidence ratio of primary scattering were shown in Fig. 4. As the thickness of the front detector unit increases, both the primary scattering ratio and coincidence ratio of primary scattering were decreased. This may be caused by the increasing of probability of multiple scattering of neutrons in the front detector layer when the thickness of the front detector unit increases. When the thickness increases from 1 to 8 cm, the primary scattering ratio decreases from 96.5% to 85.6%, and the coincidence ratio of primary scattering decreases from 98.3% to 92.9%, indicating that the thickness of front detector unit has a great influence on the primary scattering ratio and the coincidence ratio of primary scattering.

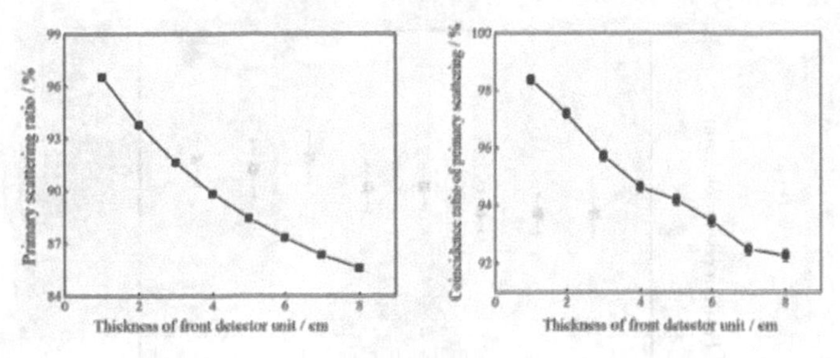

Fig. 4. The thickness of front detector unit Vs. Primary scattering ratio and coincidence ratio of primary scattering

The influences of thickness of front detector on absolute detection efficiency were shown in Fig. 5. The absolute detection efficiency increases with the increase of the thickness. When the thickness of front detector is about 7cm, the absolute detection efficiency basically reaches saturation at about 2.6×10^{-6}.

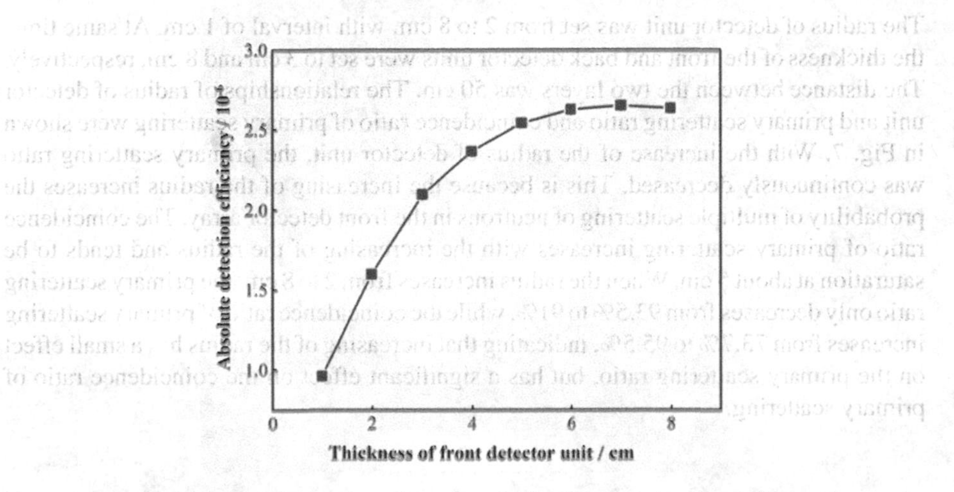

Fig. 5. The thickness of front detector unit Vs. Absolute detection efficiency

The influences of thickness of front detector unit on position resolution of neutron source were also studied and shown in Fig. 6. The position resolution was slightly deteriorated with the increase of thickness. When the thickness increases from 1 to 8 cm, the position resolutions were remained between 500 and 600 mm. Although with the increase of the thickness of front detector unit, the error of scattering point position is introduced to impact the position resolution, but this effect is very limited because the thickness of the front detector unit is quite small when compared to 50 cm distance between two layers.

Considering the simulation results comprehensively, the thickness of the front detector unit is selected as 3 cm. At this circumstance, the primary scattering ratio is bigger than 90%, and the absolute detection efficiency is about 2.2×10^{-6}.

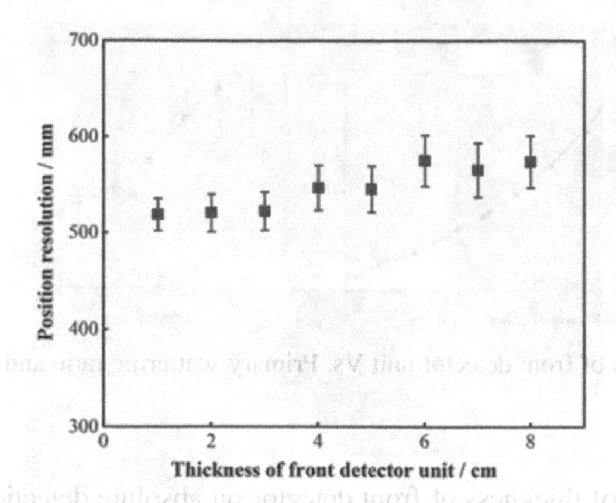

Fig. 6. Front detector thickness unit Vs. Position resolution

3.2 The Radius of the Detector Unit

The radius of detector unit was set from 2 to 8 cm, with interval of 1 cm. At same time, the thickness of the front and back detector units were set to 3 cm and 8 cm, respectively. The distance between the two layers was 50 cm. The relationships of radius of detector unit and primary scattering ratio and coincidence ratio of primary scattering were shown in Fig. 7. With the increase of the radius of detector unit, the primary scattering ratio was continuously decreased. This is because the increasing of the radius increases the probability of multiple scattering of neutrons in the front detector array. The coincidence ratio of primary scattering increases with the increasing of the radius and tends to be saturation at about 5 cm. When the radius increases from 2 to 8 cm, the primary scattering ratio only decreases from 93.5% to 91%, while the coincidence ratio of primary scattering increases from 73.7% to 95.5%, indicating that increasing of the radius has a small effect on the primary scattering ratio, but has a significant effect on the coincidence ratio of primary scattering.

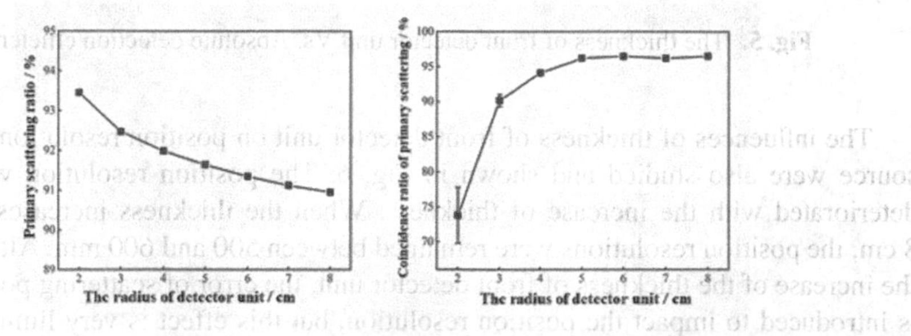

Fig. 7. The radius of detector unit Vs. Primary scattering ratio and coincidence ratio of primary scattering

The absolute detection efficiency under different radius was shown in Fig. 8. It can be seen that the absolute detection efficiency increases with the increasing of radius. When the detector radius increases from 2 cm to 8 cm, the absolute detection efficiency was increased from 1.78×10^{-8} to 1.07×10^{-5}, which were increased about 600 times. This indicates that the radius has a significant influence on the absolute detection efficiency.

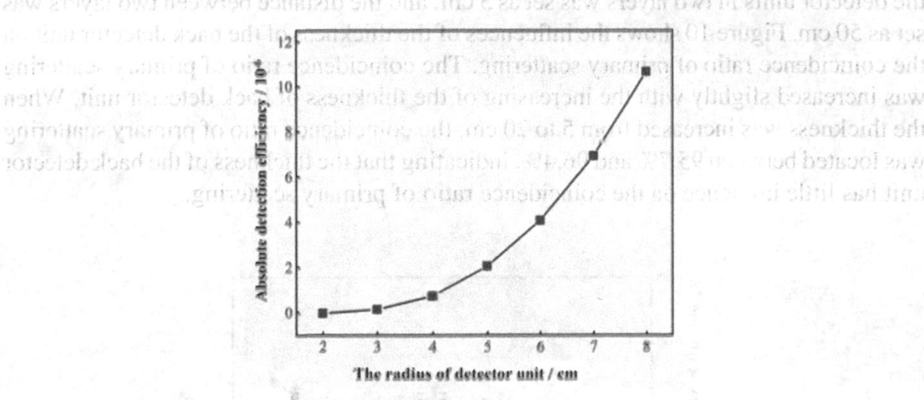

Fig. 8. The radius of detector unit Vs. Absolute detection efficiency

The influence of radius on position resolution of neutron source was shown in Fig. 9. With the increasing of the radius, the position resolutions of neutron source were got worse and worse. When the radius increases from 3 cm to 8 cm, the position resolution of neutron source was increased from 305 mm to 766 mm, indicating that the radius has a significant effect on the position resolution.

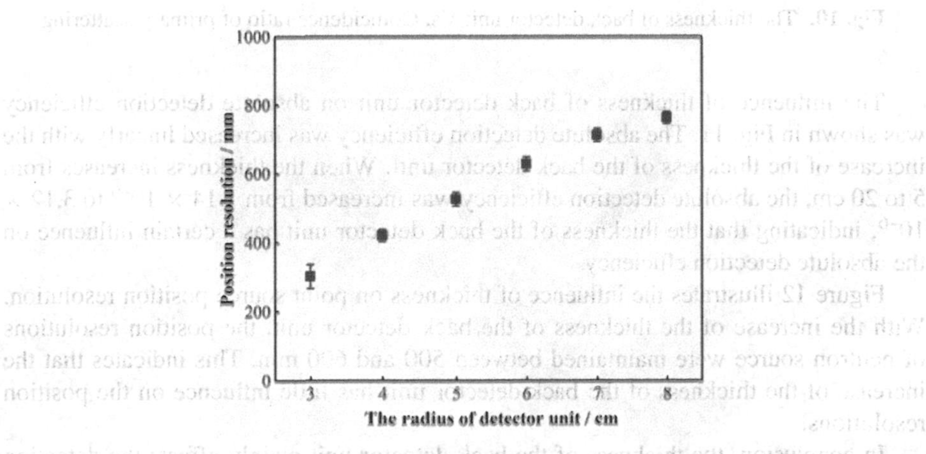

Fig. 9. The radius of detector unit Vs. Position resolution

Considering the simulation results comprehensively, when the radius of the detector unit was 5 cm, the better absolute detection efficiency and coincidence ratio of primary scattering can be obtained. Besides, the NSC also has better position resolution of neutron source. So, the radius of the detector unit was determined to be 5 cm.

3.3 The Thickness of the Back Detector Unit

In the simulation, the thickness of the back detector unit was from 5 to 20 cm with the interval of 1 cm, while the thickness of front detector unit was set as 3 cm, the radius of the detector units in two layers was set as 5 cm, and the distance between two layers was set as 50 cm. Figure 10 shows the influences of the thickness of the back detector unit on the coincidence ratio of primary scattering. The coincidence ratio of primary scattering was increased slightly with the increasing of the thickness of back detector unit. When the thickness was increased from 5 to 20 cm, the coincidence ratio of primary scattering was located between 95.7% and 96.4%, indicating that the thickness of the back detector unit has little influence on the coincidence ratio of primary scattering.

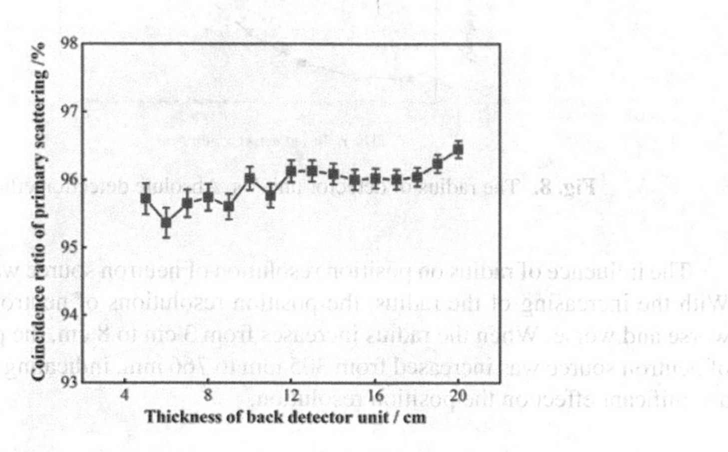

Fig. 10. The thickness of back detector unit Vs. Coincidence ratio of primary scattering

The influence of thickness of back detector unit on absolute detection efficiency was shown in Fig. 11. The absolute detection efficiency was increased linearly with the increase of the thickness of the back detector unit. When the thickness increases from 5 to 20 cm, the absolute detection efficiency was increased from 1.14×10^{-6} to 3.12×10^{-6}, indicating that the thickness of the back detector unit has a certain influence on the absolute detection efficiency.

Figure 12 illustrates the influence of thickness on point source position resolution. With the increase of the thickness of the back detector unit, the position resolutions of neutron source were maintained between 500 and 600 mm. This indicates that the increase of the thickness of the back detector unit has little influence on the position resolutions.

In conclusion, the thickness of the back detector unit mainly affects the detection efficiency. As the thickness of the back detector unit increased, the detection efficiency increases linearly. Considering the absolute detection efficiency and economic cost, the thickness of the detector is determined to be 8 cm.

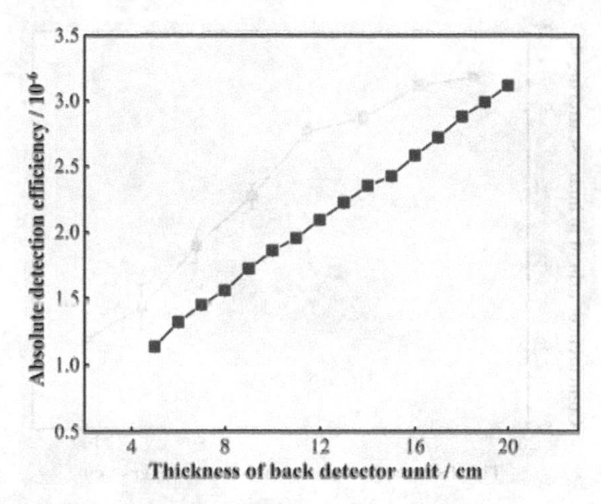

Fig. 11. The thickness of back detector unit Vs. Absolute detection efficiency

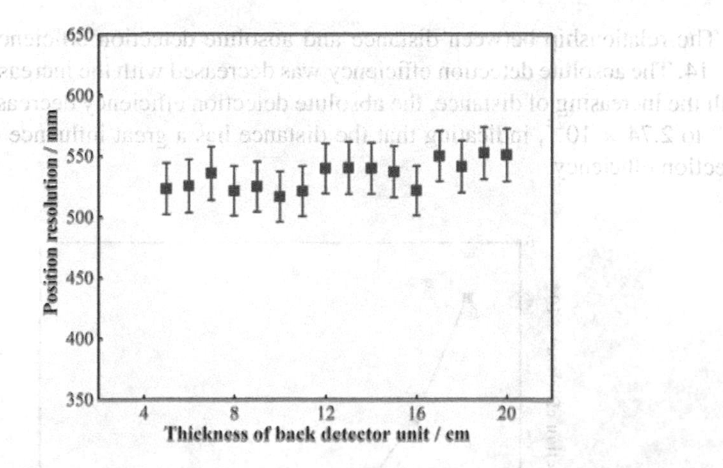

Fig. 12. The thickness of back detector unit Vs. Position resolution

3.4 The Distance Between the Two Layers

The distance between two detector layers was ranged from 30 to 100 cm with interval of 10 cm, while the thickness of the front and back detector unit were 3 and 8 cm, and the radius of the detectors unit was 5 cm. The influence of distances on the coincidence ratio of primary scattering was simulated and shown in Fig. 13. With the increasing of distance, the coincidence ratio of primary scattering was always decreased. When the distance increased from 30 to 100 cm, the coincidence ratio of primary scattering was decreased from 96.8% to 90.2%, indicating that the distance has a great influence on the coincidence ratio of primary scattering.

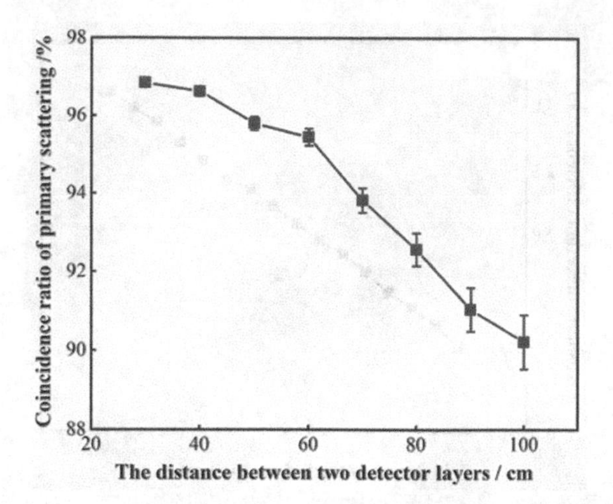

Fig. 13. The distance between two layers Vs. Coincidence ratio of primary scattering

The relationship between distance and absolute detection efficiency was given in Fig. 14. The absolute detection efficiency was decreased with the increasing of distance. With the increasing of distance, the absolute detection efficiency decreases from 5.13×10^{-6} to 2.74×10^{-7}, indicating that the distance has a great influence on the absolute detection efficiency.

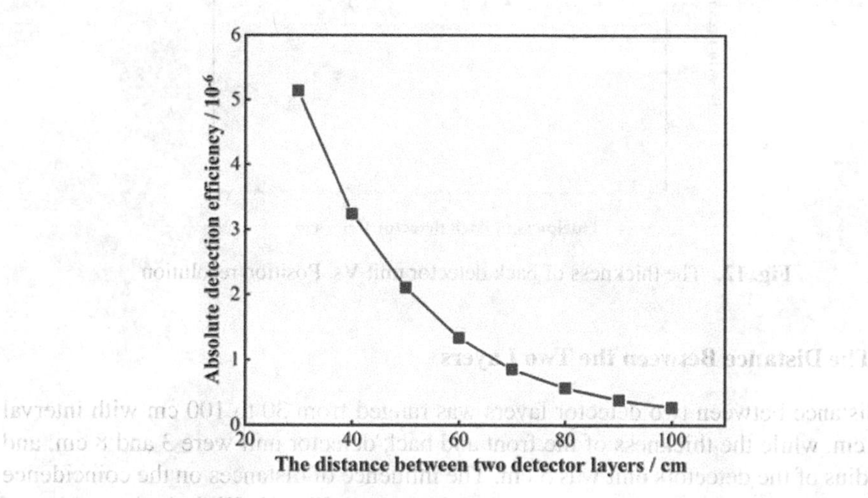

Fig. 14. The distance between two layers Vs. Absolute detection efficiency

The influence of distance between two layers on position resolution of neutron source was given in Fig. 15. With the increase of distance, the position resolution becomes better and better. When the plane distance was increased to 100 cm, the position resolution was changed from 750 to 300 mm, indicating that the plane distance has an obvious influence on the position resolution.

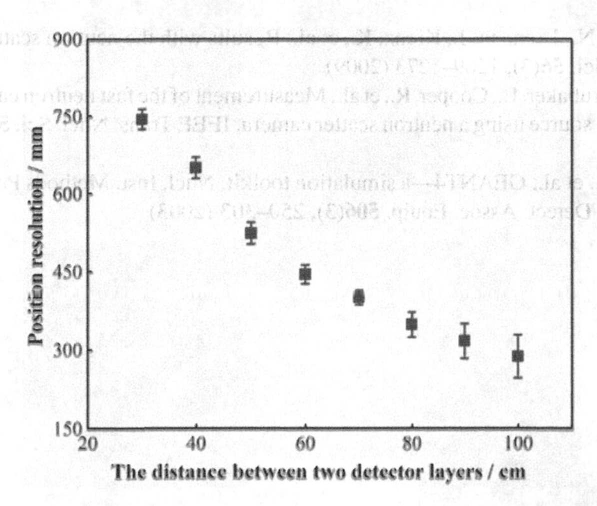

Fig. 15. The distance between two layers Vs. Position resolution

Based on the simulations, it can be seen that the distance mainly affects the coincidence ratio of primary scattering, absolute detection efficiency and point source position resolution. When the distance was 50 cm, the NSC not only has a high coincidence ratio of primary scattering and absolute detection efficiency, but also has a good position resolution. So, the distance was determined to be 50 cm.

4 Conclusions

In this paper, the effects of the thickness and radius of detector units and distance between two layers on image reconstruction were studied by the Geant4 to optimize the geometric parameter of the NSC. The simulation results show that the thickness of front detectors mainly affects the primary scattering ratio and the coincidence ratio of primary scattering, the radius of detector unit and distance between two layers mainly affect the first scattering coincidence ratio,

detection efficiency and position resolution of neutron source. Based on the simulation results, the thickness of the front detector, the radius of the detector, the thickness of the back detector and the distance between two layers were determined to be 3 cm, 5 cm, 8 cm and 50 cm, respectively.

References

1. Steinberger, W.M., Ruch, M.L., Giha, N., et al.: Imaging special nuclear material using a handheld dual particle imager. Sci. Rep. **10**(1), 1–11 (2020)
2. Bravar, U., Woolf, R.S., Bruillard, P.J., et al.: Calibration of the fast neutron imaging telescope (FNIT) prototype detector. Trans. Nucl. Sci. **56**(5), 2947–2954 (2009)
3. Pirard, B., Woolf, R.S., Bravar, U., et al.: Test and simulation of a fast neutron imaging telescope. Nucl. Instrum. Methods Phys. Res. **603**(3), 406–414 (2009)

4. Mascarenhas, N., Brennan, J., Krenz, K., et al.: Results with the neutron scatter camera. IEEE Trans. Nucl. Sci. **56**(3), 1269–1273 (2009)
5. Brennan, J., Brubaker, E., Cooper, R., et al.: Measurement of the fast neutron energy spectrum of an am-241-Be source using a neutron scatter camera. IEEE Trans. Nucl. Sci. **58**(5), 2426–2430 (2011)
6. Agostinelli, S., et al.: GEANT4—a simulation toolkit. Nucl. Inst. Methods Phys. Res. Sect. A: Accel. Spectr. Detect. Assoc. Equip. **506**(3), 250–303 (2003)

The Design of the Robust Controller for Active Magnetic Bearings on Active Disturbance Rejection Technology

Qian Shi[1], Yichen Yao[2], and Suyuan Yu[2](✉)

[1] China National Intellectual Property Administration, Beijing, China
[2] Department of Energy and Power Engineering, Tsinghua University, Beijing, China
suyuan@mail.tsinghua.edu.cn

Abstract. Active magnetic bearings (AMBs) have advantages of no friction, no lubrication or sealing requirements, long lifespan, low maintenance cost, and, especially, active controllability of dynamic characteristics. Thus, AMBs are now widely used in helium-turbine circle of the high temperature gas-cooled reactor and many other high-speed rotating machinery. The design of controller is the core problem of AMBs. The AMB force has high nonlinearity and the AMBs-rotor system may be influenced by external disturbance during operation, which increase the threshold of the controller robustness and make it hard to design. Based on the AMB-rigid rotor system model, this paper adopts lumped uncertainties to describe nonlinear error and external load disturbance and then the plant model of the decentralized controller is obtained. Then, a linear active disturbance rejection controller (LADRC) is designed to compensate the model error. The LADRC contains a proportional-differential controller and a three-order external state observer. The adjustable parameters of the LADRC can be selected according to pole assignment. In order to verify the effectiveness of the LADRC, levitation experiments, rotation experiments, and re-levitation experiments are carried out with traditional PID controller as comparison.

Keywords: active magnetic bearings · active disturbance rejection controller · lumped uncertainty

1 Introduction

Active magnetic bearings (AMBs) have the advantages of no friction, no lubrication or sealing requirements, long lifespan, low maintenance and active vibration control and have widely application in helium-turbine circle of the high temperature gas-cooled reactor and many other high-speed rotating machinery [1, 2]. In fact, the dynamic characteristics of AMBs mainly depend on the controller. A suitable controller is the core problem of AMBs design. As a matter of fact, the AMB-rotor system is open-loop unstable because of the high nonlinearity of electromagnetic bearing force. Besides, the AMBs-rotor system is always influenced by external disturbance, which makes it harder

C. Liu (Ed.): PBNC 2022, SPPHY 283, pp. 1147–1158, 2023.
https://doi.org/10.1007/978-981-99-1023-6_99

to design a suitable controller to achieve moderate rotor displacement response during normal operation.

In order to solve this problem, many researches have been carried out. The proportional-integral-differential (PID) controller is widely used in industrial applications because of its simplicity and robustness. Markus Hutterer designed PID controller through linear quadratic regulator method and validated the controller on a turbomolecular pump [3]. Tianhao Zhou proposed a robust PD control via eigenstructure assignment and evaluated the closed-loop sensitivity to change of the bias current [4]. Sun zhe applied a PID controller on a prototype of a 27000 rpm/150 kw blower and analyzed the nonlinearity of the system [5]. Chenzi Liu proposed a simple lead-lag controller, which parameters were determined through backprogation neural network [6].

Moreover, many advanced controllers have also been developed. Alexander presented a μ-synthesis-based controller to robustly minimize the difference between the tool reference and the estimated tool position in tooltip tracking spindle [7]. Syed Muhammad amrr proposed a robust control law based on high-order sliding mode control scheme, and carried out numerical analysis [8]. Xuan Yao proposed a dual-loop neural network sliding mode control to achieve large-motion rotor tracking, and validated its effectiveness through simulations [9].

These advanced controllers show better performance compared with PID controllers, but they have more complex structure and may have difficulties in parameter adjustment. Besides, they are model-based controller and may have deteriorated performance on an imprecise model.

Nowadays, linear active disturbance rejection controllers (LADRCs) have been attractive because it can achieve robust performance on an imprecise model and can simply realize parameter adjustment [10–14].

In this paper, a plant model of the decentralized controller is established based on the AMB-rigid rotor system model. This model adopts lumped uncertainty to describe nonlinear error and external load disturbance. Furthermore, a LADRC is designed, which can attenuating the impact of model nonlinear and external disturbance. The LADRC contains a proportional-differential controller and a three-order external state observer. The adjustable parameters of the LADRC can be selected according to pole assignment. In order to verify the effectiveness of the LADRC, levitation experiments, rotation experiments, and re-levitation experiments are carried out with traditional PID controller as comparison.

This paper is organized as follows. First, in Sect. 2, the plant model of the decentralized controller is established. Then, the LADRC is designed in Sect. 3. Experiments are developed in Sect. 4. Finally, Sect. 5 concludes this paper.

2 Description of the AMB-rotor System

2.1 The Rotor Model

The structure of the AMB-rigid rotor system is shown in Fig. 1. x_c, y_c denote the displacement of the centroid of rotor; α, β are the angular displacement of the rotor around x and y axes. $l_{bA}, l_{bB}, l_{sA}, l_{sB}$ show the distance between the A/B bearing/sensor and the

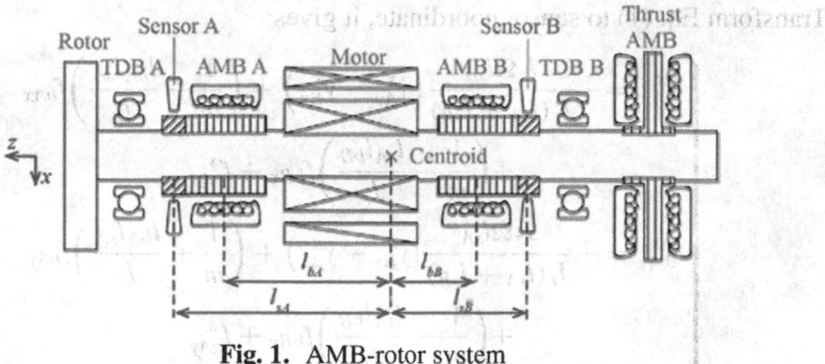

Fig. 1. AMB-rotor system

centroid. In AMB-rigid rotor system, the coupling between the axial DOF and radial DOF is negligible. Thus, axial DOF is not mentioned in this model and merely radial 4 DOF and axial angle γ are considered. There are three popular coordinates to illustrate the motion of rotor, which are centroid coordinate $\mathbf{y}^c = \left[x_c\ \alpha\ y_c\ \beta\right]^T$, sensor coordinate $\mathbf{y}^s = \left[y^s_{Ax}\ y^s_{Ay}\ y^s_{Bx}\ y^s_{By}\right]^T$ and bearing coordinate $\mathbf{y}^b = \left[y^b_{Ax}\ y^b_{Ay}\ y^b_{Bx}\ y^b_{By}\right]^T$. $\mathbf{y}^c$ describes the displacement of rotor centroid, $\mathbf{y}^s$ and $\mathbf{y}^b$ are rotor displacement at sensor/bearing. The relation between them can be expressed as $\mathbf{y}^s = \mathbf{T}^{sc}_s \mathbf{y}^c$, $\mathbf{y}^s = \mathbf{T}^{cb}_b \mathbf{y}^b$, in which

$$\mathbf{T}^{cb}_b = \begin{bmatrix} 1 & 0 & 1 & 0 \\ 0 & -l_{bA} & 0 & l_{bB} \\ 0 & 1 & 0 & 1 \\ l_{bA} & 0 & -l_{bB} & 0 \end{bmatrix}, \mathbf{T}^{sc}_s = \begin{bmatrix} 1 & 0 & 0 & l_{sA} \\ 0 & -l_{sA} & 1 & 0 \\ 1 & 0 & 0 & -l_{sB} \\ 0 & l_{sB} & 1 & 0 \end{bmatrix}$$

Define bearing force as $\mathbf{f}^b_b = \left[f_{bAx}\ f_{bAy}\ f_{bBx}\ f_{bBy}\right]^T$ under bearing coordinate, and then define external disturbance as $\mathbf{f}^c_g = \left[f^c_{gx}\ f^c_{g\alpha}\ f^c_{gy}\ f^c_{g\beta}\right]^T$ under rotor centroid coordinate. The equation of motion of the rotor can be written as:

$$\begin{cases} m\ddot{x}_c = f_{bAx} + f_{bBx} + f^c_{gx} \\ J_r\ddot{\alpha} + J_z\dot{\gamma}\dot{\beta} = -l_{bA}f_{bAy} + l_{bB}f_{bBy} + f^c_{g\alpha} \\ m\ddot{y}_c = f_{bAy} + f_{bBy} + f^c_{gy} \\ J_r\ddot{\beta} + J_z\dot{\gamma}\dot{\beta} = l_{bA}f_{bAx} - l_{bB}f_{bBx} + f^c_{g\beta} \end{cases} \tag{1}$$

Transform Eq. (1) to sensor coordinate, it gives:

$$\begin{cases} \ddot{y}^s_{Ax} = -\dfrac{J_z\Omega l_{sA}}{J_r(l_{sA}+l_{sB})}\left(\dot{y}^s_{Ay}-\dot{y}^s_{By}\right)+\left(\dfrac{1}{m}+\dfrac{l_{bA}l_{sA}}{J_r}\right)f_{bAx} \\ \qquad +\left(\dfrac{1}{m}-\dfrac{l_{sA}l_{bB}}{J_r}\right)f_{bBx}+f^s_{gAx} \\ \ddot{y}^s_{Ay} = -\dfrac{J_z\Omega l_{sA}}{J_r(l_{sA}+l_{sB})}\left(\dot{y}^s_{Bx}-\dot{y}^s_{Ax}\right)+\left(\dfrac{1}{m}+\dfrac{l_{bA}l_{sA}}{J_r}\right)f_{bAy} \\ \qquad +\left(\dfrac{1}{m}-\dfrac{l_{sA}l_{bB}}{J_r}\right)f_{bBy}+f^s_{gAy} \\ \ddot{y}^s_{Bx} = -\dfrac{J_z\Omega l_{sB}}{J_r(l_{sA}+l_{sB})}\left(\dot{y}^s_{By}-\dot{y}^s_{Ay}\right)+\left(\dfrac{1}{m}-\dfrac{l_{bA}l_{sB}}{J_r}\right)f_{bAx} \\ \qquad +\left(\dfrac{1}{m}+\dfrac{l_{bB}l_{sB}}{J_r}\right)f_{bBx}+f^s_{gBx} \\ \ddot{y}^s_{By} = -\dfrac{J_z\Omega l_{sB}}{J_r(l_{sA}+l_{sB})}\left(\dot{y}^s_{Ax}-\dot{y}^s_{Bx}\right)+\left(\dfrac{1}{m}-\dfrac{l_{bA}l_{sB}}{J_r}\right)f_{bAy} \\ \qquad +\left(\dfrac{1}{m}+\dfrac{l_{bB}l_{sB}}{J_r}\right)f_{bBy}+f^s_{gBy} \end{cases} \tag{2}$$

For slender rotor, the rotor shape usually satisfies $\left|\frac{1}{m}+\frac{l_{bA}l_{sA}}{J_r}\right| \gg \left|\frac{1}{m}-\frac{l_{sA}l_{bB}}{J_r}\right|, \left|\frac{1}{m}+\frac{l_{bB}l_{sB}}{J_r}\right| \gg \left|\frac{1}{m}-\frac{l_{sB}l_{bA}}{J_r}\right|$, and $J_r \gg J_z$. Then, Eq. (2) can be simplified as

$$\ddot{y}^s = k_{stru}f_b + f^s_g \tag{3}$$

where $k_{stru} = \frac{1}{m}+\frac{l_{bA}l_{sA}}{J_r}$ in AMB A and $k_{stru} = \frac{1}{m}+\frac{l_{bB}l_{sB}}{J_r}$ in AMB B.

2.2 The AMB Model

The AMB always works under differential-driven mode, shown in Fig. 2. There are two electromagnets placed in one direction, since each electromagnet can only achieve attractive force. The electromagnetic force generated by the single electromagnet can be expressed as

$$f_p = k\frac{i_a^2}{s^2}, k = \frac{\mu_0 A N^2}{4}\cos\theta \tag{4}$$

where i_a is the current, s is the gap, μ_0 is the magnetic field constant in vacuum, N is the number of coils turns, A is the cross-section area of the pole and θ is the angle of the electromagnet. Then, the net AMB force in one direction is the differential force of a pair of electromagnets. The net force can be linearized around the neighborhood of the operating point [1, 15], and can be expressed as

$$f_b = k_i i_c + k_s y^s \tag{5}$$

where $i_c = i_{a1} - i_{a2}$, k_i and k_s are the current-force factor (in N/A) and the displacement-force factor (in N/m) respectively.

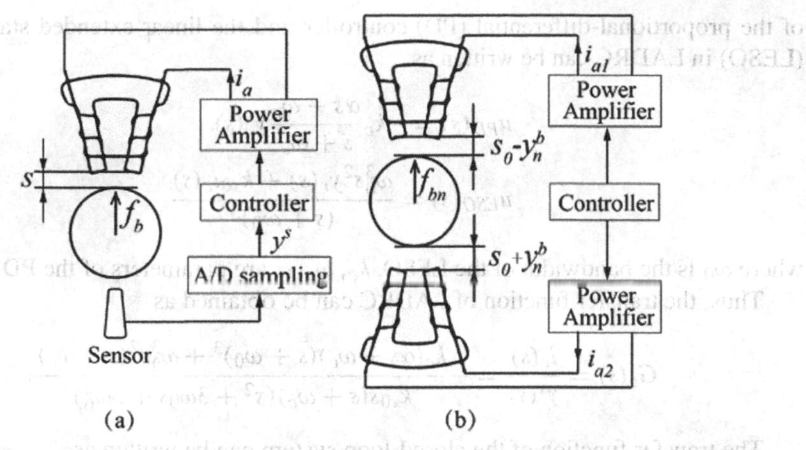

Fig. 2. Differential-driven mode in AMBs. (a) AMB force of a radial magnet. (b) AMB force of a pair of magnets.

2.3 The AMB-rotor Model

Substituting Eq. (5) into (3), the decentralized AMBs-rotor model can be written as:

$$\ddot{y}^s = k_{stru}k_i i_c + k_{stru}k_s y^s + f_g^s \tag{6}$$

It is noticed that the precise position of the rotor centroid is hard to determine due to its complex shape and its heterogeneous material, which indicate k_{stru} has uncertainty. This parameter variation can be described as $k_{stru}=(1+\lambda)k_{stru,0}$. Then, Eq. (6) can be rewritten as

$$\ddot{y}^s = k_{x0}(1+\lambda)\left(i_c + \frac{k_s}{k_i}y^s\right) + f_g^s \tag{7}$$

where $k_{x0}=k_i k_{stru,0}$.

3 LADRC Design

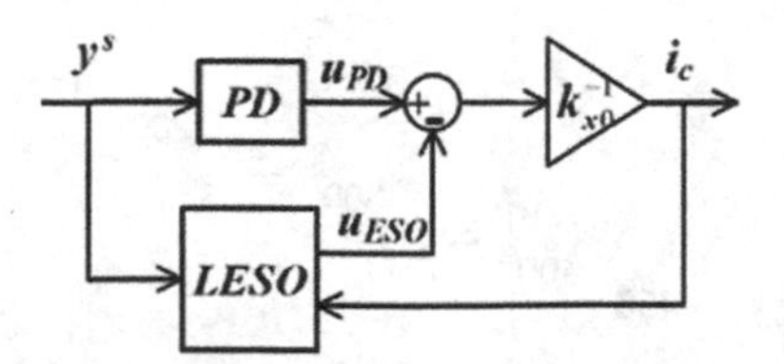

Fig. 3. The structure of LADRC.

The structure of LADRC is shown in Fig. 3. The LADRC contains a proportional-differential controller and a three-order external state observer. The transfer function of the proportional-differential (PD) controller and the linear extended state observer (LESO) in LADRC can be written as

$$\begin{aligned} u_{PD}(s) &= -k_c \frac{\alpha s + \omega_c}{s + \omega_c} y^s(s) \\ u_{ESO}(s) &= \frac{\omega_0^3 s^2 y_r(s) + k_{x0} i_c(s)}{(s + \omega_0)^3} \end{aligned} \tag{8}$$

where ω_0 is the bandwidth of the LESO, k_c, ω_c, α are parameters of the PD controller.

Thus, the transfer function of LADRC can be obtained as

$$G_c(s) = \frac{i_c(s)}{y^s(s)} = -\frac{k_c(\alpha s + \omega_c)(s + \omega_0)^3 + \omega_0^3 s^2 (s + \omega_c)}{k_{x0} s (s + \omega_c)(s^2 + 3\omega_0 s + 3\omega_0^2)}. \tag{9}$$

The transfer function of the closed-loop system can be written as

$$G_{cl}(s) = \frac{y^s(s)}{f_g^s(s)} = \frac{k_i}{k_i s^2 - k_{x0}(1 + \lambda)[k_i G_c(s) - k_s]}. \tag{10}$$

Therefore, the closed-loop system can achieve robust stability through proper pole assignment. If designing $\omega_0 = 50$, $\omega_c = 500$, $\alpha = 3.25$, $k_c = 25$ with nominal parameter $k_{x0} = 20$, the root locus of the closed-loop system with parameter variation can be calculated. Consider that the range of λ is $\lambda \in [0.5, 1.5]$. The roots of the closed-loop system (10) is shown in Fig. 4. The designed LADRC shows good robustness. The mapping from $f_g^s(s)$ to $y^s(s)$ under the LADRC is shown in Fig. 5. This demonstrates the capability of the LADRC to suppress the disturbance $f_g^s(s)$ under the plant with

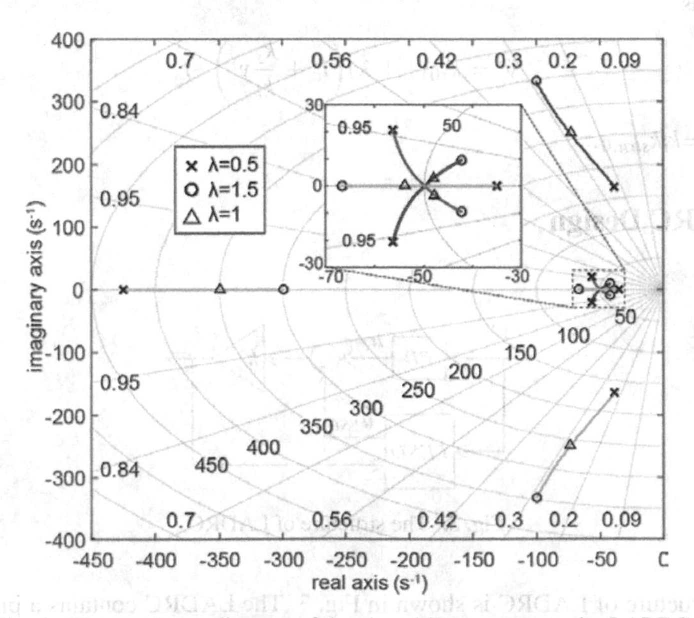

Fig. 4. The root locus diagram of the closed-loop system under LADRC.

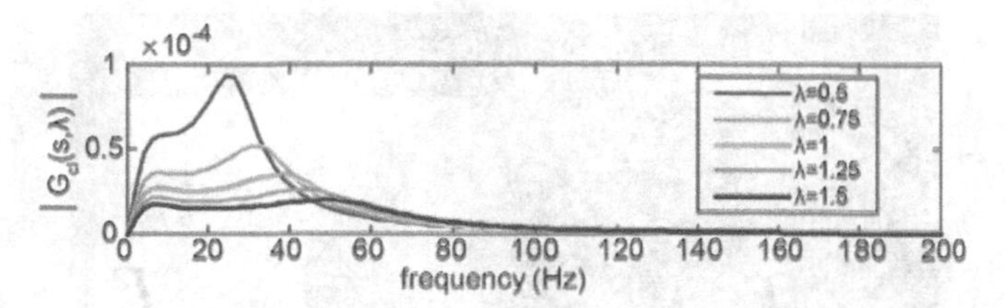

Fig. 5. The mapping from f_g^s to y^s under LADRC.

lumped uncertainty. In fact, $f_g^s(s)$ mainly contains a static load and unbalanced force. Unbalanced force causes an auto-balance effect after rigid critical speed; thus, it does not require compensation within a high-frequency range. The peak is larger when λ is smaller.

4 Experimental Results

This section contains two type of experiments, levitation experiments, and rotation experiments. In levitation experiments, the transient responses are analyzed. In rotation experiments, the analysis of rotor displacement responses within working speed range are given.

4.1 Description of the Experimental Platform

In order to verify the effectiveness of the LADRC, verification experiments are carried out on the AMBs-supported permanent magnet synchronous motor platform at Tsinghua University [16], pictured in Fig. 6. The rotor of the platform is 0.4866 m long with a total mass of 13.98 kg and is horizontally supported by two radial AMBs and one axial AMB. The radial clearance of the AMBs is 0.4 mm and the clearance of touchdown bearings (TDBs) is 0.2 mm. Radial and axial displacements of the rotor were measured by five inductive sensors. Ten pulse width modulated amplifiers power the magnet coils to generate the expected bearing force.

For comparison, a traditional PID controller, the most popular controller in industrial practice, is involved in this section. This PID controller is well-designed and performs well in various working conditions.

4.2 Levitation Experiments

The transient response at AMB A of levitation experiments are shown in Fig. 7 and Fig. 8 since AY channel has the worst results of the four. Sub-figure (a) gives the rotor trajectory. In this sub-figure, the transient response is divided into 3 phases. The first phase ends when the differential signal reaches its maximum. In this phase, the proportional signal plays the most important role. It starts at its maximum absolute value and its proportion in the total control signal gradually reduce. This phase is very short and the end position of the rotor during this phase mostly depends on the PD controller. The PID and LADRC nearly have similar displacement response in this phase. The second phase ends when the

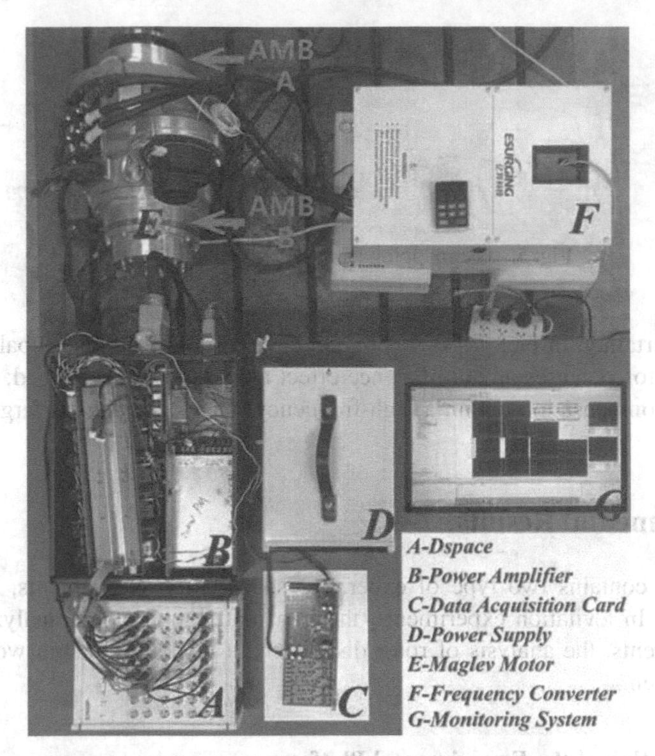

Fig. 6. The experimental platform.

proportional signal line intersects with the compensation signal line (I in PID and ESO in LADRC) at the first time. During this phase, the compensation signal increase gradually and cannot be ignored. The displacement response of the rotor results from the combined influence of three signals and is more complex. Under PID, this phase is about 0.042 s and the rotor trajectory in Fig. 7(a) displays fluctuation. Under LADRC, this phase is 0.012 s and there is less fluctuation in rotor trajectory because ESO signal has faster response. The third phase ends when the displacement response remains within 5% of the radial clearance of the TDBs (0.2 mm). The compensation signal plays important role in this phase. Since ESO signal has faster response, this phase is shorter under LADRC controller. The total startup period is 0.14 s under PID and is 0.05 s under LADRC, which shows LADRC have better performance in transient response. This indicator is significant for AMB controllers.

4.3 Rotation Experiments

Rotation experiments are of great significance not only due to being the most common working conditions but also because synchronous excitation caused by residual unbalance can help to analyze the frequency domain characteristics of the system and the control strategy.

The displacement response of rotation experiments from 0–200 Hz (up to 12000 r/min) under PID and LADRC are given in Fig. 9. During the rigid rotation speed range,

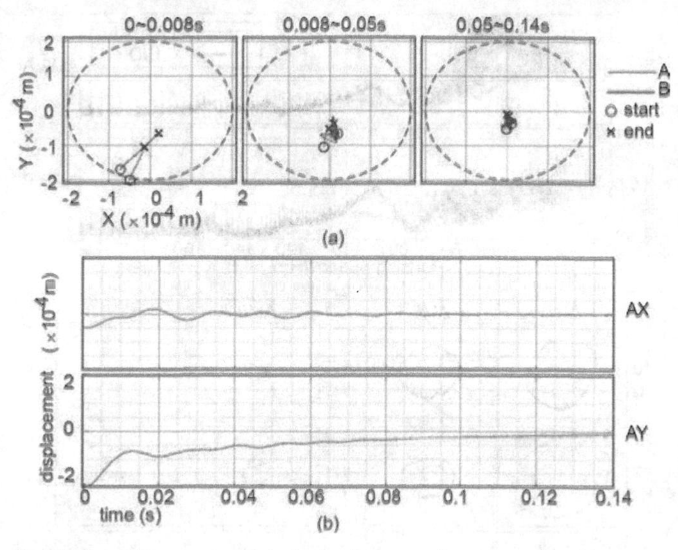

Fig. 7. The transient response in levitation experiments under PID controller

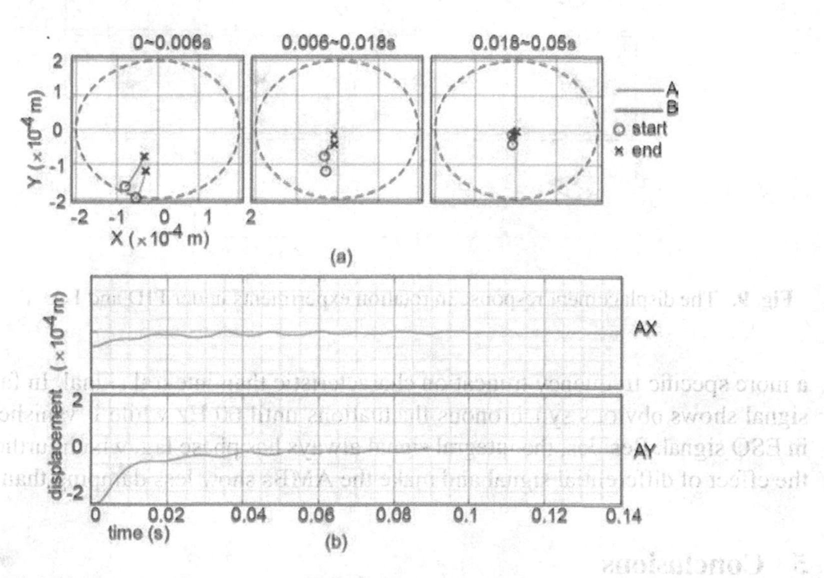

Fig. 8. The transient response in levitation experiments under LADRC

both of the two controllers show good displacement response. The amplitude of vibration remains within half of the TDBs clearance. There are four peaks within the speed range, shown in Fig. 9 (a), which correspond to the four vibration modes of the system.

However, LADRC shows better performance at the four peaks and under 100 Hz. It seems that the proposed LADRC have better damping property. When it comes to low rotating speed range, it indicates that ESO signal shows better performance in isolation of low-frequency vibration. There is a feedback loop in ESO, so that ESO signal has

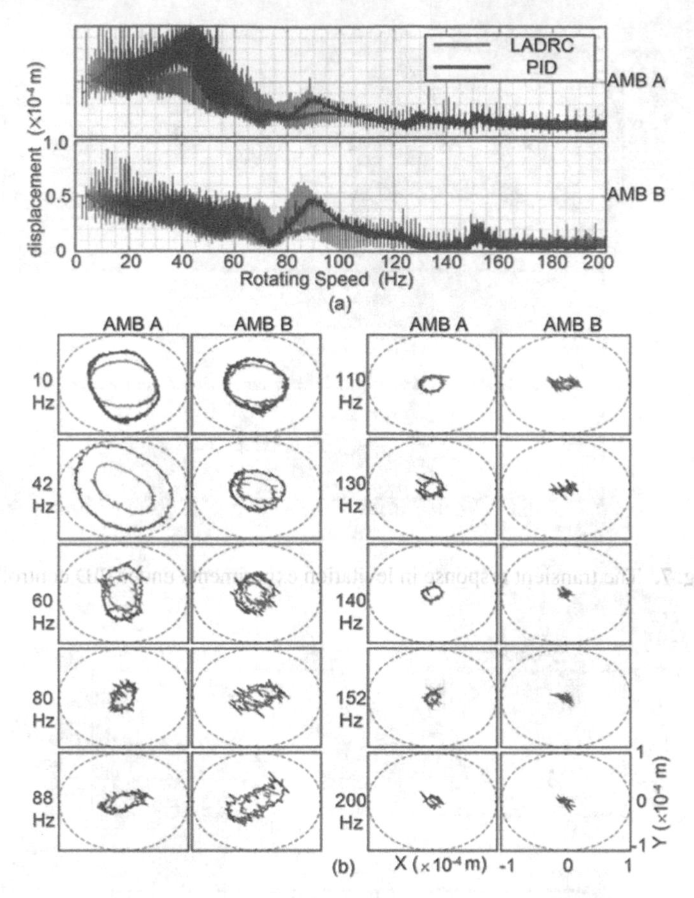

Fig. 9. The displacement response in rotation experiments under PID and LADRC controller

a more specific frequency truncation characteristic than integral signal. In fact, integral signal shows obvious synchronous fluctuations until 60 Hz while it vanishes at 20 Hz in ESO signal. Besides, the integral signal always has phase lag, which further suppress the effect of differential signal and make the AMBs show less damping than designed.

5 Conclusions

In view of the high nonlinearity, parameter perturbation and external disturbance existing in AMB-rotor system, a LADRC is designed in this paper to suppress the rotor displacement response while attenuating the impact of uncertainties. A plant model of the decentralized controller is established with lump uncertainties including the model error of the AMB force, the external load, and related parametric perturbation. Then, a LADRC is designed. The LADRC contains a proportional-differential controller and a three-order external state observer. The adjustable parameters of the LADRC can be selected according to pole assignment. In order to verify the effectiveness of the LADRC, levitation experiments and rotation experiments are carried out with traditional PID controller as comparison.

References

1. Schweitzer, G., Maslen, E.H.: Magnetic Bearings: Theory, Design, and Application to Rotating Machinery. Springer, Heidelberg (2009)
2. Shi, Z., Yang, X., et al.: Design aspects and achievements of active magnetic bearing research for htr-10gt. Nucl. Eng. Des. **238**, 1121–1128 (2008)
3. Hutterer, M., Wimmer, D., Schrödl, M.: Stabilization of a magnetically levitated rotor in the case of a defective radial actuator. IEEE Trans. Mechatron. **66**(12), 9383–9393 (2019)
4. Zhou, T., Zhu, C.: Robust proportional-differential control via eigenstructure assignment for active magnetic bearings-rigid rotor systems. IEEE Trans. Ind. Electron. **69**(7), 6572–6585 (2022)
5. Zhang, X., Sun, Z., et al.: Nonlinear dynamic characteristics analysis of active magnetic bearing system based on cell mapping method with a case study. Mech. Syst. Signal Process. **117**, 116–137 (2019)
6. Liu, C., Deng, Z., Xie, L., Li, K.: The design of the simple structure-specified controller of magnetic bearings for the high-speed srm. IEEE/ASME Trans. Mechatron. **20**(4), 1798–1808 (2014)
7. Pesch, A.H., Smirnov, A., Pyrhonen, O., Sawicki, J.T.: Magnetic bearing spindle tool tracking through mu-synthesis robust control. IEEE/ASME Trans. Mechatron. **20**, 1448–1457 (2015)
8. Amrr, S.M., Alturki, A.: Robust control design for an active magnetic bearing system using advanced adaptive smc technique. IEEE Access **9**, 155662–155672 (2021)
9. Yao, X., Chen, Z., Jiao, Y.: A dual-loop control approach of active magnetic bearing system for rotor tracking control. IEEE Access **7**, 121760–121768 (2019)
10. Dong, S.: Comments on active disturbance rejection control. IEEE Trans. Ind. Electron. **54**(6), 3428–3429 (2007)
11. Han, J.: From pid to active disturbance rejection control. IEEE Trans. Ind. Electron. **56**(3), 900–906 (2009)
12. Huang, Y., Xue, W.: Active disturbance rejection control: methodology and theoretical analysis. ISA Trans. **53**(4), 963–976 (2014)
13. Zhao, Z.-L., Guo, B.-Z.: A novel extended state observer for output tracking of mimo systems with mismatched uncertainty. IEEE Trans. Autom. Control **63**(1), 211–218 (2018)
14. Zhang, X., Sun, L., Zhao, K., Sun, L.: Nonlinear speed control for pmsm system using sliding-mode control and disturbance compensation techniques. IEEE Trans. Power Electron. **28**(3), 1358–1365 (2013)
15. Chen, M., Knospe, C.R.: Feedback linearization of active magnetic bearings: current-mode implementation. IEEE/ASME Trans. Mechatron. **10**(6), 632–639 (2005)
16. Yao, Y.C., Sha, H.L., Su, Y.X., Ren, G.X., Yu, S.Y.: Identification of system parameters and external forces in amb-supported pmsm system. Mech. Syst. Signal Process. **166**, 18 (2022)

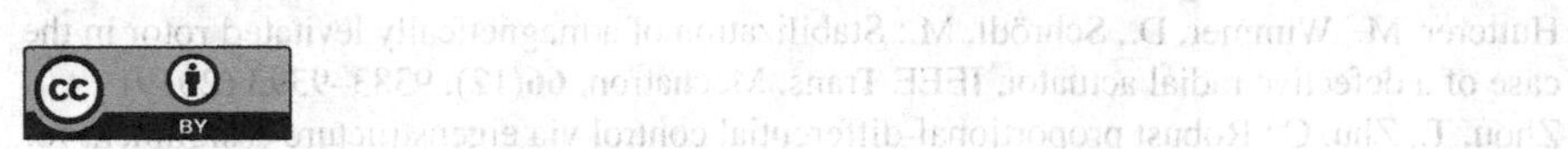

Author Index

C. Liu (Ed.): PBNC 2022, SPPHY 283, pp. 1159–1163, 2023.
https://doi.org/10.1007/978-981-99-1023-6

Z